Organic Phosphorus in the Environment

Ciascun uomo durante la sua esistenza ripercorre la storia di tutta l'umanità dal neolitico all'era moderna. Per questa ragione, l'uomo non può non partire dalla conoscenza della terra e dei suoi frutti.

Every man, during his life, travels again over the history of all humanity, from the Neolithic to the modern age. For this reason, he must know the value of the soil and all that it yields.

Lucien Muller

Organic Phosphorus in the Environment

Edited by

Benjamin L. Turner

Smithsonian Tropical Research Institute
Ancon
Republic of Panama

Emmanuel Frossard

Institute of Plant Sciences
Swiss Federal Institute of Technology (ETH)
Lindau
Switzerland

and

Darren S. Baldwin

Murray–Darling Freshwater Research Centre
Albury, New South Wales
Australia

CABI Publishing

CABI Publishing is a division of CAB International

CABI Publishing
CAB International
Wallingford
Oxfordshire OX10 8DE
UK

Tel: +44 (0)1491 832111
Fax: +44 (0)1491 833508
E-mail: cabi@cabi.org
Website: www.cabi-publishing.org

CABI Publishing
875 Massachusetts Avenue
7th Floor
Cambridge, MA 02139
USA

Tel: +1 617 395 4056
Fax: +1 617 354 6875
E-mail: cabi-nao@cabi.org

A catalogue record for this book is available from the British Library, London, UK.

Library of Congress Cataloging-in-Publication Data

Organic Phosphorus Workshop (2003 : Ascona, Switzerland)

Organic phosphorus in the environment / edited by Benjamin L. Turner, Emmanuel Frossard, Darren S. Baldwin.

p. cm.

Papers presented at the Organic Phosphorus Workshop, held July 2003, Ascona, Switzerland.

Includes bibliographical references and index.

ISBN 0-85199-822-4 (alk. paper)

1. Soils--Phosphorus content--Congresses. 2. Water--Phosphorus content--Congresses. 3. Organophosphorus compounds--Environmental aspects--Congresses. 4. Organophosphorus compounds--Toxicology--Congresses. I. Turner, Benjamin L. II. Frossard, Emmanuel. III. Baldwin, Darren S. IV. Title.

S592.6.P5O74 2003
631.4′17--dc22 2004006554

ISBN 0 85199 822 4

Typeset in Melior by Servis Filmsetting Ltd, Manchester
Printed and bound in the UK by Biddles Ltd, King's Lynn

Contents

Contributors

Al-Shehri, Abdulrahman M., *Biological Sciences Department, College of Science, King Khalid University, Abha, Saudi Arabia.*

Baldwin, Darren S., *Murray–Darling Freshwater Research Centre, Cooperative Research Centre for Freshwater Ecology, PO Box 921, Albury, NSW 2640, Australia. E-mail: darren.baldwin@csiro.au*

Barberis, Elisabetta, *Università di Torino, DIVAPRA Chimica Agraria, via Leonardo da Vinci 44, Grugliasco, 10095 Torino, Italy.*

Beattie, James K., *School of Chemistry, University of Sydney, Sydney, NSW 2006, Australia.*

Bennett, G. Lee, *Department of Chemistry and Biochemistry, Florida State University, Tallahassee, FL 32306, USA. Current address: AstraZeneca Pharmaceuticals, Wilmington, DE 19803, USA.*

Bünemann, Else K., *Institute of Plant Sciences, Swiss Federal Institute of Technology (ETH), Eschikon 33, PO Box 185, CH-8315, Switzerland. Current address: School of Earth and Environmental Sciences, University of Adelaide, SA 5005, Australia.*

Cade-Menun, Barbara J., *Geological and Environmental Sciences Department, Building 320, Room 118, Stanford University, Stanford, CA 94305-2115, USA. E-mail: bjcm@pangea.stanford.edu*

Celi, Luisella, *Università di Torino, DIVAPRA Chimica Agraria, via Leonardo da Vinci 44, Grugliasco, 10095 Torino, Italy. E-mail: luisella.celi@unito.it*

Condron, Leo M., *Agriculture and Life Sciences, PO Box 84, Lincoln University, Canterbury 8150, New Zealand. E-mail: condronl@lincoln.ac.nz*

Cooper, William T., *Department of Chemistry and Biochemistry, Florida State University, Tallahassee, FL 32306, USA. E-mail: cooper@chemmail.chem.fsu.edu*

Ellwood, Neil T.W., *Dipartimenti di Scienze Geologiche, Universita Roma Tre, Largo San Leonardo Murialdo, 00146 Roma, Italy.*

Frossard, Emmanuel, *Institute of Plant Sciences, Swiss Federal Institute of Technology (ETH), Eschikon 33, PO Box 185, CH-8315 Lindau, Switzerland.*

George, Tim S., *CSIRO Plant Industry, PO Box 1600, Canberra, ACT 2601, Australia.*

Heath, Robert T., *Department of Biological Sciences and Water Resources Research Institute, Kent State University, Kent, OH 44242, USA. E-mail: rheath@kent.edu*

Hens, Maarten, *CSIRO Plant Industry, PO Box 1600, Canberra, ACT 2601, Australia.*

Howitt, Julia A., *Water Studies Centre and School of Chemistry, Monash University, Clayton, Victoria 3800, Australia.*

Joner, Erik J., *Norwegian Forest Research Institute, Hogskoleveien 12, N-1432 Aas, Norway.*

Llewelyn, Jennifer M., *Department of Chemistry and Biochemistry, Florida State University, Tallahassee, FL 32306, USA. Current address: AstraZeneca Pharmaceuticals, Wilmington, DE 19803, USA.*

McKelvie, Ian D., *Water Studies Centre and School of Chemistry, Monash University, Clayton 3800, Victoria, Australia. E-mail: ian.mckelvie@sci.monash.edu.au*

Mitchell, Alison M., *Murray–Darling Freshwater Research Centre, Cooperative Research Centre for Freshwater Ecology, PO Box 921, Albury, NSW 2640, Australia. E-mail: alison.mitchell@csiro.au*

Mousain, Daniel, *Unité Mixte de Recherche Rhizosphère et Symbiose, INRA-ENSAM, 2 Place Pierre Viala, 34060 Montpellier, France.*

Neff, Jason, *US Geological Survey, MS980 Denver Federal Center, Denver, CO 80225, USA.*

Nziguheba, Generose, *Laboratory of Soil and Water Management, Kasteelpark Arenberg 20, 3001 Heverlee, Belgium. E-mail: gn@akad.sun.ac.za*

Oberson, Astrid, *Institute of Plant Sciences, Swiss Federal Institute of Technology (ETH), Eschikon 33, PO Box 185, CH-8315 Lindau, Switzerland. E-mail: astrid.oberson@ipw.agrl.ethz.ch*

Parton, William J., *Natural Resource Ecology Laboratory, Colorado State University, Fort Collins, CO 80523, USA. E-mail: billp@nrel.colostate.edu*

Quiquampoix, Hervé, *Unité Mixte de Recherche Rhizosphère et Symbiose, INRA-ENSAM, 2 Place Pierre Viala, 34060 Montpellier, France. E-mail: quiquamp@ensam.inra.fr*

Reichert, Peter, *Department of Systems Analysis, Integrated Assessment and Modelling (SIAM), Swiss Federal Institute for Environmental Science and Technology (EAWAG), PO Box 611, 8600 Dübendorf, Switzerland. E-mail: reichert@eawag.ch*

Richardson, Alan E., *CSIRO Plant Industry, PO Box 1600, Canberra, ACT 2601, Australia. E-mail: alan.richardson@csiro.au*

Salters, Vincent J.M., *National High Magnetic Field Laboratory and Department of Geological Sciences, Florida State University, Tallahassee, FL 32306, USA.*

Simpson, Richard J., *CSIRO Plant Industry, PO Box 1600, Canberra, ACT 2601, Australia.*

Stenson, Alexandra C., *Department of Chemistry and Biochemistry, Florida State University, Tallahassee, FL 32306, USA. Current address: Vertex Pharmaceuticals, Cambridge, MA 02139, USA.*

Tiessen, Holm, *Department of Tropical Agronomy, Göttingen University, Grisebachstrasse 6, 37077 Göttingen, Germany.*

Turner, Benjamin L., *Smithsonian Tropical Research Institute, 2072, Balboa, Ancon, Republic of Panama.*

Vitousek, Peter M., *Department of Biological Sciences, Stanford University, Stanford, CA 94305, USA.*

Wehrli, Bernhard, *Swiss Federal Institute of Technology (ETH), 8092 Zürich, Switzerland.*

Whitton, Brian A., *School of Biological and Biomedical Sciences, University of Durham, Durham DH1 3LE, UK. E-mail: b.a.whitton@durham.ac.uk*

Preface

Phosphorus is essential for life, yet is frequently the element that most limits biological productivity in ecosystems. Indeed, from polar regions to the humid tropics it is possible to find organisms in both terrestrial and aquatic ecosystems limited by the availability of phosphorus. Most organisms take up phosphorus as inorganic phosphate, but organic forms of phosphorus (molecules containing both phosphorus and carbon) often dominate in soils and aquatic systems. It is not surprising, therefore, that many organisms possess complex adaptations enabling them to access phosphorus from organic compounds. Understanding these mechanisms may provide answers to fundamental questions in plant and microbial ecology, but organic phosphorus remains poorly understood and currently represents the greatest gap in our knowledge of the global phosphorus cycle. For example, a major fraction of the organic phosphorus in soils, surface waters and aquatic sediments remains unidentified, which precludes even basic research into its dynamics in the environment.

Despite the lack of information on organic phosphorus in the environment, recent advances in technology have revealed novel insights into processes operating in a variety of ecosystems. However, this information remains dispersed in many disciplines, including chemistry, ecology, microbiology, soil science and limnology. The Organic Phosphorus Workshop, held during July 2003 in Ascona, Switzerland, brought together an international group of scientists working on organic phosphorus in a diverse range of environments, from soils of the Caucasina mountains, to particles in the depths of the Pacific Ocean. The strength of the meeting was derived in part from a mutual interest in the mechanisms that regulate organic phosphorus dynamics in the environment and the analytical procedures used to investigate them. Advances in both these areas have benefits that transcend disciplinary boundaries. Mechanisms also link scales, one of the key research issues in contemporary biogeochemistry and ecosystem modelling.

Drawing on the latest research and opinion, this book contains state-of-the-art reviews of organic phosphorus characterization and transformations in the environment. It is aimed at scientists with an interest in nutrient dynamics in all fields of environmental science. The seventeen chapters are organized into three main sections. *Section 1* describes analytical techniques used to characterize organic phosphorus compounds in environmental samples, with chapters dedicated to state-of-the-art procedures involving chromatographic separation, nuclear magnetic resonance spectroscopy and mass spectrometry. *Section 2* addresses processes that control organic phosphorus behaviour in terrestrial and aquatic environments, including the origins, stabilization, hydrolysis (biotic and abiotic) and biological utilization of organic phosphorus. *Section 3* integrates these processes at the ecosystem level, with chapters synthesizing information on organic phosphorus in soils and aquatic systems, its interaction with other nutrient cycles and its transfer between terrestrial and aquatic environments. Two chapters in this section describe state-of-the-art models used to investigate the behaviour of organic phosphorus in both terrestrial and aquatic environments.

The final chapter summarizes the importance of organic phosphorus in the environment and identifies key areas towards which future research effort should be directed. Understanding organic phosphorus in the environment almost certainly holds the key to addressing some of the most important problems facing the modern world, including climate change, environmental pollution, declining agricultural productivity, and threats to biodiversity. By bringing together the most up-to-date information on organic phosphorus in the environment, we hope this book goes some way towards facilitating research on these fundamental issues.

Benjamin L. Turner
Smithsonian Tropical Research Institute, Ancon, Republic of Panama

Emmanuel Frossard
Institute of Plant Sciences, Swiss Federal Institute of Technology (ETH), Lindau, Switzerland

Darren S. Baldwin
Murray–Darling Freshwater Research Centre, Albury, New South Wales, Australia

A Note on Nomenclature

Few aspects of biogeochemistry can claim a more confusing nomenclature than phosphorus speciation. A cursory glance through the literature soon reveals a startling multitude of terms and associated acronyms, which undoubtedly causes dismay to students and scientists new to the subject. Indeed, judging by the heated debate generated on the subject of terminology at the Organic Phosphorus 2003 meeting in Ascona, even those experienced in phosphorus research find the matter to be a source of considerable frustration. More often than not, such terminology issues are rather academic, but become worrisome when they lead to errors in interpretation that are perpetuated in the literature. Perhaps the most notable example is the assumption that molybdate-reactive phosphorus is equivalent to free phosphate, despite the possible (and often likely) inclusion of acid-labile colloidal and organic phosphates in the reactive fraction. Similarly, the molybdate-unreactive phosphorus fraction is often termed 'organic', despite the possible presence of inorganic polyphosphates.

From the outset, it was clear that some standardization of nomenclature was essential in this multidisciplinary volume. However, marked differences in convention between the soil and aquatic sciences made this a challenging task. The result is that some chapters may contain terminology that differs slightly from the 'standard' in that particular discipline, but we feel that this is a small price to pay for a more unified approach.

Aspects of particular note are as follows. The terms inorganic and organic phosphorus are reserved for general use. Unless specific compounds have been determined, phosphorus fractions in water samples are described by operational terms such as reactive and filterable, in preference to terms such as inorganic and dissolved. Following the convention of the International Union of Pure and Applied Chemistry, the term phosphate describes inorganic orthophosphate unless specified otherwise, for example in the terms glucose 6-phosphate or organic phosphate. The term inorganic phosphate is avoided unless used in a general sense, because although it is often used to refer to inorganic orthophosphate, it also includes polyphosphates and a range of mineral phosphates. The term organic phosphorus is reserved for general use, although it is sometimes used to describe an operationally defined phosphorus fraction (e.g. in sequential fractionation schemes). However, the reader should be aware that this may also include inorganic polyphosphates, as demonstrated by parallel analyses using nuclear magnetic resonance spectroscopy. Further discussion of terminology can be found in Chapter 1 (McKelvie) and Chapter 14 (Mitchell and Baldwin).

One compound around which there is notable ambiguity is *myo*-inositol hexakisphosphate. This commonly occurring organic phosphate is also known as phytic acid, although the term phytate is used for salt forms, and phytin is sometimes used to refer specifically to the calcium–magnesium salt. To avoid ambiguity, the term *myo*-inositol hexakisphosphate is used throughout the book, unless reference is made to a specific salt form (e.g. calcium phytate).

Finally, it is worth noting that we strived to minimize the use of abbreviations throughout the text. Frequent abbreviations make tiresome reading, particularly for non-specialists, and can make certain topics rather indigestible (e.g. some of the analytical chemistry). Notably, abbreviated terms for elements and chemicals have been avoided unless associated with a numerical value, such as the concentration of a compound or element. It is hoped that the limited use of abbreviations makes chapters easier to read and more accessible to non-specialists, while at the same time minimizing the confusion that can arise from complex terminology.

Acknowledgements

The editors thank Paolo Demaria and Mathimaran Natarajan (Institute of Plant Sciences, ETH Zürich) for their efforts behind the scenes that helped make the Organic Phosphorus 2003 workshop such a wonderful success. We also thank The Centro Stefano Franscini (CSF), the Swiss Federal Institute of Technology (ETH Zürich), The Swiss National Foundation for Science (SNF), the Swiss Center for International Agriculture (ZIL), the COST office of the Swiss Office for Education and Science (OFES), and Division 4 of the International Union of Soil Science for funding to support the meeting. Finally, we thank all those who gave their time to peer-review chapters in this volume.

Ben Turner thanks the Agricultural Research Service of the United States Department of Agriculture and the Soil and Water Science Department of the University of Florida for support during the planning and production of this book.

1 Separation, Preconcentration and Speciation of Organic Phosphorus in Environmental Samples

Ian D. McKelvie

Water Studies Centre and School of Chemistry, Monash University, Clayton 3800, Victoria, Australia

Introduction

'Organic phosphorus' is a term of convenience that embraces a plethora of organic compounds containing phosphorus, including nucleic acids, phospholipids, inositol phosphates, phosphoamides, phosphoproteins, sugar phosphates, phosphonic acids, organophosphate pesticides, humic-associated organic phosphorus compounds, and organic condensed phosphates in dissolved, colloidal and particle-associated forms. In aquatic systems, the role of organic phosphorus has largely been ignored, because techniques for its determination are not readily accessible, and consequently it is often included in the non-reactive, non-bioavailable component of total filterable or total phosphorus. Interest in organic phosphorus derives from its relative abundance in aquatic and terrestrial systems, and the potential for its release and eventual transformation to bioavailable forms. It has long been recognized that organic phosphorus is a potentially important source of phosphorus for phytoplankton (Chu, 1946). During the last two decades, strong evidence has emerged that organic phosphorus is at least as abundant as (and often greatly in excess of) phosphate in many water and sediment systems, and that this component of the phosphorus pool may be much less refractory than was previously assumed. Furthermore, soluble organic phosphorus is much more bioavailable to phytoplankton and bacterioplankton than hitherto believed (Cotner and Wetzel, 1992). For example, a study of sediments in Tokyo Bay, Japan showed that *myo*-inositol hexakisphosphate (phytic acid) was hydrolysed to phosphate within days under typical aerobic marine conditions (Suzumura and Kamatani, 1993). Similarly, organic phosphorus was released from freshwater sediments that had been subjected to increased salinity, and that this organic phosphorus was hydrolysed within hours to a more bioavailable form (Gardolinski *et al.*, 2004). More recently, it was shown that bioavailable phosphorus in the euphotic zone of the North Pacific contained a substantial component of soluble organic phosphorus (Bjorkman and Karl, 2003).

For many scientists involved in the analysis of phosphorus in water, sediment and soil, the only available techniques depend on the colorimetric determination of molybdate-reactive phosphorus, and that which can be made reactive during a digestion step. Molybdate-reactive phosphorus in filtered samples is generally equated to 'inor-

 Organic Phosphorus in the Environment (eds B.L. Turner, E. Frossard and D.S. Baldwin)

ganic' phosphorus. Thus, organic phosphorus is loosely considered to be the molybdate-unreactive phosphorus fraction (i.e. the fraction comprising the difference between the total filterable[1] phosphorus and the filterable reactive phosphorus). However, by this definition, condensed phosphorus (polyphosphate, an inorganic phosphorus form) would quite erroneously be considered as part of the filterable organic phosphorus fraction. Therefore there is a strong case for ensuring that condensed phosphorus is measured in addition to total and reactive phosphorus in the filterable fraction, so that the true filterable organic phosphorus concentration can be determined. However, this provides only a gross estimate of filterable organic phosphorus, and there is a strong imperative for the development of new or improved techniques for the measurement of organic phosphorus in particulate and filterable forms, and for individual organic phosphorus moieties. This chapter provides a broad, and by no means comprehensive, review of approaches that have been used historically and that are currently in use to measure organic phosphorus, and to isolate, quantify and characterize individual organic phosphorus species.

Organic Phosphorus in Sediments and Soils

The forms of phosphorus in sediments and soils can be operationally defined by chemical extraction schemes. The major inorganic phosphorus components are considered to comprise those: (i) adsorbed by exchange sites; (ii) associated with iron, aluminium and manganese oxides; (iii) associated with carbonate; (iv) associated with calcium as apatite; or (v) bound in a crystalline mineral form (e.g. silicates). The organic phosphorus fraction, on the other hand, is either considered as a whole, or operationally divided into: (i) labile organic substances; (ii) organic phosphorus associated with the humic/fulvic fraction; (iii) acid-soluble organic components; or (iv) a residual or refractory group consisting of phosphate esters and phosphonates (Barbanti *et al.*, 1994).

Extraction of Phosphorus from Sediments and Soils

Sequential extraction or chemical fractionation techniques have been widely used in the characterization of various phosphorus fractions in soils and sediments, with an emphasis on the more bioavailable or plant-available inorganic forms (Condron *et al.*, 2005). Early extraction procedures (Chang and Jackson, 1957; Williams *et al.*, 1976b) focused on inorganic phosphorus associated with iron, aluminium and calcium, using various acid, base or salt extraction steps. Organic phosphorus was considered to be the residual or refractory phosphorus-containing fraction that remained after all other extractions had been performed.

The SEDEX procedure has emerged as the benchmark extraction scheme for sediments (Ruttenberg, 1992). It uses only magnesium chloride, acetate buffer and citrate-bicarbonate reagents at pH values between 4 and 8 to leach sediments of the inorganic phosphorus fractions, before ashing at 550°C and a final extraction with 1 M HCl to determine the so-called residual organic phosphorus. However, extraction of organic phosphorus from sediments and soils should be performed in a manner that as far as possible avoids hydrolysis or oxidation, and this is often incompatible with the procedures used for inorganic phosphorus extraction. Extraction of organic phosphorus has conventionally involved an initial extraction with 1 M HCl for an extended period (e.g. 16 h); after centrifugation and removal of the acid, a further extraction

[1] The terms 'dissolved' and 'soluble' are conventionally used to denote that portion of species that are operationally separated by filtration through a defined pore size (usually 0.45 or 0.2 μm). However, it is widely recognized that the filterable fraction also includes high molecular mass colloidal species. Use of the term 'filterable' is therefore preferable to either 'dissolved' or 'soluble', which imply the presence of pure solution.

Table 1.1. Sequential chemical extraction steps for isolation of phytate (de Groot and Golterman, 1993). (NTA, nitriloacetic acid; EDTA, ethylenediaminetetraacetic acid.)

Fraction designation	Abbreviation	Extractant
Iron-bound phosphate	Fe(OOH)≈P	0.02 M Ca-NTA/dithionite, pH 7.8–8.0
Calcium-bound phosphate	$CaCO_3$≈P	0.05 M Na-EDTA, pH about 8.0
Acid-soluble organic phosphate	ASOP	0.5 M HCl or 0.25 M H_2SO_4 (30 min)
Sodium hydroxide-extractable phosphorus	$NaOH_{extr}$-P	2.0 M NaOH (90°C, 30 min)

with 1 M NaOH (4 h) is performed, and this fraction is quantified as the organic phosphorus fraction (Williams *et al.*, 1976a). In attempting to minimize the hydrolysis of organic phosphorus that necessarily accompanies these procedures, various authors have used a range of progressively milder acid and base conditions, shorter extraction times and less extreme temperatures (Broberg and Persson, 1988). Analysis of these extracts for both total phosphorus and reactive phosphorus can be used to provide estimates of organic phosphorus, because the condensed phosphorus forms should arguably be hydrolysed to reactive phosphorus in the initial acid extraction step.

Golterman (1988), however, has argued that strongly acidic and basic conditions are intrinsically unsuitable for extraction because of the likelihood of hydrolytic breakdown. As an alternative, he recommended the use of chelating extractants such as nitriloacetic acid or ethylenediaminetetraacetic acid (EDTA), which can be used at near-neutral pH. Use of this approach is exemplified in a study by de Groot and Golterman (1993), which focused on the isolation of *myo*-inositol hexakisphosphate with an extraction scheme that used complexation and reduction to remove inorganic phosphorus prior to the characterization and quantification of organic phosphorus (Table 1.1). The sodium hydroxide-extractable phosphorus was subsequently digested and used to determine the organic phosphorus content, or further acidified to precipitate the humic-acid phosphorus, leaving the fulvic-acid phosphorus fraction in solution. Further hydrolysis with phytase was then performed to confirm the presence of inositol phosphates. This illustrates the advantages of so-called sequential chemical extraction for the separation and broad classification of the various forms of organic and inorganic phosphorus associated with sediment. Similar extractions involving 0.5 M NaOH and chelation of metals by Chelex 100 or EDTA have been used widely for the extraction of organic phosphorus from soil (Condron *et al.*, 2005).

Some components of organic phosphorus may also be extracted using organic solvents. Phospholipids, for example, which are used as biomarkers for microbial communities, can be extracted using methanol-chloroform-water (White *et al.*, 1979), by pressurized hot solvent extraction (Macnaughton *et al.*, 1997), cold methanol–water, or a surfactant such as Triton X-100 (Amini, 2001).

Preconcentration of Organic Phosphorus in Water

Unlike sediments, the concentrations of organic phosphorus compounds in water can be sufficiently low to make quantitative analysis difficult, especially if characterization by chromatography or nuclear magnetic resonance spectroscopy is to be performed. A number of different preconcentration approaches have been used to overcome this problem. Consideration should always be given to the potential for modification or degradation of organic phosphorus that may occur during such procedures.

Reverse osmosis and ultrafiltration

A number of authors have described the use of ultrafiltration, sometimes combined with reverse osmosis techniques, to concentrate and size-fractionate organic phosphorus in

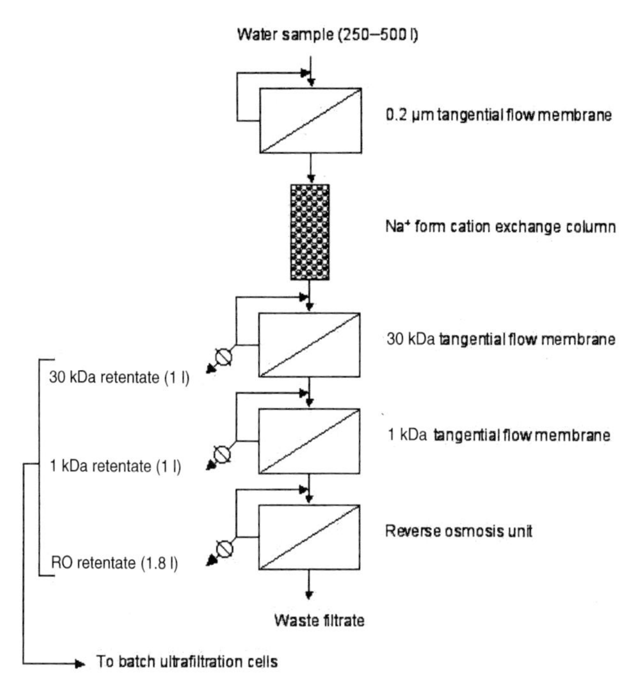

Fig. 1.1. Ultrafiltration and reverse osmosis system used for preconcentration of organic phosphorus. Reprinted from Nanny and Minear (1997) with permission from Elsevier Science.

waters (Cooper *et al.*, 1991; Nanny *et al.*, 1994). For example, Stewart *et al.* (1991) used a reverse osmosis module to perform a 40-fold concentration of organic phosphorus in water from the Lough Neagh catchment, with a recovery efficiency of 93%. Similarly, Nanny and Minear (1997) used Na^+ form Dowex 50×8 cation-exchange columns to remove ferric ions and avoid precipitation of calcium carbonate and magnesium during the preconcentration of filtered lake and river water samples by reverse osmosis/ultrafiltration. Samples were then concentrated and size-fractionated by passing them through a series of ultrafiltration and reverse osmosis membranes with successively smaller cut-off sizes (e.g. 30,000 Da, 1000 Da, or reverse osmosis), in such a manner that the retentate was continuously recycled while the filtrate underwent further filtration (see Fig. 1.1). In this manner, sample volumes of 250–500 l could be concentrated to final volumes of 1–2 l, which were further fractionated using batch ultrafiltration pressure cells, with 10,000, 5000, 1000 and 500 Da membrane cut-offs. Other authors described the use of ultrafiltration alone for the batch preconcentration of water samples using membranes typically with 500, 10,000 and 100,000 Da cut-off sizes (Eisenreich and Armstrong, 1977; Matsuda *et al.*, 1985).

Lyophilization

Lyophilization (freeze-drying) is used as a means of preconcentration, although this process can involve some physical or chemical modification of the sample (Galanos and Kapoulas, 1965; Nanny and Minear, 1997). For example, Minear (1972) achieved preconcentration of between 100- and 240-fold for water samples that had previously been treated to remove calcium and magnesium ions. Lyophilization can also be used to remove solvent *after* a separation process, and has been used as part of a clean-up and concentration step preparatory to chromatographic separation of inositol phosphate esters. A sample containing inositol phosphates was loaded on to an amino high-performance liquid chromatography column, washed with water and then eluted with ammonia, which was then removed by lyophilization (Heathers *et al.*, 1989). Rotary evaporation at less than 40°C has also been used to achieve 50-fold preconcentration of organic phosphorus in water samples (Hino, 1989).

Precipitation and coprecipitation

Significant preconcentration of filterable organic phosphorus was demonstrated by Stevens and Stewart (1982), who precipitated both filterable organic and reactive phosphorus by adding lanthanum (20 mg/l) and adjusting the pH to 6.5. Under these conditions, they were able to isolate approximately 44 mg/l of precipitate containing both filterable reactive phosphorus and filterable organic phosphorus from water samples as large as 100 l.

Magnesium-induced coprecipitation (Karl and Tien, 1992) involves the precipitation of brucite ($Mg(OH)_2$) from seawater by the addition of base. It was observed that phosphorus compounds, including organic phosphates, coprecipitate with the brucite, and this was used as the basis for preconcentration of phosphorus prior to analysis of total phosphorus in the collected precipitate. Nucleic acids have also been determined quantitatively by precipitation with cetyltrimethylammonium bromide, with DNA and RNA determined at μg/l levels by fluorometric and colorimetric procedures (Karl and Bailiff, 1989). Iron and barium salts are also commonly used as precipitants to isolate inositol phosphates after selective oxidation of organic phosphorus (Cosgrove, 1980).

Solid phase extraction in waters

Solid phase extraction procedures are now commonplace as a means of sample clean-up and preconcentration in chromatography. Espinosa *et al.* (1999) used strong anion exchange (SAX) solid phase cartridges to concentrate and isolate phosphorus-containing species from soil leachate prior to separation by high-performance liquid chromatography. Using this approach, 1000-fold preconcentration of a series of different organic phosphates was achieved for an original sample volume of 2 l. Similarly, Suzumura and Ingall (2001) used both solvent extraction and solid phase extraction cartridges of silica gel or aminopropyl-bonded silica gel to extract and clean up phospholipids prior to high-temperature combustion acid hydrolysis and photometric detection of the reactive phosphorus thus formed.

A technique for the extraction of total filterable phosphorus from seawater using iron (III) hydroxide-coated acrylic fibres was described by Lee *et al.* (1992). These fibres had previously been shown to adsorb adenosine 5′-triphosphate, DNA, glucose 1-phosphate and glucose 6-phosphate, and were used to preconcentrate large volumes (>1000 l) of seawater. While workers such as Baldwin *et al.* (1995, 1996) have demonstrated that hydrolysis of organic phosphorus can occur at mineral surfaces, only a small amount of hydrolysis (<3%) was observed in these studies.

Non-specific Determination of Organic Phosphorus in Waters and Extracts

Peroxydisulphate oxidation

Thermal oxidation using peroxydisulphate has been widely used for determination of organic phosphorus (Menzel and Corwin, 1965). Arguably only organic phosphorus should be oxidized by controlled digestion with alkaline peroxydisulphate. However, the timing of digestion and concentration of peroxydisulphate used are apparently critical, because this forms sulphuric acid when it decomposes, and under these acidic conditions condensed phosphorus will also be hydrolysed, as in the commonly used acid peroxydisulphate digestion.

Others have used a high-temperature (500°C) ashing technique that includes magnesium sulphate or magnesium nitrate as oxidant (Solórzano and Sharp, 1980;

Cembella *et al.*, 1986), and for total filterable phosphorus determination this is followed by hydrolysis with hydrochloric acid. Monaghan and Ruttenberg (1999) subsequently investigated claims that high-temperature ashing caused errors through volatilization losses (Ormaza-González and Statham, 1991), but found these to be insignificant.

Ultraviolet photo-oxidation

Ultraviolet photo-oxidation may be used to convert organic phosphorus to phosphate prior to detection, and this was the subject of a comprehensive review (Golimowski and Golimowska, 1996). Ultraviolet photo-oxidation can be performed in batch mode, using a high-wattage ultraviolet source and a quartz reactor vessel (Armstrong *et al.*, 1966; Henriksen, 1970), or in a continuous flow mode using either quartz or Teflon photoreactors (McKelvie *et al.*, 1989). More commonly, batch ultraviolet radiation systems have used high-wattage ultraviolet lamps ($\geq$1000 W) and lengthy irradiation times. Under these conditions, condensed phosphates may be hydrolysed, but this is an artefact of the elevated temperatures that are reached and the gradual acidification of the sample that occurs as peroxydisulphate decomposes to sulphuric acid. Solórzano and Strickland (1968) noted that ultraviolet photo-oxidation alone is insufficient to convert condensed phosphates to phosphate, and suggested that ultraviolet photo-oxidation provides a basis for discrimination between the organic and condensed phosphorus fractions.

Photo-oxidation of organic phosphorus may be performed by ultraviolet irradiation of the untreated sample by relying on the dissolved oxygen present as the source of oxidant. However, it is more usual that oxidants such as hydrogen peroxide, potassium peroxydisulphate, or ozone are added to enhance oxidation. When hydrogen peroxide is exposed to ultraviolet light, it forms hydroxyl radicals:

$$H_2O_2 + h\nu \rightarrow 2\ OH^{\bullet}$$

Hydroxyl radicals are among the strongest oxidizing agents present in aqueous systems (Hoigné and Bader, 1978), and can initiate radical chain reactions with organic substances present, resulting in mineralization of the sample. Ultraviolet photolysis of peroxydisulphate also produces hydroxyl radicals and oxygen by the following route:

$$S_2O_8^{2-} + h\nu \rightarrow 2\ SO_4^{\bullet}$$

$$SO_4^{\bullet} + H_2O \rightarrow HSO_4^{-} + OH^{\bullet}$$

$$S_2O_8^{2-} + OH^{\bullet} \rightarrow HSO_4^{-} + SO_4^{\bullet} + \tfrac{1}{2}\ O_2$$

$$SO_4^{\bullet} + OH^{\bullet} \rightarrow HSO_4^{-} + \tfrac{1}{2}\ O_2$$

Photo-oxidation of organic phosphorus can be achieved in batch mode, and the resultant phosphate is determined photometrically as phosphomolybdenum blue. However, the oxidation may be more conveniently performed using similar digestion and phosphate detection conditions, but with a flow-through photoreactor in flow injection (McKelvie *et al.*, 1989) or segmented continuous flow analysis systems (Aminot and Kérouel, 2001). Both approaches exhibit high sensitivity and rapid sample throughput (30–60 analyses/h). The residence time of sample in the flow injection photoreactor is $<$90 s (McKelvie *et al.*, 1989), but this is sufficient to oxidize even refractory compounds such as 2-aminoethylphosphonic acid, which requires $>$2 h ultraviolet exposure in a batch system (Cembella *et al.*, 1986).

Organic phosphorus can also be converted to phosphate using long wavelength ultraviolet light with titanium dioxide as a catalyst. Excitation of an electron from the valence band (*v*) into the conduction band (*c*) creates an electron-hole pair, which may then react with oxygen adsorbed to the titanium dioxide surface to form radicals such as $O_2^{\bullet-}$ and $OH^{\bullet}$. This approach was shown to very effectively mineralize organic phosphates (Low and Matthews, 1990; Matthews *et al.*, 1990). However, determining filterable organic phosphorus as the difference between molybdate-reactive phosphorus before and after photo-oxidation will in some instances lead to an underestimation of the true value. This may occur because the filterable reactive phosphorus measurement will include some acid-labile organic phosphates, and subtracting these from the filterable organic plus filterable reactive phosphorus concentration will lead to error. Selective measurement of phosphate by ion chromatography or weak anion exchange chromatography (Baldwin, 1998), as an alternative to the use of filterable reactive phosphorus, would overcome this problem.

Selective oxidation: hypobromite oxidation

Hypobromite oxidation with sodium hypobromite has been widely used as a means of isolating the congeners of inositol phosphates. The technique oxidizes all natural organic matter and organic phosphorus with the exception of the inositol phosphates, and has been widely used as a sample clean-up method for sediment and soil extracts preparatory to their separation by ion-exchange chromatography or gel filtration (Cosgrove, 1962; Eisenreich and Armstrong, 1977). Modified alkaline extraction and bromination conditions were claimed to provide improved recovery of inositol phosphates from the humic-acid fraction (Hong and Yamane, 1980).

Model compounds

The paucity of certified reference materials for organic phosphorus poses something of a challenge for researchers developing new methods of analysis. Until such time as certified reference materials become more readily available, it is advisable that spike recovery tests be included in any validation of a new method or analytical procedure. Kérouel and Aminot (1996) suggested that in any determination of filterable organic phosphorus in fresh or marine waters, several model organic phosphorus compounds should be used. These should include one or more labile (phosphoenolpyruvate, glycerophosphate, or riboflavin 5′-phosphate) and refractory (aminophosphonic acid, *myo*-inositol hexakisphosphate, or choline phosphate) organic phosphorus compounds.

Characterization and Selective Measurement of Organic Phosphorus Species

Enzymatic determinations

Enzymatic reactions can be used as a means of both characterizing and releasing classes of organic phosphorus present in waters, soils and sediments. For example, phosphate monoesters can be quantified by the use of a phosphomonoesterase such as alkaline phosphatase. Alternately, the high specificity of some enzymes for particular substrates can be used as the basis for determination of specific organic phosphorus compounds, or as part of a post-separation quantification step.

Enzymatic hydrolysis

Strickland and Solórzano (1966) were among the earliest workers to use alkaline phosphatase as a means of assessing the fraction of organic phosphorus that might become bioavailable through enzymatic hydrolysis by exocellular algal phosphatase released under conditions of phosphate depletion. In this approach, enzymatic hydrolysis of organic phosphorus yielded phosphate that was detected as molybdate-reactive phosphorus. Herbes *et al.* (1975) extended this approach to include alkaline phosphatase, phosphodiesterase and alkaline phosphatase in sequence, and phytase. They found that, while no inorganic phosphorus was liberated by alkaline phosphatase or phosphodiesterase in concert with alkaline phosphatase, significant amounts were released by phytase. Phytase hydrolysis of different fractions obtained using gel filtration suggested that the low molecular mass fraction of the phytase-hydrolysable organic phosphorus was *myo*-inositol hexakisphosphate of sedimentary origin, while the high molecular mass fraction comprised lipid or fulvic acid-associated inositol phosphates from soils.

Francko and Heath (1979) used both gel filtration and ion-exchange chromatography to fractionate organic phosphorus and inorganic phosphorus, and performed functional characterization using either alkaline phosphatase hydrolysis or low-intensity ultraviolet irradiation. Their studies suggested that phosphate was released from two distinct classes of compounds, namely organic phosphorus substrates that underwent hydrolysis in the presence of alkaline phosphatase, and phosphate adsorbed to iron (III) oxyhydroxides or organic colloids that were solubilized by ultraviolet-induced photoreduction of iron (III). Later work on preconcentrated waters showed that alkaline phosphatase and phosphodiesterase in concert hydrolysed approximately 14% of the filterable organic phosphorus, and caused a similar degree of hydrolysis in substrates with molecular masses ranging between 500 and 100,000 Da (Cooper *et al.*, 1991). Similar application has been made of enzyme-induced inorganic phosphorus release in soils and soil waters (Pant *et al.*, 1994; Shand and Smith, 1997; Turner *et al.*, 2002) and sediments (Feuillade and Dorioz, 1992). However it has been observed that some commercial enzyme preparations (e.g. alkaline phosphatase) are inhibited by free phosphate, and it is arguable that the extent of phosphate release observed *in vitro* may be less than that which would occur naturally. Chróst *et al.* (1986) addressed this by determining the net phosphate release during a 2–5 day incubation of sterilized seawater samples that were buffered at the natural pH.

Shan *et al.* (1994) used alkaline phosphatase from *Escherichia coli* in immobilized enzyme reactors in a flow injection analysis system to perform rapid on-line enzymatic hydrolysis of organic phosphorus and detection of the released phosphate as molybdate-reactive phosphorus. This approach has several advantages compared with the conventional batch reaction approach:

- an immobilized enzyme reactor can be used for several hundred enzymatic hydrolysis experiments before activity declines;
- phosphate inhibition is minimized because hydrolysis products are rapidly transported away from the active sites;
- the sample reaction time (3 min) is extremely rapid compared with the previously described batch methods, which take hours or even days to perform (Shan *et al.*, 1993).

However, this approach proved unsuccessful when applied to soil waters because of adsorption of iron hydroxides or natural organic matter to the enzyme reactor, causing rapid loss of enzyme activity.

Immobilized phytase (from *Aspergillus ficuum*) has been applied in a similar manner to determine phosphate release from waters (McKelvie *et al.*, 1995), but attempts to immobilize phosphodiesterase alone or co-immobilized with alkaline phosphatase proved unsuccessful because of loss of phosphodiesterase activity. Alkaline phosphatase, phytase and phosphodiesterase are relatively non-specific enzymes, and their use in applications such as those described above provides information on the broad functional classes of organic phosphorus compounds present.

Enzymatic determination of specific organic phosphorus compounds

The high substrate specificity of some enzymes can be used as the basis for highly selective determinations of organic phosphorus species. For example, Amini (2001) determined phosphatidyl choline using co-immobilized phospholipase C, alkaline phosphatase and choline oxidase (Fig. 1.2a). A stoichiometric product of this sequence is hydrogen peroxide, and this was determined using the luminol chemiluminescence reaction, using the flow injection manifold shown (Fig. 1.2b) to provide rapid, sensitive and reproducible indirect measurement of phosphatidyl choline (Amini, 2001).

Marko-Varga and Gorton (1990) described an immobilized enzyme system for post-anion-exchange detection of inositol phosphates. This sequence used immobilized alkaline phosphatase to dephosphorylate lower inositol phosphates, and the inositol thus produced underwent further reactions with co-immobilized inositol dehydrogenase, lactate dehydrogenase and lactate oxidase to produce hydrogen peroxide. A third enzyme reactor containing horseradish peroxidase was used to catalyse the reaction between hexacyanoferrate (II) and hydrogen peroxide, which was monitored amperometrically. Monitoring of the hydrogen peroxide produced by this enzyme sequence has also been performed using chemiluminescence (Gudermann and Cooper, 1986).

Chromatographic Separations

Separation and quantification of organic phosphorus species have been performed using all of the commonly used chromatographic techniques. However, most organic phosphorus compounds do not contain strongly ultraviolet-absorbing chromophores, so post-separation fraction collection and measurement of total phosphorus has been a common means of quantification in liquid chromatography. A number of authors have also described online post-column detection or derivatization systems for sensitive detection of organic phosphorus. Similarly, it is often necessary to convert organic phosphorus species to more volatile derivatives if gas chromatography is to be used for separation.

Size exclusion chromatography/gel filtration

Gel filtration chromatography has been widely used to separate high and low molecular mass phosphorus in sediments, soil solutions, and preconcentrated surface water. Steward and Tate (1971) showed, for example, that Sephadex G50 could be used to separate high molecular mass phosphorus that corresponded with the molecular weight of *myo*-inositol hexakisphosphate. Others used Sephadex gel filtration separations to demonstrate the presence and importance of inositol phosphates in preconcentrated lake water (Eisenreich and Armstrong, 1977), algal cultures (Minear, 1972), river water (Stevens and Stewart, 1982) and marine water (Matsuda *et al.*, 1985).

A focus of the early work in this area was directed towards discriminating between the high-molecular-mass phosphorus fraction and the more bioavailable low-molecular-mass phosphorus fraction in aquatic systems, because it was recognized that the formation of phosphomolybdate was not selective for phosphate, but also involved the hydrolysis of organic phosphates (Lean, 1973). Other authors reported a proportion of the high-molecular-mass phosphorus in lake waters to be molybdate-reactive (Downes and Paerl, 1978), with the implication that this fraction might also be bioavailable. White and Payne (1980) found that reactive high-molecular-mass phosphorus comprised 70–80% of total filterable reactive phosphorus in eutrophic lake water, and that this was available to phosphorus-starved algae, although it was not as readily available as free phosphate.

In most of these studies, gel filtration was performed on large columns, and the organic phosphorus concentration of each eluting fraction was determined by performing total or reactive phosphorus analysis off-line. As an alternative, McKelvie *et al.* (1993) used small, high-resolution Sephadex G25 columns connected to an on-line ultraviolet photo-oxidation flow injection system for continuous detection of eluting inorganic and organic phosphorus species (Fig. 1.3a). The advantages of this approach are the rapid separations and the low detection limit for phosphorus species of <10 μg/l. Consequently, water samples or sediment extracts can be analysed without preconcentration. Figure 1.3b shows typical separations of pore water and overlying water for a eutrophic coastal lagoon sediment, in which the predominance of organic phosphorus (corresponding to the elution time for *myo*-inositol hexakisphosphate) in the immediate subsurface pore waters is evident (Fig. 1.3c; M. Hindle and I.D. McKelvie, unpublished data). However, anion exclusion and both reversible and specific adsorption phenomena are common in gel filtration separations (Downes and Paerl, 1978; Stevens and Stewart, 1982), and can only be eliminated by careful control of elu-

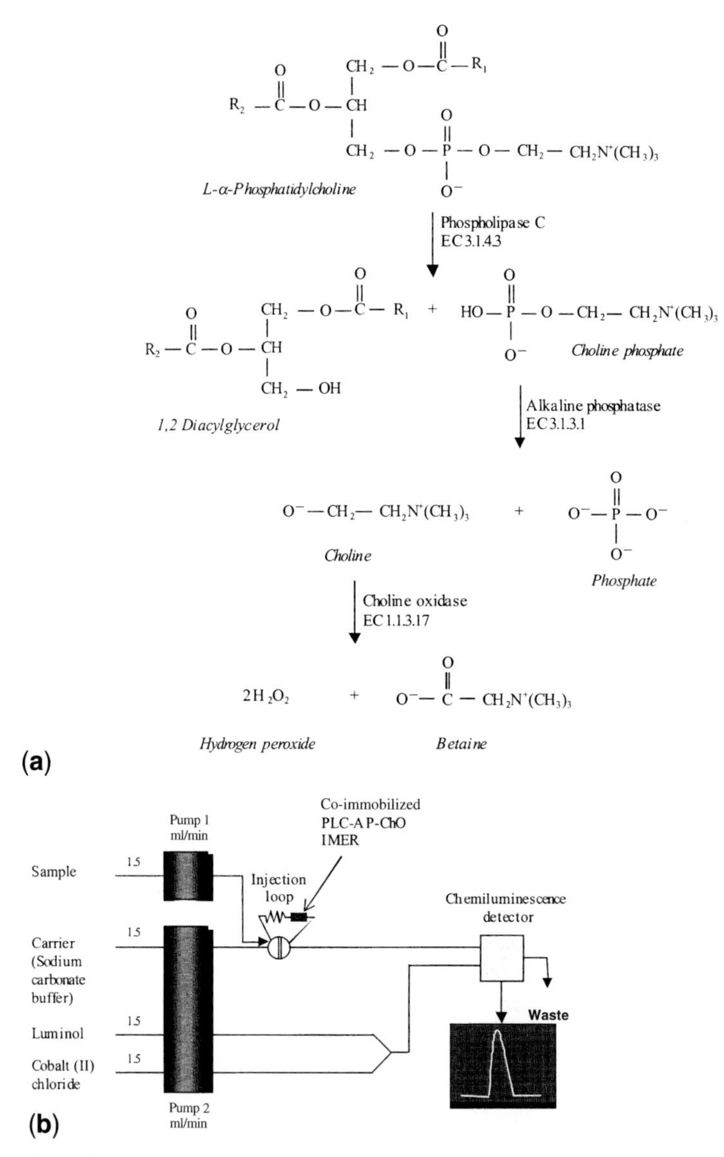

Fig. 1.2. (a) Scheme for enzymatic determination of phosphatidyl choline (PC) using phospholipase C, alkaline phosphatase and choline oxidase. (b) Schematic diagram of the flow injection manifold including a co-immobilized phospholipase C-alkaline phosphatase–choline oxidase immobilized enzyme reactor (PLC-AP-ChO IMER) for the conversion of phosphatidyl choline to hydrogen peroxide. The hydrogen peroxide produced was detected using a chemiluminescence detector.

tion conditions (McKelvie *et al.*, 1993). Post-separation conditioning of the column with EDTA (Fig. 1.3a) was found to give improved reproducibility of separations (M. Hindle and I.D. McKelvie, unpublished data). One suggestion is that adsorptive behaviour of this type is due to complexation by metal ions bound to the gel, which might be overcome by adding a complexing agent to the eluent (Town and Powell, 1992). To some extent, gel filtration separations have been superseded by ion-exchange separations; although, if the

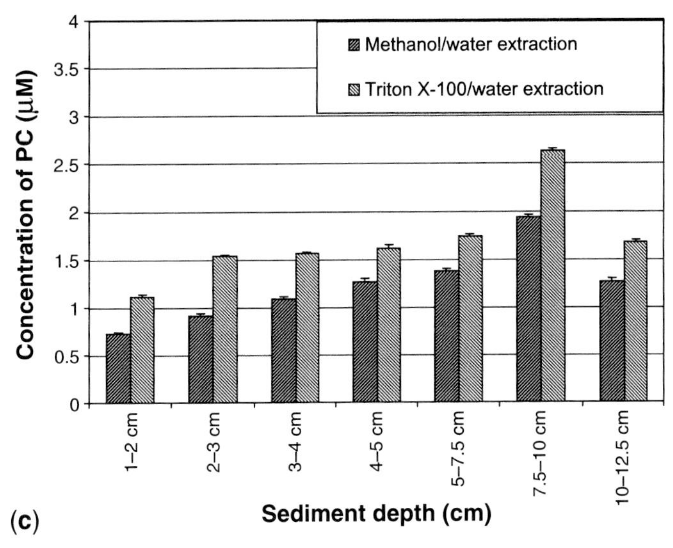

Fig. 1.2. (*continued*) (c) Phosphatidyl choline concentrations in Yarra River sediment extracts (methanol/water and Triton X-100/water). The bar indicates the range for duplicate measurements of each sample (Amini, 2001).

limitations of gel filtration are recognized, there is still considerable information to be obtained from this technique. For example, Hens and Merckx (2001) recently used gel filtration to study the partitioning of phosphate between the aqueous and colloid phases in soil solutions.

Ion-exchange chromatography

The ionizable phosphate group present in organic phosphorus species provides the basis for separation by anion exchange chromatography. This approach has been applied to the separation of phosphate, inositol phosphates, organic condensed phosphates, nucleic acids, nucleoside phosphonates, phospholipids and other organic phosphorus species.

Anion-exchange chromatography has been used extensively for the separation of inositol phosphate in soils (Cosgrove, 1969a), sediments (Sommers *et al.*, 1972; Weimer and Armstrong, 1977; Suzumura and Kamatani, 1995) and waters (Minear *et al.*, 1988) (Fig. 1.4). Many early reports of ion-exchange separations used large columns (typically 15–60 × 1 cm) of Dowex AG resins, long elution times, and required the collection and off-line digestion and subsequent spectrophotometric detection of reactive phosphorus. Most separations typically required gradient elution with, for example, hydrochloric acid to elute strongly retained species such as *myo*-inositol hexakisphosphate (Cosgrove, 1969b).

More recently, ion-exchange separations have been performed using ion-exchange chromatography, high-performance liquid chromatography, or ion chromatography, to achieve better resolution and more rapid sample throughput (Minear *et al.*, 1988; Skoglund *et al.*, 1997; Espinosa *et al.*, 1999) (Figs 1.4 and 1.5). In a number of cases, post-column derivatization (Rounds and Nielsen, 1993) or post-column photo-oxidation followed by molybdate-reactive phosphorus measurement was employed, thus avoiding the need for time-consuming fraction collection and off-line digestion and spectrophotometry (Clarkin *et al.*, 1992; Nanny *et al.*, 1995). A similar approach involving on-line thermal digestion was also described for the separation and detection of nucleoside H-phosphonates (Baba *et al.*, 1990).

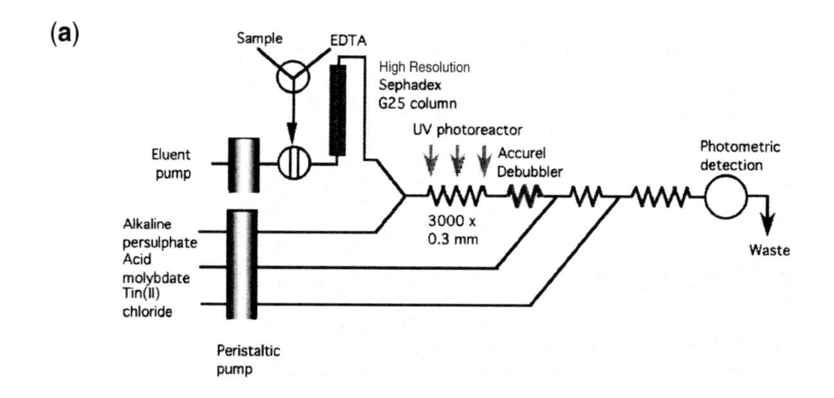

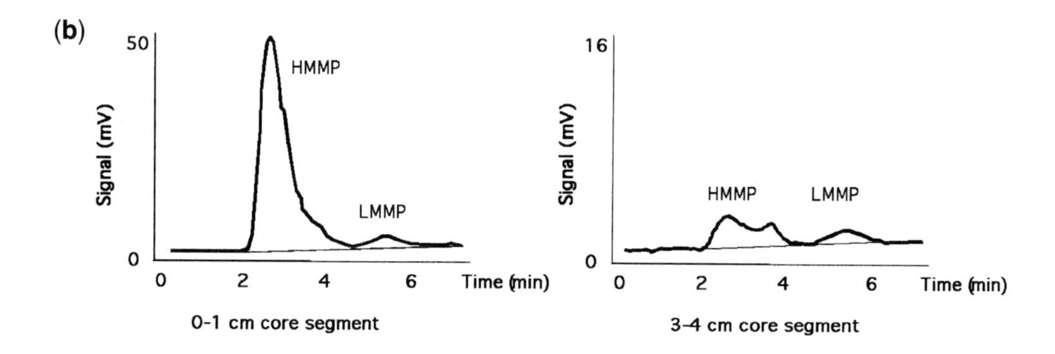

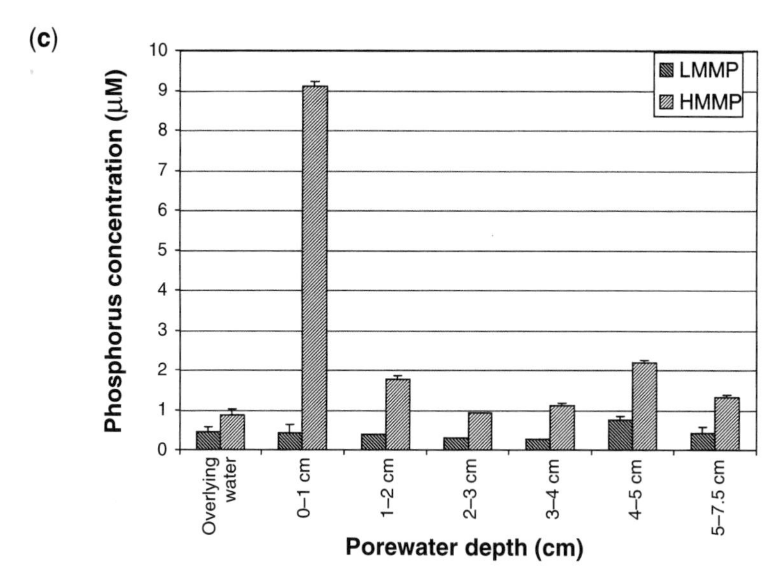

Fig. 1.3. (a) Low pressure separation of high (HMMP) and low (LMMP) molecular mass organic phosphorus species by gel filtration with continuous on-line detection of organic phosphorus species. (b) Typical gel filtration chromatograms for high and low molecular mass phosphorus obtained using system shown in (a). (c) Graph showing relative concentrations of high and low molecular mass phosphorus in sediment pore and overlying water (M. Hindle and I.D. McKelvie, unpublished data). Error bars are the mean of three replicates.

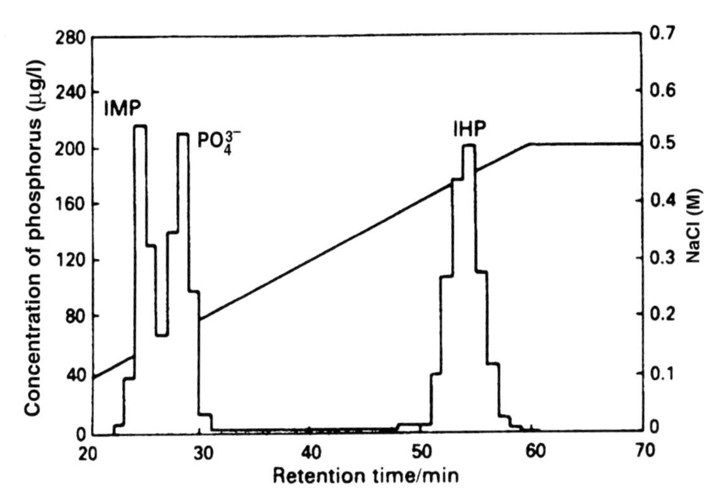

Fig. 1.4. Gradient elution ion-exchange chromatography of an inositol phosphate standard mixture. Concentrations (mg/l): phosphate (PO_4^{3-}) = 25, inositol monophosphate (IMP) = 28, *myo*-inositol hexakisphosphate (IHP) = 34. Reproduced from Minear *et al.* (1988) with permission from The Royal Society for Chemistry.

Other detection methods employed have included the use of fluorescence for the determination of bisphosphonates (Lovdahl and Pietrzyk, 1999), while ion-exchange has been used for separation of organic phosphorus moieties prior to their derivatization and determination by gas chromatography (Heathers *et al.*, 1989). Ion-exchange separations have been performed on untreated waters and sediment extracts, preconcentrated samples, or on samples that were subjected to hypobromite oxidation to isolate the inositol phosphates (Suzumura and Kamatani, 1995). Some organic phosphorus anions, including *myo*-inositol hexakisphosphate, are highly charged and are strongly retained by ion-exchange stationary phases. For this reason, gradient elution using high eluent salt concentrations (e.g. 0.5 M) is often required. On-column hydrolysis may be a problem for some organic phosphorus species, although the incorporation of sodium EDTA in the eluent has been found to minimize this. Adsorption of organic phosphorus to ion-exchange columns may be quite pronounced, and it is often necessary to re-equilibrate columns or even pass through a cleaning solution of EDTA in order to maintain satisfactory recovery and precision.

Partition chromatography

While ion-exchange high-performance liquid chromatography is used extensively for the separation of organic phosphorus species, a number of authors have shown that reversed-phase partition chromatography has a definite role in the separation of even quite charged or polar organic phosphates. Phospholipids are commonly separated and quantified using either reversed-phase (Lesnefsky *et al.*, 2000; Guan *et al.*, 2001) or normal-phase (Aries *et al.*, 2001) high-performance liquid chromatography, and this approach may often be used as a tool in studying sediment microbial communities (Martz *et al.*, 1983; Nichols *et al.*, 1987; Rütters *et al.*, 2002a,b; Zink *et al.*, 2003). Reversed-phase chromatography has also been used in the study of inositol phosphates. For example, Graf and Dintzis (1982) performed sample clean-up and preconcentration by ion exchange before separation of *myo*-inositol hexakisphosphate on a C18 reversed-phase column. A major advantage of this approach is that the elution times of the highly charged *myo*-inositol hexakisphosphate are quite short.

The chiral separation of *myo*-inositol 1,4,5-trisphosphate and some isomers using

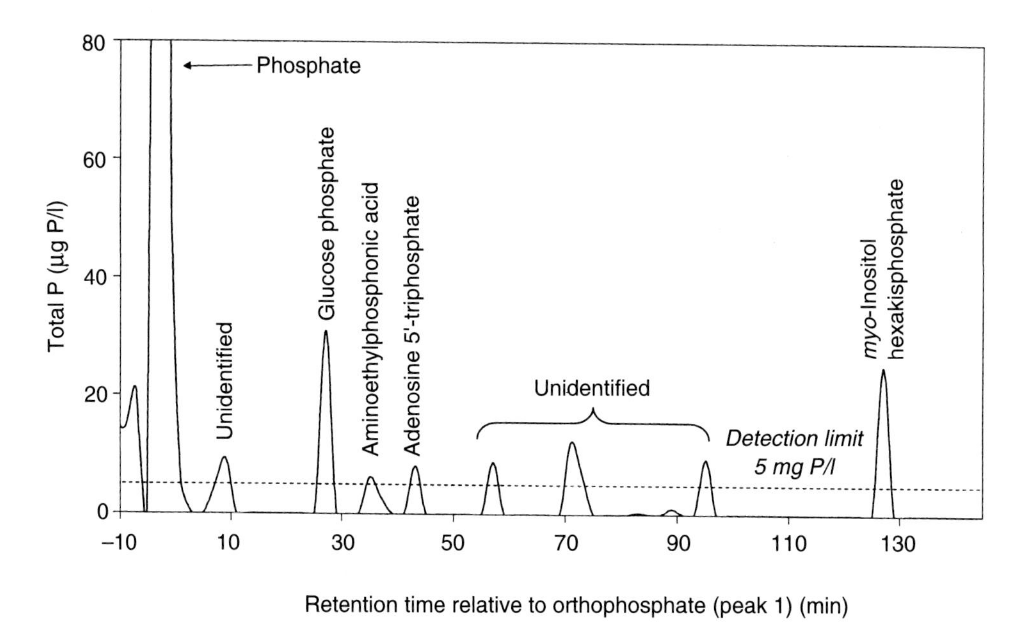

Fig. 1.5. High-performance liquid chromatographic separation of phosphorus compounds determined in soil leachate using a strong anion exchange stationary phase. Reprinted from Espinosa *et al.* (1999) with permission from the American Society of Agronomy.

a β-cyclodextrin-bonded column under reversed-phase conditions was described by Burmester and Jastorff (1996). The use of ion-pair reversed-phase chromatography is also advantageous for the separation of highly charged species and may avoid some of the adsorption problems encountered using gel filtration or ion-exchange chromatography, especially of highly charged species. Reversed-phase ion-pair chromatographic separations have been described for inositol phosphates (Patthy *et al.*, 1990; Shayman and Barcelon, 1990; Hamada, 2002) and some common nucleotides and sugar phosphates (Patthy *et al.*, 1990).

Inositol phosphates isomers have also been separated by micellar mobile high-performance liquid chromatography (Brando *et al.*, 1990). This involved addition of the surfactant hexadecyltrimethylammonium hydroxide ($HDTMA^+OH^-$) to the mobile phase to form micellar ion association complexes that were separated on a reversed-phase column. Enhanced sensitivity and selectivity can be achieved by use of high-performance liquid chromatography in conjunction with a selective detection system, such as electrospray mass spectrometry (Lytle *et al.*, 2000).

Gas chromatography

Gas chromatography with either flame photometric or mass spectrometric detection is commonly used for determination of organophosphate pesticides (Santos and Galceran, 2002). Analysis of derivatized samples may also be a useful approach in some instances. To separate the *myo*- and *chiro*-isomers of inositol tetrakis-, tris-, bis- and monophosphates, Heathers *et al.* (1989) dephosphorylated preconcentrated samples using magnesium and alkaline phosphatase. The resultant solution was then lyophilized and reacted with bis-(trimethylsylyl)trifluoroacetomide to form the hexatrimethylsilyl derivatives. These were separated using gas chromatography to allow quantification of the *myo*- and *chiro*-inositols. While this approach was developed for biological samples, it should be equally applicable to environmental samples given sufficient preconcentration. A similar procedure was described by Steward and Tate (1971), who prepared inositol acetates by reaction of dephosphorylated inositol phosphates with acetic anhydride.

Capillary electrophoresis

The high separation efficiency of capillary electrophoresis makes it an attractive technique for the fast separation and determination of organic phosphorus compounds. Unfortunately, most organic phosphorus compounds are poor ultraviolet chromophores, and sensitivity is often limited compared with other techniques. Alternative detection techniques must therefore be employed if this approach is to be useful for separation and identification of organic phosphorus compounds in waters and sediments. Henshall *et al.* (1992) used capillary electrophoresis with indirect photometric detection to determine inositol phosphates in physiological samples, while Buscher *et al.* (1995) separated all six inositol phosphate esters using electrospray mass spectrometry for detection. The limits of detection achieved using this approach were in the micromolar range. Capillary electrophoresis has also been applied to the determination of individual organophosphate species, such as chlorpyrifos (Guardino *et al.*, 1998) and bisphosphonates (Lovdahl and Pietrzyk, 1999).

Spectrometric Techniques

Nuclear magnetic resonance spectroscopy, especially ^{31}P, offers a powerful means of characterizing and identifying individual organic phosphorus moieties in soils (Lookman *et al.*, 1997; Cade-Menun *et al.*, 2002) and sediments (Carman *et al.*, 2000). Phosphate, phosphonates, phosphate monoesters and diesters, pyrophosphate and polyphosphates all exhibit signals at different chemical shifts, which is the basis for their identification and quantification, although this may be complicated by overlapping signals, especially in the phosphate monoester region (Turner *et al.*, 2003). In the case of water samples, there is usually a need for preconcentration (Nanny and Minear, 1997), although this may cause interference from the corresponding increase in paramagnetic ion concentrations. A detailed description of ^{31}P nuclear magnetic resonance techniques can be found elsewhere in this book (see Cade-Menun, Chapter 2).

Gas chromatography–mass spectrometry combines high efficiency separation with sensitive and highly selective detection technique, and is a powerful tool for characterization of organic phosphorus (Llewelyn *et al.*, 2002). Similarly, the use of electrospray mass spectrometry provides greatly enhanced detection capability for high-performance liquid chromatography (Lytle *et al.*, 2001) and particularly capillary electrophoresis (Buscher *et al.*, 1995). For this reason, mass spectrometry-based detection methods can only gain in popularity (see Cooper *et al.*, Chapter 3 this volume).

The Future

> I think that instead of losing time using inappropriate fractionation and culture techniques, (which probably give irreproducible results) our efforts should first be directed towards the development of satisfactory techniques and culture methods.
>
> Golterman (1988)

The same comment might well be applied to some of the techniques currently used for phosphorus analysis. There is a real need for improved measurement techniques for determining both total organic phosphorus and the concentrations of individual organic phosphorus moieties. Given the tendency for molybdate colorimetry to underestimate organic phosphorus and overestimate inorganic phosphorus, this may entail a shift from the often uncritical reliance on this detection chemistry to techniques with greater specificity. Such techniques may be based on chromatographic or electrophoretic separations combined with post-column molybdate-reactive chemistry, specific enzymatic reactions, or mass spectrometry.

Application of flow analysis or other automation techniques to unit operations such as extraction, digestion or photo-oxidation may also prove beneficial for better control and reproducibility of processing conditions and hence the quality of the final result. Another important area requiring attention is the thorny issue of sampling. Alteration of the sample is implicit in the act of sampling, and it is not a matter of whether or not alteration occurs, but to what degree. Changes in temperature, moisture, light and redox regime may all affect the speciation of organic phosphorus. Thus, the application of *in situ* techniques that minimize perturbation of the sample is highly desirable. Use of passive, diffusive sampling devices, such as peepers or diffusive gradients in thin films (Zhang *et al.*, 1998) may well prove to be suitable for the sampling of organic phosphorus in sediments, soils and waters. Application of these devices to date has focused on sampling and measurement of the most bioavailable ionic forms of metals or nutrients (e.g. filterable reactive phosphorus), but there is certainly a role for their use in the study of organic nutrients.

References

Amini, N. (2001) Development of enzymatic flow injection systems for the determination of phospholipid species in environmental samples. MSc thesis, Monash University, Clayton, Australia.

Aminot, A. and Kérouel, R. (2001) An automated photo-oxidation method for the determination of dissolved organic phosphorus in marine and fresh water. *Marine Chemistry* 76, 113–126.

Aries, E., Doumenq, P., Artaud, J., Molinet, J. and Bertrand, J.C. (2001) Occurrence of fatty acids linked to non-phospholipid compounds in the polar fraction of a marine sedimentary extract from Carteau Cove, France. *Organic Geochemistry* 32, 193–197.

Armstrong, F.A.J., Williams, P.N. and Strickland, J.D.H. (1966) Photooxidation of organic matter in seawater by ultraviolet radiation, analytical and other applications. *Nature* 211, 481.

Baba, Y., Tsuhako, M. and Yoza, N. (1990) Rapid and sensitive determination of nucleoside H-phosphonates and inorganic H-phosphoates by high-performance liquid chromatography coupled with flow-injection analysis. *Journal of Chromatography* 507, 103–111.

Baldwin, D.S. (1998) Reactive 'organic' phosphorus revisited. *Water Research* 32, 2265–2270.

Baldwin, D.S., Beattie, J.K., Coleman, L.M. and Jones, D.R. (1995) Phosphate ester hydrolysis facilitated by mineral phases. *Environmental Science and Technology* 29, 1706–1709.

Baldwin, D.S., Beattie, J.K. and Jones, D.R. (1996) Hydrolysis of organic phosphorus compounds by iron-oxide impregnated filter papers. *Water Research* 30, 1123–1126.

Barbanti, A., Begamini, M.C., Frascari, F., Miserocchi, S. and Rosso, G. (1994) Ecosystem processes: critical aspects of sedimentary phosphorus chemical fractionation. *Journal of Environmental Quality* 23, 1093–1102.

Bjorkman, K.M. and Karl, D.M. (2003) Bioavailability of dissolved organic phosphorus in the euphotic zone at Station ALOHA, North Pacific Subtropical Gyre. *Limnology and Oceanography* 48, 1049–1057.

Brando, C., Hoffman, T. and Bonvini, E. (1990) High-performance liquid chromatographic separation of inositol phosphate isomers employing a reversed-phase column and a micellar mobile phase. *Journal of Chromatography* B 529, 65–80.

Broberg, O. and Persson, G. (1988) Particulate and dissolved phosphorus forms in freshwater: composition and analysis. *Hydrobiologia* 170, 61–90.

Burmester, A. and Jastorff, B. (1996) Enantioseparation in the synthesis of *myo*-inositol phosphates by high-performance liquid chromatography using a β-cyclodextrin-bonded column. *Journal of Chromatography* A 749, 25–32.

Buscher, B.A.P., van der Hoeven, R.A.M., Tjaden, U.R., Andersson, E. and van der Greef, J. (1995) Analysis of inositol phosphates and derivatives using capillary zone electrophoresis–mass spectrometry. *Journal of Chromatography* A 712, 235–243.

Cade-Menun, B.J., Liu, C.W., Nunlist, R. and McColl, J.G. (2002) Soil and litter phosphorus-31 nuclear magnetic resonance spectroscopy: extractants, metals and P relaxation times. *Journal of Environmental Quality* 31, 457–465.

Carman, R., Edlund, G. and Damberg, C. (2000) Distribution of organic and inorganic phosphorus compounds in marine and lacustrine sediments: a ^{31}P NMR study. *Chemical Geology* 163, 101–114.

Cembella, A.D., Antia, N.J. and Taylor, F.J.R. (1986) The determination of total phosphorus in seawater by nitrate oxidation of the organic component. *Water Research* 20, 1197–1199.

Chang, S.C. and Jackson, M.L. (1957) Fractionation of soil phosphorus. *Soil Science* 84, 133–144.

Chróst, R.Z., Siuda, W., Albrecht, D. and Overbeck, J. (1986) A method for determining enzymatically hydrolyzable phosphate (EHP) in natural waters. *Limnology and Oceanography* 31, 662–667.

Chu, S.P. (1946) The utilization of organic phosphorus by phytoplankton. *Journal of the Marine Biological Association of the UK* 26, 285–295.

Clarkin, C.M., Minear, R.A., Kim, S. and Elwood, J.W. (1992) An HPLC postcolumn reaction system for phosphorus-specific detection in the complete separation of inositol phosphate congeners in aqueous samples. *Environmental Science and Technology* 26, 199–204.

Condron, L.M., Turner, B.L. and Cade-Menun, B.J. (2005) Chemistry and dynamics of soil organic phosphorus. In: Sims, J.T. and Sharpley, A.N. (eds) *Phosphorus: Agriculture and the Environment.* ASA/CSSA/ SSSA, Madison, Wisconsin.

Cooper, J.E., Early, J. and Holding, A.J. (1991) Mineralization of dissolved organic phosphorus from a shallow eutrophic lake. *Hydrobiologia* 209, 89–94.

Cosgrove, D.J. (1962) Forms of inositol hexaphosphate in soils. *Nature* 194, 1265–1266.

Cosgrove, D.J. (1969a) The chemical nature of soil organic phosphorus. II. Characterization of the supposed DL-*chiro*-inositol hexaphosphate component of soil phytate as D-*chiro*-inositol hexaphosphate. *Soil Biology and Biochemistry* 1, 325–327.

Cosgrove, D.J. (1969b) Ion-exchange chromatography of inositol polyphosphates. *Annals of the New York Academy of Sciences* 165, 677–686.

Cosgrove, D.J. (1980) *Inositol Phosphates: Their Chemistry, Biochemistry and Physiology.* Elsevier Scientific, Amsterdam, 197 pp.

Cotner, J.B. and Wetzel, R.G. (1992) Uptake of dissolved inorganic and organic phosphorus compounds by phytoplankton and bacterioplankton. *Limnology and Oceanography* 37, 232–243.

de Groot, C.J. and Golterman, H.L. (1993) On the presence of organic phosphate in some Camargue sediments: evidence for the importance of phytate. *Hydrobiologia* 252, 117–126.

Downes, M.T. and Paerl, H.W. (1978) Separation of two dissolved reactive phosphorus fractions in lake water. *Journal of the Fisheries Research Board of Canada* 35, 1636–1639.

Eisenreich, S.J. and Armstrong, D.E. (1977) Chromatographic investigations of inositol phosphate esters in lake waters. *Environmental Science and Technology* 11, 497–501.

Espinosa, M., Turner, B.L. and Haygarth, P.M. (1999) Preconcentration and separation of trace phosphorus compounds in soil leachate. *Journal of Environmental Quality* 28, 1497–1504.

Feuillade, M. and Dorioz, J.M. (1992) Enzymatic release of phosphate in sediments of various origins. *Water Research* 26, 1195–1201.

Francko, D.A. and Heath, R.T. (1979) Functionally distinct classes of complex phosphorus compounds in lake water. *Limnology and Oceanography* 24, 463–473.

Galanos, D.S. and Kapoulas, V.M. (1965) Preparation and analysis of lipid extracts from milk and other tissues. *Biochimica et Biophysica Acta: Lipids and Lipid Metabolism* 98, 278–292.

Gardolinski, P.C.F.C., Worsfold, P.J. and McKelvie, I.D. (2004) Seawater induced release and transformation of organic and inorganic phosphorus from river sediments. *Water Research* 38, 688–692.

Golimowski, J. and Golimowska, K. (1996) UV-photooxidation as a pretreatment step in inorganic analysis of environmental samples. *Analytica Chimica Acta* 325, 111–133.

Golterman, H.L. (1988) Reflections on fractionation and bioavailability of sediment bound phosphorus. *Archiv für Hydrobiologie* 30, 1–4.

Graf, R. and Dintzis, F.R. (1982) High-performance liquid chromatographic method for the determination of phytate. *Analytical Biochemistry* 119, 413–417.

Guan, Z., Grunler, J., Piao, S. and Sindelar, P.J. (2001) Separation and quantitation of phospholipids and their ether analogues by high-performance liquid chromatography. *Analytical Biochemistry* 297, 137–143.

Guardino, X., Obiols, J., Rosell, M.G., Farran, A. and Serra, C. (1998) Determination of chlorpyrifos in air, leaves and soil from a greenhouse by gas-chromatography with nitrogen-phosphorus detection, high-performance liquid chromatography and capillary electrophoresis. *Journal of Chromatography* A 823, 91–96.

Gudermann, T.W. and Cooper, T.G. (1986) A sensitive bioluminescence assay for myoinositol. *Analytical Biochemistry* 158, 59–63.

Hamada, J.S. (2002) Scale-up potential of ion-pair high-performance liquid chromatography method to produce biologically active inositol phosphates. *Journal of Chromatography* A 944, 241–248.

Heathers, G.P., Juehne, T., Rubin, L.J., Corr, P.B. and Evers, A.S. (1989) Anion exchange chromatographic separation of inositol phosphates and their quantification by gas chromatography. *Analytical Biochemistry* 176, 109–116.

Henriksen, A. (1970) Determination of total nitrogen, phosphorus and iron in freshwater by photo-

oxidation with ultraviolet radiation. *Analyst* 95, 601–608.
Hens, M. and Merckx, R. (2001) Functional characterization of colloidal phosphorus species in the soil solution of sandy soils. *Environmental Science and Technology* 35, 493–500.
Henshall, A., Harrold, M.P. and Tso, J.M.Y. (1992) Separation of inositol phosphates by capillary electrophoresis. *Journal of Chromatography* A 608, 413–419.
Herbes, S.E., Allen, H.E. and Mancy, K.H. (1975) Enzymatic characterization of soluble organic phosphorus in lake water. *Science* 187, 432–434.
Hino, C. (1989) Characterization of orthophosphate release from dissolved phosphorus by gel filtration and several hydrolytic enzymes. *Hydrobiologia* 174, 49–55.
Hoigné, J. and Bader, H. (1978) Ozone and hydroxyl radical-initiated oxidations of organic and organometallic trace impurities in water. In: Brinckman, F.E. and Bellama, J.M. (eds) *Organometallics and Organometalloids: Occurrence and Fate in the Environment.* American Chemical Society, Washington, DC, pp. 292–313.
Hong, J.K. and Yamane, I. (1980) Examination of the conventional method and a proposal for a new method of determining inositol phosphate in the humic acid fraction. *Soil Science and Plant Nutrition (Tokyo)* 26, 497–505.
Karl, D.M. and Bailiff, M.D. (1989) The measurement and distribution of dissolved nucleic acids in aquatic environments. *Limnology and Oceanography* 34, 543–558.
Karl, D.M. and Tien, G. (1992) MAGIC: a sensitive and precise method for measuring dissolved phosphorus in aquatic environments. *Limnology and Oceanography* 37, 105–116.
Kérouel, R. and Aminot, A. (1996) Model compounds for the determination of organic and total phosphorus dissolved in natural waters. *Analytica Chimica Acta* 318, 385–390.
Lean, D.R.S. (1973) Movement of phosphorus between its biologically important forms in lake water. *Journal of the Fisheries Research Board of Canada* 30, 1525–1536.
Lee, T., Barg, E. and Lal, D. (1992) Techniques for extraction of dissolved inorganic and organic phosphorus from large volumes of sea water. *Analytica Chimica Acta* 260, 113–121.
Lesnefsky, E.J., Stoll, M.S.K., Minkler, P.E. and Hoppel, C.L. (2000) Separation and quantitation of phospholipids and lysophospholipids by high-performance liquid chromatography. *Analytical Biochemistry* 285, 246–254.
Llewelyn, J.M., Landing, W.M., Marshall, A.M. and Cooper, W.T. (2002) Electrospray ionization transform ion cyclotron resonance mass spectrometry of dissolved organic phosphorus species in a treatment wetland after selective isolation and concentration. *Analytical Chemistry* 74, 600–606.
Lookman, R., Grobet, P., Merckx, R. and van Reimsdijk, W.H. (1997) Application of ^{31}P and ^{27}Al MAS NMR for phosphate speciation studies in soil and aluminium hydroxides: promises and constraints. *Geoderma* 80, 369–388.
Lovdahl, M.J. and Pietrzyk, D.J. (1999) Anion-exchange separation and determination of bisphosphonates and related analytes by post-column indirect fluorescence detection. *Journal of Chromatography* A 850, 143–152.
Low, G.K.-C. and Matthews, R. (1990) Flow-injection determination of organic contaminants in water using an ultraviolet-mediated titanium dioxide film reactor. *Analytica Chimica Acta* 231, 13–20.
Lytle, C.A., Gan, Y.D. and White, D.C. (2000) Electrospray ionization/mass spectrometry compatible reversed-phase separation of phospholipids: piperidine as a post-column modifier for negative ion detection. *Journal of Microbiological Methods* 41, 227–234.
Lytle, C.A., Fuller, M.E., Gan, Y.D.M., Peacock, A., DeFlaun, M.F., Onstott, T.C. and White, D.C. (2001) Utility of high-performance liquid chromatography/electrospray/mass spectrometry of polar lipids in specifically Per-C-13 labeled Gram-negative bacteria DA001 as a tracer for acceleration of bioremediation in the subsurface. *Journal of Microbiological Methods* 44, 271–281.
Macnaughton, S.J., Jenkins, T.L., Wimpee, M.H., Cormier, M.R. and White, D.C. (1997) Rapid extraction of lipid biomarkers from pure culture and environmental samples using pressurized accelerated hot solvent extraction. *Journal of Microbiological Methods* 31, 19–27.
Marko-Varga, G. and Gorton, L. (1990) Post-column derivatization in liquid chromatography using immobilized enzyme reactors and amperometric detection. *Analytica Chimica Acta* 234, 13–29.
Martz, R.F., Sebacher, D.I. and White, D.C. (1983) Biomass measurement of methane forming bacteria in environmental samples. *Journal of Microbiological Methods* 1, 53–61.
Matsuda, O., Ohmi, K. and Sasada, K. (1985) Dissolved organic phosphorus in sea water, its molecular weight fractionation and availability to phytoplankton. *Journal of the Faculty of Applied Biological Science, Hiroshima University* 24, 33–42.
Matthews, R.W., Abdullah, M. and Low, G.K.-C.

(1990) Photocatalytic oxidation for total organic carbon analysis. *Analytica Chimica Acta* 233, 171–179.

McKelvie, I.D., Hart, B.T., Cardwell, T.J. and Cattrall, R.W. (1989) Spectrophotometric determination of dissolved organic phosphorus in natural waters using in-line photo-oxidation and flow injection. *Analyst* 114, 1459–1463.

McKelvie, I.D., Hart, B.T., Cardwell, T.J. and Cattrall, R.W. (1993) Speciation of dissolved phosphorus in environmental samples by gel filtration and flow-injection analysis. *Talanta* 40, 1981–1993.

McKelvie, I.D., Hart, B.T., Cardwell, T.J. and Cattrall, R.W. (1995) Use of immobilized phytase and flow injection for the determination of dissolved phosphorus species in natural waters. *Analytica Chimica Acta* 316, 277–289.

Menzel, D.W. and Corwin, N. (1965) The measurement of total phosphorus in seawater on the liberation of organically bound fractions by persulfate oxidation. *Limnology and Oceanography* 10, 280–282.

Minear, R.A. (1972) Characterization of naturally occurring dissolved organophosphorus compounds. *Environmental Science and Technology* 6, 431–437.

Minear, R.A., Segars, J.E. and Elwood, J.W. (1988) Separation of inositol phosphates by high-performance ion-exchange chromatography. *Analyst* 113, 645–649.

Monaghan, E.J. and Ruttenberg, K.C. (1999) Dissolved organic phosphorus in the coastal ocean: reassessment of available methods and seasonal phosphorus profiles from the Eel River Shelf. *Limnology and Oceanography* 44, 1702–1714.

Nanny, M.A. and Minear, R.A. (1997) Characterization of soluble unreactive phosphorus using ^{31}P nuclear magnetic resonance spectroscopy. *Marine Geology* 139, 77–94.

Nanny, M.A., Seungdo, K., Gadomski, J.E. and Minear, R.A. (1994) Aquatic soluble unreactive phosphorus: concentration by ultrafiltration and reverse osmosis membranes. *Water Research* 28, 1355–1365.

Nanny, M.A., Seungdo, K. and Minear, R.A. (1995) Aquatic soluble unreactive phosphorus: HPLC studies on concentrated water samples. *Water Research* 29, 2138–2148.

Nichols, P.D., Mancuso, C.A. and White, D.C. (1987) Measurement of methanotroph and methanogen signature phosopholipids for use in assessment of biomass and community structure in model systems. *Organic Geochemistry* 11, 451–461.

Ormaza-González, F.I. and Statham, P.J. (1991) Determination of dissolved inorganic phosphorus in natural waters at nanomolar concentrations using a long capillary cell detector. *Analytica Chimica Acta* 244, 63–70.

Pant, H.K., Edwards, A.C. and Vaughan, D. (1994) Extraction, molecular fractionation and enzyme degradation of organically associated phosphorus in soil solutions. *Biology and Fertility of Soils* 17, 196–200.

Patthy, M., Balla, T. and Aranyi, P. (1990) High-performance reversed-phase ion-pair chromatographic study of myo-inositol phosphates: separation of myo-inositol phosphates, some common nucleotides and sugar phosphates. *Journal of Chromatography* A 523, 201–216.

Rounds, M.A. and Nielsen, S.S. (1993) Anion-exchange high-performance liquid chromatography with post-column detection for the analysis of phytic acid and other inositol phosphates. *Journal of Chromatography* A 653, 148–152.

Ruttenberg, K.C. (1992) Development of a sequential extraction method for different forms of phosphorus in marine sediments. *Limnology and Oceanography* 37, 1460–1482.

Rütters, H., Sass, H., Cypionka, H. and Rullkotter, J. (2002a) Microbial communities in a Wadden Sea sediment core–clues from analyses of intact glyceride lipids, and released fatty acids. *Organic Geochemistry* 33, 803–816.

Rütters, H., Sass, H., Cypionka, H. and Rullkotter, J. (2002b) Phospholipid analysis as a tool to study complex microbial communities in marine sediments. *Journal of Microbiological Methods* 48, 149–160.

Santos, F.J. and Galceran, M.T. (2002) The application of gas chromatography to environmental analysis. *Trends in Analytical Chemistry* 21, 672–685.

Shan, Y., McKelvie, I.D. and Hart, B.T. (1993) Characterisation of immobilised *Escherichia coli* alkaline phosphatase reactors in flow injection analysis. *Analytical Chemistry* 65, 3053–3060.

Shan, Y., McKelvie, I.D. and Hart, B.T. (1994) Determination of alkaline phosphatase hydrolysable-phosphorus in natural water systems by enzymatic flow injection. *Limnology and Oceanography* 39, 1993–2000.

Shand, C.A. and Smith, S. (1997) Enzymatic release of phosphate from model substrates and P compounds in soil solution from a peaty podzol. *Biology and Fertility of Soils* 24, 183–187.

Shayman, J.A. and Barcelon, F.S. (1990) Ion-pair chromatography of inositol polyphosphates with N-methylimipramine. *Journal of Chromatography* 528, 143–154.

Skoglund, E., Carlsson, N.-G. and Sandberg, A.-S. (1997) Determination of isomers of inositol mono- to hexaphosphates in selected foods and intestinal contents using high-performance ion

chromatography. *Journal of Agricultural and Food Chemistry* 45, 431–436.
Solórzano, L. and Sharp, J.H. (1980) Determination of total dissolved phosphorus and particulate phosphorus in natural waters. *Limnology and Oceanography* 25, 754–758.
Solórzano, L. and Strickland, J.D. (1968) Polyphosphate in seawater. *Limnology and Oceanography* 13, 515–518.
Sommers, L.E., Harris, R.F., Williams, J.D.H., Armstrong, D.E. and Syers, J.K. (1972) Fractionation of organic phosphorus in lake sediments. *Soil Science Society of America Proceedings* 36, 51–54.
Stevens, R.J. and Stewart, B.M. (1982) Concentration, fractionation and characterization of soluble organic phosphorus in river water entering Loch Neagh. *Water Research* 16, 1507–1519.
Steward, J.H. and Tate, M.E. (1971) Gel chromatography of soil organic phosphorus. *Journal of Chromatography* 60, 75–82.
Stewart, B.M., Jordan, C. and Burns, D.T. (1991) Reverse osmosis as a concentration technique for soluble organic phosphorus in fresh water. *Analytica Chimica Acta* 244, 267–274.
Strickland, J.D. and Solórzano, L. (1966) Determination of monoesterase hydrolyzable phosphorus and phosphomonoesterase activity in seawater. In: Barnes, J. (ed.) *Some Contemporary Studies in Marine Science.* Allen and Unwin, London, pp. 665–674.
Suzumura, M. and Ingall, E.D. (2001) Concentrations of lipid phosphorus and its abundance in dissolved and particulate organic phosphorus in coastal seawater. *Marine Chemistry* 75, 141–149.
Suzumura, M. and Kamatani, A. (1993) Isolation and determination of inositol hexaphosphate in sediments from Tokyo Bay. *Geochimica et Cosmochimica Acta* 57, 2197–2202.
Suzumura, M. and Kamatani, A. (1995) Origin and distribution of inositol hexaphosphate in estuarine and coastal sediments. *Limnology and Oceanography* 40, 1254–1261.
Town, R.M. and Powell, H.J.J. (1992) Elimination of adsorption effects in gel permeation chromatography of humic substances. *Analytica Chimica Acta* 256, 81–86.
Turner, B.L., McKelvie, I.D. and Haygarth, P.M. (2002) Characterisation of water-extractable soil organic phosphorus by phosphatase hydrolysis. *Soil Biology and Biochemistry* 34, 29–37.
Turner, B.L., Mahieu, N. and Condron, L.M. (2003) Phosphorus-31 nuclear magnetic resonance spectral assignments of phosphorus compounds in soil NaOH–EDTA extracts. *Soil Science Society of America Journal* 67, 497–510.
Weimer, W.C. and Armstrong, D.E. (1977) Determination of inositol phosphate esters in lake sediments. *Analytica Chimica Acta* 94, 35–47.
White, D.C., Davis, W.M., Nickels, J.S., King, J.D. and Bobbie, R.J. (1979) Determination of the sedimentary microbial biomass by extractible lipid phosphate. *Oecologia* 40, 51–62.
White, E. and Payne, G. (1980) Distribution and biological availability of reactive high molecular weight phosphorus in natural waters in New Zealand. *Canadian Journal of Fisheries and Aquatic Sciences* 37, 664–669.
Williams, J.D.H., Jaquet, J.-M. and Thomas, R.L. (1976a) Forms of phosphorus in the surficial sediment of Lake Erie. *Journal of the Fisheries Research Board of Canada* 33, 413–429.
Williams, J.D.H., Syers, J.K. and Walker, T.W. (1976b) Fractionation of soil inorganic phosphorus by a modification of Chang and Jackson's procedure. *Soil Science Society of America Proceedings* 31, 736–739.
Zhang, H., Davison, W., Gadi, R. and Kobayashi, T. (1998) In-situ measurement of dissolved phosphorus in natural waters using DGT. *Analytica Chimica Acta* 370, 29–38.
Zink, K.-G., Wilkes, H., Disko, U., Elvert, M. and Horsfield, B. (2003) Intact phospholipids-microbial 'life markers' in marine deep subsurface sediments. *Organic Geochemistry* 34, 755–769.

2 Using Phosphorus-31 Nuclear Magnetic Resonance Spectroscopy to Characterize Organic Phosphorus in Environmental Samples

Barbara J. Cade-Menun

Geological and Environmental Sciences Department, Building 320, Room 118, Stanford University, Stanford, CA 94305-2115, USA

General Introduction

Nuclear magnetic resonance (NMR) spectroscopy is a non-destructive, non-invasive technique that uses the magnetic resonance of a nucleus to identify the chemical forms of that nucleus in a sample. The technique was developed by physicists in the 1940s, and was first used to study soil organic matter by Barton and Schnitzer (1963), with a ^{1}H NMR study of methylated humic acid. ^{31}P NMR spectroscopy was first used on soil extracts by Newman and Tate (1980), and since then has advanced our knowledge of organic phosphorus in soils and environmental samples more than any other technique. Phosphorus NMR spectroscopy may be used on solid or extracted samples, with the advantage that all phosphorus species in a sample can be characterized simultaneously without the need for complex clean-up and chromatographic separation procedures. However, the heterogeneous physical and chemical properties of soils, relatively low phosphorus concentrations, and the natural association of phosphorus with paramagnetic ions such as iron and manganese, make ^{31}P NMR analysis of soil samples more complicated than studies of purified compounds. Phosphorus NMR experiments on environmental samples must be conducted carefully in order to obtain acceptable spectral resolution and reliable, quantitative results.

This chapter reviews the use of ^{31}P NMR spectroscopy for soil, water and other environmental samples. After a brief overview of the principles of NMR spectroscopy, the requirements for a successful NMR experiment are described. Finally, literature on ^{31}P NMR spectroscopy in soils and environmental samples is reviewed, followed by suggestions for future research needs.

Basic Principles of Nuclear Magnetic Resonance Spectroscopy

This section provides a short summary of the principles of NMR spectroscopy. More detailed descriptions are available in textbooks (e.g. Wilson, 1987; Canet, 1996) and review articles (e.g. Wilson, 1991; Preston, 1993, 1996, 2001; Knicker and Nanny, 1997; Randall *et al.*, 1997; Veeman, 1997; Lens and Hemminga, 1998).

Nuclear magnetic resonance spectroscopy is based on the concept that some nuclei have magnetic properties. In these nuclei, the sum of the number of neutrons and protons is an odd number (e.g. ^{1}H, ^{13}C, ^{15}N, ^{27}Al, ^{31}P). Each of these nuclei has a positive charge and

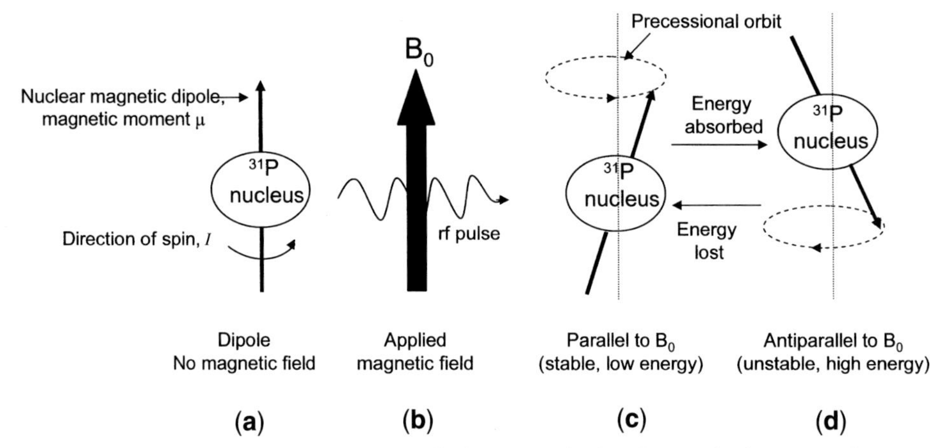

Fig. 2.1. Response of phosphorus nuclei to an applied magnetic field, and to a radio frequency (rf) wave. (a) Nuclear magnetic dipole without applied magnetic field. (b) Applied magnetic field, B_0, with applied rf wave. (c) Alignment of dipole with B_0 (stable, low energy). (d) Alignment against B_0 (unstable, high energy).

a half-integer spin, I (i.e. $I=1/2$, 3/2, etc.). This spinning generates a small magnetic field so that the nucleus behaves as a magnetic dipole, analogous to a bar magnet with north and south poles, and possesses a magnetic moment, μ (Fig. 2.1a). Each nucleus also has a gyromagnetic ratio, γ, which is a fundamental nuclear constant and is different for every nucleus. The variables μ, γ, and I are related by the equation $\mu=\gamma Ih/2\pi$, where h is Planck's constant.

When a nucleus with $I=½$ and positive γ, such as phosphorus ($I=½$; $\gamma=10.829\times10^7$ rad/T/sec (17.25 MHz/T); $\mu=1.9581$), is placed in an external magnetic field of strength B_0 (Fig. 2.1b), it can align itself in one of two ways. The first is parallel to B_0 (Fig. 2.1c), which is the stable low-energy configuration. The other possible alignment is antiparallel to B_0 (Fig. 2.1d), which has a higher energy and is less stable. At equilibrium, slightly more of the nuclei will be in the lower energy alignment than the higher energy alignment. In these alignments, the spinning nucleus is not oriented exactly parallel or antiparallel, but precesses in an orbit with a frequency of $\omega_0=\gamma B_0$, where ω_0 is the Larmor frequency in Hz, and varies depending on the nucleus in question and the strength of the magnetic field (B_0).

When a radio-frequency pulse with an angular velocity equal to ω_0 is applied to the sample in the magnetic field B_0 (Fig. 2.1b), nuclei resonate and absorb energy, 'flipping' from the lower energy state to the higher energy state. After the radio-frequency pulse, the nuclei relax and emit energy, which is detected as an emission signal and recorded as a peak. However, a resonating nucleus only experiences a portion of the magnetic field, due to shielding by an electron cloud from the molecule in which the nucleus is located in a sample. This electron shielding affects the amount of energy required for resonance. Thus, different nuclei absorb and subsequently emit energy in different ways, and show peaks at different positions in a spectrum, depending on their chemical bonds. The intensity of the peaks is proportional to the number of each type of nuclei that are emitting energy, which allows NMR spectroscopy to quantitatively identify different phosphorus forms in a sample. Figure 2.2 shows solution ^{31}P NMR spectra for several extracted soil and environmental samples. These include: forest floor material from a cedar forest in Washington; an alkaline agricultural soil from Idaho; a marine sediment from the Santa Barbara basin, California; a sewage sludge sample from England; and a humic acid extracted from an Australian soil. The groups of biological phosphorus

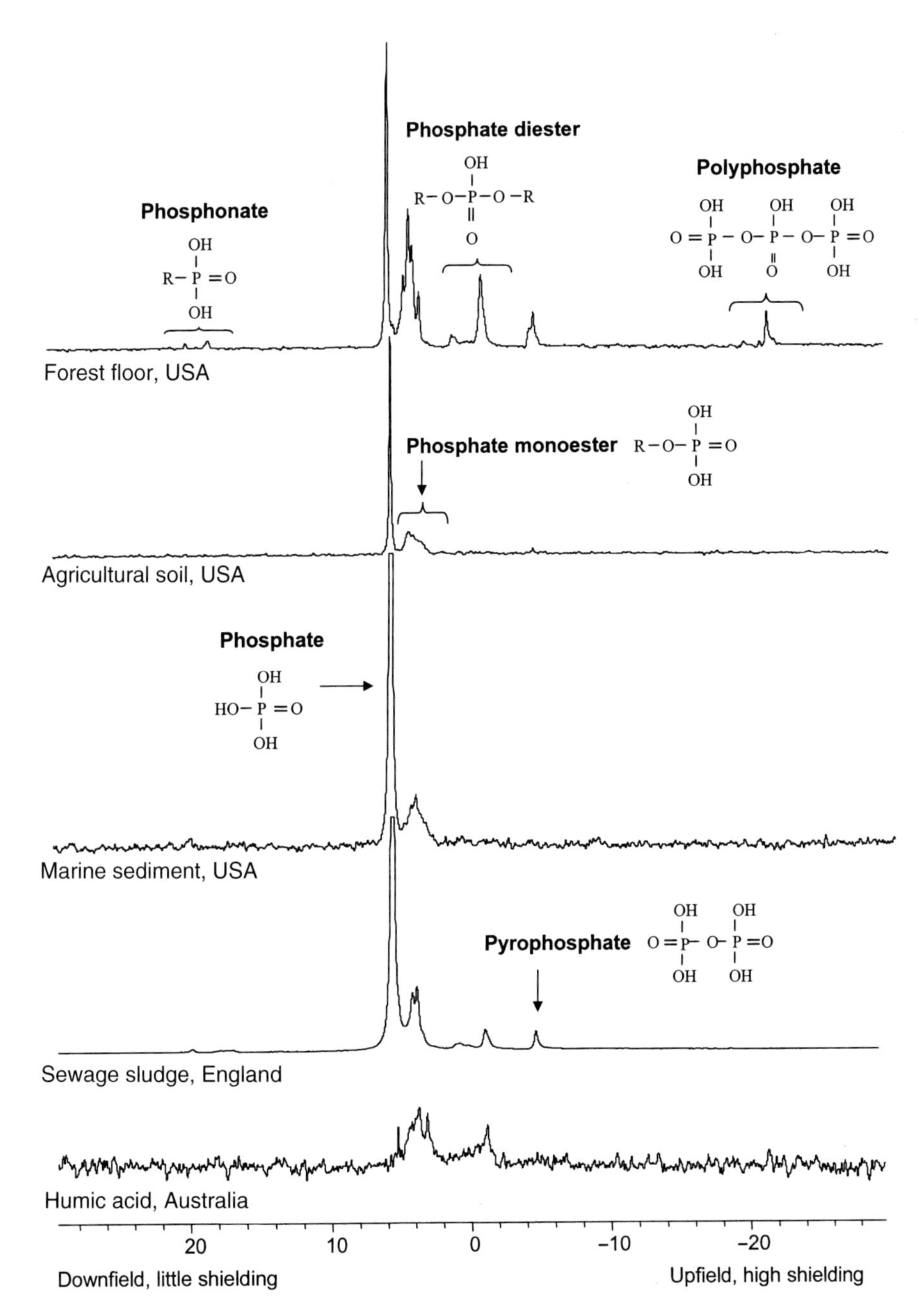

Fig. 2.2. Solution ^{31}P nuclear magnetic resonance spectra of a range of environmental samples. These include: forest floor material from a cedar forest in Washington (sample from B. Cade-Menun); an alkaline agricultural soil from Idaho (D. Strawn, University of Idaho); a marine sediment from the Santa Barbara Basin, California (A. Paytan, Stanford University); a sewage sludge sample from England (M. Smith, Bournemouth University); and a humic acid extracted from Australian soil (N. Mathers, Griffiths University). The humic acid was treated with Chelex, while the other samples were extracted with NaOH–EDTA (Cade-Menun and Preston, 1996). Spectra were generated by B. Cade-Menun on a Varian Unity INOVA 500-MHz spectrometer equipped with a 10-mm broadband probe, using a 90° pulse, 0.68 s acquisition, 4.32 s pulse delay and 25°C temperature.

compounds (both organic and inorganic) that may be visible in a ^{31}P NMR spectrum are phosphonates, phosphate, phosphate monoesters, phosphate diesters, pyrophosphate and polyphosphate. The general chemical structures of these compound groups are shown in Fig. 2.2. Phosphorus nuclei in phosphonates, with a carbon-phosphorus bond, are less shielded than phosphorus nuclei in polyphosphates, which are phosphate groups linked by energy-rich phosphoanhydride bonds. Therefore, it takes less energy for the resonance of phosphorus nuclei in phosphonates, and they appear to the far left (downfield) on a ^{31}P NMR spectrum, whereas polyphosphates appear to the far right (upfield).

In order for the excited nuclei to relax back to equilibrium after a radio-frequency pulse, they exchange energy with their surroundings (spin–lattice relaxation) or with each other (spin–spin relaxation). These types of relaxation are governed by the exponential time constants T_1 and T_2, respectively. Relaxation is important for quantitative data: if the nuclei do not relax back to equilibrium between radio-frequency pulses, the system becomes saturated, some nuclei can no longer resonate, and the peak intensities will not reflect all the nuclei within each phosphorus form in the sample. T_1 is dependent on the γ of the nucleus and the mobility of the lattice. The more mobile the lattice, the greater the interaction it will have with the excited nuclei. Therefore, T_1 is shorter in solutions than in solids. T_1 can also be reduced by the presence of paramagnetic ions such as iron and manganese, which are commonly associated with phosphorus in soils. The unpaired electrons in paramagnetic ions readily transfer energy away from excited nuclei. In spin–spin relaxation (governed by T_2), nuclei exchange energy with neighbouring nuclei of the same type, but in a different excitation state. Nuclei that are in the lower energy level become excited, while nuclei that were excited relax to the lower energy level. This decreases the average time that a nucleus remains in the excited state, and can result in line broadening. T_2 is more important in solids; T_1 and T_2 are equal in solutions.

As previously mentioned, electron shielding influences the portion of the applied magnetic field B_0 that a nucleus will experience. The effect of the shielding factor (α) means that $B_{nucleus} = B_0 - \sigma B_0$. The actual value of σ depends on the orientation of the molecule containing the resonating nucleus relative to B_0. The electron cloud around molecules is anisotropic, meaning that it is not uniform in all directions, so shielding will vary depending on the position of the molecule within the sample and within the magnet. Because molecules in solutions are in rapid motion, σ is averaged out to a single value. However, molecules in solids are fixed in all possible orientations. This chemical shift anisotropy results in spectral line broadening. In addition, dipoles in solids are more likely to interact with one another than dipoles in solutions. These dipole–dipole interactions can also broaden spectral lines. Nuclear spins may also interact through their surrounding electrons in spin–spin (also called indirect or scalar) coupling. These adjacent spins may be aligned with or against the resonating phosphorus nuclei, splitting the energy of the spin of the phosphorus nuclei. In ^{31}P NMR, spin–spin coupling with protons occurs most frequently, and may broaden spectral lines and affect peak intensities, because phosphorus nuclei near to protons will resonate slightly differently than those away from protons. However, because phosphorus is never directly bonded to a proton, the effect should be minimal for most phosphorus nuclei.

Requirements for a Successful ^{31}P Nuclear Magnetic Resonance Spectroscopy Experiment

As ^{31}P is the only naturally occurring phosphorus isotope (100% natural abundance), all phosphorus within a sample should be detected by NMR spectroscopy. This high natural abundance, and large gyromagnetic ratio (γ), make NMR spectroscopy an ideal tool for the study of phosphorus in environ-

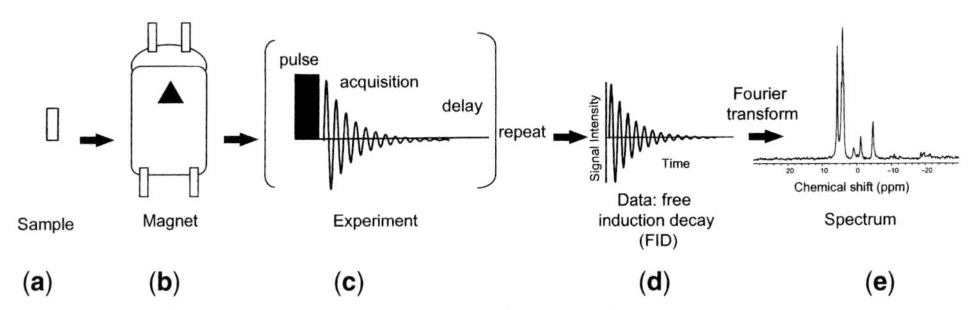

Fig. 2.3. Outline of a conventional one-dimensional nuclear magnetic resonance experiment. (a) A sample in a nuclear magnetic resonance tube; (b) a magnet into which the sample is placed; (c) the outline of a simple experiment; (d) the free induction decay (FID), which is Fourier-transformed to a spectrum (e).

mental samples. However, phosphorus concentrations in environmental samples are low relative to those required for NMR spectroscopy, and the natural association of phosphorus with paramagnetic ions can complicate ^{31}P NMR studies of environmental samples, particularly soils. Care must be taken to obtain the best spectral resolution and quantitative results. Figure 2.3 outlines the components of a typical ^{31}P NMR experiment. A sample (a) in an NMR tube or rotor is placed in a magnet (b). An experiment, consisting of a radio-frequency pulse, acquisition of a signal and a delay, is repeated as many times as necessary (c). The data are collected as a free induction decay (FID) plot (d), which is Fourier-transformed to a ^{31}P NMR spectrum (e). There are important considerations at each of these steps to maximize the information obtained from the experiment, which will be discussed below.

Experimental set-up

The first choice is whether to use a solid sample (solid-state ^{31}P NMR) or an extract (solution ^{31}P NMR, also called liquid-state ^{31}P NMR). Solid-state ^{31}P NMR allows samples to be examined directly with minimal preparation or alteration and requires only small sample sizes. However, the technique is limited by the low natural concentrations of phosphorus, because a minimum concentration of 1 mg P/g is needed for signal detection (Magid *et al.*, 1996). Spectral resolution is also poor in solid-state ^{31}P NMR spectra compared with solution ^{31}P NMR for a variety of reasons, including the presence of paramagnetics and chemical shift anisotropy. This difference in resolution can be seen in Fig. 2.4, which shows the ^{31}P NMR spectra of two algal species grown in culture. The algal samples were first analysed by solid-state ^{31}P NMR (shown in the spectra on the left), and were then extracted with sodium hydroxide plus ethylenediaminetetraacetic acid (EDTA) and analysed by solution ^{31}P NMR (spectra on the right). Note the differences in scale on the bottom of the spectra: the solid spectra are plotted from 100 to −100 ppm, while the solution spectra are plotted from 20 to −20 ppm (see below for an explanation of chemical shift). This puts the solution spectra within the range noted by the arrows in the solid spectra. The solid-state spectra also show artefacts called spinning side bands, which are marked with an asterisk (*, see below). The solid-state spectra show three broad peaks, each of which overlaps the chemical shifts of several phosphorus nuclei. In contrast, peaks in the solution spectra are narrow, allowing phosphorus nuclei to be more easily identified.

Sample preparation for solid-state ^{31}P NMR can be as simple as drying and grinding (Kolowith and Berner, 2002). Solid-state soil studies have also used dried extracts of soils, in an attempt to remove paramagnetic ions with reagents such as dithionite (Lookman *et al.*, 1996; Delgado *et al.*, 2000), DTPA (diethylenetriamine pentaacetate) (Sutter *et*

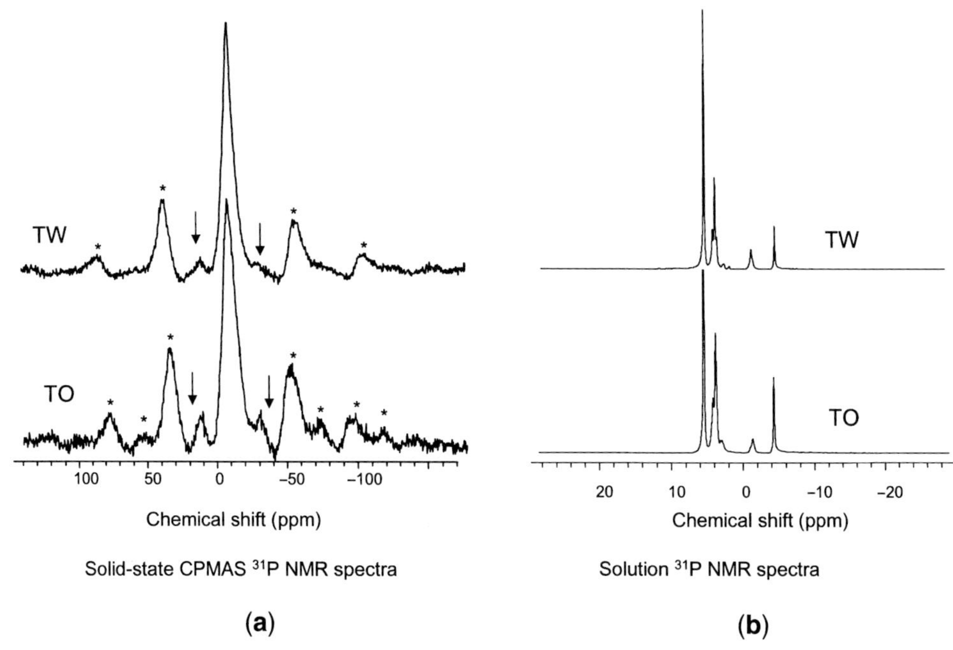

Fig 2.4. A comparison of solid-state and solution ^{31}P nuclear magnetic resonance (NMR) spectra for the same samples. (a) Solid-state magic angle spinning (MAS) ^{31}P NMR spectra of cultures of the algae *Thalassiosira oceanica* (TO) and *Thalassiosira weissflogii* (TW). (b) Solution ^{31}P NMR spectra of the same algal cultures, extracted after solid-state analysis. The arrows on the solid-state spectra indicate the range of peaks for phosphorus species; other peaks are spinning side bands, marked with an asterisk (*). A. Paytan (Stanford University) provided the algal samples. J. Stebbins (Stanford University) produced the solid-state spectra on a 400-MHz Varian VRX/Unity spectrometer with a MAS probe and a spinning speed of 6.8 kHz. B. Cade-Menun produced the solution spectra on a Varian Unity INOVA 500-MHz spectrometer equipped with a 10-mm broadband probe, using a 90° pulse, 0.68 s acquisition, 4.32 s pulse delay and 25°C temperature.

al., 2002), or oxalate (Lookman *et al.*, 1997). The sample is then packed into a rotor that must be able to withstand the spinning speeds used for ^{31}P NMR.

Solution ^{31}P NMR uses extracts of soils and environmental samples. As with any extraction procedure for soil organic phosphorus, there is always the risk of hydrolysis (Turner *et al.*, 2003b). There is little agreement on the most suitable extractant, but those in use include sodium hydroxide (Newman and Tate, 1980; Hawkes *et al.*, 1984), Bu_4NOH (Emsley and Niazi, 1983), the cation exchange resin Chelex in water (Adams and Byrne, 1989; Condron *et al.*, 1996), sodium hydroxide plus Chelex (Gressel *et al.*, 1996), sodium hydroxide plus sodium fluoride (Sumann *et al.*, 1998), and sodium hydroxide plus EDTA (Cade-Menun and Preston, 1996). These extractants solubilize different concentrations and forms of phosphorus from soils (Fig. 2.5). Chelex and EDTA are both used to release phosphorus from paramagnetic ions, thereby reducing line broadening and improving spectral quality. Chelex is removed after extraction, which removes paramagnetic ions from solution, but may also remove polyphosphate (Cade-Menun and Preston, 1996; Gressel *et al.*, 1996; Cade-Menun *et al.*, 2002). Sodium hydroxide plus EDTA extracts more phosphorus than Chelex, but iron and manganese remain in solution complexed with EDTA (Cade-Menun and Preston 1996; Dai *et al.*, 1996). This hastens relaxation, but may increase line broadening (Cade-Menun *et al.*, 2002).

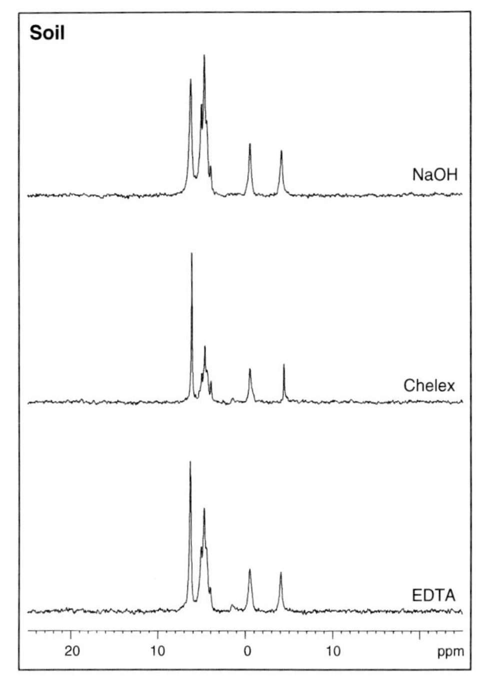

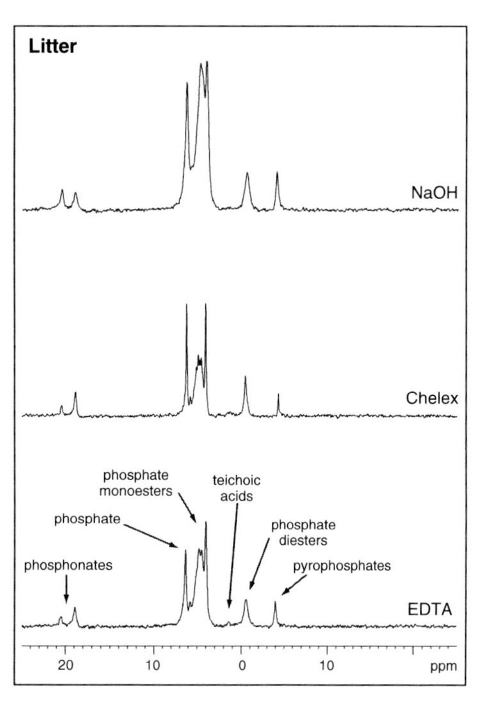

Fig. 2.5. A comparison of the effect of extractants on solution ^{31}P nuclear magnetic resonance spectra (Cade-Menun *et al.*, 2002). The samples were a soil and a litter sample collected under stands of cedar in Washington, USA. The NaOH samples were extracted with 0.25 M NaOH, the Chelex samples were extracted with 0.25 M NaOH plus 6:1 Chelex to soil or litter (weight basis) and the EDTA samples were extracted with a 1:1 mix of 0.5 M NaOH plus 0.1 M Na_2EDTA. Spectra were collected on a DRX-500 spectrometer with a 10-mm broadband probe, using a 90° pulse, 3.2 s pulse delay and 32°C temperature. See Cade-Menun *et al.* (2002) for more details.

Pretreatments prior to extraction have been used to lower the paramagnetic concentration or to increase the phosphorus concentration. Such pretreatments include acid (Tate and Newman, 1982; Hinedi *et al.*, 1988; Adams and Byrne, 1989; Makarov *et al.*, 2002a) and dithionite (Ingall *et al.*, 1990; Carman *et al.*, 2000). Post-extraction treatments include dialysis (Rubæk *et al.*, 1999; Amelung *et al.*, 2001) or exchange resins such as Chelex (Bishop *et al.*, 1994; Robinson *et al.*, 1998; Preston and Trofymow, 2000; Rheinheimer *et al.*, 2002) and Sephadex (Pant *et al.*, 1999, 2002).

Solutions are concentrated after extraction by either lyophilization (e.g. Cade-Menun and Preston, 1996; Guggenberger *et al.*, 1996a,b), a stream of nitrogen at 40°C (Zech *et al.*, 1987; Trasar-Cepeda *et al.*, 1989; Makarov 1998), or rotary evaporation (e.g. Adams and Byrne, 1989; Pant *et al.*, 1999; Turrion *et al.*, 2000, 2001). However, given the risk of hydrolysis with higher temperatures (Cade-Menun *et al.*, 2002; Turner *et al.*, 2003b), lyophilization may be the safest method. Water samples have been concentrated by tangential-flow ultrafiltration (Nanny and Minear, 1994a,b, 1997a,b; Kolowith *et al.*, 2001).

Samples need to be redissolved for solution ^{31}P NMR, with the final volume determined by the size of the NMR probe. For a 10-mm probe (the most common size used for environmental ^{31}P NMR), the final volume should be 2–3 ml, and the sample should not be too viscous, to prevent line broadening from spin–spin coupling. Samples may be redissolved in water, sodium deuteroxide (e.g. Sumann *et al.*, 1998), deuterium oxide plus sodium hydroxide (e.g.

Cade-Menun *et al.*, 2002) or dimethylsulphoxide (Carman *et al.*, 2000). Deuterium oxide is used for a signal lock in the spectrometer (see below), while sodium hydroxide or sodium deuteroxide are used to adjust the pH to >10 for optimal spectral resolution (Nanny and Minear, 1994b; Carman *et al.*, 2000; Crouse *et al.*, 2000). Redissolved samples may be filtered or centrifuged prior to decanting into NMR tube. To prevent hydrolysis, samples should be analysed as soon as possible after dissolution (Turner *et al.*, 2003b).

Experimental considerations

Solution ^{31}P nuclear magnetic resonance spectroscopy

High field super-conducting magnets are used for solution ^{31}P NMR of soils and environmental samples. The field strength (B_0) of a magnet is usually designated in terms of the frequency of ^1H resonance. Thus, a 300-MHz spectrometer has a 7.05-tesla magnet, a resonance frequency of 300 MHz for ^1H, and a resonance frequency of 121.4 MHz for ^{31}P. As B_0 increases, the spectral resolution or signal-to-noise ratio will increase by $B_0^{3/2}$. Therefore, an NMR experiment using a 500-MHz magnet should produce better signal-to-noise than a 300-MHz magnet for the same sample and experimental parameters. A broad-band probe is required to obtain the resonance frequency for ^{31}P. Because signal-to-noise is proportional to the number of phosphorus nuclei in the sample tube, a 10-mm probe that holds an NMR tube with a diameter of 10 mm is preferable for ^{31}P NMR studies of environmental samples. The increase in volume by using a 10-mm tube rather than a 5-mm tube increases the phosphorus concentration in the tube by a factor of four, requiring one-quarter the number of scans for the same sample in the larger tube, and thus reducing the length of the NMR experiment.

After the sample is inserted into the probe in the magnet, it must be locked and shimmed, and the probe must be tuned. To achieve the best signal intensity and resolution, the magnetic field B_0 must be as homogeneous as possible. Spectrometers use the deuterium in a deuterated solvent such as deuterium oxide as a frequency lock to detect, and correct for, minor fluctuations in B_0. Deuterium oxide can be added to the sample solution during sample preparation, or in a capillary tube in the NMR tube with the sample (Nanny and Minear, 1997a). The deuterium lock is also used to manually homogenize B_0 prior to the start of each NMR experiment, in a process called shimming. Each sample should also be manually tuned after it is in the probe, to optimize the probe circuitry.

After locking and shimming, the experimental parameters must be set. The range of peaks seen in the spectrum depends on the spectral window. The phosphorus forms of interest in environmental studies generally fall between 25 and −25 ppm (Tables 2.1 and 2.2; see below for an explanation of chemical shift), so the spectral window (also known as the sweep width) is set to 50 ppm, centred at 0 ppm. For a 500-MHz spectrometer (202.4 MHz for ^{31}P) 50 ppm equals 10,120 Hz. Next, the number of data points is chosen. NMR signals are recorded digitally as discrete data points, and sufficient data points must be collected to quantitatively represent the signal, usually between 8000 and 64,000 for environmental ^{31}P NMR. The acquisition time, or period over which data are recorded after the radio-frequency pulse, is determined by the spectral width and the number of data points. With a sweep width of 10,120 Hz, and 8000 data points, each point represents 10,120/8000 = 1.265 Hz. The acquisition time is the inverse of this, or 0.79 s. For 16,000 data points, acquisition time will be 1.58 s. Thus, the length of time required for each scan, and ultimately the entire NMR experiment, will be determined by the sweep width and the number of data points. Acquisition times used for soil and environmental ^{31}P NMR experiments range from 0.17 s for 8000 data points (Shand *et al.*, 1999) to 1.31 s for 32,000 data points (Zhang *et al.*, 1999).

The length of time in which the radio-frequency pulse excites the nuclei is the pulse width, which is measured in μs but usually expressed in terms of pulse angles. Pulse angles between 30° and 90° are used for environmental ^{31}P NMR, with 90° being the most common. Shorter pulse angles are sometimes used to decrease the delay times between pulses, because if the nuclei are not fully excited, they will take less time to return to equilibrium (see below for more discussion of relaxation). However, with shorter pulse widths there is the risk that not all nuclei will be fully excited, particularly the less abundant nuclei, which could make the results less accurate.

A delay between radio-frequency pulses is required to allow the excited nuclei to relax back to equilibrium (Fig. 2.3c), which must be long enough to allow complete relaxation of phosphorus species if quantitative results are desired (Cade-Menun *et al.*, 2002). Different phosphorus species may have different relaxation times. The general rule, established mathematically because relaxation is an exponential function, is that a delay time of $5 \times T_1$ will allow a return to 99.3% of equilibrium (Canet, 1996). Cade-Menun *et al.* (2002) determined T_1 values for soil phosphorus compounds in different extractants. They suggested that delay times of 1–2 s are adequate for most samples, with longer delay times required for samples from which all paramagnetic ions have been removed (e.g. Chelex-treated samples), or for samples naturally low in paramagnetics, such as calcareous soils and manures (Turner *et al.*, 2003a). To shorten delay times, lanthanide shift reagents may be added to the sample (Nanny and Minear, 1994a), or pulse angles of less than 90° may be used (Wilson, 1991), although this may also make quantification less accurate (see above). Delay times used for environmental ^{31}P NMR have included 20 s (Newman and Tate 1980), 10 s (Hupfer *et al.*, 1995), 7 s (Kemme *et al.*, 1999), 5 s (Condron *et al.*, 1985), 2.5 s (Turner and McKelvie, 2002), 2.0 s (Gressel *et al.*, 1996), 1.5 s (Dai *et al.*, 1996), 1.0 s (Carman *et al.*, 2000) and 0.2 s (Solomon *et al.*, 2002).

About half of the environmental ^{31}P NMR experiments reported in the literature used proton decoupling to remove scalar coupling of protons to phosphorus. This is done with an additional radio-frequency pulse at the Larmor frequency for protons, which is inverse-gated, or switched off, during the delay between pulses in order to suppress the nuclear Overhauser enhancements that can distort relative signal area. A detailed comparison of peak intensities with and without proton decoupling has not been done for soil or environmental ^{31}P NMR, although Turner *et al.* (2003b) showed splitting of a phosphonate peak without decoupling, and a single peak with decoupling. Spinning is seldom used for solution ^{31}P NMR of environmental samples, although it may decrease line broadening (Wilson, 1991).

The number of scans is the number of times the NMR experiment (Fig. 2.3c) is repeated. This is determined to some extent by the signal-to-noise ratio required for adequate spectral resolution. The signal-to-noise is determined by $n^{1/2}I_S/I_N$, where n is number of scans, and I_S and I_N are the intensities of the signal and noise. This shows that doubling the number of scans will increase signal-to-noise by $2^{1/2}$, or 1.4. The number of scans needed is also governed by the time required for each scan, the concentration of phosphorus in the probe (and therefore the diameter of the NMR tube used), and the cost and availability of NMR spectrometer time. The number of scans used for environmental ^{31}P NMR have ranged from 500 (Pant *et al.*, 1999) to 110,000 (Gressel *et al.*, 1996).

Due to the risk of hydrolysis of some organic phosphorus compounds, temperature should be controlled at 20–25°C, especially for long ^{31}P NMR experiments (Crouse *et al.*, 2000; Cade-Menun *et al.*, 2002; Turner *et al.*, 2003b). Temperature will also change the relaxation of nuclei, affecting chemical shift and relative signal intensity (Crouse *et al.*, 2000; Turner *et al.*, 2003b).

Solid-state ^{31}P nuclear magnetic resonance spectroscopy

Most solid-state ^{31}P NMR of soils and environmental samples uses cross-polarization magic angle spinning (CPMAS) NMR techniques. Magic angle spinning involves spinning the sample rapidly at the magic angle of 54°7′ to B_0. This is called 'magic' because at this angle the interactions that cause line broadening, chemical shift anisotropy and dipole–dipole interactions are averaged to zero (Wilson, 1991). Cross-polarization uses a pulse sequence to transfer magnetization from protons to phosphorus nuclei. This improves the signal-to-noise ratio and shortens T_1 relaxation times, which can be long in solid samples. With cross-polarization magic angle spinning, a 90° pulse must be applied (Wilson, 1991). A few solid-state ^{31}P NMR studies of soil and environmental samples have been conducted using Bloch decay experiments and magic angle spinning, with and without high-power decoupling (Bleam *et al.*, 1989; McDowell *et al.*, 2002b; Sutter *et al.*, 2002). Bloch decay uses a single pulse sequence similar to that shown in Fig. 2.3c for solution ^{31}P NMR spectroscopy. There have been no detailed comparisons of cross-polarization and Bloch decay experiments for ^{31}P NMR studies of environmental samples.

In contrast to solution ^{31}P NMR, high field magnets are not advantageous for solid-state ^{31}P NMR spectroscopy, because increasing the magnetic field increases the strength of spinning sidebands (Fig. 2.4). These appear at multiples of the spinning speed, distorting the relative peak intensities and possibly obscuring features of the spectrum (Preston, 2001). Solid-state ^{31}P NMR studies of environmental samples predominantly use 300- or 400-MHz spectrometers (121 or 162 MHz for ^{31}P), spin rates of 5–10 kHz, and spectral widths as wide as 400 ppm (200 to −200 ppm). Experiments may collect as many as 170,000 scans, and may last as long as 48 h. An external standard of 85% H_3PO_4 is used, and hydrolysis caused by high temperatures is not a concern.

Post-experimental processing

The end result of an NMR experiment is the emission signal, which is detected and recorded as a free-induction decay, or FID (Fig. 2.3d). A free-induction decay is a time-domain signal, because it is recorded as a function of intensity over time. The transformation of a free-induction decay to a spectrum (Fig. 2.3e) is known as processing.

Processing software is used to Fourier transform the free-induction decay. This converts the data to a frequency-domain form with signal intensity (*y*-axis) recorded as a function of frequency (*x*-axis). Frequency is expressed as chemical shift, which is the measure of the position of a resonance signal relative to a standard (usually 85% phosphoric acid) and is defined by:

$$\frac{V_s - V_r}{V_r} \times 10^6$$

where V_s and V_r are the frequencies of the sample and reference standard. Chemical shift values are dimensionless and are expressed in parts per million (ppm) with the external standard set at 0 ppm. Chemical shift is independent of the field strength of the magnet, allowing the results obtained from different spectrometers to be compared. Chemical shift of some phosphorus compounds, particularly phosphate, is affected by pH due to the dissociation of protons (e.g. Adams and Byrne, 1989), so that the chemical shift of phosphate in an alkaline soil extract will be downfield from that of phosphate in the reference (6 ppm *vs.* 0 ppm). Carman *et al.* (2000) also observed chemical shift differences when using dimethyl sulphoxide instead of deuterium oxide for the signal lock.

Other important processing tools are phasing, baseline correction and line-broadening. Phasing removes artefacts from Fourier transformation and should produce symmetric peaks flanked by flat baselines. Baseline correction will also remove baseline artefacts. Line-broadening uses an exponential multiplication factor, in Hz, to reduce noise and improve the signal-to-noise ratio. Examples of the same spectrum with different line-broadening are shown in Fig. 2.6. If the line-broadening is too low, noise reduction is inefficient; if too high, useful data may be lost. Line-broadening of 10–20 Hz is common for solution ^{31}P NMR of environmental samples, while 50–200 Hz line-broadening may be needed for solid-state samples.

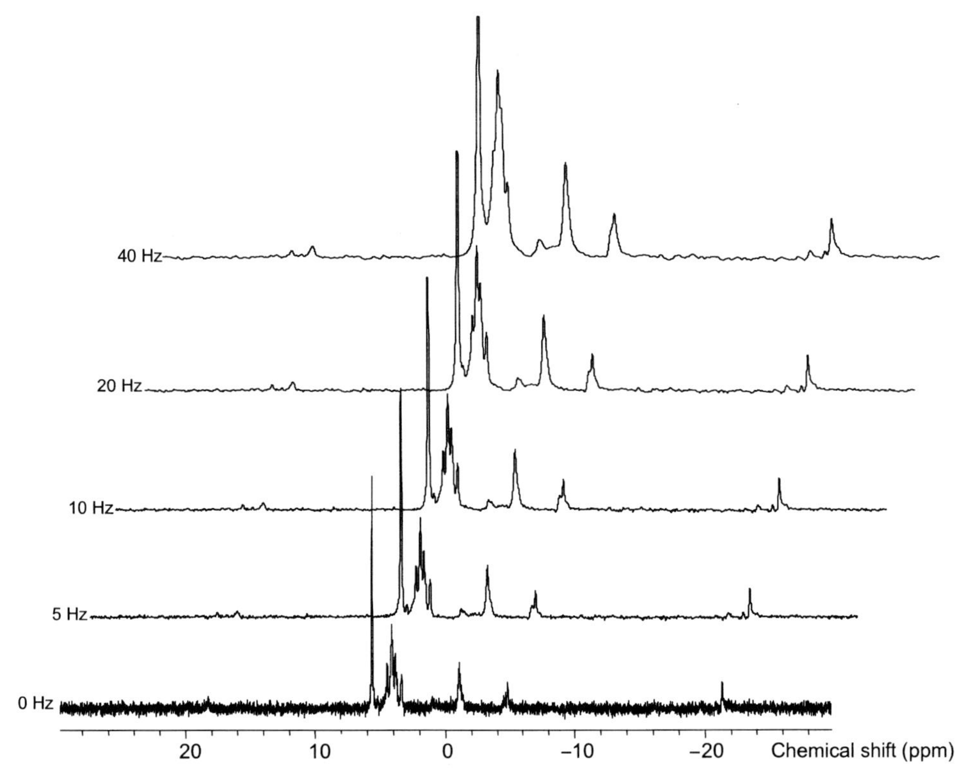

Fig. 2.6. Solution ^{31}P nuclear magnetic resonance spectra of a single experiment, showing the effect of line broadening on spectral resolution and the signal-to-noise ratio.

Both an automatic peak-picking routine in the processing software and visual inspection are used to identify peaks. Chemical shifts are then compared with literature reports. Table 2.1 shows some of the biological phosphorus compounds identified in ^{31}P NMR studies of environmental samples, while Table 2.2 shows peak shifts for phosphorus minerals determined by solid-state spectroscopy. Although this chapter focuses on organic phosphorus, the overlap of peak shifts for biological and mineral phosphorus compounds must be noted. This may complicate the identification of phosphorus species in solid-state spectroscopy, because both biological and mineral phosphorus forms may be present. To further confirm peak shifts, standards such as methylene diphosphonic acid may be added directly to the sample or included as capillary tube inserts in solution ^{31}P NMR (e.g. Koopmans *et al.*, 2003).

The area under each peak is proportional to the peak intensity or amount of the total sample phosphorus found in each phosphorus species. Using an integration routine in the processing software, the entire spectrum is integrated (Fig. 2.7), and the integral is then divided into regions representing each peak. The height of the integral for each peak is determined as a percentage of the total integral, which is the percentage of total sample phosphorus in each phosphorus species. Nuclei that are equivalent magnetically, such as the two phosphorus nuclei in pyrophosphate (P_2O_7), will show only a single peak. It can be difficult to determine peak intensities if peaks overlap, but spectral deconvolution can separate broad peaks that may include more than one compound, such as those in the phosphate monoester region of solution (Turner *et al.*, 2003d), or solid-state ^{31}P NMR spectra (Lookman *et al.*, 1996; McDowell *et al.*, 2002a,b).

Table 2.1. Chemical shift references for biological phosphorus compounds in solution ^{31}P nuclear magnetic resonance spectroscopy. Bold type indicates general peak shift ranges; regular type indicates specific chemical shift assignments.

Chemical shift (ppm)	Compound	Reference
20	**Phosphonates**	Various
20	Aminoethyl phosphonates	Turner *et al.* (2003b)
18	Phosphonolipids	Cade-Menun *et al.* (2002)
12 to 14	Aromatic phosphonic acid esters	Turner *et al.* (2003b)
7.4	Aromatic diesters	Turner *et al.* (2003b)
5.7 to 6.1	**Phosphate**	Various
6 to 3	**Phosphate monoesters**	Various
5.85, 4.92, 4.55, 4.43	*myo*-inositol hexakisphosphate (phytic acid)	Turner *et al.* (2003b)
5.4	Glucose 6-phosphate	Pant *et al.* (1999)
4.78 to 4.32	Mononucleotides	Turner *et al.* (2003b)
4.71	Ethanolamine phosphate	Turner *et al.* (2003b)
4.05	Choline phosphate	Turner *et al.* (2003b)
3.6	Polynucleotides	Pant *et al.* (1999)
2.5 to −1.0	**Phosphate diesters**	Various
2.5 to 1.2	Teichoic acids	Makarov *et al.* (2002a,b)
1.75	Phosphatidyl ethanolamine	Turner *et al.* (2003b)
1.57	Phosphatidyl serine	Turner *et al.* (2003b)
0.54	RNA	Turner *et al.* (2003b)
0	DNA	Makarov *et al.* (2002a,b)
−0.37	DNA	Turner *et al.* (2003b)
−4	Polyphosphate terminal phosphate group	Turner *et al.* (2003b)
−5	**Pyrophosphate**	Various
−10	Adenosine di- or triphosphate α-phosphate	Turner *et al.* (2003b)
−19 to −21	**Polyphosphates**	Various
−19.68	Adenosine triphosphate β-phosphate	Turner *et al.* (2003b)

Table 2.2. Chemical shift references for phosphorus compounds from solid-state ^{31}P nuclear magnetic resonance spectroscopic studies.

Chemical shift (ppm)	Compound	Reference
9	Dicalcium phosphate dihydrate ($Ca_2HPO_4 \cdot 2H_2O$)	McDowell *et al.* (2002a,b)
3	Hydroxyapatite [$Ca_5(PO_4)_3OH$]	Frossard *et al.* (1994a)
3	Octacalcium phosphate [$Ca_8H_2(PO_4)_6 \cdot 5H_2O$]	Frossard *et al.* (1994a)
−0.4	$Na_4P_2O_7$	Condron *et al.* (1997)
−2	Monetite ($CaHPO_4$)	Frossard *et al.* (1994a)
−5	Crandallite [($CaAl_3(OH)_5(PO_4)_2$]	Bleam *et al.* (1989)
−8	$Na_3HP_2O_7$	Condron *et al.* (1997)
−9.9	$Na_2H_2P_2O_7$	Condron *et al.* (1997)
−10.2	Brazilianite [$NaAl_3(OH)_4(PO_4)_2$]	Bleam *et al.* (1989)
−11	Wavellite [$Al_3(OH)_3(PO_4)_2 \cdot 5H_2O$]	Bleam *et al.* (1989)
−13.2	Metavariscite ($AlPO_4 \cdot 2H_2O$)	Duffy and van Loon (1995)
−16	Senegalite [$Al_2(OH)_3(PO_4) \cdot H_2O$]	Bleam *et al.* (1989)
−19	Variscite ($AlPO_4 \cdot 2H_2O$)	Bleam *et al.* (1989)
−20	Lazulite [$MgAl_2(OH)_2(PO_4)_2$]	Bleam *et al.* (1989)
−21.7	Metaphosphate ($Na_6P_6O_{18}$)	Condron *et al.* (1997)
−25	Berlinite ($AlPO_4$)	Bleam *et al.* (1989)
−30	Augelite [$Al_2(OH)_3PO_4$]	Bleam *et al.* (1989)

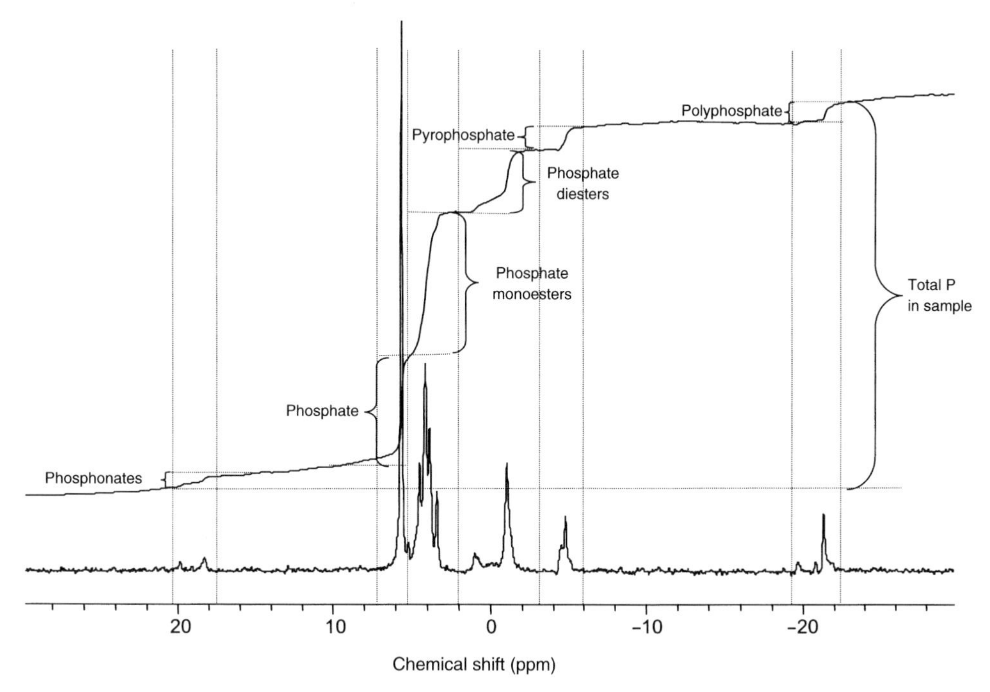

Fig. 2.7. Peak areas are determined by integration. The integral is calculated using processing software, and the height corresponding to each peak is calculated as a percentage of the total integral. This corresponds to the percentage of total sample phosphorus in each phosphorus species or functional group.

^{31}P NMR Studies of Environmental Samples

This section summarizes the literature for ^{31}P NMR spectroscopy of environmental samples. Due to space constraints, ^{31}P NMR studies of compounds with chemical shifts outside the spectral width of biological phosphorus compounds, such as pesticides (see Condron *et al.*, 1997 for a recent review), or biological ^{31}P NMR studies such as *in vivo* NMR (e.g. Lee and Ratcliffe, 1993) or mycorrhizal studies (e.g. Martin *et al.*, 1983) are not included. Space also limits a detailed discussion of all environmental ^{31}P NMR papers. Instead, they are grouped into tables based on the type of material studied or the type of NMR experiment. Some papers may appear in more than one table.

Papers listed in Table 2.3 are primarily methodological and range from the first report of soil ^{31}P NMR (Newman and Tate, 1980) to those testing extractants, relaxation times, and signal assignments. Next grouped are the studies using whole-soil extractions to survey phosphorus forms in agricultural soils (Table 2.4), forest soils (Table 2.5) and natural ecosystems (Table 2.6). Some studies have also used ^{31}P NMR to monitor the transformation of phosphorus in laboratory incubations (Table 2.6), while others have separated soil components into humic acids or particle sizes (Table 2.7). Environmental ^{31}P NMR papers include studies of phosphorus forms in manure, compost, and sludge or sludge-amended soils (Table 2.8), while aquatic ^{31}P NMR papers include studies of lake water, lake and river sediments, wetland and estuary sediments, and ocean water and sediments (Table 2.9). Solid-state ^{31}P NMR studies of environmental samples are listed in Table 2.10, and include studies of waste materials, soil, ocean water, and shale.

Table 2.3. Solution ^{31}P nuclear magnetic resonance (NMR) methodology papers for environmental samples.

Reference	Analysis
Newman and Tate (1980)	First soil phosphorus NMR; extractants; peak assignments; relaxation; storage
Emsley and Niazi (1983)	Extractants
Condron *et al.* (1985)	Extractants
Adams and Byrne (1989)	Extractants; peak assignments
Hinedi *et al.* (1989a)	Extractants
Adams (1990)	Extractants
Cade-Menun and Preston (1996)	Extractants; peak assignments
Dai *et al.* (1996)	Peak assignments
Gressel *et al.* (1996)	^{31}P- and ^{13}C-NMR on same soil extract
Pant *et al.* (1999)	Extractants, peak assignments
Zhang *et al.* (1999)	Extractants
Crouse *et al.* (2000)	pH and temperature optimization
Cade-Menun *et al.* (2002)	Extractants, relaxation times, temperature
Makarov *et al.* (2002b)	Peak assignments
Turner and McKelvie (2002)	Preconcentration of inositol phosphate
Turner *et al.* (2003b)	Peak assignments, temperature, storage
Turner *et al.* (2003c)	Quantification with deconvolution

Table 2.4. Studies of phosphorus in agricultural soils using whole-soil extractions and solution ^{31}P nuclear magnetic resonance spectroscopy.

Reference	Crop	Location
Newman and Tate (1980)	Grassland, pasture	New Zealand
Tate and Newman (1982)	Grassland climosequence	New Zealand
Emsley and Niazi (1983)	Grassland, crops	England
Hawkes *et al.* (1984)	Grassland	England
Condron *et al.* (1985)	Pasture	New Zealand
Zech *et al.* (1985)	Tree and vegetable crops	Mexico
Condron *et al.* (1990)	Grassland and cultivated soils	Canada
Bedrock *et al.* (1994)	Grassland, virgin peat	Scotland
Guggenberger *et al.* (1996a)	Fertilized cropland	Denmark
Guggenberger *et al.* (1996b)	Tropical pasture	Colombia
Leinweber *et al.* (1997)	Grassland, crop soils	Germany
Escudey *et al.* (1997)	Crops, pasture; P fertilizer	Chile
Pant *et al.* (1999)	Cultivated soils	Canada
Zhang *et al.* (1999)	Maize	Canada
Solomon and Lehmann (2000)	Maize, beans	Tanzania
Möller *et al.* (2000)	Cabbage, ^{13}C and ^{31}P	Thailand
Chapuis-Lardy *et al.* (2001)	Cultivated and pasture cerrado	Brazil
Solomon *et al.* (2002)	Maize, sorghum, tea	Ethiopia
Rheinheimer *et al.* (2002)	Crops; tillage systems	Brazil
Koopmans *et al.* (2003)	Grasslands; manure, fertilizer	The Netherlands
Toor *et al.* (2003)	Dairy manure leachate	New Zealand
Turner *et al.* (2003a)	Croplands, sagebrush	USA
Turner *et al.* (2003d)	Grassland	England and Wales
Tchienkoua and Zech (2003)	Tea, eucalyptus	West Cameroon

Table 2.5. Studies of phosphorus in forest soils using whole-soil extractions and solution ^{31}P nuclear magnetic resonance spectroscopy.

Reference	Plant community	Location
Emsley and Niazi (1983)	Conifer	England
Zech *et al.* (1985)	*Acacia*, *Vitex*; land use	Mexico
Zech *et al.* (1987)	*Alnus*, *Pinus*; land use	Germany
Adams and Byrne (1989)	*Eucalyptus*; fire	Australia
Adams (1990)	*Eucalyptus*; fire, logging	Australia
Forster and Zech (1993)	Evergreen tropical rainforest	Liberia
Parfitt *et al.* (1994)	Fertilized *Pinus*	New Zealand
Makarov *et al.* (1995)	Aspen, P-pollution	Russia
Guggenberger *et al.* (1996a)	Beech, fir, spruce; land use	Germany
Condron *et al.* (1996)	*Pinus*, on grassland	New Zealand
Dai *et al.* (1996)	Spruce, fir	USA
Gressel *et al.* (1996)	Mixed conifer; ^{13}C and ^{31}P	USA
Cade-Menun *et al.* (2000a)	Cedar, hemlock, fir	Canada
Cade-Menun *et al.* (2000b)	Cedar, hemlock; burning	Canada
Möller *et al.* (2000)	*Pinus*; reforestation	Thailand
Preston and Trofymow (2000)	Conifers; chronosequence	Canada
Solomon and Lehmann (2000)	*Acacia*; deforestation	Tanzania
Taranto *et al.* (2000)	*Banksia*; pasture soil	Australia
Turrión *et al.* (2000)	*Juniperus*; deforestation	Kyrgyzstan
Chapuis-Lardy *et al.* (2001)	Cerrado forest	Brazil
Turrión *et al.* (2001)	Oak, chestnut	Spain
Solomon *et al.* (2002)	Mixed native forest; plantations	Ethiopia

Table 2.6. Studies utilizing solution ^{31}P nuclear magnetic resonance spectroscopy to investigate phosphorus in natural ecosystems, or phosphorus transformations in the laboratory.

Reference	Description	Location
Natural ecosystems		
Trasar-Cepeda *et al.* (1989)	Soil catenas; high soil organic matter	Spain
Gil-Sotres *et al.* (1990)	Heather, furze	Spain
Makarov (1998)	Alpine meadows and heath	Russia
Sumann *et al.* (1998)	Native grasslands, climate	Central North America
Amelung *et al.* (2001)	Steppe	Russia
Kristiansen *et al.* (2001)	Abandoned ant-hills	Denmark
Phosphorus transformations		
Hinedi *et al.* (1988)	Incubations of sewage, soil	USA
Quiquampoix and Ratcliffe (1992)	Protein adsorption on clays	England
Bishop *et al.* (1994)	Enzymatic mineralization	Chile
Miltner *et al.* (1998)	Incubations of litter, minerals	Germany
Makarov *et al.* (2002a)	Incubations, climosequences	Russia

Table 2.7. Studies investigating phosphorus in humic acids or particle size separations with solution ^{31}P nuclear magnetic resonance spectroscopy.

Reference	Material	Location
Humic acids		
Ogner (1983)	Raw humus	Norway
Bedrock *et al.* (1994)	Peat, soil	Scotland
Bedrock *et al.* (1995)	Peat, soil	Scotland
Francioso *et al.* (1996)	Peat	Ireland
Guggenberger *et al.* (1996b)	Tropical pasture	Colombia
Makarov (1996)	Various soils	Russia
Makarov *et al.* (1996)	Organic mountain soils	Russia
Makarov *et al.* (1997)	Alpine soil; toposequence	Russia
Francioso *et al.* (1998)	Peat	Ireland
Shand *et al.* (1999)	Peat, soil	Scotland
Fan *et al.* (2000)	Forest soil	California
Mahieu *et al.* (2000)	Corn, rice soils	Philippines
Tikhova *et al.* (2000)	Various soils	Russia
Particle size separates		
Leinweber *et al.* (1997)	Clay; cropland	Germany
Sumann *et al.* (1998)	Clay; natural grassland	North America
Rubæk *et al.* (1999)	Sand, silt, clay; forest, cropland	Germany, Denmark
Guggenberger *et al.* (2000)	Sand, silt, clay; cropland	Denmark
Solomon and Lehmann (2000)	Sand, silt, clay; forest, cropland	Tanzania
Turrión *et al.* (2000)	Sand, silt, clay; forest, pasture	Kyrgyzstan
Amelung *et al.* (2001)	Sand, silt, clay; steppe	Russia
Solomon *et al.* (2002)	Sand, silt, clay; cropland, forest	Ethiopia
Makarov *et al.* (2004)	Sand, silt, clay; forest, meadow, heath	Caucasus

Table 2.8. Studies using solid-state or solution ^{31}P nuclear magnetic resonance spectroscopy to investigate phosphorus in manure, sludge and sludge-amended soils and other waste-materials.

Reference	Material	State	Location
Manure and waste material			
Preston *et al.* (1986)	Fish and crab scrap	Solid/solution	Canada
Leinweber *et al.* (1997)	Pig, chicken manure	Solution	Germany
Preston *et al.* (1998)	Backyard compost, ^{13}C and ^{31}P	Solution	Canada
Kemme *et al.* (1999)	Pig manure	Solution	The Netherlands
Crouse *et al.* (2000)	Turkey manure	Solution	USA
Crouse *et al.* (2002)	Poultry manure	Solution	USA
Gigliotti *et al.* (2002)	Urban compost, pig slurry	Solution	Italy
Frossard *et al.* (2002)	Composted organic wastes	Solid	Switzerland
Sludge and sludge-amended soils			
Florentz and Granger (1983)	Activated sludge	Solution	France
Hinedi *et al.* (1988)	Sludges, soils	Solution	USA
Hinedi *et al.* (1989a)	Digested sludge, dairy manure	Solution	USA
Hinedi *et al.* (1989b)	Digested sludge	Solid	USA
Hinedi and Chang (1989)	Sludge-amended soil	Solid	USA
Uhlmann *et al.* (1990)	Activated sludge	Solution	Germany
Jing *et al.* (1992)	Activated sludge	Solution	USA
Röske and Schönborn (1994)	Activated sludge	Solution	Germany
Frossard *et al.* (1994a)	Activated and digested sludges	Solid	France
Duffy and van Loon (1995)	Sludge, Al-hydroxide	Solid	Canada
Gigliotti *et al.* (2002)	Sludge	Solution	Italy

Table 2.9. ^{31}P nuclear magnetic resonance spectroscopy studies of aquatic systems, including: freshwater, estuary, wetland and ocean studies.

Reference	Sample type	State	Location
Lakes and rivers			
Nanny and Minear (1994a,b)	Lake water, filterable phosphorus, shift reagents	Solution	USA
Hupfer *et al.* (1995)	Lake sediments	Solution	Switzerland
Baldwin (1996)	Lake, pond, billabong sediments	Solution	Australia
Nanny and Minear (1997a,b)	Lake water, filterable phosphorus	Solution	USA
Carman *et al.* (2000)	Lake sediments	Solution	Sweden
Khoshmanesh *et al.* (2002)	Wetland, river sediment bacteria	Solution	Australia
Selig *et al.* (2002)	Lake water, filterable and particulate phosphorus	Solution	Germany
Watts *et al.* (2002)	Lake sediment bacteria	Solution	Canada
Hupfer *et al.* (2004)	Lake sediments	Solution	Europe
Wetlands			
Robinson *et al.* (1998)	Marsh; muck farmland	Solution	USA
Delgado *et al.* (2000)	Calcareous marsh	Solid	Spain
Pant and Reddy (2001)	*Typha* detrital material	Solution	USA
Pant *et al.* (2002)	Benthic floc	Solution	USA
Estuaries			
Sundareshwar *et al.* (2001)	Estuary sediments	Solution	USA
Halls (2002)	Estuary sediments	Solution	USA
Oceans			
Ingall *et al.* (1990)	Marine sediments	Solution	
Clark *et al.* (1998, 1999)	Marine water, algal cultures	Solid	Pacific Ocean
Clark *et al.* (1999)	Marine water, filterable organic phosphorus	Solid	Pacific Ocean
Carman *et al.* (2000)	Marine sediments	Solution	Sweden
Kolowith *et al.* (2001)	Marine water, filterable organic phosphorus	Solid	Various oceans
Paytan *et al.* (2003)	Sediment trap material	Solution	Various oceans

Table 2.10. Solid-state cross-polarization magic angle spinning ^{31}P nuclear magnetic resonance spectroscopic studies of environmental samples.

Reference	Material	Location
Preston *et al.* (1986)	Fish and crab scrap	Canada
Hinedi and Chang (1989)	Sludge-amended soil	USA
Hinedi *et al.* (1989b)	Sewage sludge, soil	USA
Bleam *et al.* (1989)	Al phosphates	USA
Frossard *et al.* (1994a)	Sewage sludge	France
Frossard *et al.* (1994b)	Fertilizer materials	France
Duffy and van Loon (1995)	Sewage sludge, Al hydroxide	Canada
Lookman *et al.* (1996)	High-P soil	Belgium
Lookman *et al.* (1997)	Soil, Al hydroxides	Belgium
Clark *et al.* (1998, 1999)	Marine water, algal cultures	Pacific Ocean
Crouse *et al.* (2000)	Turkey manure	USA
Delgado *et al.* (2000)	Calcareous marsh	Spain
Kolowith *et al.* (2001)	Marine water, dissolved organic P	Various oceans
Sutter *et al.* (2002)	Synthetic apatites	USA
Kolowith and Berner (2002)	Black shale	USA
McDowell *et al.* (2002a)	Arable soils	England
McDowell *et al.* (2002b)	Arable soils	England

Future Research Needs

In the more than 20 years of ^{31}P NMR studies of soils and environmental samples, the technique has been considerably improved and refined. However, only a fraction of the potential of NMR has been tapped (Preston, 1996). Although spectroscopy can be used for more detailed studies, such as of phosphorus transformations, structures or reaction kinetics, the majority of environmental ^{31}P NMR papers are surveys of phosphorus forms in samples. ^{31}P NMR can also be linked to NMR studies of other nuclei, such as ^{13}C or ^{1}H (e.g. Gigliotti *et al.*, 2002), with the potential for two-dimensional NMR spectroscopy (e.g. Veeman, 1997). In addition, despite the improvement in extraction techniques in recent years, we still need to determine whether extraction is altering the phosphorus forms in samples in any way, such as by hydrolysis (Turner *et al.*, 2003b). We should also explore the use of different extractants: those currently in use are designed to extract all of the biological phosphorus forms in a sample, but additional information could be gained by also using specific extractants for phosphorus forms, such as methanol and chloroform for phospholipids (Makarov *et al.*, 2002b; Watts *et al.*, 2002). Finally, there is a real need to standardize results among laboratories. The use of different extractants and parameters make it difficult to compare the results of different studies. One option is to select a standard method for extraction, then compare the results from other extractants to this standard, perhaps with commercially available reference material. Laboratories that regularly conduct ^{31}P NMR experiments using unusually short parameters such as delay times, pulse angles and pulse widths, should also include experiments with longer parameters, to demonstrate that their results are quantitative.

Acknowledgements

I am grateful to Leo Condron, Emmanuel Frossard, Ludwig Haumaier, Corey Liu, Kelly Kryc and Ben Turner for helpful comments.

References

Adams, M.A. (1990) Phosphatase activity and phosphorus fractions in Karri (*Eucalyptus diversicolor* F. Muell.) forest soils. *Biology and Fertility of Soils* 14, 200–204.

Adams, M.A. and Byrne, L.T. (1989) ^{31}P NMR analysis of phosphorus compounds in extracts of surface soils from selected Karri (*Eucalyptus diversicolor* F. Muell.) forests. *Soil Biology and Biochemistry* 21, 523–528.

Amelung, W., Rodionov, A., Urusevskaja, I.S., Haumaier, L. and Zech, W. (2001) Forms of organic phosphorus in zonal steppe soils of Russia assessed by ^{31}P NMR. *Geoderma* 103, 335–350.

Baldwin, D.S. (1996) The phosphorus composition of a diverse series of Australian sediments. *Hydrobiologia* 335, 63–73.

Barton, D.H.R. and Schnitzer, M. (1963) A new experimental approach to the humic acid problem. *Nature* 198, 217–218.

Bedrock, C.N., Cheshire, M.V., Chudek, J.A., Goodman, B.A. and Shand, C.A. (1994) Use of ^{31}P NMR to study the forms of phosphorus in peat soils. *Science of the Total Environment* 152, 1–8.

Bedrock, C.N., Cheshire, M.V., Chudek, J.A., Fraser, A.R., Goodman, B.A. and Shand, C.A. (1995) Effect of pH on precipitation of humic acid from peat and mineral soils on the distribution of phosphorus forms in humic and fulvic acid fractions. *Communications in Soil Science and Plant Analysis* 26, 1411–1425.

Bishop, M.L., Chang, A.C. and Lee, R.W.K. (1994) Enzymatic mineralization of organic phosphorus in a volcanic soil in Chile. *Soil Science* 157, 238–243.

Bleam, W.F., Pfeffer, P.E. and Frye, J.S. (1989) ^{31}P solid-state nuclear magnetic resonance spectroscopy of aluminum phosphate minerals. *Physics and Chemistry of Minerals* 16, 455–464.

Cade-Menun, B.J. and Preston, C.M. (1996) A comparison of soil extraction procedures for ^{31}P NMR spectroscopy. *Soil Science* 161, 770–785.

Cade-Menun, B.J., Berch, S.M., Preston, C.M. and Lavkulich, L.M. (2000a) Phosphorus forms and related soil chemistry of Podzolic soils on northern Vancouver Island. I. A comparison of two forest types. *Canadian Journal of Forest Research* 30, 1714–1725.

Cade-Menun, B.J., Berch, S.M., Preston, C.M. and Lavkulich, L.M. (2000b) Phosphorus forms and related soil chemistry of Podzolic soils on northern Vancouver Island. II. The effects of clear-cutting and burning. *Canadian Journal of Forest Research* 30, 1726–1741.

Cade-Menun, B.J., Liu, C.W., Nunlist, R. and McColl, J.G. (2002) Soil and litter phosphorus-31 nuclear magnetic resonance spectroscopy: extractants, metals and phosphorus relaxation times. *Journal of Environmental Quality* 31, 457–465.

Canet, D. (1996) *Nuclear Magnetic Resonance: Concepts and Methods.* John Wiley & Sons, New York, 270 pp.

Carman, R., Edlund, G. and Damberg, C. (2000) Distribution of organic and inorganic phosphorus compounds in marine and lacustrine sediments: a ^{31}P NMR study. *Chemical Geology* 163, 101–114.

Chapuis-Lardy, L., Brossard, M. and Quiquampoix, H. (2001) Assessing organic phosphorus status of Cerrado oxisols (Brazil) using ^{31}P NMR spectroscopy and phosphomonoesterase activity measurement. *Canadian Journal of Soil Science* 81, 591–601.

Clark, L.L., Ingall, E.D. and Benner, R. (1998) Marine phosphorus is selectively remineralized. *Nature* 393, 426.

Clark, L.L., Ingall, E.D. and Benner, R. (1999) Marine organic phosphorus cycling: novel insights from nuclear magnetic resonance. *American Journal of Science* 299, 724–737.

Condron, L.M., Goh, K.M. and Newman, R.H. (1985) Nature and distribution of soil phosphorus as revealed by a sequential extraction method followed by ^{31}P nuclear magnetic resonance analysis. *Journal of Soil Science* 36, 199–207.

Condron, L.M., Frossard, E., Tiessen, H., Newman, R.H. and Stewart, J.W.B. (1990) Chemical nature of organic phosphorus in cultivated and uncultivated soils under different environmental conditions. *Journal of Soil Science* 41, 41–50.

Condron, L.M., Davis, M.R., Newman, R.H. and Cornforth, I.S. (1996) Influence of conifers on the forms of phosphorus in selected New Zealand grassland soils. *Biology and Fertility of Soils* 21, 37–42.

Condron, L.M., Frossard, E., Newman, R.H., Tekely, P. and Morel, J.-L. (1997) Use of ^{31}P NMR in the study of soils and the environment. In: Nanny, M.A., Minear, R.A. and Leenheer, J.A. (eds) *Nuclear Magnetic Resonance Spectroscopy in Environmental Chemistry.* Oxford University Press, Oxford, pp. 247–271.

Crouse, D.A., Sierzputowska-Gracz, H. and Mikkelson, R.L. (2000) Optimization of sample pH and temperature for phosphorus-31 nuclear magnetic resonance spectroscopy of poultry manure extracts. *Communications in Soil Science and Plant Analysis* 31, 229–240.

Crouse, D.A., Sierzputowska-Gracz, H., Mikkelson, R.L. and Wollum, A.G. (2002) Monitoring phosphorus mineralization from poultry manure using phosphatase assays and phosphorus-31 nuclear magnetic resonance spectroscopy. *Communications in Soil Science and Plant Analysis* 33, 1205–1217.

Dai, K.H., David, M.B., Vance, G.F. and Krzyszowska, A.J. (1996) Characterization of phosphorus in a spruce-fir Spodosol by phosphorus-31 nuclear magnetic resonance spectroscopy. *Soil Science Society of America Journal* 60, 1943–1950.

Delgado, A., Ruíz, J.R., del Campillo, M.D., Kassem, S. and Andreu, L. (2000) Calcium- and iron-related phosphorus in calcareous and calcareous marsh soils: sequential chemical fractionation and ^{31}P nuclear magnetic resonance study. *Communications in Soil Science and Plant Analysis* 31, 2483–2499.

Duffy, S.J. and van Loon, G.W. (1995) Investigations of aluminium hydroxyphosphates and activated sludge by ^{27}Al and ^{31}P MAS NMR. *Canadian Journal of Chemistry* 73, 1645–1659.

Emsley, J. and Niazi, S. (1983) The analysis of soil phosphorus by ICP and ^{31}P NMR spectroscopy. *Phosphorus and Sulfur* 16, 303–312.

Escudey, M., Galindo, G., Förster, J.E., Salazar, I., Page, A.L. and Chang, A. (1997) ^{31}Phosphorus-nuclear magnetic resonance analysis in extracts of a phosphorus-enriched volcanic soil of Chile. *Communications in Soil Science and Plant Analysis* 28, 727–737.

Fan, T.W.-M., Higashi, R.M. and Lane, A.N. (2000) Chemical characterization of a chelator-treated soil humate by solution-state multinuclear two-dimensional NMR with FTIR and pyrolysis-GCMS. *Environmental Science and Technology* 34, 1636–1646.

Florentz, M. and Granger, P. (1983) Phosphorus-31 nuclear magnetic resonance of activated sludge: use for the study of the biological removal of phosphates from wastewater. *Environmental Technology Letters* 4, 9–14.

Forster, J.C. and Zech, W. (1993) Phosphorus status of a soil catena under Liberian evergreen rain forest: results of ^{31}P NMR spectroscopy and phosphorus adsorption experiments. *Zeitschrift für Pflanzenernährung und Bodenkunde* 156, 61–66.

Francioso, O., Sanchez-Cortes, S., Tugnoli, V., Ciavatta, C., Sitti, L. and Gessa, C. (1996) Infrared, Ramen and nuclear magnetic resonance (^{1}H, ^{13}C, and ^{31}P) spectroscopy in the study of fractions of peat humic acids. *Applied Spectroscopy* 50, 1165–1174.

Francioso, O., Ciavatta, C., Tugnoli, V., Sanchez-Cortes, S. and Gessa, C. (1998) Spectroscopic characterization of pyrophosphate incorporation during extraction of peat humic acids. *Soil Science Society of America Journal* 62, 181–187.

Frossard, E., Tekely, P. and Grimal, J.Y. (1994a) Characterization of phosphate species in urban sewage sludges by high resolution solid-state ^{31}P NMR. *European Journal of Soil Science* 45, 403–408.

Frossard, E., Tekely, P. and Morel, J.L. (1994b) Chemical characterization and agronomic effectiveness of phosphorus applied as a polyphosphate-chitosan complex. *Fertilizer Research* 37, 151–158.

Frossard, E., Skrabal, P., Sinaj, S., Bangerter, F. and Traore, O. (2002) Forms and exchangeability of inorganic phosphate in composted solid organic wastes. *Nutrient Cycling in Agroecosystems* 62, 103–113.

Gigliotti, G., Kaiser, K., Guggenberger, G. and Haumaier, L. (2002) Differences in the chemical composition of dissolved organic matter from waste material of different sources. *Biology and Fertility of Soils* 36, 321–329.

Gil-Sotres, F., Zech, W. and Alt, H.G. (1990) Characterization of phosphorus fractions in surface horizons of soils from Galicia (NW Spain) by ^{31}P NMR spectroscopy. *Soil Biology and Biochemistry* 22, 75–79.

Gressel, N., McColl, J.G., Preston, C.M., Newman, R.H. and Powers, R.F. (1996) Linkages between phosphorus transformations and carbon decomposition in a forest soil. *Biogeochemistry* 33, 97–123.

Guggenberger, G., Christensen, B.T., Rubaek, G. and Zech, W. (1996a) Land-use and fertilization effects on P forms in two European soils: resin extraction and ^{31}P NMR analysis. *European Journal of Soil Science* 47, 605–614.

Guggenberger, G., Haumaier, L., Thomas, R.J. and Zech, W. (1996b) Assessing the organic phosphorus status of an Oxisol under tropical pastures following native savanna using ^{31}P NMR spectroscopy. *Biology and Fertility of Soils* 23, 332–339.

Guggenberger, G., Christensen, B.T. and Rubæk, G.H. (2000). Isolation and characterization of labile organic phosphorus pools in soils from the Askov long-term field experiments. *Zeitschrift für Pflanzenernährung und Bodenkunde* 163, 151–155.

Halls, J.N. (2002) A spatial sensitivity analysis of land use characteristics and phosphorus levels in small tidal creek estuaries of North Carolina, USA. *Journal of Coastal Research* Special Issue 36, 340–351.

Hawkes, G.E., Powlson, D.S., Randall, E.W. and Tate, K.R. (1984) A ^{31}P nuclear magnetic resonance study of the phosphorus species in alkali extracts of soils from long-term field experiments. *Journal of Soil Science* 35, 35–45.

Hinedi, Z.R. and Chang, A.C. (1989) Solubility and phosphorus-31 magic angle spinning nuclear magnetic resonance of phosphorus in sludge-amended soils. *Soil Science Society of America Journal* 53, 1057–1061.

Hinedi, Z.R., Chang, A.C. and Lee, R.W.K. (1988) Mineralization of phosphorus in sludge-amended soils monitored by phosphorus-31 nuclear magnetic resonance. *Soil Science Society of America Journal* 52, 1593–1596.

Hinedi, Z.R., Chang, A.C. and Lee, R.W.K. (1989a) Characterization of phosphorus in sludge extracts using phosphorus-31 nuclear magnetic resonance spectroscopy. *Journal of Environmental Quality* 18, 323–329.

Hinedi, Z.R., Chang, A.C. and Yesinowski, J.P. (1989b) Phosphorus-31 magic angle spinning nuclear magnetic resonance of wastewater sludges and sludge-amended soil. *Soil Science Society of America Journal* 53, 1053–1056.

Hupfer, M., Gächter, R. and Rüegger, H. (1995) Polyphosphate in lake sediments: ^{31}P NMR spectroscopy as a tool for its identification. *Limnology and Oceanography* 40, 610–617.

Hupfer, M., Rübe, B. and Schmieder, P. (2004) Origin and diagenesis of polyphosphate in lake sediments: a ^{31}P-NMR study. *Limnology and Oceanography* 49, 1–10.

Ingall, E.D., Schroeder, P.A. and Berner, R.A. (1990) The nature of organic phosphorus in marine sediments: new insights from ^{31}P NMR. *Geochimica et Cosmochimica Acta* 54, 2617–2620.

Jing, S.R., Benefield, L.D. and Hill, W.E. (1992) Observations relating to enhanced phosphorus removal in biological systems. *Water Research* 26, 213–223.

Kemme, P.A., Lommen, A., De Jonge, L.H., Van der Klis, J.D., Jongbloed, A.W., Mroz, Z. and Beynen, A.C. (1999) Quantification of inositol phosphates using ^{31}P nuclear magnetic resonance spectroscopy in animal nutrition. *Journal of Agricultural and Food Chemistry* 47, 5116–5121.

Khoshmanesh, A., Hart, B.T., Duncan, A. and Beckett, R. (2002) Luxury uptake of phosphorus by sediment bacteria. *Water Research* 36, 774–778.

Knicker, H. and Nanny, M.A. (1997) Nuclear magnetic resonance spectroscopy: basic theory and background. In: Nanny, M.A., Minear, R.A. and Leenheer, J.A. (eds) *Nuclear Magnetic Resonance Spectroscopy in Environmental Chemistry.* Oxford University Press, Oxford, pp. 3–15.

Kolowith, L.C. and Berner, R.A. (2002) Weathering of phosphorus in black shales. *Global Biogeochemical Cycles* 16, 1140.

Kolowith, L.C., Ingall, E.D. and Benner, R. (2001) Composition and cycling of marine organic phosphorus. *Limnology and Oceanography* 46, 309–320.

Koopmans, G.F., Chardon, W.F., Dolfing, J., Oenema, O., van der Meer, P. and van Riemsdijk, W.H. (2003) Wet chemical and phosphorus-31 nuclear magnetic resonance analysis of phosphorus speciation in a sandy soil receiving long-term fertilizer or animal manure applications. *Journal of Environmental Quality* 32, 287–295.

Kristiansen, S.M., Amelung, W. and Zech, W. (2001) Phosphorus forms as affected by abandoned anthills (*Formica polyctena* Förster) in forest soils: sequential extraction and liquid-state ^{31}P NMR spectroscopy. *Zeitschrift für Pflanzenernährung und Bodenkunde* 164, 49–55.

Lee, R.B. and Ratcliffe, R.G. (1993) Nuclear magnetic studies of the location and function of plant nutrients *in vivo*. *Plant and Soil* 155/156, 45–55.

Leinweber, P., Haumaier, L. and Zech, W. (1997) Sequential extractions and ^{31}P NMR spectroscopy of phosphorus forms in animal manures, whole soils and particle-size separates from a densely populated livestock area in northwest Germany. *Biology and Fertility of Soils* 25, 89–94.

Lens, P.N.L. and Hemminga, M.A. (1998) Nuclear magnetic resonance in environmental engineering: principles and applications. *Biodegradation* 9, 393–409.

Lookman, R., Grobet, P., Merckx, R. and van Riemsdijk, W.H. (1997) Application of ^{31}P and ^{27}Al MAS NMR for phosphate speciation studies in soil and aluminium hydroxides: promises and constraints. *Geoderma* 80, 369–388.

Lookman, R., Geerts, H., Grobet, P., Merckx, R. and Vlassak, K. (1996) Phosphate speciation in excessively fertilized soil: a ^{31}P and ^{27}Al MAS NMR spectroscopy study. *European Journal of Soil Science* 47, 125–130.

Magid, J., Tiessen, H. and Condron, L.M. (1996) Dynamics of organic phosphorus in soils under natural and agricultural ecosystems. In: Piccolo, A. (ed.) *Humic Substances in Terrestrial Ecosystems*. Elsevier, Amsterdam, pp. 429–466.

Mahieu, N., Olk, D.C. and Randall, E.W. (2000) Analysis of phosphorus in two humic acid fractions of intensively cropped lowland rice soils by ^{31}P NMR. *European Journal of Soil Science* 51, 391–402.

Makarov, M.I. (1996) Forms of P compounds in humic and fulvic acids in some types of soils. *Moscow University Soil Science Bulletin* 51, 15–22.

Makarov, M.I. (1998) Organic phosphorus compounds in alpine soils of the northwestern Caucasus. *Eurasian Soil Science* 31, 778–786.

Makarov, M.I., Guggenberger, G., Alt, H.G. and Zech, W. (1995) Phosphorus status of Eutric Cambisols polluted by P-containing immisions: results of ^{31}P NMR spectroscopy and chemical analysis. *Zeitschrift für Pflanzenernährung und Bodenkunde* 158, 293–298.

Makarov, M.I., Guggenberger, G., Zech, W. and Alt, H.G. (1996) Organic phosphorus species in humic acids of mountain soils along a toposequence in the Northern Caucasus. *Zeitschrift für Pflanzenernährung und Bodenkunde* 159, 467–470.

Makarov, M.I., Malysheva, T.I., Haumaier, L., Alt, H.G. and Zech, W. (1997) The forms of phosphorus in humic and fulvic acids of a toposequence of alpine soils in the northern Caucasus. *Geoderma* 80, 61–73.

Makarov, M.I., Haumaier, L. and Zech, W. (2002a) The nature and origins of diester phosphates in soils: a ^{31}P NMR study. *Biology and Fertility of Soils* 35, 136–146.

Makarov, M.I., Haumaier, L. and Zech, W. (2002b) Nature of soil organic phosphorus: an assessment of peak assignments in the diester region of ^{31}P NMR spectra. *Soil Biology and Biochemistry* 34, 1467–1477.

Makarov, M.I., Haumaier, L., Zech, W. and Malysheva, T.I. (2004) Organic phosphorus compounds in particle-size fractions of mountain soils in the northwest Caucasus. *Geoderma* 118, 101–114.

Martin, F., Canet, D., Rolin, D., Marchal, J.P. and Larher, F. (1983) Phosphorus-31 nuclear magnetic resonance study of polyphosphate metabolism in intact ectomycorrhizal fungi. *Plant and Soil* 71, 469–476.

McDowell, R.W., Brookes, P.C., Mahieu, N., Poulton, P.R., Johnston, A.E. and Sharpley, A.N. (2002a) The effect of soil acidity on potentially mobile phosphorus in a grassland soil. *Journal of Agricultural Science* 139, 27–36.

McDowell, R.W., Condron, L.M., Mahieu, N., Brookes, P.C., Poulton, P.R. and Sharpley, A.N. (2002b) Analysis of potentially mobile phosphorus in arable soils using solid-state nuclear magnetic resonance. *Journal of Environmental Quality* 31, 450–456.

Miltner, A., Haumaier, L. and Zech, W. (1998) Transformations of phosphorus during incubation of beech leaf litter in the presence of oxides. *European Journal of Soil Science* 49, 471–475.

Möller, A., Kaiser, K., Amelung, W., Niamskul, C.,

Udomsri, S., Puthawong, M., Haumaier, L. and Zech, W. (2000) Forms of organic C and P extracted from tropical soils as assessed by liquid-state ^{13}C- and ^{31}P NMR spectroscopy. *Australian Journal of Soil Research* 38, 1017–1035.

Nanny, M.A. and Minear, R.A. (1994a) Use of lanthanide shift reagents with ^{31}P FT NMR spectroscopy to analyze concentrated lake samples. *Environmental Science and Technology* 28, 1521–1527.

Nanny, M.A. and Minear, R.A. (1994b) Organic phosphorus in the hydrosphere: characterization via ^{31}P Fourier transform nuclear magnetic resonance spectroscopy. In: Baker, L.A. (ed.) *Environmental Chemistry of Lakes and Reservoirs.* Advances in Chemistry Series 237. American Chemical Society, Washington, DC, pp. 161–191.

Nanny, M.A. and Minear, R.A. (1997a) Characterization of soluble unreactive phosphorus using ^{31}P nuclear magnetic resonance spectroscopy. *Marine Geology* 139, 77–94.

Nanny, M.A. and Minear, R.A. (1997b) ^{31}P FT NMR of concentrated lake water samples. In: Nanny, M.A., Minear, R.A. and Leenheer, J.A. (eds) *Nuclear Magnetic Resonance Spectroscopy in Environmental Chemistry.* Oxford University Press, Oxford, pp. 221–246.

Newman, R.H. and Tate, K.R. (1980) Soil phosphorus characterisation by ^{31}P-nuclear magnetic resonance. *Communications in Soil Science and Plant Analysis* 11, 835–842.

Ogner, G. (1983) ^{31}P NMR spectra of humic acids: a comparison of four different raw humus types in Norway. *Geoderma* 29, 215–219.

Pant, H.K. and Reddy, K.R. (2001) Hydrologic influence on stability of organic phosphorus in wetland detritus. *Journal of Environmental Quality* 30, 668–674.

Pant, H.K., Warman, P.R. and Nowak, J. (1999) Identification of soil organic phosphorus by ^{31}P nuclear magnetic resonance spectroscopy. *Communications in Soil Science and Plant Analysis* 30, 757–772.

Pant, H.K., Reddy, K.R. and Dierberg, F.E. (2002) Bioavailability of organic phosphorus in a submerged aquatic vegetation-dominated treatment wetland. *Journal of Environmental Quality* 31, 1748–1756.

Parfitt, R.L., Tate, K.R., Yeates, G.W. and Beets, P.N. (1994) Phosphorus cycling in a sandy Podsol under *Pinus radiata. New Zealand Journal of Forestry Science* 24, 253–267.

Paytan, A., Cade-Menun, B.J., McLaughlin, K. and Faul, K.L. (2003) Selective phosphorus regeneration of sinking marine particles: evidence from ^{31}P NMR. *Marine Chemistry* 82, 55–70.

Preston, C.M. (1993) The NMR user's guide to the forest. *Canadian Journal of Applied Spectroscopy* 38, 61–69.

Preston, C.M. (1996) Application of NMR to soil organic matter analysis: history and prospects. *Soil Science* 161, 144–166.

Preston, C.M. (2001) Carbon-13 solid-state NMR of soil organic matter: using the technique effectively. *Canadian Journal of Soil Science* 81, 255–270.

Preston, C.M. and Trofymow, J.A. (2000) Characterization of soil P in coastal forest chronosequences of southern Vancouver Island: effects of climate and harvesting disturbance. *Canadian Journal of Soil Science* 80, 633–647.

Preston, C.M., Ripmeester, J.A., Mathur, S.P. and Lévesque, M. (1986) Application of solution and solid-state multinuclear NMR to a peat-based composting system for fish and crab scrap. *Canadian Journal of Spectroscopy* 31, 63–69.

Preston, C.M., Cade-Menun, B.J. and Sayer, B.G. (1998) Characterization of Canadian backyard composts: chemical analysis and ^{13}C CPMAS and ^{31}P NMR spectroscopy. *Compost Science and Utilization* 6, 53–66.

Quiquampoix, H. and Ratcliffe, R.G. (1992) A ^{31}P NMR study of the adsorption of bovine serum albumin on montmorillonite and the paramagnetic cation Mn^{2+}: modification of conformation with pH. *Journal of Colloid Interface Science* 148, 343–352.

Randall, E.W., Mahieu, N. and Ivanova, G.I. (1997) NMR studies of soil, organic matter and nutrients: spectroscopy and imaging. *Geoderma* 80, 307–325.

Rheinheimer, D.S., Anghinoni, I. and Flores, A.F. (2002) Organic and inorganic phosphorus as characterized by phosphorus-31 nuclear magnetic resonance in subtropical soils under management systems. *Communications in Soil Science and Plant Analysis* 33, 1853–1871.

Robinson, J.S., Johnston, C.T. and Reddy, K.R. (1998) Combined chemical and ^{31}P NMR spectroscopic analysis of phosphorus in wetland organic soils. *Soil Science* 163, 705–713.

Röske, I. and Schönborn, C. (1994) Influence of the addition of precipitants on the biological phosphorus elimination in a pilot plant. *Water Science and Technology* 30, 323–332.

Rubæk, G.H., Guggenberger, G., Zech, W. and Christensen, B.T. (1999) Organic phosphorus in soil size separates characterized by phosphorus-31 nuclear magnetic resonance and resin extraction. *Soil Science Society of America Journal* 63, 1123–1132.

Selig, U., Hübener, T. and Michalik, M. (2002) Dissolved and particulate phosphorus forms in a eutrophic shallow lake. *Aquatic Sciences* 64, 97–105.

Shand, C.A., Cheshire, M.V., Bedrock, C.N., Chapman, P.J., Fraser, D.A. and Chudek, J.A. (1999) Solid–phase ^{31}P NMR spectra of peat and mineral soils, humic acids and soil solution components: influence of iron and manganese. *Plant and Soil* 214, 153–163.

Solomon, D. and Lehmann, J. (2000) Loss of phosphorus from soil in semi-arid northern Tanzania as a result of cropping: evidence from sequential extraction and ^{31}P NMR spectroscopy. *European Journal of Soil Science* 51, 699–708.

Solomon, D., Lehmann, J., Mamo, T., Fritzsche, F. and Zech, W. (2002) Phosphorus forms and dynamics as influenced by land use changes in the subhumid Ethiopian highlands. *Geoderma* 105, 21–48.

Sumann, M., Amelung, W., Haumaier, L. and Zech, W. (1998) Climatic effects on soil organic phosphorus in the North American Great Plains identified by phosphorus-31 nuclear magnetic resonance. *Soil Science Society of America Journal* 62, 1580–1586.

Sundareshwar, P.V., Morris, J.T., Pellechia, P.J., Cohen, H.J., Porter, D.E. and Jones, B.C. (2001) Occurrence and implications of pyrophosphate in estuaries. *Limnology and Oceanography* 46, 1570–1577.

Sutter, B., Taylor, R.E., Hossner, L.R. and Ming, D.W. (2002) Solid state phosphorus nuclear magnetic resonance of iron-, manganese-, and copper-containing synthetic hydroxyapatites. *Soil Science Society of America Journal* 66, 455–463.

Taranto, M.T., Adams, M.A. and Polglase, P.J. (2000) Sequential fractionation and characterisation (^{31}P NMR) of phosphorus-amended soils in *Banksia integrifolia* (L.f.) woodland and adjacent pasture. *Soil Biology and Biochemistry* 32, 169–177.

Tate, K.R. and Newman, R.H. (1982) Phosphorus fractions of a climosequence of soils in New Zealand tussock grassland. *Soil Biology and Biochemistry* 14, 191–196.

Tchienkoua, M. and Zech, W. (2003) Chemical and spectral characterization of soil phosphorus under three land uses from an Andic Palehumult in West Cameroon. *Agriculture, Ecosystems and Environment* 100, 193–200.

Tikhova, V.D., Shakirov, M.M., Fadeeva, V.P. and Dergacheva, M.I. (2000) ^{31}P NMR study of soil humic acids. *Russian Journal of Applied Chemistry* 73, 1278–1281.

Toor, G.S., Condron, L.M., Di, H.J., Cameron, K.C. and Cade-Menun, B.J. (2003) Characterization of organic phosphorus in leachate from a grassland soil. *Soil Biology and Biochemistry* 35, 1319–1325.

Trasar-Cepeda, M.C., Gil-Sotres, F., Zech, W. and Alt, H.G. (1989) Chemical and spectral analysis of organic P forms in acid high organic matter soils in Galicia (NW Spain). *Science of the Total Environment* 81/82, 429–436.

Turner, B.L. and McKelvie, I.D. (2002) A novel technique for the pre-concentration and extraction of inositol hexakisphosphate from soil extracts with determination by phosphorus-31 nuclear magnetic resonance. *Journal of Environmental Quality* 31, 466–470.

Turner, B.L., Cade-Menun, B.J. and Westermann, D.T. (2003a) Organic phosphorus composition and potential bioavailability in calcareous soils of the western United States. *Soil Science Society of America Journal* 67, 1168–1179.

Turner, B.L., Mahieu, N. and Condron, L.M. (2003b) Phosphorus-31 nuclear magnetic resonance spectral assignments of phosphorus compounds in NaOH–EDTA extracts. *Soil Science Society of America Journal* 67, 497–510.

Turner, B.L., Mahieu, N. and Condron, L.M. (2003c) Quantification of *myo*-inositol hexakisphosphate in alkaline soil extracts by solution ^{31}P NMR spectroscopy and spectral deconvolution. *Soil Science* 168, 469–478.

Turner, B.L., Mahieu, N. and Condron, L.M. (2003d) The phosphorus composition of temperate pasture soils determined by NaOH–EDTA extraction and solution ^{31}P NMR spectroscopy. *Organic Geochemistry* 34, 1199–1210.

Turrión, M.-B., Glaser, B., Solomon, D., Ni, A. and Zech, W. (2000) Effects of deforestation on phosphorus pools in mountain soils of the Alay Range, Khyrgyzia. *Biology and Fertility of Soils* 31, 134–142.

Turrión, M.-B., Gallardo, J.F., Haumaier, L., González, M.-I. and Zech, W. (2001) ^{31}P NMR characterization of phosphorus fractions in natural and fertilized forest soils. *Annals of Forest Science* 58, 89–98.

Uhlmann, D., Röske, I., Hupfer, M. and Ohms, G. (1990) A simple method to distinguish between polyphosphate and other phosphate fractions of activated sludge. *Water Research* 24, 1355–1360.

Veeman, W.S. (1997) Nuclear magnetic resonance: a simple introduction to the principles and applications. *Geoderma* 80, 225–242.

Watts, E.E., Dean, P.A.W. and Martin, R.R. (2002) ^{31}P nuclear magnetic resonance study of sediment microbial phospholipids. *Canadian Journal of Analytical Sciences and Spectroscopy* 47, 127–133.

Wilson, M.A. (1987) *NMR Techniques and Applications in Geochemistry and Soil Chemistry.* Pergamon, New York, 353 pp.

Wilson, M.A. (1991) Analysis of functional groups in soil by nuclear magnetic resonance spectroscopy. In: Smith, K.A. (ed.) *Soil Analysis: Modern Instrumental Techniques* (2nd edn). Marcel Dekker, New York, pp. 601–645.

Zech, W., Alt, H.G., Zucker, A. and Kögel, I. (1985) ^{31}P NMR spectroscopic investigations of NaOH-extracts from soils with different land use in Yucatan (Mexico). *Zeitschrift für Pflanzenernährung und Bodenkunde* 148, 626–632.

Zech, W., Alt, H.G., Haumaier, L. and Blasek, R. (1987) Characterization of phosphorus fractions in mountain soils of the Bavarian Alps by ^{31}P NMR spectroscopy. *Zeitschrift für Pflanzenernährung und Bodenkunde* 150, 119–123.

Zhang, T.Q., Mackenzie, A.F. and Sauriol, F. (1999) Nature of organic phosphorus as affected by long-term fertilization under continuous corn (*Zea mays* L.): a ^{31}P NMR study. *Soil Science* 164, 662–670.

3 Organic Phosphorus Speciation in Natural Waters by Mass Spectrometry

William T. Cooper,[1] Jennifer M. Llewelyn,[1] G. Lee Bennett,[1] Alexandra C. Stenson[1] and Vincent J.M. Salters[2]

[1]*Department of Chemistry and Biochemistry, Florida State University, Tallahassee, FL 32306, USA;* [2]*National High Magnetic Field Laboratory and Department of Geological Sciences, Florida State University, Tallahassee, FL 32306, USA*

Introduction

Characterizing the molecular speciation of individual organic phosphorus compounds in natural environments (i.e. the chemical formulas and structures of organic phosphorus compounds as they exist in nature) is one of the more difficult analytical challenges in modern geochemistry. While some speciation information is available for a few types of organic phosphorus compounds (e.g. inositol phosphates), little is known about the hundreds of others that are thought to play significant roles in a variety of environmental processes. This lack of information is due in great part to the limitations of current analytical techniques, which do not have the sensitivity to detect individual organic phosphorus compounds at the levels at which they normally exist in natural environments, nor the selectivity to distinguish them from the overwhelming background of the other natural organic compounds present. For example, while mass spectrometry has been one of the principal techniques used to develop a molecular understanding of many environmental processes, this is not true for organic phosphorus speciation, for which low concentrations, complex backgrounds, and polar, non-volatile target analytes make the application of traditional mass spectrometry techniques difficult, if not impossible. Indeed, there are surprisingly few studies of organic phosphorus speciation that have employed mass spectrometry techniques, either alone or in combination with modern separation techniques, such as gas chromatography, liquid chromatography or capillary electrophoresis.

The molecular properties of organic phosphorus compounds are no doubt responsible to a great degree for this resistance to using one of the most powerful techniques available in analytical chemistry. From the mass spectroscopist's perspective, organic phosphorus is dominated by the phosphate group(s) which imparts polarity and mass. Until about 1990 it was very difficult to make gas-phase ions from these relatively high-molecular-weight, low-volatility molecules. However, the dramatic developments in 'soft' ionization techniques for large biomolecules that have occurred over the past 15 years now make mass spectrometry of molecules such as nucleotides, proteins, peptides, and even whole cells, almost routine. In tandem with these new ionization techniques, large-molecule mass spectrometry has evolved as well, and there are now spectrometers capable of providing elemental composition, fine structure and even sequence information of biomolecules. It should be noted that the importance of the breakthrough technologies in large-molecule ionization and their applications in biomolecular mass spectrometry was recognized by the award of a Nobel Prize to John Fenn (electrospray ionization) and Koichi Tanaka (matrix-assisted laser desorption ionization) who developed the technology (Nobel Foundation, 2002).

This chapter begins with a review of basic mass spectrometry techniques, with an emphasis on the more 'advanced' techniques that have been developed for studies of large biomolecules. Readers interested in a more detailed discussion of these techniques should consult any of the many literature reviews and books that have appeared recently (e.g. Dass, 2001). Biomolecular mass spectrometry is evolving rapidly, and many of the methods are directly applicable to organic phosphorus speciation. A discussion of the new soft ionization techniques that make large-molecule mass spectrometry possible (electrospray and matrix-assisted laser desorption ionization) will follow. The chapter concludes with a description of our work on phosphorus speciation in the Florida Everglades. These studies utilized high-resolution phosphorus-specific mass spectrometry for quantitative analyses, and ultrahigh-resolution mass spectrometry experiments for qualitative identification of individual compounds in the dissolved organic phosphorus fraction in waters from a treatment wetland in the Everglades. The latter experiments included both single- and multi-stage Fourier-transform ion cyclotron resonance (FT–ICR) mass spectrometry with electrospray ionization. Ultrahigh-resolution single-stage FT–ICR mass spectrometry with a mass resolving power $(m/\Delta m) > 10^6$ was used to identify prominent peaks that contained phosphate. Several of the peaks that exhibited distinct spatial trends in the Everglades nutrient removal site were then further analysed by a multi-stage tandem mass spectrometry scheme. This procedure allows the chemical formulas and structures of these high-molecular-weight organic phosphorus compounds to be estimated with a great degree of confidence.

Mass Spectrometry: a Brief Review of Current Technology

Mass spectrometry includes a collection of techniques that produce information about the masses and abundances of gas-phase ions. These are some of the most powerful analytical methods available for determining the elemental composition and chemical formulas of molecules, since molecular mass is the single most important piece of information necessary for characterizing chemical structures. In addition, mass spectrometry is an important quantitative tool, because mass spectrometers are in general sensitive and respond to most compounds in a predictable way over large concentration ranges. Mass spectrometry can also provide element-specific detection by monitoring a single mass that corresponds to the element of interest (e.g. phosphorus at 30.974 Da). This application is useful in quantifying organic phosphorus compounds that normally exist in complex matrices dominated by other non-phosphorus organic compounds. It should be noted that in the mass spectrometry community the currently accepted unit for atomic or molecular mass is the dalton (Da), which is exactly 1/12th the mass of a ^{12}C atom.

While mass spectrometers may differ greatly in their design, they all produce spectra as a result of the same sequence of four processes:

1. *Generation* of gas-phase atoms or molecules;
2. *Ionization* of these species;
3. *Separation* based on mass-to-charge (m/z) ratio;
4. *Detection* of the mass-sorted ions.

A mass spectrum is thus a plot of intensities of ions as a function of their mass-to-charge ratios. To get the mass of an ion its charge must be known, but this can usually be determined from the spectrum. Many molecules will fragment during the ionization step because of the energies used in that process, and their spectra are then composed of a molecular ion and a series of fragment ions. While such fragmentation will produce complicated spectra if many compounds are present, it can nevertheless provide valuable information about the sub-structures present in the original molecules. Indeed, the technique of tandem mass spectrometry is designed to deliberately fragment molecules so that they can be reconstructed from the ionic pieces that result.

In this brief review of mass spectrometry we focus on the mass analysis and ionization steps of the mass spectrometry process. It is these two areas where dramatic developments in technology have made large-molecule mass spectrometry possible.

Table 3.1. Chemical formulas from exact masses in electrospray ionization Fourier-transform ion cyclotron resonance mass spectra (adapted from Fievre *et al*., 1997).

	Number of possible formulas within a given mass tolerance		
Observed ion	10 ppm[a]	5 ppm[a]	1 ppm[a]
543.0931	49	25	6
559.1371	52	33	6
575.1023	58	29	6

[a] Here, 'ppm' (parts per million) refers to the relative uncertainty in determining the molecular mass. It is calculated from $(\Delta M/M) \times 10^6$, where ΔM is the absolute uncertainty and M the measured mass.

Mass analysers

Before describing the various mass analysers currently available, a brief discussion of mass resolving power and mass accuracy is appropriate. These are the most useful criteria for comparing the capabilities of mass spectrometers.

Mass resolving power refers to the ability of the analyser to distinguish ions of similar mass-to-charge ratio. It is quantitatively defined by equation (3.1):

$$R = m/\Delta m \tag{3.1}$$

where R is the resolution, m is mass, and Δm is the difference in masses resolved.

In equation (3.1), m is the mass at which two adjacent peaks are resolved by one mass unit ($\Delta m = 1$). 'Resolved' is normally taken to mean overlap at 10% of the height of the peaks. Thus, an analyser with a resolution of 500 can provide unit mass resolution at mass 500 Da. That is, molecules separated by one mass unit can be distinguished. However, that analyser can only distinguish molecules that differ by two mass units at 1000 Da. This illustrates the need for high-resolution analysers to characterize large molecules such as organic phosphates. Furthermore, unit mass resolution is not always sufficient when complex mixtures such as natural dissolved organic matter are being analysed. As an example, consider two molecules we have identified in fulvic acid extracts from the Suwannee River in northern Florida (Stenson *et al*., 2002).

$$C_{21}H_{30}O_8;\ \text{molecular weight} = 410.1941$$

$$C_{22}H_{34}O_7;\ \text{molecular weight} = 410.2305$$

$$M = 410;\ \Delta m = 0.0364;\ R_{required} = 11{,}300$$

Mass accuracy is also a very important characteristic of mass spectrometers used for qualitative identification of unknown compounds. The greater the mass accuracy, the greater the confidence in the assignment of exact mass and chemical formula. Again, greater mass accuracy is needed for larger molecules because of the increasing chemical formula possibilities as mass increases. Table 3.1 illustrates this, again using data from our studies of Suwannee River fulvic acids (Fievre *et al*., 1997).

Magnetic sector analysers

For decades, the most widely used mass spectrometers employed magnetic sector analysers for sorting ions by mass, or more correctly, mass-to-charge ratio, commonly referred to as m/z. Magnetic sectors use a magnetic field to deflect the trajectory of

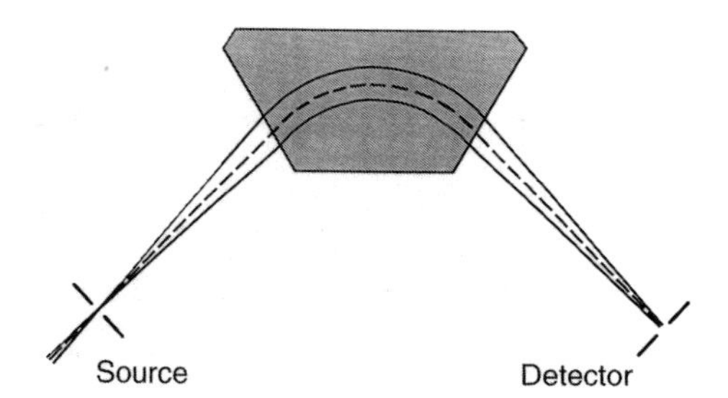

Fig. 3.1. Schematic representation of a magnet sector analyser. For a given source potential *V* and magnetic field *B*, only ions of one *m/z* will be deflected along the correct radius *r* and strike the detector.

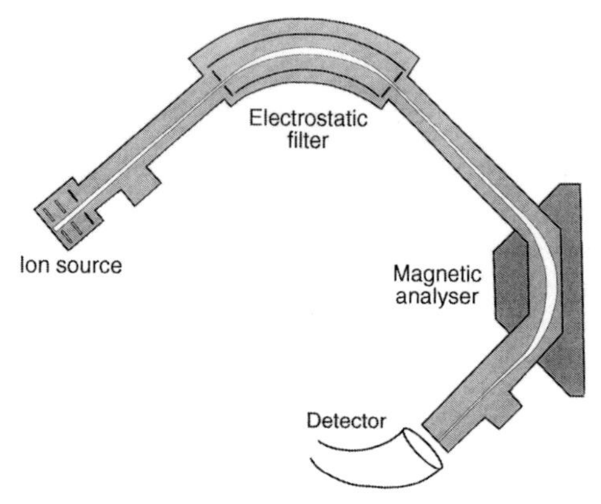

Fig. 3.2. High-resolution double-focusing mass analyser with Nier–Johnson geometry. The electrostatic sector filters ions according to kinetic energy before magnetic sorting according to *m/z*. The combination of electric and magnet fields produces narrow peaks and high spectral resolution.

moving ions after they have been injected into the magnetic field by acceleration through an electrostatic potential (Fig. 3.1). Since the trajectory of a charged particle in a magnetic field is dependent on the momentum of the particle, ions are thus sorted by mass-to-charge ratio according to the following equation:

$$\frac{m}{z} = \frac{B^2r^2}{2V} \tag{3.2}$$

where m/z is the mass-to-charge ratio, B is the magnetic field strength, V is the electrostatic potential through which the ion has been accelerated, and r is the radius of curvature of the ion trajectory. Ions of different m/z values thus follow different radii and can be distinguished in that way. However, in normal operation only one detector is used at a fixed r, and different ions are brought to focus on the detector by scanning either the magnetic field or the accelerating voltage. Equation (3.2) also expresses the relationship between magnetic field, accelerating voltage and m/z when r is a constant.

Magnetic sector analysers are considered to be medium-resolution instruments with $R \leq 2000$. They are also an important component in the high-resolution double-focusing spectrometers discussed next.

High-resolution double-focusing mass spectrometers

The relatively low resolution provided by magnetic sector instruments is actually not due to limitations in the mass sorting, but rather the dispersion in kinetic energies of ions leaving the ionization source. In equation (3.2) it is assumed that all ions are accelerated through the same electrostatic field and thus enter the magnetic field with the same kinetic energy. In practice there is a range of ion energies produced, which results in broader peaks and lower mass resolving power. These variations in kinetic energy can be overcome by coupling the magnetic sector with a second electrostatic sector that sorts ions by their kinetic energies (Fig. 3.2). By taking only those ions that fall within a narrow range of kinetic energies (normally accomplished by including a narrow slit at the exit of the electrostatic analyser), very narrow dispersions in ion trajectories are achieved in the magnetic mass analyser, resulting in very high mass resolving power, frequently approaching 20,000. Although slow and not very sensitive, double-focusing mass spectrometers are now routinely used when high mass resolving power and mass accuracy are required.

Quadrupole and ion trap mass spectrometers

These two low-resolution spectrometers sort ions according to different principles,

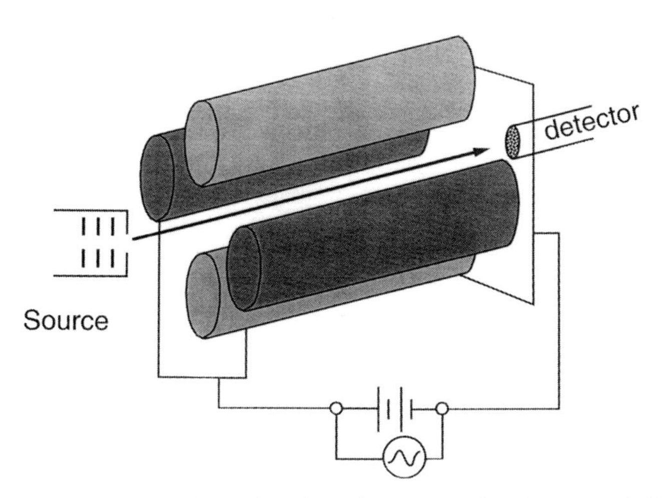

Fig. 3.3. Schematic representation of a quadrupole analyser. A complex electrostatic field is produced by application of direct and alternating current fields to conducting rods. For given applied direct and alternating current voltages, only ions of a specific *m/z* reach the detector; all others strike the rods. A mass spectrum is obtained by ramping the direct current and alternating current fields.

but have similar characteristics. *Quadrupole analysers* consist of four parallel, cylindrical, electrical-conducting rods (Fig. 3.3). Each pair of opposite rods is connected electrically, and direct current and radio frequency alternating current voltages are applied to both pairs. This combination creates a complex field inside the rods that affects ion trajectories in a complex but predictable way. In essence, the quadrupolar field acts as a mass-to-charge 'notch' filter, allowing only one *m/z* to reach the detector for particular applied direct and alternating current potentials. A mass spectrum is obtained by varying the field through ramping of the direct and alternating current potentials, allowing ions of different *m/z* to be detected.

Ion trap analysers operate on a different principle than magnetic sector or quadrupole analysers, both of which manipulate ion trajectories. Ion traps store ions in a three-dimensional electrostatic field that is formed by appropriate application of voltages to electrodes in the trap (Fig. 3.4). Ions of a broad distribution of *m/z* values can be stored in the trap as they precess at a frequency that is dependent on their *m/z* ratio. A mass spectrum is obtained by increasing the magnitude of the direct and alternating current fields so that the motions of higher *m/z* ions become unstable and they are ejected from the trap along an axial vector where a detector is located. This is referred to as the mass-selective instability mode of mass analysis (Stafford *et al.*, 1984).

In general, for the analysis of large biomolecules and complex environmental samples, these low-resolution mass spectrometers have been supplanted by the more sophisticated, higher mass resolution instruments described next. However, they do find some use in structural characterization using two-dimensional mass spectrometry techniques, as well as in selective ion monitoring of unique molecular fragments. For example, characterization of the basic molecular properties of individual organic phosphorus compounds found in complex organic matter matrices can be accomplished by coupling high-resolution separation methods (e.g. high-performance liquid chromatography) to phosphorus-specific mass spectrometry detection. This approach will be demonstrated later in this chapter.

Time-of-flight mass analysers

Time-of-flight mass analysers, which have been used for many decades, operate on a

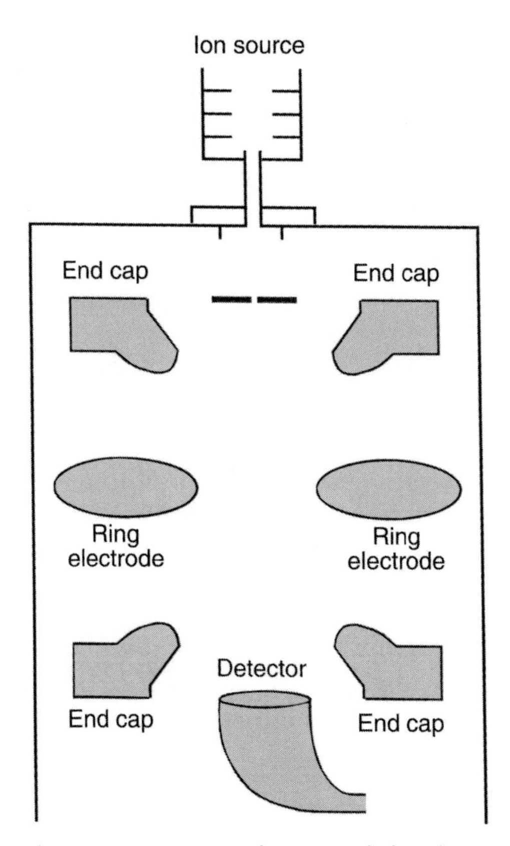

Fig. 3.4. Ion trap mass analyser. Ions of a broad distribution of *m/z* values are stored in the trap by application of radio frequency voltages to the end caps and ring electrode. A mass spectrum is obtained by increasing the magnitude of the radio frequency between the end caps so that the motions of selected ions become unstable and they are ejected.

simple principle. They are in essence velocity spectrometers in that ions are separated by *m/z* based upon different velocities acquired in an electrostatic field. Ions are accelerated from the ionization source into a field-free drift tube. Since all ions are accelerated through the same electrostatic field, they all have the same nominal kinetic energy. Thus, their velocities will vary inversely with mass-to-charge ratio. A mass spectrum is obtained by monitoring ions arriving at the detector as a function of time. The defining relationships that equate drift time to mass-to-charge ratio are given in equation (3.3):

$$t = \frac{L}{v} = L\sqrt{\left(\frac{1}{2V}\right)\left(\frac{m}{z}\right)} \quad (3.3)$$

where t is the time for the ion to reach the detector, L the length of the tube, v the ion velocity, and V the accelerating voltage.

Early time-of-flight analysers did not have particularly good resolution, generally less than about 500, but their speed and virtual unlimited mass range made them popular for many applications. However, significant developments in time-of-flight technology occurred in the 1990s that increased their resolution. In addition, these analysers are particularly suited for laser ionization techniques that produce temporally narrow bursts of ions. This combination of increased resolution, very high mass range, and compatibility with matrix-assisted laser desorption ionization (MALDI) has led to a rebirth in time-of-flight usage, particularly for large biomolecular characterization.

The most significant effect that limits resolution in time-of-flight analysers is the dispersion in kinetic energies produced during injection into the field-free region. Ions of a particular mass travelling faster than ions of the same mass arrive at the detector sooner, producing a spread in the peak for that molecule. While a number of approaches can minimize this effect, the most successful by far has been the reflectron, a series of electrical lenses that compensate for variations in ion velocity. Ions traverse the first field-free region and enter the reflectron, where their velocity is reduced and then reversed. Ions are then injected back into a second field-free region and finally detected. Faster-moving ions penetrate farther into the lens system and thus spend less time in the field-free region than slower-moving ions of the same mass. This extra time spent in the reflectron compensates for their faster velocity, and all ions of the same mass arrive at the detector in a narrow time window.

Great improvements in the resolution of time-of-flight spectrometers followed the introduction of the reflectron (Mamyrin, 1994). Further enhancements in resolution were realized when time-delay extraction

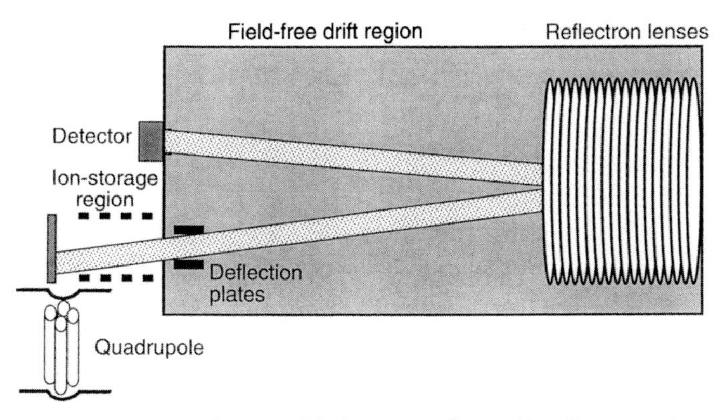

Fig. 3.5. Schematic representation of a time-of-flight mass analyser with reflectron and external ion storage for time-delay extraction. Ions that enter the field-free drift region migrate to the detector at a rate that is dependent on their *m/z* ratio. The reflectron lenses compensate for variations in kinetic energies of the injected ions; these variations would otherwise produce broadened peaks and loss of spectral resolution.

was introduced. This technique involves storing ions in an external field for a fixed time period, then introducing them into the time-of-flight reflectron in a very narrow burst. Figure 3.5 is a schematic diagram of a reflectron time-of-flight analyser with external ion storage for time-delay extraction.

Time-of-flight reflectron mass spectrometry with some form of time-delay extraction has become popular for large-molecule characterizations. Resolutions of up to 15,000 can be routinely achieved over very broad mass ranges, extending up to about 300,000 Da. As noted previously, time-of-flight analysers are particularly compatible with laser-based ionization techniques that produce very short bursts of ions. They are also very fast, with mass spectra often obtainable in about 25 μs. Finally, time-of-flight analysers have been paired with a quadrupole to produce a hybrid two-dimensional mass spectrometry system that has found widespread use in protein analyses. This device will be discussed in more detail in the section on two-dimensional mass spectrometry.

Fourier-transform ion cyclotron resonance mass spectrometry

Fourier-transform ion cyclotron resonance (FT–ICR) mass spectrometry makes use of the circular motion of ions when they are constrained in a magnetic field. The velocity and thus angular frequency of this cyclotron motion is given by equation (3.4).

$$\omega = B\frac{z}{m} \qquad (3.4)$$

where ω = ion cyclotron frequency; B = magnetic field strength; m/z = mass-to-charge ratio of ion. It is important to note that in ICR mass spectrometry, frequency is measured, rather than arrival of ions at a detector. Frequency is inherently the most precise physical parameter that can be measured and largely accounts for the power of ICR mass spectrometry.

Figure 3.6 is a schematic representation of an ICR trap. Although ions are confined in the *x–y* plane by the magnetic field, it is necessary to further trap the ions in the *z* plane through application of a small potential to the end caps of the trap. The first step in the FT–ICR experiment is to excite ions coherently to a larger and detectable radius. This is accomplished by applying a broadband radio frequency voltage pulse at the transmitter plates. When the radio frequency matches the cyclotron frequency of a particular ion, the ion absorbs energy and increases in velocity and radius. Ions of a particular *m/z* ratio will form a coherent ion packet, and the packet generates an 'image

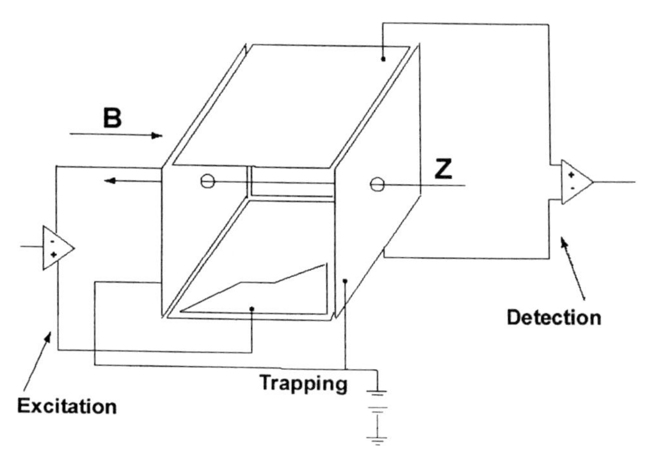

Fig. 3.6. Schematic representation of an ion cyclotron resonance ion trap. Ions are trapped in the *x–y* plane by the magnetic field (*B*) and held in the *z* plane by a trapping potential applied to the end plates. Excitation occurs by applying broadband radio frequency voltage pulse to the transmitter plates, followed by detection of the image current on the same plates. Figure adapted from Fig. 7 in Marshall *et al.* (1998).

current' each time it comes into proximity with the detector plate. This image current decays over time as the ion packet loses coherence. The frequency of this transient current is characteristic of the m/z ratio of the ions. However, at this point the experiment has produced only a time-domain signal. The frequency information, and thus the mass-to-charge ratios of ions present, is extracted by performing a Fourier transformation on the time-domain signal, producing frequencies of ions that are converted to m/z via a straightforward second-order calibration equation. Figure 3.7 contains both time-domain and frequency-domain (converted to mass) information generated with an FT–ICR spectrometer.

Several mass spectral parameters are improved with increasing magnetic fields (Marshall *et al.*, 1998). This recognition led to the development of the Ion Cyclotron Resonance Mass Spectrometry Facility at the National High Magnetic Field Laboratory. This laboratory, operated by Florida State University in Tallahassee, USA, is sponsored by the National Science Foundation and the State of Florida and is responsible for developing and operating state-of-the-art high magnetic field facilities for research in physics, biology, bioengineering, chemistry, geochemistry, biochemistry and materials science. The facility is specifically devoted to development and applications of high-field FT–ICR mass spectrometry.

Four mass spectral parameters that are particularly relevant to organic phosphorus speciation in natural environments are discussed briefly here. First, *mass resolving power, R*, increases linearly with increasing magnetic field (Sommer *et al.*, 1951). High mass resolving power is necessary to analyse ions in complex mixtures where several ions per nominal mass may be present. It has also been observed that peak coalescence decreases quadratically with increasing magnetic field. *Peak coalescence* is a phenomenon in which ions of similar cyclotron frequencies couple, appearing in the mass spectrum as a single peak. This makes unique determination of the exact mass of an ion impossible, even when tandem mass spectrometry is utilized. Thirdly, the *number of ions* that can be trapped during a single time event increases quadratically with increasing magnetic field, increasing signal-to-noise ratios. Finally, greater *ion kinetic energy* can be achieved, thus improving fragmentation efficiency in tandem mass spectrometry experiments.

Fourier-transform ion cyclotron resonance mass spectrometry at 9.4 Tesla now offers the highest resolving power and mass

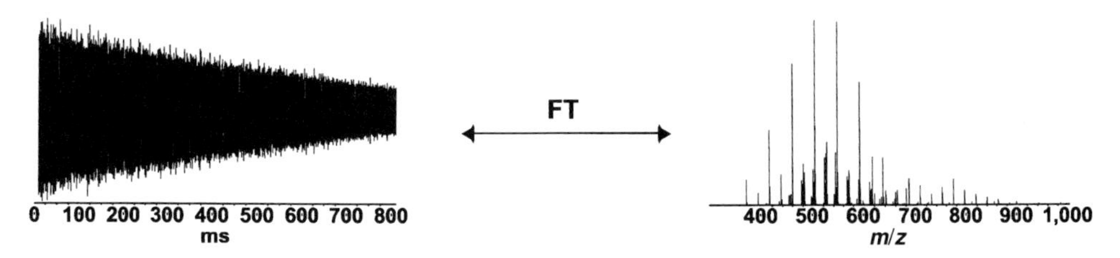

Fig. 3.7. Data (time) domain signal produced by 9.4-Tesla Fourier-transform ion cyclotron resonance mass spectrometer (left), and frequency domain signal (i.e. mass spectrum) after fast Fourier transformation and frequency-to-mass calculation (right). The sample was a mixture of poly(ethylene) glycol polymers used for internal calibration.

accuracy of any currently available mass spectrometry technique (Marshall *et al.*, 1998). At the National High Magnetic Field Laboratory, a mass resolving power of greater than 1 million has been achieved and the isotopic fine structure of a 15.8-kDa protein has been determined using the 9.4-Tesla instrument (Shi *et al.*, 1999). For biogeochemical studies of dissolved organic matter, resolving powers of 500,000 have been achieved with mass accuracies of less than 1 part per million (ppm) (Stenson *et al.*, 2003). The unsurpassed ability of high-field FT–ICR mass spectrometry to identify individual compounds in very complex mixtures was recently demonstrated by Hughey *et al.* (2002). In that report, over 11,000 individual compounds in a sample of crude petroleum were assigned chemical formulas based on exact masses from an FT–ICR spectrum. For this reason, FT–ICR at 9.4 Tesla is now referred to as 'ultrahigh-resolution' mass spectrometry to distinguish it from current high-resolution double-focusing and time-of-flight instruments.

The need for such resolution and mass accuracy in environmental and geochemical studies can be understood from an evaluation of Fig. 3.8, which is an expanded view of a spectrum of fulvic acids isolated from the Suwannee River in Florida. From this spectrum it is clear that in natural samples there are ions that differ by as little as 0.0364 Da. It is also clear that there are regular patterns in the mass spacing of ions, and these mass differences can be attributed to specific atomic variations in molecules (Stenson *et al.*, 2002). For example, a difference of 2.0157 Da represents the addition of a H_2 group (normally to saturate a double bond), a difference of 1.0034 Da represents substitution of a ^{13}C atom for a ^{12}C atom, and 0.9953 Da the substitution of –NH–(amine) for $-CH_2-$. The smallest mass difference observed, 0.0364 Da, arises from the addition of oxygen and loss of methane, a molecular change that is apparently due to the specific biochemical pathways used by microbes as they degrade plant matter (Filley *et al.*, 2002).

Tandem mass spectrometry

While these individual mass spectrometry techniques are themselves very useful in determining masses and formulas of organic molecules, their combination results in a very powerful technique referred to as tandem mass spectrometry. Tandem mass spectrometers normally combine two mass analysers and a field-free region where chemical reactions are allowed to occur. The first analyser acts as a mass filter, allowing in principle only a single molecular ion to be transferred to the reaction section. Here the original *precursor ion* is fragmented by any one of a number of processes, and the resulting fragments, or *product ions*, are then transferred to the second analyser where mass analysis is carried out. This type of tandem mass spectrometry is often referred to as MS/MS. Instruments in which this process is repeated, with product ions from the first tandem serving as precursor ions for an additional tandem spectrometer, is referred to as MS/MS/MS/MS (Biemann, 1990).

All the mass analysers described so far

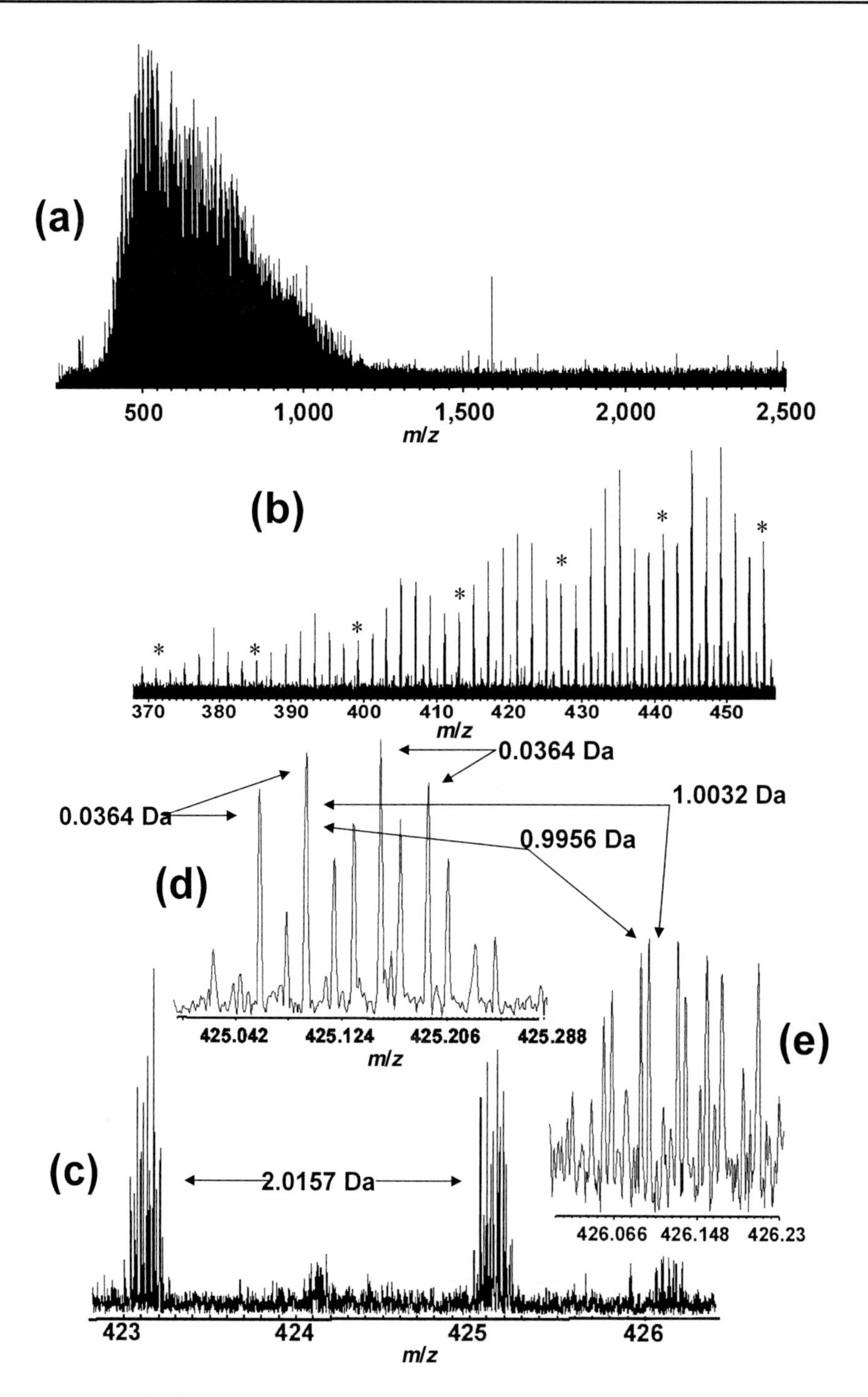

Fig. 3.8. Positive mode electrospray ionization Fourier-transform ion cyclotron resonance mass spectrum of Suwannee River fulvic acid mixture. (a) Entire spectrum over 225–3000 *m/z* region; (b) and (c) after scale expansion to highlight the 370–450 and 423–426 *m/z* regions; (d) and (e) after further expansion to highlight 0.3 mass windows. Reprinted from Stenson *et al.* (2002) with permission from the American Chemical Society.

have been used in tandem mass spectrometery. Sector analysers were the first to be employed in this way, and the variations made possible by combining magnetic and electrostatic sectors led to a designation system in which magnetic sectors were represented by *B*, electrostatic sectors by *E*, and quadrupoles by *Q*. For example, the first tandem instruments were high-resolution double-focusing instruments with an additional field-free collision chamber inserted between the electrostatic and magnetic sectors; hence, an *EB* configuration. Virtually every combination of *E* and *B* sectors has been utilized, however, including two double-focusing spectrometers with collision chambers in both (i.e. an *EBEB* configuration).

The most popular two-dimensional mass spectrometry configuration at present is the *QQQ*, or triple-sector quadrupole, represented schematically in Fig. 3.9. Three scan modes are possible with this configuration: product ion scan, precursor ion scan, and constant neutral loss scan. Product ion scan is the most widely used, and involves using Q_1 to selectively transmit one precursor ion to Q_2 where it is fragmented, normally by collisions with an inert gas such as helium. This type of fragmentation is referred to as collision-induced dissociation, or CID. Q_2 is operated in radio frequency mode only, and thus stores ions of a broad *m*/*z* range until they are transmitted to Q_3 for mass analysis of the product ions.

In *precursor scan mode*, the functions of Q_1 and Q_3 are reversed. Q_3 is set to a particular *m*/*z*, while Q_1 scans. This allows identification of all precursor ions that yield a particular product. A relevant example of such an application is the identification of phosphopeptides (Neubauer and Mann, 1999). Since phosphopeptides will lose PO_3 fragments in the collision-induced dissociation step, Q_3 is set at *m*/*z* 79 and Q_1 scanned. This allows identification of peptides that contain phosphate. Figure 3.10 includes the product and precursor ion spectra of such phosphopeptides.

Neutral loss scan mode involves introducing a fixed mass offset to Q_1 and Q_3. Q_1 and Q_3 are scanned in tandem, maintaining this offset. In this way precursor ions that lose a fragment corresponding to the mass offset between the two quadrupoles can be identified.

Triple-sector quadrupole tandem spectrometers gained popularity because they were cheaper, simpler to operate, and provided unit mass resolution below about 1000 Da. However, their restricted mass range and low resolution precluded their use for studies of larger molecules. This led to the development of hybrid tandem instruments, where hybrid refers to the coupling of two or more types of analysers. One of the more commercially successful of these hybrids has been the *Q*-time-of-flight that combines a normal and radio frequency-only quadrupole for precursor isolation with a time-of-flight mass analyser. *Q*-time-of-flight instruments have been used with MALDI ionization for numerous biological applications. Figure 3.11 contains the spectrum of the tryptic digest of a tetraphosphorylated peptide, from which the number of phosphorylation sites were determined (Bennett *et al.*, 2002).

Two-dimensional Fourier-transform ion cyclotron resonance mass spectrometry

An experiment similar to the tandem mass spectrometry just described can be carried out by FT–ICR mass spectrometry. However, FT–ICR tandem mass spectrometry is different from other forms of tandem mass spectrometry, because the entire process of precursor isolation, precursor reaction and product mass analysis can be done in a single cell. The technique is thus not truly tandem, but is better described as two-dimensional.

In an ICR cell, two-dimensional spectrometry begins with application of the stored waveform inverse Fourier transform (SWIFT) excitation. This technique removes all but a single chosen ion from the trap (Marshall *et al.*, 1985) and is performed by first determining which ions are to be ejected and their cyclotron frequency. Inverse Fourier transform then produces an excitation waveform that excites selected ions radially until they come into contact

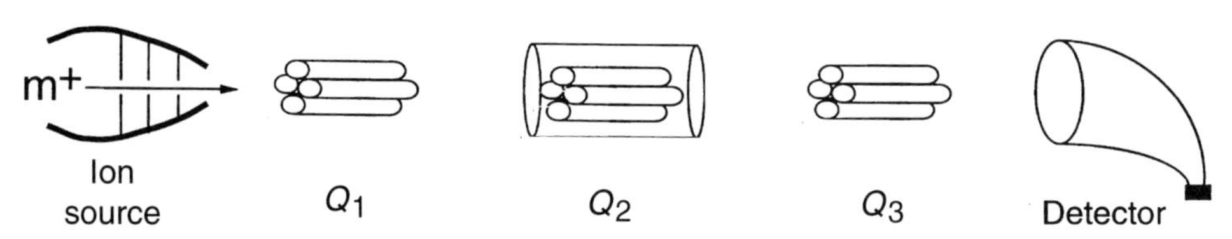

Fig. 3.9. Triple quadrupole configuration for tandem mass spectrometry. Quadrupole Q_1 serves as a mass filter and delivers one ion to the middle quadrupole Q_2, which is the fragmentation chamber and is operated in radio frequency mode only to store and transmit ions formed in fragmentation step. Mass analysis of product ions then occurs in quadrupole Q_3.

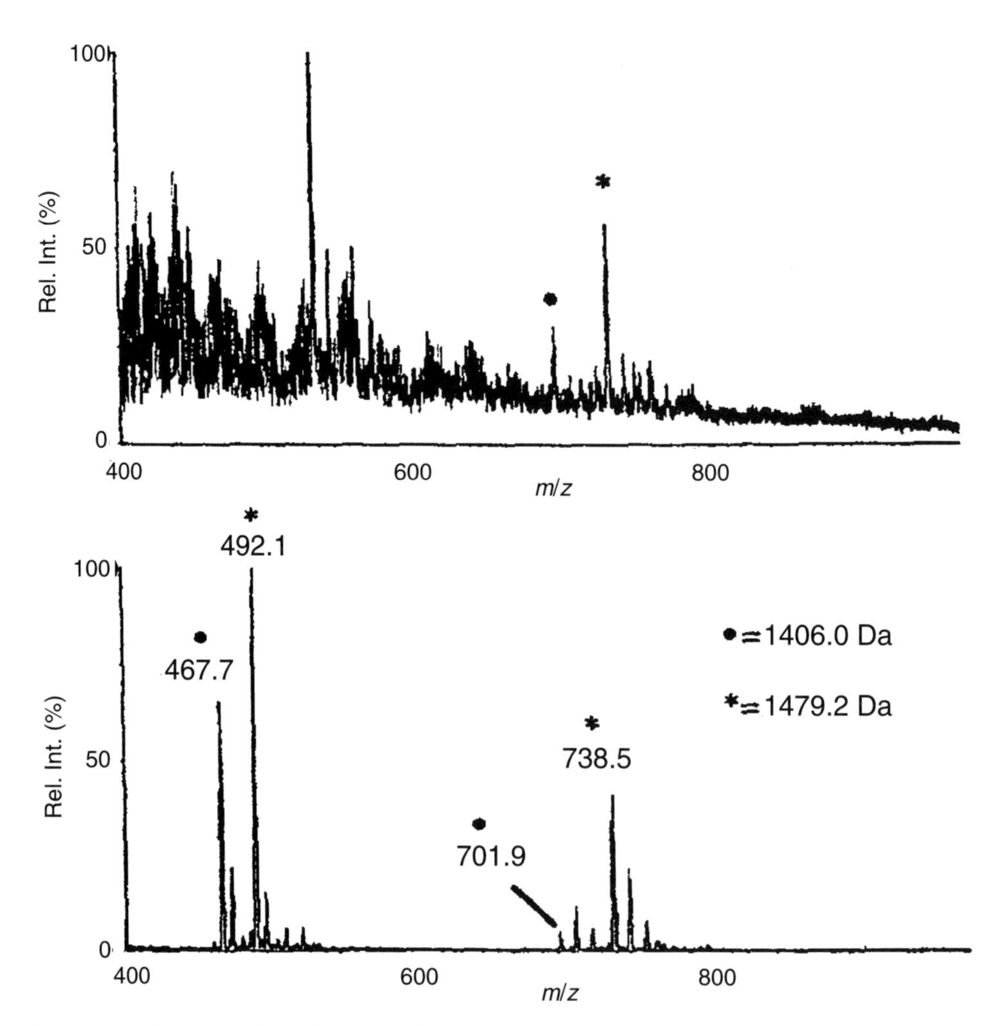

Fig. 3.10. Triple quadrupole two-dimensional mass spectrometry of a phosphopeptide after phosphorylation with kinase. Top: Q_1 scan of the entire peptide digest. Bottom: Q_1 precursor ion scan that identifies phosphopeptide ions that lose *m/z* 79 (PO_3). This is accomplished by holding Q_3 at *m/z*=79 and scanning Q_1; any ion in the Q_1 scan that loses 79 Da will then give rise to a signal in Q_3 and can thus be identified as a phosphorylation product (here marked with an asterisk, *). Note that precursor ions are doubly-charged as well, indicating neutral loss of PO_3. Reprinted from Neubauer and Mann (1999) with permission from the American Chemical Society.

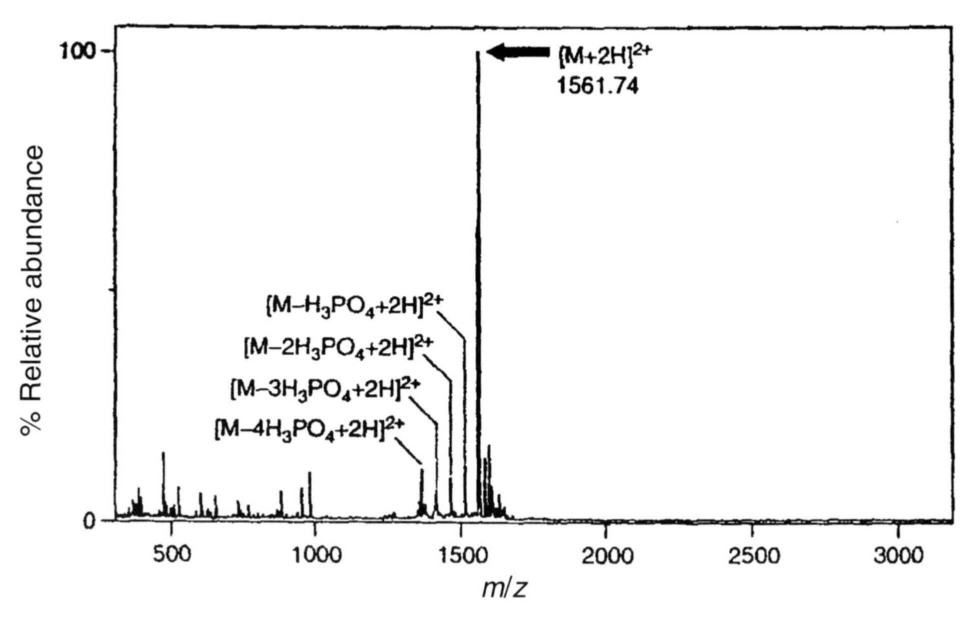

Fig. 3.11. Matrix-assisted laser desorption ionization Q–time-of-flight two-dimensional mass spectrometry of peptide containing phosphoserine. Doubly charged peptide, $[M+2H]^{+} = 1561.74$, isolated and fragmentation pattern observed. Note successive loss of up to four phosphoric acid groups. Reprinted from Bennett *et al.* (2002) with permission from John Wiley and Sons.

with the surface of the ion trap, eliminating them from the cell. After removal of all ions except that chosen, this single *precursor ion* is fragmented and the *product* or *daughter ions* produced from fragmentation mass are analysed as previously described.

There are two common methods of fragmenting the precursor ion:

1. Dissociation by collisions with a neutral target, or sustained off-resonance irradiation collision-induced dissociation (SORI–CID) (Amster, 1996).
2. Photodissociation by infrared multiphoton dissociation (Gauthier *et al.*, 1991).

SORI–CID is achieved by applying a radio-frequency excitation of a few kHz, above or below the cyclotron frequency (off-resonance) of the selected ion. Thus, the ion packet is accelerated quickly to large and small radii 90° out of phase with the excitation frequency. During this time, neutral target gas molecules are introduced, increasing the pressure in the cell. The ions undergo multiple collisions with the target gas, and fragmentation normally occurs. In infrared multiphoton dissociation, selected ions are fragmented with a laser pulse that is applied directly to ions in the trap. Ions are heated by absorbing multiple infrared photons until the threshold dissociation limit is achieved. If the frequency and power of the applied radiation is sufficient, fragmentation can take place. SORI-CID and infrared multiphoton dissociation tend to be complementary with regard to the types of molecules they will fragment. Figure 3.12 demonstrates a FT–ICR two-dimensional mass spectrometry experiment with the spectra of a humic acid molecule and its collision-induced dissociation products.

Ionization

These impressive advances in mass spectrometry would have gone largely unnoticed if it were not for parallel developments in large-molecule ionization techniques that

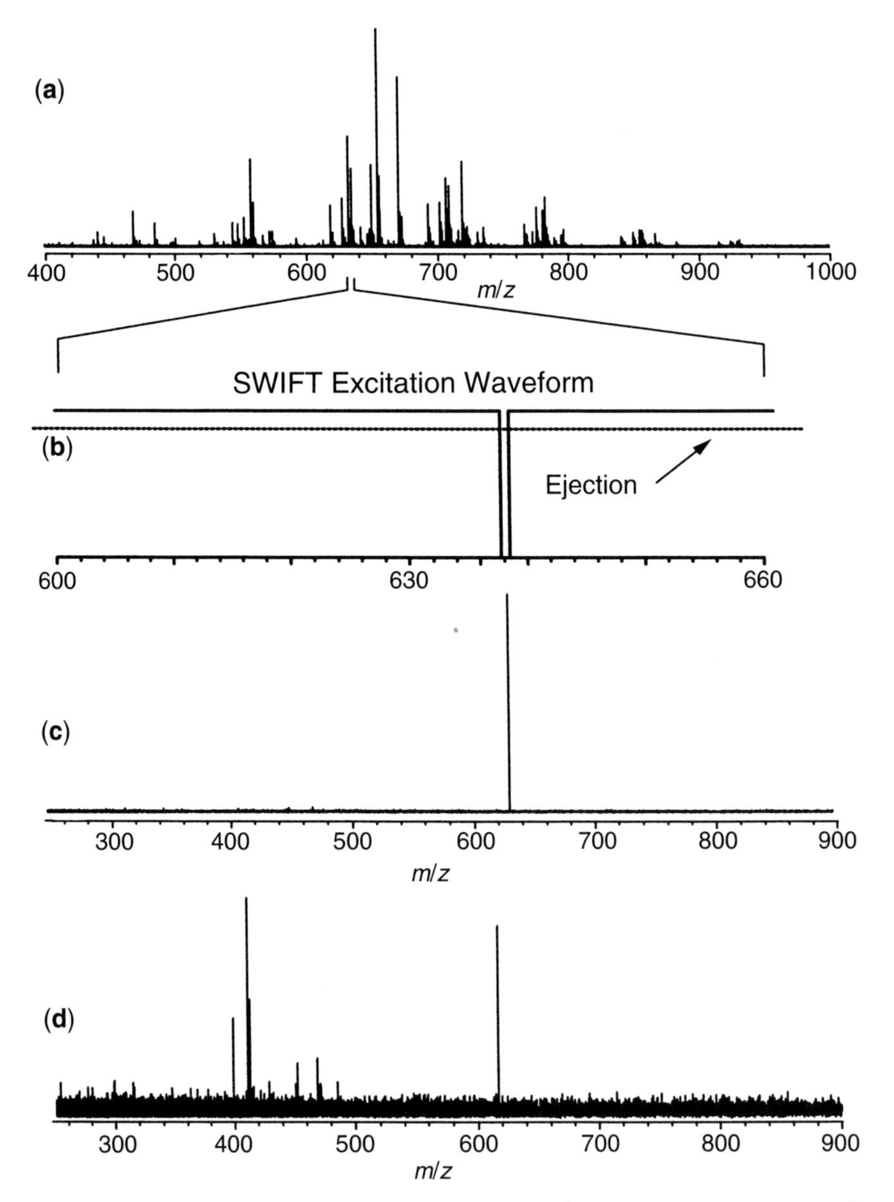

Fig. 3.12. Series of electrospray ionization Fourier-transform ion cyclotron resonance mass spectra obtained in a two-dimensional mass spectrometry experiment. Proceeding from top to bottom: (a) full mass spectrum of a fulvic acid mixture; (b) stored waveform inverse Fourier transform (SWIFT) waveform ejection from the ion cyclotron resonance cell of ions of all but a narrow *m/z* range; (c) the resulting isolated parent ion mass spectrum; and (d) the product ion mass spectra produced by collision-induced dissociation. Reprinted from Fievre *et al.* (1997) with permission from the American Chemical Society.

occurred in the mid- to late 1980s. Traditional electron impact and chemical ionization cannot produce intact, gas-phase ions of molecules above about 300 Da. While a few techniques for producing large intact ions were available (e.g. fast atom bombardment, Californium-252 plasma desorption), they were unreliable and hard to implement.

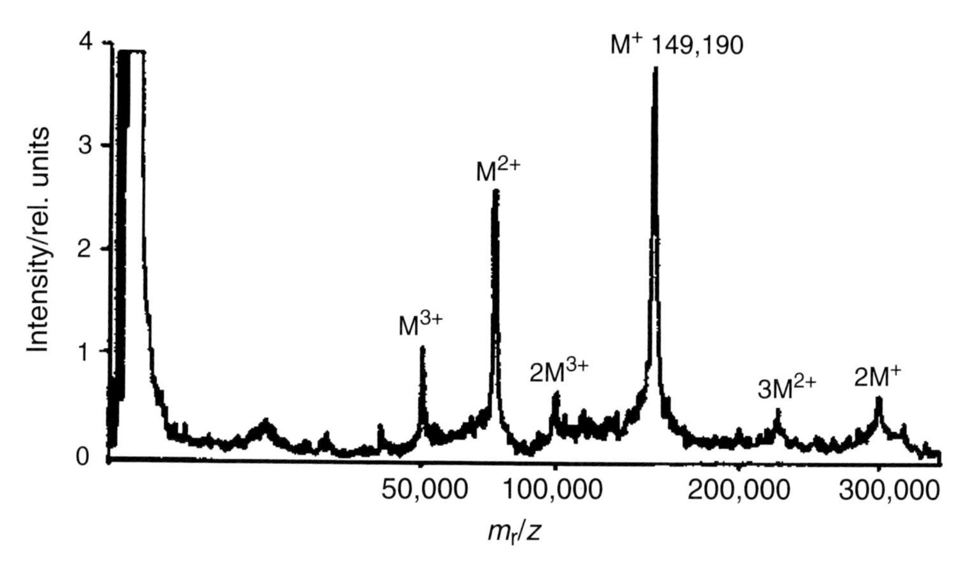

Fig. 3.13. Matrix-assisted laser desorption ionization (MALDI) time-of-flight spectrum of mouse anticlonal antibody, $M = 149{,}000$, showing formation of singly ($^{+}$) and multiply charged ($^{2+}$, $^{3+}$) monomers (M), dimers (2M), and trimers (3M). Reprinted from Hillenkamp and Karas (1990) with permission from Academic Press.

Then, within a short period, both MALDI (matrix-assisted laser desorption ionization) and electrospray ionization were introduced and found widespread use in biomolecular mass spectrometry. As noted earlier, Koichi Tanaka (MALDI) and John Fenn (electrospray ionization) were awarded the 2002 Nobel Prize in Chemistry for these innovations. Both MALDI and electrospray ionization are considered to be 'soft ionization' techniques that induce little, if any, molecular fragmentation.

Matrix-assisted laser desorption ionization (MALDI)

Matrix-assisted laser desorption ionization is, as its name implies, a desorption technique that uses a laser to impart energy to a substance for volatization and ionization. While direct laser desorption ionization is possible, it usually imparts too much energy to molecules and fragmentation occurs. Almost simultaneously, Karas and Hillenkamp (1988) and Tanaka *et al.* (1988) reported indirect laser desorption approaches in which laser energy is transmitted directly to a matrix in which the target analyte is mixed. Karas and Hillenkamp (1988) used a nicotinic acid matrix, while Tanaka *et al.* (1988) used platinum in glycerol. Short, intense laser pulses volatilize the matrix, taking analyte molecules with it. Ionized, gas-phase matrix molecules then transfer charge (usually as protons) to analyte molecules in a very gentle process that results in gas-phase ionized analytes. While the exact mechanism of the charge transfer process is not yet understood, it is thought to result from acid–base exchange of protons. However, it is established that strong analyte–matrix solvation forces are necessary to successfully ionize a target analyte, and a variety of matrices have been used for different classes of molecules.

MALDI is particularly suited for use with time-of-flight analysers because pulsed lasers match well the ion introduction requirements of these instruments. In addition, the high mass ranges and short analysis times of these analysers make MALDI–time-of-flight the method of choice for many large biomolecular applications. Figure 3.13 is an example of a MALDI spectrum of a monoclonal antibody with a molecular weight of 149,000 (Hillenkamp and Karas, 1990). This spectrum is instructive in that it contains

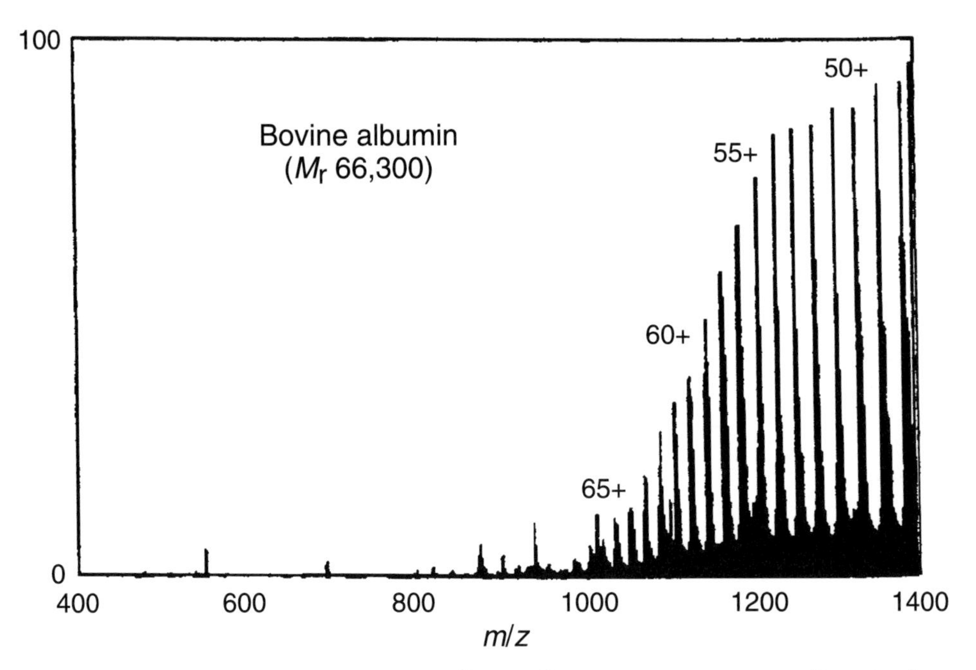

Fig. 3.14. Electrospray ionization mass spectrum of bovine albumin protein (M=66,300), illustrating the ability of electrospray ionization to produce multiply charged, large biomolecular ions. Reprinted from Smith *et al.* (1990) with permission from the American Chemical Society.

monomers (M), dimers (2M), and trimers (3M) of the antibody, as well as multiply-charged forms of each (e.g. M^{2+}, $3M^{2+}$). Such multiple peaks for biomolecules are common. It should be noted that MALDI is finding increasing use with FT–ICR spectrometers because of their inherently greater mass resolving power (Vollmer *et al.*, 1996).

Electrospray ionization

John Fenn is generally credited for first developing the application of electrospray ionization with mass spectrometry (Fenn *et al.*, 1989). Electrospray ionization is an atmospheric pressure ionization technique, in contrast to MALDI, which is normally carried out *in vacuo*. When used for mass spectrometry, atmospheric pressure ionization involves ionization at pressure, solvent removal and charge transfer to analyte molecules, and then introduction into the mass spectrometer via some sort of ion guide system. In electrospray ionization, the liquid solution containing the analyte(s) is pumped slowly through a capillary tip held at a potential of 1–3 kV relative to a counter electrode in the source. The electrostatic field generated by the applied potential disperses the emerging solution into a fine mist of charged droplets. These droplets are rapidly desolvated, leaving behind charged analyte molecules which are then introduced into the mass spectrometer. Both positive and negative ions can be formed, depending on the polarity of the capillary tip.

Electrospray ionization mass spectra are characterized by intact molecular ions that often have multiple charges. Figure 3.14 is a dramatic illustration of such a spectrum, where the large protein bovine albumin has acquired more than 60 positive charges (Smith *et al.*, 1990). High charge states like those seen in Fig. 3.14 are desirable because they 'fold down' the mass-to-charge range of interest to reasonable levels because of the high z. Mass analysers with limited mass ranges (e.g. quadrupoles, simple time-of-flight) can then be used.

Inductively coupled plasma

The inductively coupled plasma ion source was developed to accomplish exactly the opposite of the two 'soft' ionization methods just described. Molecules are reduced to their atomic (i.e. elemental) components through the application of intense energy. Masses corresponding to elements of interest (e.g. 30.974 for phosphorus) are then specifically monitored.

Inductively coupled plasma has become the ionization method of choice for elemental mass spectrometry. It was initially developed as the excitation source for multi-element optical spectrometers, because at typical plasma temperatures of 5000–10,000°C virtually all elements on the periodic chart emit detectable light. Most molecules are also atomized at these temperatures, which makes inductively coupled plasma ideal for mass spectrometry monitoring of elemental composition as well. Fassell and co-workers introduced the first inductively coupled plasma interfaced to a mass spectrometer in 1980 (Houk *et al.*, 1980). Elemental mass spectrometry normally requires only low-resolution analysers because unit mass resolution is typically required (i.e. the mass difference between elements, which is always equal to or greater than 1 Da).

Organic Phosphorus Speciation in a Treatment Wetland

We noted earlier that mass spectrometry had not been used to any extent in speciation studies of organic phosphorus in natural environments. Here we define 'phosphorus speciation' as the determination of actual molecular formulas and structures of molecules containing phosphorus, and not the more commonly accepted, broad classes of phosphorus such as filterable reactive phosphorus, total filterable phosphorus, filterable organic phosphorus, etc. While a few studies have focused on particular classes of organic phosphorus, such as the inositol phosphates, we are not aware of any attempt to broadly speciate organic phosphorus in any natural environment.

The following is a summary of our efforts to better understand organic phosphorus speciation and cycling in the surface waters of a treatment wetland using modern mass spectrometry techniques. We employed inductively coupled plasma high-resolution mass spectrometry with phosphorus-specific detection to quantitatively monitor classes of organic phosphorus as they moved through various stages of the wetland. We also performed high-resolution electrospray ionization FT–ICR experiments on filterable organic phosphorus isolated from the wetland to qualitatively identify individual organic phosphorus compounds that appeared to be resistant to removal. This work was part of a larger effort to restore the Florida Everglades ecosystem to its historical condition.

The Everglades Experimental Nutrient Removal (ENR) Project

The Florida Everglades have been historically characterized as an oligotrophic, phosphorus-limited wetland. The unperturbed Everglades is typified by extensive sawgrass (*Cladium jamaicense* Crantz) marsh communities, interspersed with an attached and floating periphyton mat (primarily calcareous mats of cyanobacterial–diatom assemblages), and relatively unvegetated sloughs. However, during recent decades, the Everglades has been impacted by human intrusion in many ways, including canal cutting, wetland reclamation and nutrient loading. Previously covering the lower third of the Florida peninsula, the Everglades now comprise only three water conservation areas and the Everglades National Park (McCormick and O'Dell, 1996; McCormick *et al.*, 1996).

Originally, the water conservation areas were designed as buffer zones for waters destined for the southern Everglades and Everglades National Park. However, phosphorus impacts in these areas have shown signs of spreading to the pristine Everglades. Various governmental agencies realized this problem and began to acquire lands that are critical to the ongoing restoration. As a part of this programme the South Florida Water Management District constructed artificial wetlands to filter the nutrient-rich water before it enters the water conservation areas. These treatment wetlands are referred to as stormwater treatment areas.

The Everglades Experimental Nutrient Removal (ENR) Project was the stormwater treatment area chosen to study the effectiveness of various biological and hydrological practices in maximizing phosphorus removal. The ENR, now an operational treatment area, is the largest constructed treatment wetland in North America. It is a single-pass treatment system with a buffer cell and two flow-ways, each consisting of two treatment cells (Fig. 3.15). The wetland receives inflow from drainage canals of the Everglades Agricultural Area and pumps outflow into Water Conservation Area 2A. The purpose of the ENR is to convert readily available nutrients, principally phosphorus, to organic forms; the assumption being that organic forms of phosphorus are less bioavailable and thus have less of an impact on the environment. The exact sources of the phosphorus entering the ENR have not been fully determined, but probably originate from a combination of the direct application of fertilizers and the leaching of organic-rich soils. In addition, the exact composition and relative abundances of different organic phosphorus species produced in the ENR have not been determined.

Quantitative organic phosphorus determination by inductively coupled plasma high-resolution mass spectrometry

Inductively coupled plasma high-resolution mass spectrometry

Interest in inductively coupled plasma mass spectrometry analysis of organic phosphorus resulted from our desire to characterize organic phosphorus species using liquid chromatography and capillary electrophoresis separations. However, because of the low concentrations of filterable organic phosphorus found in the ENR, a sensitive and selective detection system was required. Fortunately, a Finnigan MAT Element state-of-the-art inductively coupled plasma mass spectrometer was available to us in the Geochemistry Laboratory at the National High Magnetic Field Laboratory. Elemental mass spectrometry normally requires only low-resolution analysers because unit mass resolution should be adequate. However, there are several molecular interferences that are possible, including those mentioned previously for phosphorus. Bandura *et al.* (2002) recently showed how these interferences could be removed by converting P^+ to PO^+ in a reaction cell. However, that approach is specific to phosphorus. The high-resolution MAT Element is an instrumental approach to overcome these interferences that is applicable to most elements.

The Element is a double-focusing sector instrument with reversed Nier–Johnson (BE) geometry with high resolution (>8000 resolving power) and excellent abundance sensitivity (Giebmann and Greb, 1994). Abundance sensitivity is the ability to resolve and quantify a very small peak adjacent to a large matrix or background peak. High resolution and a large abundance sensitivity is necessary for phosphorus analyses because of interferences from NO (30.0064 Da) and O_2 (31.998 Da), both of which are present in the atmosphere and in a standard plasma, and which are very near the phosphorus peak at 30.974 Da.

Quantitative reliability of inductively coupled plasma high-resolution mass spectrometry

The quantitative reliability of inductively coupled plasma high-resolution mass spectrometry for phosphorus determinations was judged by comparing it to the standard total filterable phosphorus method (Cooper *et al.*, 1999). That method includes a persulphate oxidation step to convert all forms of phosphorus to phosphate, which is then suitable for the molybdenum blue reaction. Using the same set of standards, calibration curves for both methods were developed over a range of 0–200 μg P/l. Fig-

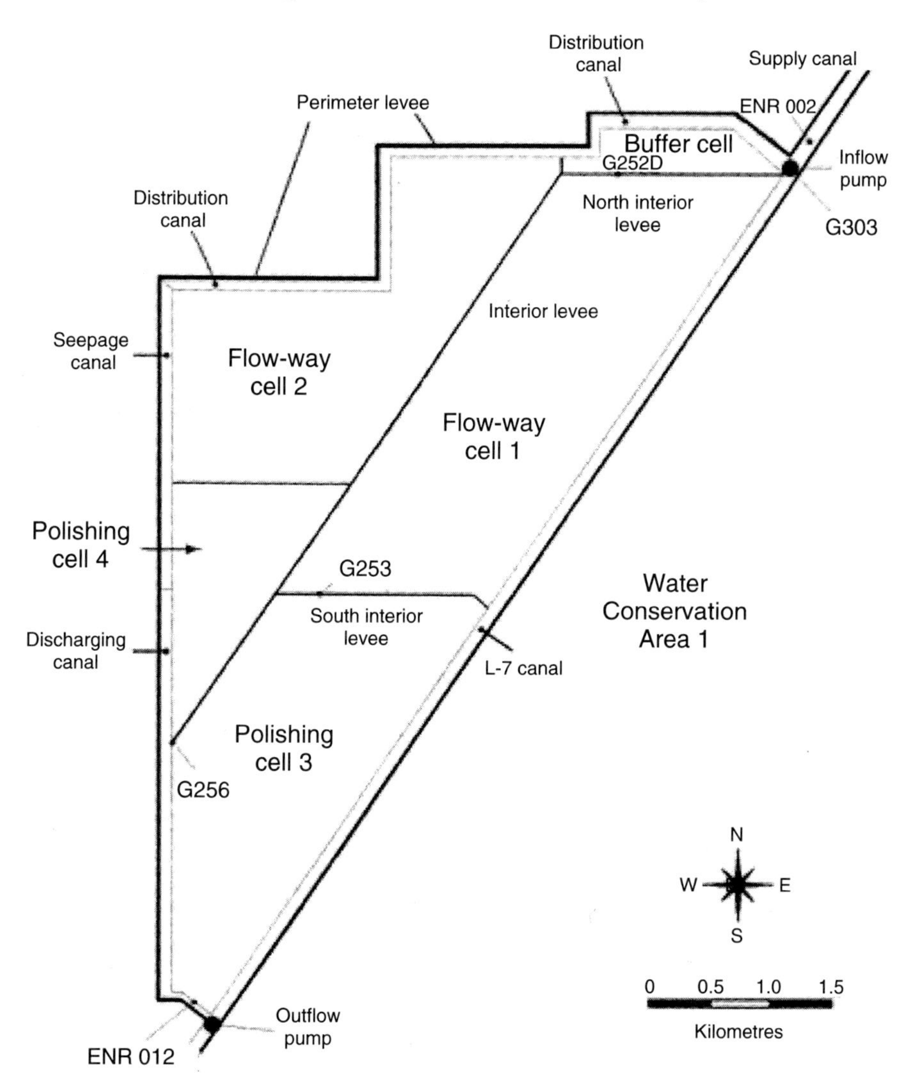

Fig. 3.15. Map of the Everglades Experimental Nutrient Removal (ENR) Project. Cells 1 and 2 are the primary treatment areas and are in a 'natural vegetative state', dominated by cattail (*Typha* spp.). Polishing Cell 3 was planted with native species, primarily sawgrass (*Cladium jamaicense* Crantz), and allowed to 'naturally vegetate'; it is now a 'mixed-marsh' community of sawgrass and cattail. Polishing Cell 4 is maintained as a submerged macrophyte community. Samples for this study were obtained from four ENR sites: G 303 (inflow), G 253 (cell 1–3 conveyance), G 256 (cell 4 outflow), and ENR 012 (total ENR outflow). Reprinted from Llewelyn *et al.* (2002) with permission from the American Chemical Society.

ures of merit for the two methods are included in Table 3.2. The inductively coupled plasma mass spectrometry method is clearly superior to the classical method over this concentration range.

The two methods were also compared using actual data on field samples generated by both methods (Cooper *et al.*, 1999). Figure 3.16 compares the results of the two methods, with a regression line of unit slope

Table 3.2. Analytical parameters for comparison of the inductively coupled plasma mass spectrometry (ICP–MS) and colorimetric methods of phosphorus determination.

Figure of merit	ICP–MS	Colorimetric
Calibration sensitivity (*m*)[a]	38.1 ± 0.152	0.004 ± 0.0004
Analytical sensitivity (γ)[b]	316	1.70
Limit of detection (LOD)[c]	~0.4 μg P/l	~3.5 μg P/l
Limit of quantitation (MQL)[d]	~1.7 μg P/l	~12 μg P/l

[a] Slope of the calibration curve.
[b] Slope of the calibration curve divided by the standard deviation of the measurement.
[c] The concentration that produces a signal equal to the average plus three standard deviations of the blank signal.
[d] The concentration that produces a signal equal to the average plus 10 standard deviations of the blank signal.

included. From this plot it appears that the colorimetric method overestimates low phosphorus concentrations due to interferences and background absorption, and underestimates higher concentrations due to incomplete reactions or colour development.

Ion-pair liquid chromatography with off-line inductively coupled plasma high-resolution mass spectrometry detection of organic phosphorus compounds

An ion-pair reversed-phase liquid chromatography method was developed that effectively separated five organic phosphate standards representative of naturally occurring, biologically derived organic phosphorus species. These particular compounds were chosen for the method development because of their ultraviolet absorption properties, variations in phosphorus content, ionization efficiency, molecular weight and hydrophobicity. Since phosphates are ionic in aqueous solutions, simple reversed-phase high-performance liquid chromatography cannot effectively separate them due to lack of partitioning into the organic stationary phase. However, by adding an appropriate ion-pairing reagent to the mobile phase (e.g. tetrabutylammonium chloride), the normally hydrophobic stationary phase becomes coated with the somewhat hydrophobic cation (tetrabutylammonium). Phosphate anions can then form ion-pairs with the exposed positive charges on the stationary phase surface, providing an ion-exchange type mechanism (Folley *et al.*, 1983; Fürst and Hallstrom, 1992). Ion-pair liquid chromatography has an advantage over pure ion exchange in that it offers an additional retention mechanism, hydrophobic partitioning, which can be exploited to optimize the separation of complex mixtures. This additional partitioning mechanism is particularly important for separating mixtures of organic phosphates, since there may be many mono- (or di-, tri-, etc.) phosphates with different structures. The optimized separation of the five standards is shown in Fig. 3.17, in which ultraviolet detection was used in the optimization experiments. The compounds included in this method development were adenosine mono-, di- and triphosphate, creatine phosphate, and phospho(enol)pyruvate. The small peak at about 4 min in Fig. 3.17 is phosphate.

Filterable organic phosphorus isolated from several sites in and adjacent to the ENR were analysed by this method, using off-line phosphorus-specific mass spectrometry detection. The organic phosphorus was isolated and concentrated by tangential ultrafiltration and lyophilization to produce concentration factors of ~25 and final total organic phosphorus concentrations of ~1 mg P/l. Column effluent was collected in 1 ml fractions after ultraviolet detection at 214 nm. The chromatographic fractions, along with a matrix blank and 1 and 10 mg/l

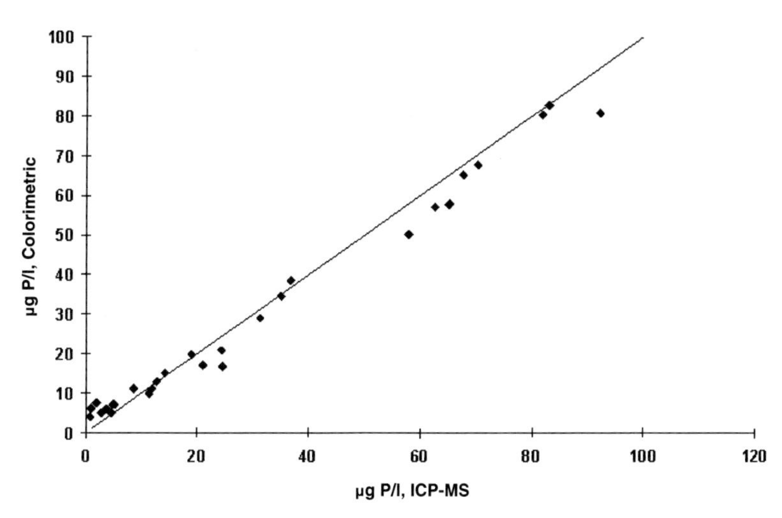

Fig. 3.16. Comparison of phosphate data determined by inductively coupled plasma mass spectrometry (ICP–MS) and colorimetric methods; solid line indicates a 1:1 relationship.

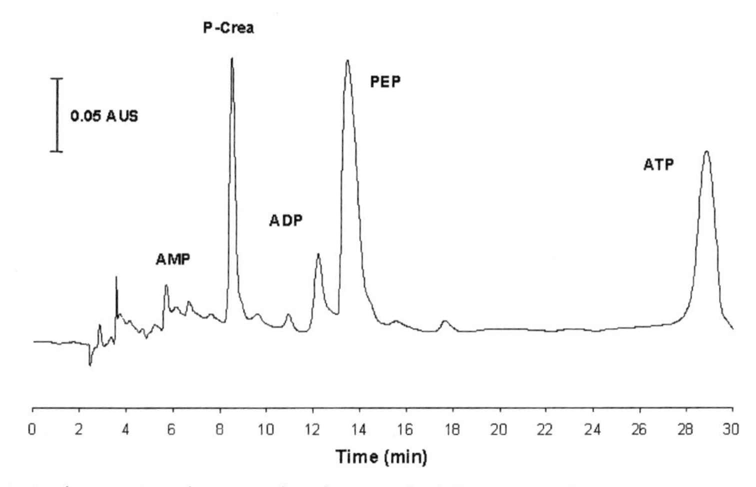

Fig. 3.17. Optimized separation of organic phosphate standards by ion-pair chromatography with ultraviolet detection (AMP, ADP, ATP = adenosine mono-, di- and triphosphate; P-Crea = creatine phosphate; PEP = phospho(enol)pyruvate; AUS = absorbance units = −log (% transmission). The small peak at ~4 min is phosphate. Conditions: 7.75 mM Tris buffer, pH 6.9; 4 mM tetrabutylammonium ion pair reagent; 22.5% acetonitrile.

phosphorus standards in eluant were adjusted to 1% HNO_3 and spiked with indium as an internal standard. The phosphorus signal of the fractions was compared to the standards, a relative phosphorus response calculated, and the points plotted as a function of elution volume to generate what we refer to as a 'phosphogram'. Such a phosphogram, along with the normal ultraviolet chromatogram, for a sample from Water Conservation Area 2A is shown in Fig. 3.18.

Based on a comparison with the phosphogram of standards, Fig. 3.12 can be interpreted as follows. The early eluting peak (6 min) not associated with ultraviolet

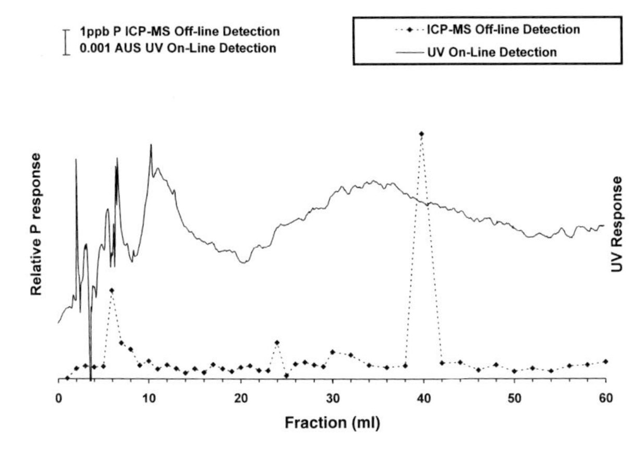

Fig. 3.18. Ion-pair chromatography of organic matter isolated from Water Conservation Area 2A in the Florida Everglades. Chromatography conditions as described in Fig. 3.11. Samples were concentrated by ultrafiltration and lyophilization by factors of ~25. Injection volume was 20 μl. On-line ultraviolet detection indicated by solid line; off-line phosphorus-specific detection by inductively coupled plasma mass spectrometry (ICP–MS) indicated by dotted line. Fractions were collected for inductively coupled plasma mass spectrometry detection at 1 min (1 ml) intervals. (AUS = absorbance units.)

absorbing compounds is likely to be free phosphate. Ultraviolet absorbing material not associated with a phosphorus species at 11 min indicates small organic molecules, probably in the hydrophilic acids class. The small phosphorus peaks at 21 and 30 min, associated with some ultraviolet-absorbing molecules, are likely to be small to moderate-sized organic phosphates such as adenosine triphosphate, or possibly some phosphorus species associated with fulvic acids. The large phosphorus peak at 40 min, overlapping the centre of the broad ultraviolet peak, is some multiply charged organic phosphate, possibly an organic polyphosphate like adenosine triphosphate or a humic-associated phosphorus compound.

Figure 3.19 contains phosphograms of filterable organic phosphorus isolated from the outlet of the ENR and from within Water Conservation Area 2A. Clearly, the filterable organic phosphorus from the ENR, which is presumably formed within the wetland and is 'younger', is different from that in Water Conservation Area 2A. In the next section we describe additional mass spectrometry experiments that attempt to qualitatively identify these different organic phosphorus compounds.

Qualitative organic phosphorus speciation by electrospray ionization and ultrahigh-resolution Fourier-transform ion cyclotron resonance mass spectrometry

Exact masses and chemical formulas

The complicated spectrum in Fig. 3.12 emphasizes the difficulty in trying to identify compounds such as organophosphates which are present at relatively low concentrations in an overwhelming background of organic carbon molecules. This is a particularly difficult problem in an oligotrophic system such as the Everglades, where filterable organic phosphorus is ~1000 times lower in concentration than organic carbon. Therefore, before any molecular identification of individual organic phosphorus compounds in the Everglades can be accomplished, the concentration of filterable organic phosphorus from sub-μg/l levels must be achieved.

We use a combination of tangential cross-flow filtration and phosphorus precipitation to get organic phosphorus fractions that are sufficiently concentrated and relatively free of interfering dissolved organic matter. Cross-flow filtration is a popular method for fractionating natural dissolved organic matter into molecular weight frac-

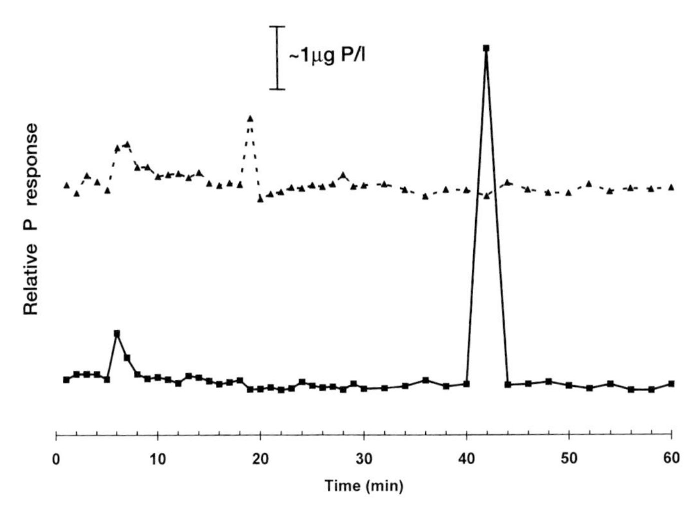

Fig. 3.19. Ion-pair chromatography with phosphorus-specific inductively coupled plasma mass spectrometry detection of organic matter isolated from the Experimental Nutrient Removal (ENR) wetland outflow (dotted line) and Water Conservation Area 2A (solid line). Chromatography conditions and sample preparation as described in Figs 3.11 and 3.12.

tions (Whitehouse *et al.*, 1990; Buesseler *et al.*, 1996; Hilger *et al.*, 1999). It can also concentrate organic species retained by the membrane (i.e. the 'retentate') by up to 20-fold. Unfortunately, this improvement is not sufficient to characterize organic phosphorus compounds in oligotrophic systems where they are present at sub-μg/l concentrations and the background organic matter component isolated by cross-flow filtration is typically 1000 times higher in concentration. This background dissolved organic matter provides a serious interference for direct mass spectral analysis by electrospray ionization. A method that further concentrates filterable organic phosphorus and isolates it from the high-level dissolved organic matter background was therefore developed (Llewelyn *et al.*, 2002). This method was adapted from that of McKercher and Anderson (1968) to isolate inositol penta- and hexakisphosphates from soil. It involves precipitation of organic phosphorus compounds with barium acetate in a mixture of 10% ethanol and 90% water, collection of the filtrate on filter paper, then redissolution of the organic phosphorus in distilled water. This new method also includes a cation removal step that produces an organic phosphorus concentrate suitable for electrospray ionization. The effectiveness of this isolation can be appreciated by comparing the spectra in Fig. 3.20.

Very low mass errors are required if chemical formulas are being assigned based on mass alone. Table 3.3 lists measured and theoretical masses of five standards used to validate the phosphorus isolation procedure. With these mass accuracies of less than one part per million (ppm), unequivocal chemical formulas can be assigned. It should be noted that the ultrahigh mass accuracy of FT–ICR mass spectrometry can be realized only when the total number of ions in the cell is minimized through some type of selective isolation procedure such as that used here.

Dissolved organic matter concentration and isolation of filterable organic phosphorus from surface waters at four sites in the ENR were carried out and the resulting concentrates analysed by ultrahigh-resolution FT–ICR with electrospray ionization. These sites represent a gradation in water residence time within the wetland and include the inflow, the conveyance linking Cells 1 and 3, the output from Cell 4, and the integrated ENR outflow (see Fig. 3.9 for reference). Detailed mass analysis of each peak in these spectra allowed several

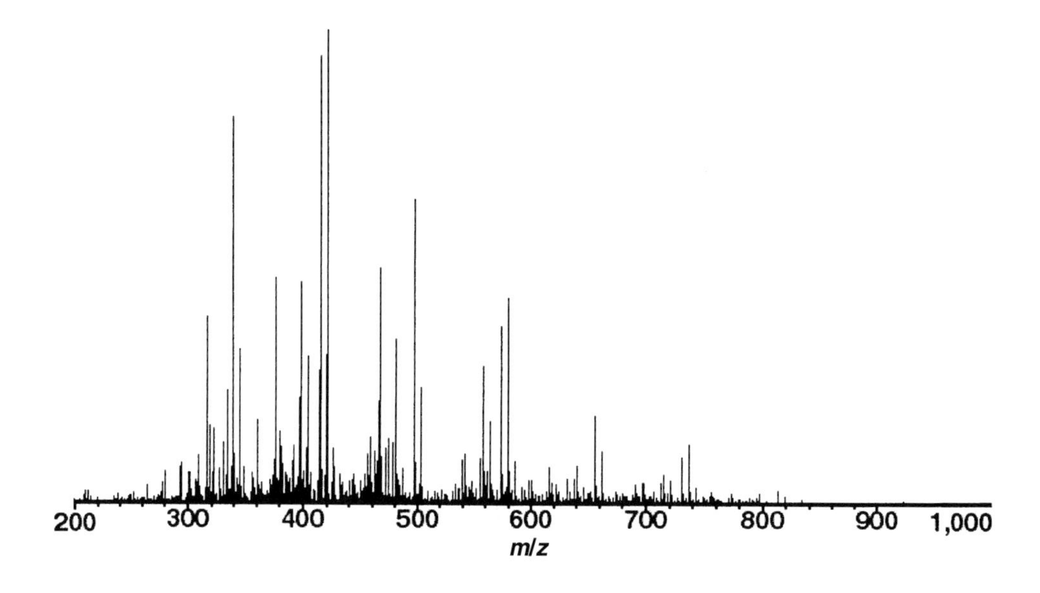

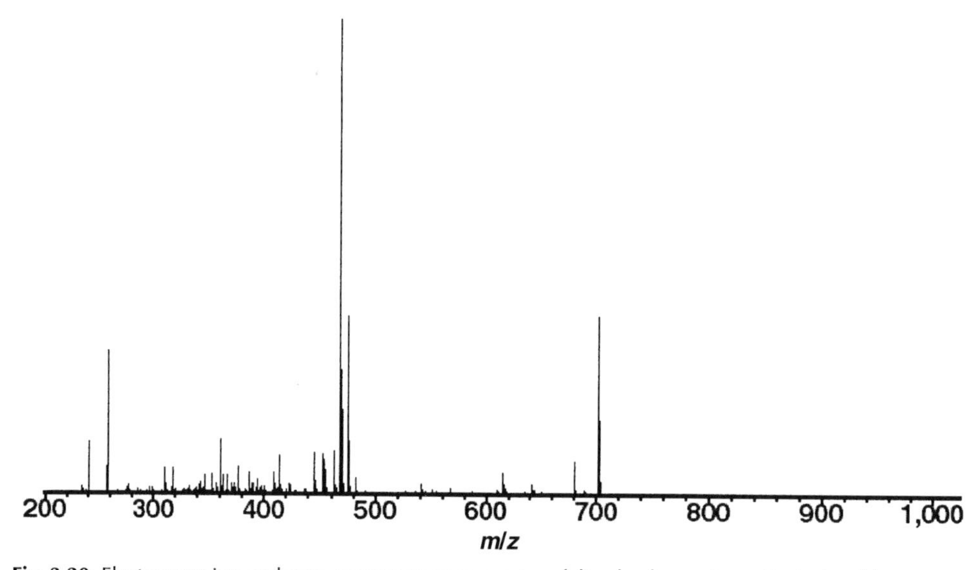

Fig. 3.20. Electrospray ion cyclotron resonance mass spectra of dissolved organic matter isolated from Experimental Nutrient Removal (ENR) wetland outflow before (top) and after (bottom) selective organic phosphorus concentration. Reprinted from Llewelyn *et al.* (2002) with permission from the American Chemical Society.

phosphorus-containing compounds to be identified. Table 3.4 lists the organic phosphorus compounds identified in the ENR inflow alone. Calculation of elemental compositions based on mass alone was made possible by the relatively low mass errors and the assumption that only carbon, hydrogen, nitrogen, oxygen, phosphorus, sodium, sulphur, and iron could be present. Because the number of chemical formulas possible within a given mass window increase with increasing mass, it is impossible to make

Table 3.3. Measured and theoretical masses of organic phosphorus standards.

Measured mass (Da)	Theoretical mass (Da)	Elemental composition	Mass error (ppm)[a]
547.9668	547.9666	2 $[C_5H_9O_8PNa_2]$	−0.36
369.0451	369.0451	$C_{10}H_{13}N_5O_7PNa$	0.00
347.0630	347.0631	$C_{10}H_{14}N_5O_7P$	+0.29
273.9833	273.9833	$C_5H_9O_8PNa_2$	0.00
261.0402	261.0402	$C_9H_{12}NO_6P$	0.00

[a] See footnote in Table 3.1 for explanation of ppm scale.

Table 3.4. Exact masses and elemental compositions of organic phosphorus molecules observed in the Experimental Nutrient Removal wetland inflow.

Measured mass (Da)	Theoretical mass (Da)	Elemental composition	Mass error (ppm)[a]
478.1186	478.1190	$C_{13}H_{24}N_6O_{10}PN_a$	+0.82
474.3182	474.3184	$C_{24}H_{43}N_4O_4N_a$	+0.44
443.3248	443.3249	$C_{24}H_{45}NO_6$	+0.22
415.2934	415.2936	$C_{22}H_{41}NO_6$	+0.34

[a] See footnote in Table 3.1 for explanation of ppm scale.

Table 3.5. Exact masses, elemental compositions and hydrogen deficiencies of organic phosphorus molecules identified at all sites in the Experimental Nutrient Removal wetland. Reprinted from Llewelyn *et al.* (2002) with permission from the American Chemical Society.

Mass (Da)	Elemental composition	Hydrogen deficiency[a]
592.15044	$C_{18}H_{30}N_6O_{13}PNa$	7
577.20237	$C_{22}H_{33}N_7O_8PNa$	10
518.13138	$C_{21}H_{23}N_6O_8P$	14
451.10521	$C_8H_{19}N_{11}O_8PNa$	5
444.11291	$C_{11}H_{17}N_{12}O_6P$	10
369.09506	$C_{12}H_{16}N_7O_5P$	9
364.12881	$C_{15}H_{25}O_8P$	4

[a] Number of double bonds and rings.

unambiguous assignments much beyond about 600 Da from these data alone.

Table 3.5 is a list of formulas of organic compounds containing phosphorus that were identified in surface waters at all four sampling sites. These would appear to be part of a refractory high-molecular-weight organic phosphorus fraction that is non-labile to phosphomonoesterase and phosphodiesterase enzymes (Cooper *et al.*, 1999). Interestingly, although all of the phosphorus-containing compounds were isolated from the high-molecular-weight (>1000 Da) organic matter fraction, all have apparent masses below 1000. While this might seem contradictory, it is in agreement with other electrospray ionization mass spectrometry analyses of humic materials, in which most of the compounds observed were below 1000 Da (Fievre *et al.*, 1997; Kujawinski *et al.*, 2002a,b; Stenson *et al.*, 2002, 2003). It may be that these compounds could be linked to larger humic-like substances through ionic interactions that are disrupted during electrospray ionization. This conclusion would be consistent with the data from ion-pair liquid chromatography high-resolution mass spectrometry that

Table 3.6. Exact masses, elemental compositions and hydrogen deficiencies of organic phosphorus molecules observed only in the Experimental Nutrient Removal wetland outflow. Reprinted from Llewelyn *et al.* (2002) with permission from the American Chemical Society.

Mass (Da)	Elemental composition	Hydrogen deficiency[a]
691.090	$C_{21}H_{28}N_6O_{16}P_2Na$	11
445.111	$C_{14}H_{20}N_7O_8P$	9
427.060	$C_{10}H_{19}N_3O_{12}PNa$	3
425.219	$C_{19}H_{32}N_5O_4P$	7
422.084	$C_{13}H_{19}N_4O_{10}P$	7
392.092	$C_{21}H_{16}N_2O_4P$	15
389.047	$C_9H_{16}N_3O_{12}P$	4
372.098	$C_{16}H_{21}O_8P$	7
323.002	$C_7H_{11}NO_{10}PNa$	3

[a] Number of double bonds and rings.

included relatively large molecules containing phosphorus.

In addition to the compounds listed in Table 3.5, there were a number of additional organic phosphorus compounds found only in water from the ENR outflow, and these are listed in Table 3.6. It might be reasonable to assume that these were compounds formed within the treatment areas, although further quantitative experiments would be needed to verify this.

Another interesting feature of the data in Tables 3.4, 3.5 and 3.6 is that almost all of the compounds identified contained nitrogen. This is actually most likely an artefact of the electrospray ionization used here. In positive ion mode electrospray ionization, nitrogen is preferentially ionized, and the presence of nitrogen in these compounds is probably the reason they can be observed, since the ionization efficiency of organic compounds containing phosphorus alone is quite low (Llewelyn *et al.*, 2002).

Structural studies and identification of organic phosphorus compounds by two-dimensional Fourier-transform ion cyclotron resonance mass spectrometry

In order to better understand the structures of the compounds identified in the ENR we initiated two-dimensional mass spectrometry experiments on selected ions. We also hoped to develop a method that would quickly identify molecules that contained phosphorus by observing product ions which had 'neutral losses' consistent with phosphorus ($\Delta m = 31$), PO_3 ($\Delta m = 79$), or H_3PO_4 ($\Delta m = 98$).

Figure 3.21 includes a SWIFT-isolated spectrum of one ion observed in an ENR sample, along with the spectrum of this same ion after SORI–CID fragmentation. From the product ion spectrum it is possible to reconstruct the precursor and verify its original chemical formula assignment. This approach is useful for determining the principal labile functional groups on a molecule, as well as assigning unequivocal chemical formulas for which the exact mass is insufficient to make such as assignment (e.g. too many formula possibilities). For example, the molecular mass of the parent ion in Fig. 3.21 is about 453 Da, and at this mass even high-resolution FT–ICR mass spectrometry (mass uncertainty 1 ppm) cannot provide an unequivocal formula for this compound. However, we can calculate very precisely the masses of fragments lost from the parent ion, and this allows us to determine the exact formulas of the fragments. In this case we see losses that correspond to water, $C_6H_{13}NO$ and $C_{12}H_{22}N_2O_3$ fragments, and we can then put these fragments back together to get the formula of the original parent molecule.

Unfortunately, neither SORI–CID nor infrared multiphoton dissociation fragmentation resulted in neutral losses from any of the ions we analysed, which could be related to phosphorus or phosphorus-containing fragments. We then carried out a similar experiment with an adenosine monophosphate standard, and again did not observe any recognizable phosphorus fragments. It thus appears that phosphonate bonds in organophosphates are too strong to be broken by these fragmentation techniques. Work is currently under way in our laboratory to find a fragmentation method that will break phosphonate bonds and thus readily identify these organic phosphorus compounds.

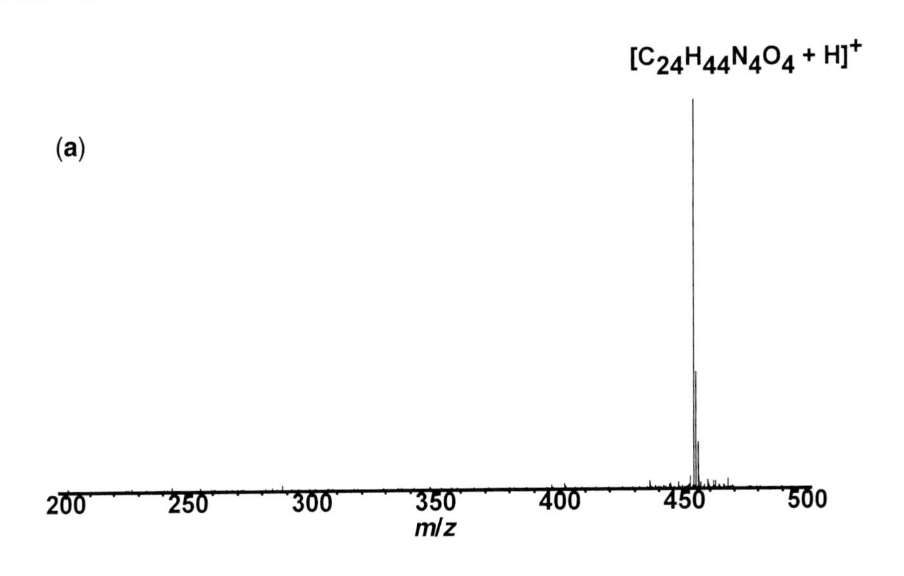

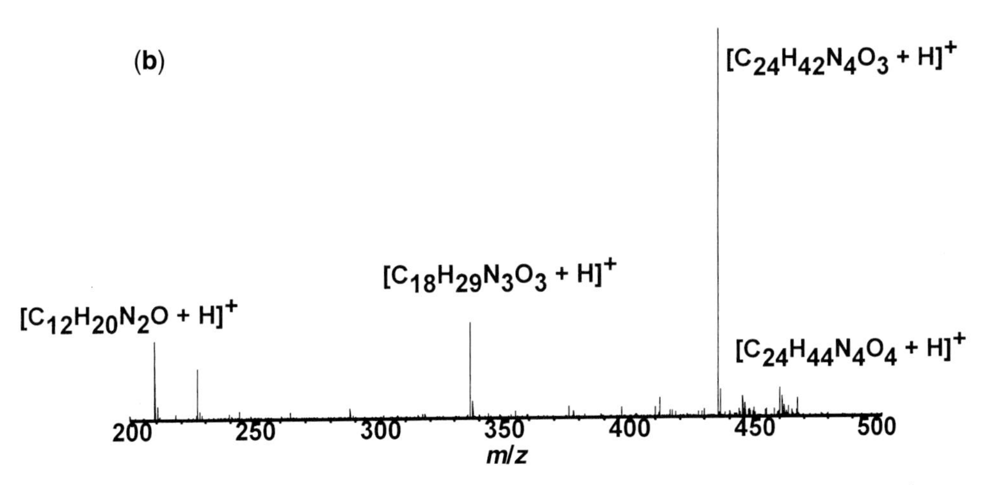

Fig. 3.21. (a) Spectrum of a stored waveform inverse Fourier-transform (SWIFT) isolated ion at nominal mass 453 *m/z*. This molecule is present in dissolved organic matter at the Experimental Nutrient Removal wetland outflow. (b) Spectrum of this ion and resulting products after fragmentation by sustained off-resonance irradiation collision-induced dissociation. Formulae in parentheses represent compositions of lost fragments. From these fragments, unambiguous determination of the elemental composition of the precursor ion at 453 *m/z* is possible.

Summary

In this chapter we have attempted to demonstrate that modern mass spectrometry techniques are poised to make a significant contribution towards improving our understanding of organic phosphorus speciation in natural environments. The results presented here are certainly not comprehensive, but they demonstrate for the first time that molecular speciation information on individual organic phosphorus compounds can be obtained, in spite of the very low levels (<1 μg P/l) at which they are found and the large background of other natural organic carbon compounds that are present.

Inductively coupled plasma high-resolution mass spectrometry is certainly a promising technique for obtaining much lower detection levels than the conventional colorimetric procedure currently used, and can be used as a phosphorus-specific detector for liquid chromatography and capillary electrophoresis speciation studies as well. We also believe that FT–ICR mass spectrometry can play an important role in determining organic phosphorus speciation, although it is obvious to us that this will always require concentration and very selective isolation of target phosphorus compounds to realize its full potential in this application. We suspect that this will also be the situation with the other new mass spectrometry technologies mentioned here, once researchers recognize their utility in these types of studies.

The future

It would appear that mass spectrometry can now begin to contribute to our understanding of organic phosphorus speciation. In particular, inductively coupled plasma mass spectrometry is already established as a sensitive, phosphorus-specific detector, and the information made possible by coupling it to some of the powerful separation techniques described by McKelvie (Chapter 1, this volume) is limited only by our imagination. While inductively coupled plasma mass spectrometry does not provide the structural information of molecular mass spectrometry, we can nevertheless learn much from combining its phosphorus-specific capabilities with intelligent separation schemes. For example, studies such as those described in Chapter 4 (Baldwin *et al.*, this volume), Chapter 5 (Quiquampoix and Mousain, this volume) and Chapter 9 (Heath, this volume) would certainly be enhanced by more molecular-specific information about the organic phosphorus involved.

The use of molecular mass spectrometry to identify individual organic phosphorus compounds in the environment is now certainly feasible, given the rapid advances in technology that have occurred in the last 10 years. However, the complexity of most natural systems and the huge number of organic molecules present will probably deter most researchers from attempting molecular-scale characterization of organic phosphorus. Before such characterizations can be made on major ecosystems, it will be necessary to refine our sampling and isolation techniques so that organic phosphorus can be selectively removed from the overwhelming background of organic matter. Indeed, we are in a situation similar to that faced by biochemists, in that extensive sample preparation and analyte isolation are necessary before these powerful mass spectrometry techniques can be used.

Finally, there is no doubt that limited access to this new generation of mass spectrometers will restrict these efforts unless at least a few are devoted to environmental and ecological studies. Most environmental researchers in this field must, like us, share instrument time with others; primarily biochemists who are generally much better funded and can thus dominate access. With a commitment to dedicating some of these new technologies to environmental research, the study of organic phosphorus in the environment could indeed become a truly molecular science.

References

Amster, I.J. (1996) Fourier transform mass spectrometry. *Journal of Mass Spectrometry* 31, 1325–1337.

Bandura, D.R., Baranov, V.I. and Tanner, S.D. (2002) Detection of ultratrace phosphorus and sulfur by quadrupole ICPMS with dynamic reaction cell. *Analytical Chemistry* 74, 1497–1502.

Bennett, K.L., Stensballe, A., Podtelejnikov, A.V., Moniatte, M. and Jenson, O.N. (2002) Phosphopeptide detection and sequencing by matrix-assisted laser desorption/ionization quadrupole time-of-flight tandem mass spectrometry. *Journal of Mass Spectrometry* 37, 179–190.

Biemann, K. (1990) Utility of exact mass measurements. In: McCloskey, J.A. (ed.) *Methods in Enzymology,* Vol. 193. Academic Press, San Diego, pp. 455–479.

Buesseler, K.O., Bauer, J.E., Chen, R.F., Eglinton, T.I., Gustafsson, O., Landing, W., Mopper, K., Moran, S.B., Santschi, P.H., Vernon, C.R. and Wells, M.L. (1996) An intercomparison of cross-flow filtration techniques used for sampling marine colloids: overview and organic carbon results. *Marine Chemistry* 55, 1–31.

Cooper, W.T., Hsieh, Y.P., Landing, W.M., Proctor, L., Salters, V.J.M. and Wang, Y. (1999) *The Speciation and Sources of Dissolved Phosphorus in the Everglades.* Final report for contract C-E8625 to the South Florida Water Management District, SFWMD Contract C-E8625. Florida State University, Tallahassee, Florida.

Dass, C. (2001) *Principles and Practice of Biological Mass Spectrometry.* Wiley Interscience, New York, 416 pp.

Fenn, J.B., Mann, M., Meng, C.K., Wong, S.F. and Whitehouse, C.M. (1989) Electrospray ionization for mass-spectrometry of large biomolecules. *Science* 246, 64–71.

Fievre, A., Solouki, T., Marshall, A.G. and Cooper, W.T. (1997) High resolution Fourier transform ion cyclotron resonance mass spectrometry of humic and fulvic acids by laser desorption/-ionization and electrospray ionization. *Energy and Fuels* 11, 554–560.

Filley, T.R., Cody, G.D., Goodell, B., Jellison, J., Noser, C. and Ostrofsky, A. (2002) Lignin demethylation and polysaccharide decomposition in spruce sapwood degraded by brown rot fungi. *Organic Geochemistry* 33, 111–124.

Folley, L.S., Power, S.D. and Poyton, R.O. (1983) Separation of nucleotides by ion-pair, reversed-phase high-performance liquid chromatography: use of Mg(II) and triethylamine as competing hetaerons in the separation of adenine and guanine derivatives. *Journal of Chromatography* 281, 199–207.

Fürst, W. and Hallstrom, S. (1992) Simultaneous determination of myocardial nucleotides, nucleosides, purine bases, and creatine phosphate by ion-pair high-performance liquid chromatography. *Journal of Chromatography* 578, 39–44.

Gauthier, J.W., Trautman, T.R. and Jacobson, D.B. (1991) Sustained off-resonance irradiation for collision-activated dissociation involving Fourier transform mass spectrometry: collision-activated dissociation technique that emulates infrared multiphoton dissociation. *Analytica Chimica Acta* 246, 211–225.

Giebmann, U. and Greb, U. (1994) High resolution ICP-MS: a new concept for elemental mass spectrometry. *Fresenius Journal of Analytical Chemistry* 350, 186–193.

Hilger, S., Sigg, L. and Barbieri, A. (1999) Size fractionation of phosphorus (dissolved, colloidal and particulate) in two tributaries to Lake Lugano. *Aquatic Sciences* 61, 337–353.

Hillenkamp, F. and Karas, M. (1990) Mass spectrometry of peptides and proteins by matrix-assisted ultraviolet laser desorption/ionization. In: McCloskey, J.A. (ed.) *Methods in Enzymology,* Vol. 193. Academic Press, San Diego, pp. 280–295.

Houk, R.S., Fassel, V.A., Flesch, G.D., Svec, H.J., Gray, A.L. and Taylor, C.E. (1980) Inductively coupled argon plasma as an ion source for mass spectrometric determination of trace elements. *Analytical Chemistry* 52, 2283–2289.

Hughey, C.A., Rodgers, R.P. and Marshall, A.G. (2002) Resolution of 11,000 compositionally distinct components in a single electrospray ionization Fourier transform ion cyclotron resonance mass spectrum of crude oil. *Analytical Chemistry* 74, 4145–4149.

Karas, M. and Hillenkamp, F. (1988) Laser desorption ionization of proteins with molecular masses exceeding 10000 daltons. *Analytical Chemistry* 60, 2299–2301.

Kujawinski, E.B., Freitas, M.A., Zang, X., Hatcher, P.G., Green-Church, K.B. and Jones, R.B. (2002a) The application of electrospray ionization mass spectrometry (ESI MS) to the structural characterization of natural organic matter. *Organic Geochemistry* 33, 171–180.

Kujawinski, E.B., Hatcher, P.G. and Freitas, M.A. (2002b) High-resolution Fourier transform ion cyclotron resonance mass spectrometry of humic and fulvic acids: Improvements and comparisons. *Analytical Chemistry* 74, 413–419.

Llewelyn, J.M., Landing, W.M., Marshall, A.G. and Cooper, W.T. (2002) Electrospray ionization Fourier transform ion cyclotron resonance mass spectrometry of dissolved organic phosphorus species in a treatment wetland after selective isolation and concentration. *Analytical Chemistry* 74, 600–606.

Mamyrin, B.A. (1994) Laser-assisted reflectron time-of-flight mass spectrometry. *International Journal of Mass Spectrometry and Ion Processes* 131, 1–19.

Marshall, A.G., Wang, T.-C.L. and Ricca, T.L. (1985) Tailored excitation for Fourier transform ion cyclotron resonance mass spectrometry. *Journal of the American Chemical Society* 107, 7893–7897.

Marshall, A.G., Hendrickson, C.L. and Jackson, G.S. (1998) Fourier transform ion cyclotron

resonance mass spectrometry: a primer. *Mass Spectrometry Reviews* 17, 1–35.
McCormick, P.V. and O'Dell, M.B. (1996) Quantifying periphyton responses to phosphorus in the Florida Everglades: a synoptic-experimental approach. *Journal of the North American Benthological Society* 15, 450–468.
McCormick, P.V., Rawlik, P.S., Lurding, K., Smith, E.P. and Sklar, F.H. (1996) Periphyton-water quality relationships along a nutrient gradient in the northern Florida Everglades. *Journal of the North American Benthological Society* 15, 433–449.
McKercher, R.B. and Anderson, G. (1968) Characterization of the inositol penta- and hexaphosphate fractions of a number of Canadian and Scottish soils. *Journal of Soil Science* 19, 302–309.
Neubauer, G. and Mann, M. (1999) Mapping of phosphorylation sites of gel-isolated proteins by nanoelectrospray tandem mass spectrometry: potentials and limitations. *Analytical Chemistry* 71, 235–242.
Nobel Foundation (2002) 2002 Nobel Prizes in Chemistry. *Nobel Foundation* website at http://www.nobel.se/chemistry/laureates/2002/index.html (accessed 21 April 2004).
Shi, S.D.-H., Hendrickson, C.L., Marshall, A.G., Siegel, M.M., Kong, F. and Carter, G.T. (1999) Structural validation of saccharomicins by high resolution and high mass accuracy Fourier transform–ion cyclotron resonance–mass spectrometry and infrared multiphoton dissociation tandem mass spectrometry. *Journal of the American Society for Mass Spectrometry* 10, 1285–1290
Smith, R.D., Loo, J.A., Edmonds, C.G., Baringa, C.J. and Udesth, H.R. (1990) New developments in biochemical mass-spectrometry: electrospray ionization. *Analytical Chemistry* 62, 882–889.
Sommer, H., Thomas, H.A. and Hipple, J.A. (1951) The measurement of e/M by cyclotron resonance. *Physical Review* 82, 697–702.
Stafford, G.C., Kelly, P.E., Syka, J.E.P., Reynolds, W.E. and Todd, J.F.J. (1984) Recent improvements in and analytical applications of advanced ion trap technology. *International Journal of Mass Spectrometry and Ion Processes* 60, 85–98.
Stenson, A.C., Landing, W.M., Marshall, A.G. and Cooper, W.T. (2002) Ionization and fragmentation of humic substances in electrospray ionization Fourier transform–ion cyclotron resonance mass spectrometry. *Analytical Chemistry* 74, 4397–4409.
Stenson, A.C., Marshall, A.G. and Cooper, W.T. (2003) Exact masses and chemical formulas of individual Suwannee River fulvic acids from ultrahigh resolution ESI FT-ICR mass spectra. *Analytical Chemistry* 75, 1275–1284.
Tanaka, K., Waki, H., Ido, H., Akita, S. and Yoshida, T. (1988) Protein and polymer analysis up to m/z 100,000 by laser ionization time-of-flight mass spectrometry. *Rapid Communications in Mass Spectrometry* 2, 151–153.
Vollmer, D.L., Rempel, D.L. and Gross, M.L. (1996) The rf-only-mode event for ion-chemistry studies in Fourier transform mass spectrometry: the reactions of propene and cyclopropane radical cations with neutral ethene. *International Journal of Mass Spectrometry and Ion Processes* 157/158, 189–198.
Whitehouse, B., Yeats, P. and Strain, P. (1990) Crossflow filtration of colloids from aquatic environments. *Limnology and Oceanography* 35, 1368–1375.

4 Abiotic Degradation of Organic Phosphorus Compounds in the Environment

Darren S. Baldwin,[1] Julia A. Howitt[1,2] and James K. Beattie[3]

[1]*Murray–Darling Freshwater Research Centre, Cooperative Research Centre for Freshwater Ecology, PO Box 921, Albury, NSW 2640, Australia;* [2]*Water Studies Centre and School of Chemistry, Monash University, Clayton, Victoria 3800, Australia;* [3]*School of Chemistry, University of Sydney, Sydney, NSW 2006, Australia*

Introduction

Organic phosphorus species can represent a large proportion of the phosphorus present in the environment (Hooper, 1973). The availability of these compounds to support microbial, plant and animal growth is dependent in part on their rate of degradation to generate free phosphate. While the importance of enzymes such as phosphatases, nucleases and phytase in the breakdown of organic phosphorus substrates is well documented (see Quiquampoix and Mousain, Chapter 5, this volume), abiotic pathways for the degradation of these compounds have received substantially less attention.

There are two principal pathways for the abiotic degradation of organic phosphorus compounds in the environment: *hydrolytic reactions* and *photolytic reactions*. The importance of either mechanism depends to a substantial degree both on the nature of the particular compound of interest (Fig. 4.1), and on its location in the environment. Some compounds, such as the phosphorodithionate pesticide malathion, are readily broken down (malathion has a half-life of about 11 h in alkaline solution; Pehkonen and Zhang, 2002). Others tend to be relatively recalcitrant; for example, the half-life of the phosphonate di-isopropyl methylphosphonate, a by-product of the manufacture of the nerve agent sarin, has been estimated to be about 500 years in groundwater (Sega *et al.*, 1998). The physical location of the compound also plays a part in determining the pathway of abiotic degradation. Photolysis can only play a significant role in the breakdown of organic phosphorus compounds in areas where light of appropriate wavelengths can penetrate, for example in surface waters. Similarly, abiotic hydrolytic reactions require the presence of free water. In this chapter we examine the role of hydrolytic and photolytic reactions in the abiotic degradation of both natural and synthetic organic phosphorus compounds in the environment.

Hydrolysis

Non-catalysed hydrolysis of phosphate esters

Hydrolysis is the addition of the elements of water across a bond. Hydrolytic cleavage of the phosphorus–oxygen bond is important in many biochemical processes and hence has generated a significant body of *in vitro* research over the last 50 years. Essentially

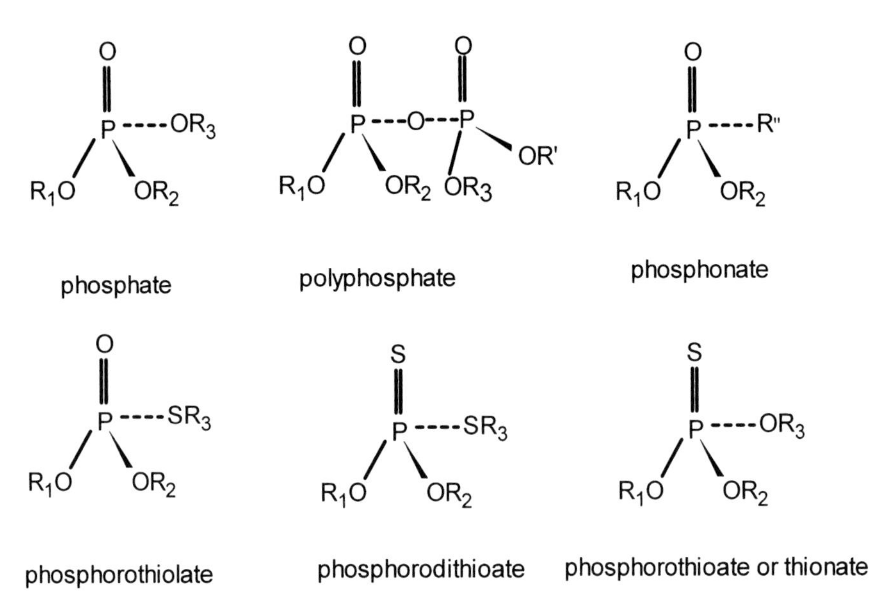

Fig. 4.1. Basic structures of common organophosphorus compounds. R_1, R_2 and R_3 can be either hydrogen or an organic group, R′ is either an organic group or another phosphate group, and R″ is an organic group bonded via carbon (after Pehkonen and Zhang, 2002).

the hydrolysis of phosphate esters can follow one of two competing pathways: a *dissociative pathway* that follows a first-order nucleophilic substitution (SN1) kinetic model, or an *associative pathway* that is described by a second-order nucleophilic substitution (SN2) kinetic model (Shen and Morgan, 1973; Wijesekera, 1992; Florian and Warshel, 1998).

In the *dissociative model* (Fig. 4.2) a phosphorus–oxygen bond is broken, yielding an alcohol or alkoxide (depending on pH) and a metaphosphate ion (PO_3^-). The metaphosphate ion is unstable and highly susceptible to nucleophilic attack. The rate of reaction is dependent on the proton dissociation constant (pK_a) of the leaving group, but independent of the basicity of the nucleophile (Wijesekera, 1992).

In the *associative model* (Fig. 4.3) a nucleophile attacks the phosphate centre to produce an unstable five coordinate oxyphosphorane species in the rate-limiting step. The oxyphosphorane then decays, losing an alkoxide leaving group. Unlike the dissociative pathway, the rate of reaction in the associative pathway is dependent on the pK_a of both the nucleophile and the leaving group.

Hydrolysis of phosphate monoesters may proceed by either the associative or dissociative pathway depending on the type of phosphate ester (including the pK_a values of stepwise deprotonation of the phosphate ester), the properties of the leaving group and the pH of the solution. The neutral diprotonated form of the phosphate monoesters (and other neutral phosphoramidates and polyphosphates) appears to react via the associative pathway, while the mono- and di-anionic forms react solely by the dissociative pathway (Shen and Morgan, 1973; Wijesekera, 1992).

Phosphate diesters tend not to be very susceptible to anionic (e.g. OH^-) nucleophilic attack. The pK_a for most phosphate diesters is low (about 1.5) and therefore, except at very low pH, most diesters are anions. Electrostatic repulsion between the anionic nucleophile and the anionic substrate is an impediment to reaction (Shen and Morgan, 1973). Phosphate triesters do not have an ionizable proton associated with the phosphate group and therefore are

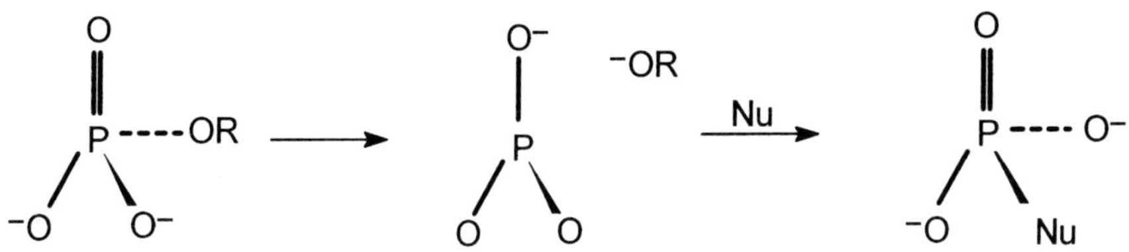

Fig. 4.2. A dissociative (SN1) reaction between a nucleophile (Nu) and a phosphate ion (after Wijesekera, 1992).

neutral at all pH values. Consequently, phosphate triesters tend to be the most reactive of all the phosphate esters (Shen and Morgan, 1973). Hydrolysis of both phosphate diesters and triesters follow the associative (oxyphosphorane) pathway.

Homogeneous metal-catalysed hydrolysis reactions

Many of the enzymes involved in the cleavage of organic phosphorus compounds require metal ions for activation. However many metal ions can also facilitate the hydrolysis of organic phosphorus compounds in the absence of enzymes. As early as 1938 it was shown that lanthanide (lanthanum and cerium) and actinide (thorium) hydroxides could accelerate the hydrolysis of α-glycerol phosphate in alkaline solution (Bamann and Mersenheimer, 1938). A similar study by Butcher and Westheimer (1955) showed that lanthanum hydroxides could accelerate the hydrolysis of three simple phosphate esters (methoxyethyl phosphate, hyrdroxyethyl phosphate and aminoethyl phosphate) by a factor of about 1000. Using ^{18}O they showed that the reaction proceeded through cleavage of the phosphorus–oxygen bond rather than at the carbon–oxygen bond.

Tetas and Lowenstein (1963) showed that simple divalent metal ions could increase the rate of hydrolysis of both adenosine triphosphate and adenosine diphosphate. At pH 5, the presence of copper increased the hydrolysis of adenosine triphosphate 60-fold while the presence of zinc increased the rate 12-fold. Other metal ions, including mercury, strontium, cobalt, calcium, nickel and beryllium, increased the rate of hydrolysis, but to a much smaller extent; barium and magnesium did not increase the rate of hydrolysis. In another study, Wolterman *et al.* (1974) showed that vanadium (III) oxide ions could increase the rate of hydrolysis of adenosine triphosphate 1000-fold. Dissolved metal ions have also been shown to facilitate the hydrolysis of artificial organic phosphorus species. For example, Smolen and Stone (1997) examined the role of cobalt (II), nickel (II), copper (II), zinc and lead (II) ions in the hydrolysis of thionate and oxonate organophosphorus pesticides at different pH values. The addition of transition metal ions increased the rate of hydrolysis of both groups by up to about a factor of 10^4, with copper (II) showing the greatest effect.

A number of mechanisms have been proposed to account for the increase in rate of hydrolysis of organic phosphorus compounds in the presence of dissolved metal ions. Butcher and Westheimer (1955) originally suggested that the formation of a complex between a phosphate and a metal ion accelerated the reaction rate by neutralizing the negative charge associated with the phosphate ligand, hence facilitating the approach of an anionic nucleophile (OH^- ion). Charge neutralization was also implicated in the million-fold acceleration of the hydrolysis of tridentate cobalt (II) coordinated tripolyphosphate relative to non-coordinated tripolyphosphate (Haight *et al.*, 1985).

Metal ions may also facilitate nucleophilic hydrolysis by shifting electron densities (Hoffman, 1980; Dixon *et al.*, 1982; Smolen and Stone, 1997). When a phosphate ester is coordinated to a metal ion,

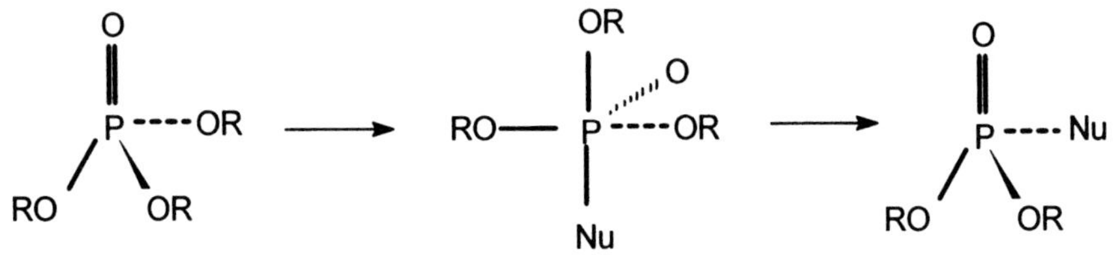

Fig. 4.3. An associative (SN2) reaction between a nucleophile (Nu) and a phosphate ion (after Wijesekera, 1992).

electron density moves away from the phosphorus centre towards the metal ion. This loss of electron density makes the phosphorus atom more susceptible to nucleophilic attack (e.g. by hydroxyl groups).

An alternative role for metal ions in the hydrolysis of organic phosphorus compounds is the coordination and induced deprotonation of water to create a very reactive nucleophile (Smolen and Stone, 1997). The effect can be further enhanced if the phosphorus compound under attack is proximally coordinated to the hydroxyl ion. For example, Jones *et al.* (1983) reported a 10^5 increase in the rate of hydrolysis of *para*-nitrophenyl phosphate when it is coordinated to a cobalt (III) complex with a proximal hydroxyl ligand.

It is of note that not all organic phosphorus species are susceptible to metal-catalysed hydrolysis. For example, Bullock *et al.* (1993) showed that a suite of 12 different metal ions had no appreciable effect on the rate of hydrolysis of phytate at elevated temperatures under either acidic or basic conditions.

Notwithstanding the impressive enhancement in the rates of hydrolysis of many organic phosphorus compounds by dissolved metal ions, because of the generally low concentration of most dissolved metal ions in the environment, homogeneous catalysis will only be an important pathway for organic phosphorus breakdown in unusual circumstances (Smolen and Stone, 1997).

Heterogeneous metal-catalysed hydrolysis

The highest concentration of most metals in the natural environment is in the solid phase (in minerals, soils and sediments) rather than in solution. These mineral phases can also play a role in the hydrolysis of organic phosphorus species (Torrents and Stone, 1994; Baldwin *et al.*, 1995); however, both equilibrium and kinetic effects on the mineral surface affect the role of solid mineral phases as catalysts. Mineral surfaces can be altered by adsorption of other cations, anions and dissolved organic matter. Generally speaking, the contribution of abiotic hydrolysis to the supply of biologically available phosphate in such complex systems is only partially understood.

Torrents and Stone (1994) examined the effect of mineral phases on the hydrolysis of three phosphorothioate esters: methylchlorpyrifos, methyl parathion and Ronnel. Titania enhanced the hydrolysis of all three, goethite and alumina enhanced the hydrolysis of the first two, while silica had no effect. These results suggest that silica-based surfaces are not effective catalytic surfaces for these hydrolysis reactions, presumably because of their strong negative charge. These conclusions are consistent with an earlier study which found that clays inhibited the enzymatic hydrolysis of various phosphate esters, presumably by adsorbing and deactivating the enzymes (Mortland and Gieseking, 1952).

Baldwin *et al.* (1995) surveyed the effect of various mineral phases on the rate of hydrolysis of the model organophosphate ester *p*-nitrophenyl phosphate. They found that, normalized for the number of independently determined phosphate adsorption sites, the manganese oxides were most effective in catalysing the hydrolysis reaction, followed by iron and titanium oxides, with a small effect for alumina. No effect was observed with silica, barium sulphate or the clay minerals kaolinite and bentonite. Given that oxides of iron and manganese are ubiquitous in most environments, it is expected that they will play an important role in the abiotic transformation of organic phosphorus compounds.

Iron oxides

Goethite (FeOOH) and other iron (III) oxides are positively charged at neutral pH and adsorb phosphate strongly. This adsorption process is of extreme importance for the phosphorus cycle in both soils and sediments (Baldwin *et al.*, 2002). Phosphate–goethite equilibria have been extensively studied (Cornell and Schwertmann, 1996). Adsorption decreases with increasing pH, but is still substantial above pH 8, approaching the point-of-zero-charge (pzc) of goethite at pH 9.4 (Boily *et al.*, 2001). The presence of calcium, ubiquitous in natural waters, further increases phosphate adsorption at high pH (Rietra *et al.*, 2001). Phosphate–goethite surface complexes are inert and desorption is slow and partly irreversible (Torrent *et al.*, 1990). Hence enhanced hydrolysis of organophosphates on the goethite surface does not lead directly to biologically available phosphate. Reduction of the iron (III) or surface biological activity is required in order to release the phosphorus (Baldwin *et al.*, 2002).

Iron oxide surfaces do facilitate the hydrolysis of organophosphate esters. (The term catalysis is not appropriate because the phosphate product is adsorbed and the oxide surface altered.) Generally the organic residue is not strongly adsorbed and its chemical and biological effects are of interest. Dannenberg and Pehkonen (1998) examined the hydrolysis of four pesticides on three iron oxides. Hazardous and persistent products can be formed and the distribution of products can depend on the reaction conditions.

Manganese oxides

Manganese oxides introduce new elements for consideration. Firstly, the various forms of manganese dioxide are negatively charged at circumneutral pH. The pH_{pzc} for α-manganese dioxide has been reported to be 4.6 (Healy *et al.*, 1966); for δ-manganese dioxide variously as 1.5 (Healy *et al.*, 1966) and 2.25 (Murray, 1974); for birnessite 1.5 and 3.0 (McKenzie, 1981); for cryptomelane 1.5 and 2.0 (McKenzie, 1981); for amorphous manganese dioxide 2.3 (Baldwin *et al.*, 2001); and for pyrolusite (β-MnO_2) 6.5 (Baldwin *et al.*, 1995). These acidic values for the pH_{pzc} of manganese dioxides, compared with neutral to alkaline values for iron oxides, are consistent with the higher oxidation state of manganese (IV) compared with iron (III).

Despite the negative surface charge, manganese oxides are still able to adsorb anions at neutral pH. Balistrieri and Chao (1990) studied the adsorption of selenium by iron and manganese oxides and reported *inter alia* that phosphate adsorption by manganese dioxide decreases with increasing pH. At pH 7, 28% of 4.8 μM phosphate remained on the surface of a suspension of 100 mg/l of amorphous manganese dioxide in 0.1 M KCl, decreasing to just a few per cent above pH 8 (Balistrieri and Chao, 1990). Yao and Millero (1996) then showed that in seawater, or 0.7 M NaCl, there was significant adsorption of phosphate on δ-manganese dioxide, again decreasing with increasing pH above pH 5. They made a number of other important observations; namely (1) that humic acid did not suppress phosphate adsorption above pH 5, (2) that calcium and magnesium ions enhanced adsorption above pH 4, and (3) that the larger surface area of δ-manganese dioxide compared to goethite means that the phosphate adsorption capacity of manganese dioxide is comparable to that of goethite despite the stronger affinity of the iron oxide for phosphate. Ouvrard *et al.* (2002) also demonstrated with chromatography experiments that phosphate adsorbs reversibly on manganese dioxide at pH 6.78. The wider significance of both of the above results is that manganese oxides are capable of acting not only as an effective sink for phosphate but also as a source of biologically available phosphate, because their adsorption is reversible under aerobic conditions.

A second novel aspect of manganese dioxide is its role as an oxidizing agent. The standard reduction potential of manganese dioxide is 1.23 V, equal to that of oxygen itself:

$$MnO_2 + 4\ H^+ + 2\ e^- = Mn^{2+} + 2\ H_2O$$

At pH 7, of course, the potential has dropped to 0.40 V, but this is still sufficient to oxidize a number of phenols, although not those with electron withdrawing groups such as *p*-nitrophenol (Stone, 1987). Hence the fate of the organic hydrolysis product, as well as that of the phosphate residue, must be considered when manganese dioxide effects the hydrolysis of organophosphates.

Suspensions of amorphous manganese dioxide do facilitate the hydrolysis of phosphate esters. Baldwin *et al.* (2001) observed an enhancement of the rate of hydrolysis of *p*-nitrophenyl phosphate, which was an order of magnitude more rapid than that observed with hydrous iron oxides. At pH 4 much of the phosphate product remained adsorbed on the surface, but at pH 8 it was all released into solution. Inman *et al.* (2001) showed that in the presence of amorphous manganese dioxide the detergent builder tripolyphosphate ($O_3POP(O)_2OPO_3^{5-}$) is hydrolysed via pyrophosphate ($O_3POPO_3^{2-}$) completely to phosphate, most of which was not adsorbed in the pH range 5–8, indicating that the phosphate produced was bioavailable. They also observed a significant enhancement of the rate of hydrolysis in the presence of calcium ions, both in natural waters and artificial systems. The inevitable presence of calcium ions in natural waters, the apparent lack of significant inhibition by humates, the large surface area of the manganese oxides, and the reversibility of phosphate adsorption suggest that manganese oxides are of central importance in the abiotic phosphate cycle.

Photolysis

Photochemical reactions are another important mechanism for the breakdown of organic phosphorus compounds in the environment. A wide range of organic phosphorus compounds are susceptible to photochemical reactions, including humic–phosphorus complexes in aquatic environments (Francko and Heath, 1982), a range of biological molecules including RNA and DNA (Golimowski and Golimowska, 1996), and organophosphorus pesticides (Pehkonen and Zhang, 2002).

Photochemical reactions arise from the absorption of light by molecules when the energy of an incoming photon corresponds to the energy difference between an occupied energy level in the molecule and an unoccupied energy level. The absorbed energy excites an electron into the higher energy level. The energies of photons in the visible and ultraviolet light are comparable to the energies of chemical bonds (100–500 kJ/mol) (Moore, 1983). The absorption of light can be described as a two-step process, the first resulting in the production of an excited state molecule:

$$M + hv \rightarrow M^*$$

M^*, the excited state of the molecule M, is a short-lived species (10^{-8}–10^{-9} s), which will then undergo one of a number of relaxation processes, the most common being relaxation to the ground state with the emission of heat or light (Skoog and Leary, 1992). Alternatively, the excited state may undergo chemical reactions to produce new species, as the atoms are more loosely bound in the excited states (Eisberg and Resnick, 1985), and these reactions are referred to as photochemical reactions.

Photochemical reactions can occur via primary or secondary mechanisms. Primary photochemical reactions are those which are the result of the molecule of interest absorbing light and forming a reactive intermediate state. Secondary photochemical reactions are those that involve the degradation of the target compound through reaction with another species formed as a result of photochemical reactions (such as oxygen radicals), or energy transfer from an excited species to the target molecule, followed by degradation (photosensitized reactions). Photochemical reactions can also be cata-

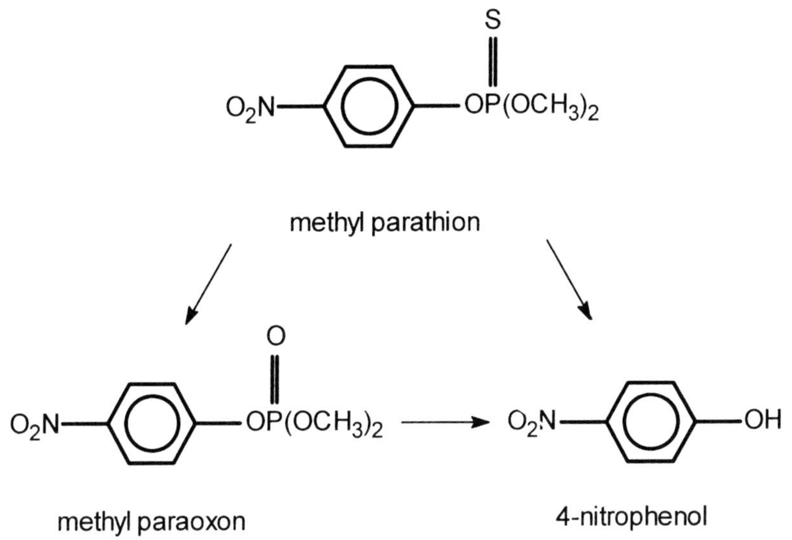

Fig. 4.4. Initial photodegradation pathways for methyl parathion (after Dzyadevych *et al.*, 2002).

lysed by the presence of species which are themselves not degraded or are transformed back into their original state by other reactions. Metal ions and metal oxides can be involved in the photocatalytic degradation of organic molecules.

Photolysis of organic phosphorus pesticides

Most studies on the photodegradation of organic phosphorus compounds in the environment have focused on organophosphorus pesticides. The widespread use of organophosphorus pesticides has led to increasing interest in the photochemical breakdown of these compounds for a range of reasons, including water treatment, identification of breakdown products, and determining the fate of these pesticides in the environment. Examples of photochemical reactions in this class of compounds will be considered here to illustrate the importance of the processes; for a more detailed overview of the field the reader is referred to a number of recent reviews (Floesser-Mueller and Schwack, 2001; Burrows *et al.*, 2002; Pehkonen and Zhang, 2002).

Direct photochemical degradation of an organophosphorus pesticide in the environment requires that it has chromophores capable of absorbing solar radiation. The huge number of structures found in organophosphorus pesticides leads to a wide range of absorbance characteristics and photochemical reactivities. Most pesticides absorb relatively short wavelength ultraviolet radiation and, as a result, direct photodegradation is of limited importance for many of these compounds (Burrows *et al.*, 2002).

The direct photolysis of organophosphates can occur via a number of mechanisms:

- *Homolysis*: degradation into two radicals
- *Heterolysis*: degradation into two components of opposite charge
- *Photoionization*: ejection of an electron, leaving a radical cation
- *Photoisomerization*: rearrangement of the molecule following excitation.

These steps will generally be followed by further reactions to give stable products or intermediates. For example, the first steps in the photodegradation of methyl parathion in deionized water are illustrated in Fig. 4.4. Irradiation with a mercury lamp results in

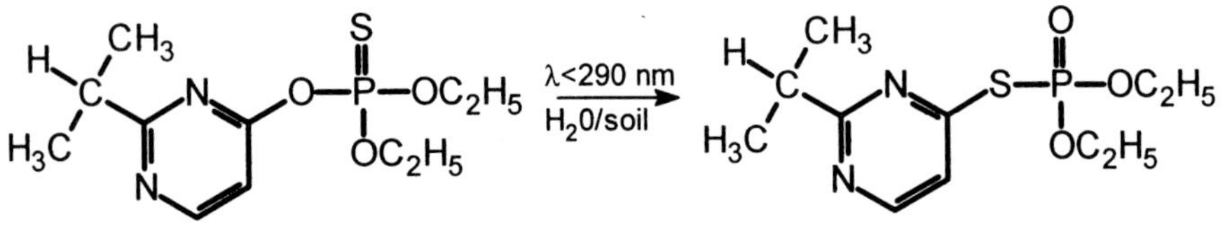

Fig. 4.5. The isomerization of diazinon in an irradiated water/soil suspension (after Mansour *et al.*, 1999).

either the replacement of sulphur with oxygen to produce the oxon, or hydrolysis of the phosphate ester bond to give the phenol and a phosphate derivative (Dzyadevych *et al.*, 2002). The phosphate derivatives were not identified in this study. In this, and many other degradation mechanisms, the photoproducts are susceptible to further photoreactions. In a highly photo-oxidative environment the end-products can be carbon dioxide and phosphate, although in a natural setting other mechanisms may be more important than direct photolysis.

The importance of environmental conditions on the products of photodegradation reactions is illustrated through comparison of the above study with work on a similar compound – parathion (Schynowski and Schwack, 1996). Parathion is different from methyl parathion only in the replacement of each of the methyl groups with an ethyl group. The compound was irradiated in a range of solvents to model various components of plant surfaces. In a saturated hydrocarbon solvent (cyclohexane), the results were similar to the methyl parathion study discussed above, with the production of paraoxon and 4-nitrophenol through photo-oxidative and cleavage reactions, respectively. However, in an unsaturated solvent, cyclohexene, the rate of degradation is accelerated and the mechanism is reductive at the nitro-group. In 2-propanol both photo-oxidative and photo-reductive mechanisms occurred. On plant cuticles the reaction resembled that in cyclohexene most closely. These results highlight the importance of studying the photochemistry of organophosphorus compounds in conditions resembling the environment of interest, as major differences in reaction mechanism may result from inappropriate choice of solvent.

Diazinon is an example of an organophosphorus compound that can undergo photoisomerization reactions (see Fig. 4.5), in addition to the formation of diazoxon and hydroxy diazinon (Mansour *et al.*, 1999). In the environment, the presence of other compounds can also result in additional mechanisms, such as photosensitization (either by interaction with a chromophore or by reaction with reactive oxygen species), or photocatalytic reactions in the presence of metal ions or on particle surfaces.

Humic substances are a class of potential photosensitizers of particular interest in the aquatic photodegradation of pesticides. Humic substances are the high-molecular-weight fraction of the organic matter found in aquatic environments and soils. They are yellow to brown in colour and have molecular weights ranging from about 500 to 100,000 Da (McKnight and Aiken, 1998). Humic substances are polymers of a diverse range of organic building blocks. The essential features are a large number of carboxylic acid functional groups (COOH), hydroxyl groups (OH), carbonyls (C=O), aromatic rings and long chains. The structures may also contain nitrogen in various forms and be bound to metals, particularly via bonds to nitrogen and oxygen (Paciolla *et al.*, 1999). Humic substances rapidly generate radicals upon irradiation with simulated sunlight (Kamiya and Kameyama, 1998). The photodegradation of a range of organophosphorus pesticides was studied in the presence and absence of this humic matter and an acceleration of the initial rate of reaction was found. The acceleration effect was influenced by the radical-generating abilities of the humic acid, with those samples with the highest radical-generating ability inducing the fastest photodegradation of the pesticides (Kamiya and Kameyama, 1998). There

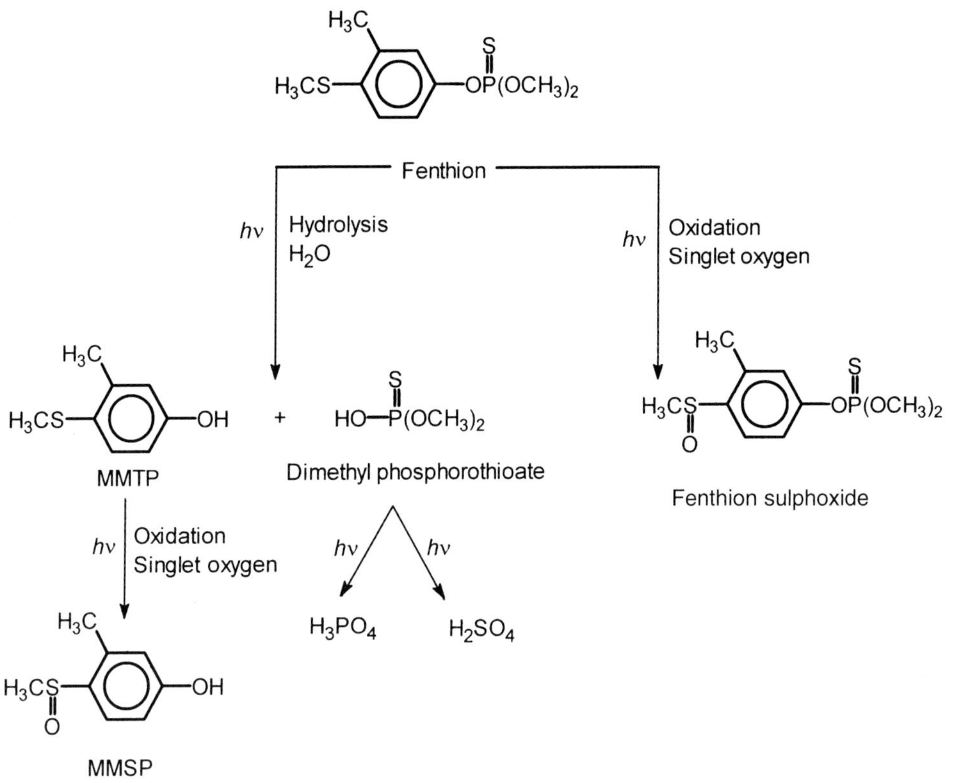

Fig. 4.6. Mechanisms involved in the photodegradation of fenthion; MMTP = 3-methyl-4-methylthiophenol, MMSP = 3-methyl-4-methylsulphoxylphenol (after Hirahara *et al.*, 2003).

was also a general tendency for pesticides containing nitrogen atoms to be more susceptible to photodegradation, and those containing hydrocarbon chains or chlorine to be less susceptible. It was also observed that the organophosphorus pesticides that had relatively low reactivity in terms of direct photolysis were the most susceptible to photosensitized reactions in the presence of humic acids. Complexation of the pesticide by the humic acid was an important factor in facilitating the radical reactions (Kamiya and Kameyama, 1998). A later study considered the interactive effects of organophosphorus pesticides, humic substances and dissolved metal ions (Kamiya and Kameyama, 2001). Nitrates of cobalt (II), chromium (III), copper (II) and manganese (II) were added to the reaction and retarded the rate of reaction. The effect increased with increasing metal concentration up to a limiting value, indicating a saturation-type effect of metal ions complexed to the humic acids. The effect was also variable depending on the pesticide involved. The paramagnetic properties of the metal ions decreased the electron spin resonance intensities of humic-generated free radicals, suggesting that the effect is a result of suppression of radical-induced pesticide degradation. Cyclodextrins increased the photodegradation of organophosphorus pesticides in humic substances, probably through trapping of both reactive radicals and pesticides within the cyclodextrin (Kamiya *et al.*, 2001).

Irradiation of fenthion (Fig. 4.6) in the presence of Rose Bengal (a photosensitizer that produces singlet oxygen) with ultraviolet-B radiation (280–320 nm) leads to photochemical degradation. A range of products are formed, including phosphate and sulphate (Hirahara *et al.*, 2003). The photoproduction of 3-methyl-4-methylthiophenol (MMTP) results from a hydrolysis reaction in which dimethyl phosphorothioate (HO-P(S)$(OCH_3)_2$) is formed as the other product. The dimethyl phosphorothioate can be further degraded to produce phosphate and sulphate. The production of MMTP was not dependent on the presence of oxygen, but the production of fenthion sulphoxide and 3-methyl-4-methylsulphoxylphenol (MMSP) were decreased when the solution was bubbled with nitrogen. The formation of fenthion sulphoxide and MMSP were shown to be diminished by the presence of singlet oxygen scavengers, but not scavengers of hydroxyl radicals or superoxide (Hirahara *et al.*, 2003). The formation of these two major reaction products is the result of singlet oxygen production by the photosensitizer, while the MMTP is the result of direct photolysis of fenthion. Photochemical degradation via multiple mechanisms is likely to be common under environmental conditions.

Photodegradation reactions on the surface of metal oxides have received attention as 'advanced oxidation processes' for the treatment of contaminated water (Pehkonen and Zhang, 2002). For example, the photodegradation of dichlorovos (Cl_2=CHOP(O)$(OCH_3)_2$) was examined in the presence of a number of metal oxides and organic sensitizers (Naman *et al.*, 2002). Ultraviolet irradiation of dichlorovos in the presence of titanium oxide leads to the formation of chloride ions and a drop in pH (assumed by the authors to indicate the formation of hydrochloric acid and dihydrogen phosphate ions). The formation of a hydroxyl radical on the oxide surface was thought to be an important factor in these reactions, and the reactions were suppressed by the removal of oxygen from the reaction vessel. Reactions also occurred in the presence of zinc oxide and a mixture of vanadium sulphides and titanium oxide.

Photochemical reactions between organic molecules and photoactive mineral surfaces will not necessarily rely on the generation of radicals. Organic molecules can adsorb to the surface of oxides and in some cases photochemical reactions may result in redox reactions occurring between metal atoms and adsorbed organics, or electron transfer between organic matter and energy bands in the oxide. Photochemical reaction between organophosphate compounds and particle surfaces are likely to be complex and varied in the environment, and further research into this area is needed.

Photolysis of natural organic phosphorus compounds

There have been few mechanistic studies on the photochemistry of natural organic phosphorus compounds. However a number of studies have implicated photodegradation in the cycling of phosphorus in aquatic environments. In the study of an acid bog lake, a large proportion of the dissolved phosphorus was found to occur as a class of high-molecular-weight compounds that were chromatographically similar to dissolved humic materials (Francko and Heath, 1982). This high-molecular-weight phosphorus released phosphate when exposed to low-intensity ultraviolet light. Daytime concentrations of ultraviolet-sensitive complex phosphorus were linearly related to co-chromatographing dissolved humic material absorbance. This is consistent with the theory that dissolved humic material had formed a complex with phosphorus that could be broken down by ultraviolet light. A diurnal variation was observed in both the concentrations of dissolved humic material and ultraviolet-sensitive phosphorus. The increase in ultraviolet-sensitive phosphorus was accompanied by an increase in iron (II), suggesting the presence of organic–iron–phosphorus complexes. The phosphate reactions were potentially linked to the reduction of iron (III) associated with humic material in the water. The high-molecular-weight phosphorus complex did not release phosphate when combined with alkaline phosphatase, providing additional evidence for an abiotic, photochemical process. The presence of ultraviolet-sensitive complex phosphorus was also observed in a later study (Francko, 1986) and the addition of dissolved humic material and ferric iron was observed to affect the processes. The addition of iron (III) to lake water labelled with ^{32}P-phosphate increased the amount of labelled phosphorus eluting with phosphate following ultraviolet irradiation (Cotner and Heath, 1990). However, phosphate release did not accompany photoreduction of iron (III) on all occasions when unamended lake water was irradiated, suggesting that phosphorus was not always associated with iron.

Light can also indirectly influence the cycling of phosphorus by changing the reactivity of phosphatases (Boavida and Wetzel, 1998). Phosphatases released by organisms can complex with humic substances in the water, leaving the enzyme inactive. Movement of the water can transport these complexes into the photic zone, where the phosphatases are released by ultraviolet irradiation. The enzymes are partially protected by the humic acid from direct photolysis. The binding capacity of the humic acid to the phosphatase depends on humic acid properties and on the source of the phosphatase. Thus, exposing humic acids in natural water systems, which may contain a reserve of stored enzyme, to sunlight can result in a release of phosphorus into the water through the action of the liberated enzyme.

Concluding Remarks

From the preceding discussion it is evident that abiotic degradation of organic phosphorus compounds is potentially an important part of the biogeochemical cycling of phosphorus in the environment. Laboratory studies have shown that both dissolved and particulate metal ions can facilitate the hydrolysis of organic phosphorus species, and that many organic phosphorus compounds are susceptible to photolysis. What is not so clear is the extent to which both natural and synthetic organic phosphorus species undergo abiotic degradation in the environment. One of the greatest difficulties faced in this area of research is differentiating between biotic and abiotic pathways. Because phosphorus may be limiting in many environments, organisms have developed a suite of enzymes to break down large phosphorus-containing species to yield free phosphate. In the environment, biotic pathways for phosphorus degradation will be operating simultaneously with abiotic processes. Even in the laboratory scrupulous attention must be paid to ensuring that microbiota are excluded from experiments. For example, Coleman (1996) traced anomalously high rates of hydrolysis of *p*-nitrophenyl phosphate in preliminary experiments on supposedly inert substrates (barium sulphate) to bacterial contamination. Consequently all her subsequent experiments were done under aseptic conditions.

Differentiating between photolytic and non-photolytic reactions (which obviously include both biotic and abiotic reactions) is relatively straightforward. One can either compare rates of breakdown in the presence and absence of light in the laboratory or, as in the study by Francko and Heath (1982), by examining diurnal differences in rates of breakdown (assuming of course that diurnal temperature differences are not significant).

Differentiating between biotic and abiotic hydrolytic reactions in environmental samples is more problematic. In many studies, the mechanism(s) responsible for organic phosphorus degradation are not important; of more importance is a measure of the overall rate of breakdown or the nature and type of the breakdown products. However, in studies where enzyme activity is used as a surrogate measure for other parameters, such as microbial biomass or the impact of pollutants on ecosystem processes, it may be important to differentiate between biotic and abiotic pathways. Because many of the hydrolytic enzymes are exo-enzymes (that is, expressed outside of the organism) simply sterilizing environmental samples will not halt enzyme activity. Indeed, some methods for inhibiting microbial growth will also affect abiotic hydrolysis reactions. Sodium azide, a common microbial growth inhibitor, will also bind to iron mineral surfaces, inhibiting abiotic hydrolysis reactions (D.S. Baldwin, unpublished results). An alternative approach that has been used successfully to differentiate between biotic and abiotic oxidation of manganese (II) in environmental samples (Johnson *et al.*, 1995) is to measure the temperature dependence of the rate of reaction. Rates of enzymatic reactions tend to increase with temperature until they reach a maximum and then decline, while abiotic reactions tend to increase linearly with temperature. Therefore, measuring the temperature dependence of rates of hydrolysis of organic phosphorus species should allow an estimate of the relative importance of biotic versus abiotic reaction pathways in any given environmental sample. As far as we are aware, this approach has not yet been applied to differentiating the two pathways for the hydrolysis of organic phosphorus species.

Acknowledgements

We thank our colleagues Dr David Jones, Dr Lynne Coleman and Dr Matthew Inman for their contributions to the understanding of abiotic pathways for the degradation of organic phosphorus, particularly metal-ion-facilitated hydrolysis of phosphate esters.

References

Balistrieri, L.S. and Chao, T.T. (1990) Adsorption of selenium by amorphous iron oxyhydroxide and manganese dioxide. *Geochimica et Cosmochimica Acta* 54, 739–751.

Baldwin, D.S., Beattie, J.K., Coleman, L.M. and Jones, D.R. (1995) Phosphate ester hydrolysis facilitated by mineral phases. *Environmental Science and Technology* 29, 1706–1709.

Baldwin, D.S., Beattie, J.K. Coleman, L.M. and Jones, D.R. (2001) Hydrolysis of an organophosphate ester by manganese dioxide. *Environmental Science and Technology* 35, 713–716.

Baldwin, D.S., Mitchell, A.M. and Olley, J. (2002) Pollutant–sediment interactions: sorption, reactivity and transport of phosphorus. In: Haygarth, P.M. and Jarvis, S.C. (eds) *Agriculture, Hydrology and Water Quality*. CAB International, Wallingford, UK, pp. 265–280.

Bamann, E. and Mersenheimer, M. (1938) Phosphatic activity of hydrogels. I. The cleavage of esters of phosphoric acid in the presence of $La(OH)_3$. *Berichte der Deutschen Chemischen Gesellschaft* 71, 1711–1720.

Boavida, M.-J. and Wetzel, R.G. (1998) Inhibition of phosphatase activity by dissolved humic substances and hydrolytic reactivation by natural ultraviolet light. *Freshwater Biology* 40, 285–293.

Boily, J.-F., Lützenkirchen, J., Balmès, O., Beattie, J. and Sjöberg, S. (2001) Modeling proton binding at the goethite (a-FeOOH)–water interface. *Colloids and Surfaces A: Physicochemical and Engineering Aspects* 179, 11–27.

Bullock, J.I., Duffin, P.A. and Nolan, K.B. (1993) In-vitro hydrolysis of phytate at 95°C and the influence of metal–ion on the rate. *Journal of the Science of Food and Agriculture* 63, 261–263.

Burrows, H.D., Canle, L., Santaballa, J.A. and Steenken, S. (2002) Reaction pathways and mechanisms of photodegradation of pesticides. *Journal of Photochemistry and Photobiology B: Biology* 67, 71–108.

Butcher, W.W. and Westheimer, F.H. (1955) The lanthanum hydroxide gel promoted hydrolysis of phosphate esters. *Journal of the American Chemical Society* 77, 2420–2420.

Coleman, L. (1996) The facilitated hydrolysis of phosphate esters by metal oxides. PhD thesis, University of Sydney, Australia.

Cornell, R.M. and Schwertmann, U. (1996) *The Iron Oxides: Structure, Properties, Reactions, Occurrence, and Uses*. VCH, Weinheim, Germany, 537 pp.

Cotner, J.B. and Heath, R.T. (1990) Iron redox effects on photosensitive phosphorus release from dissolved humic materials. *Limnology and Oceanography* 35, 1175–1181.

Dannenberg, A. and Pehkonen, S.O. (1998) Investigation of the heterogeneously catalysed hydrolysis of organophosphorus pesticides. *Journal of Agricultural and Food Chemistry* 46, 5–34.

Dixon, N., Jackson, W., Marty, W. and Sargeson, A. (1982) Base hydrolysis of pentaamminecobalt(II) complexes of urea, dimethyl sulfoxide and trimethyl phosphate. *Inorganic Chemistry* 21, 688–697.

Dzyadevych, S.V., Soldatkin, A.P. and Chovelon, J.-M. (2002) Assessment of the toxicity of methyl parathion and its photodegradation products in water samples using conductometric enzyme biosensors. *Analytica Chimica Acta* 459, 33–41.

Eisberg, R. and Resnick, R. (1985) *Quantum Physics of Atoms, Molecules, Solids, Nuclei, and Particles*, 2nd edn. John Wiley & Sons, New York, 864 pp.

Floesser-Mueller, H. and Schwack, W. (2001) Photochemistry of organophosphorus insecticides. *Reviews of Environmental Contamination and Toxicology* 172, 129–228.

Florian, J. and Warshel, A. (1998) Phosphate ester hydrolysis in aqueous solution: associative versus dissociative mechanisms. *Journal of Physical Chemistry B* 102, 719–734.

Francko, D.A. (1986) Epilimnetic phosphorus cycling: influence of humic materials and iron on coexisting major mechanisms. *Canadian Journal of Fisheries and Aquatic Sciences* 43, 302–310.

Francko, D.A. and Heath, R.T. (1982) UV-sensitive complex phosphorus: association with dissolved humic material and iron in a bog lake. *Limnology and Oceanography* 27, 564–569.

Golimowski, J. and Golimowska, K. (1996) UV-photooxidation as pretreatment step in inorganic analysis of environmental samples. *Analytica Chimica Acta* 325, 111–133.

Haight, G.P., Hambley, T.W., Hendry, P., Lawrance, G.A. and Sargeson, A.M. (1985) Rapid cleavage of tridentate cobalt(III)–co-ordinated triphosphate. *Journal of the Chemical Society: Chemical Communications* (8), 488–491.

Healy, T.W., Herring, A.P. and Fuerstenau, D.W. (1966) The effect of crystal structure on the surface properties of a series of manganese dioxides. *Journal of Colloid and Interface Science* 21, 435–444.

Hirahara, Y., Ueno, H. and Nakamuro, K. (2003) Aqueous photodegradation of fenthion by ultraviolet B irradiation: contribution of singlet oxygen in photodegradation and photochemical hydrolysis. *Water Research* 37, 468–476.

Hoffman, M.R. (1980) Trace metal catalysis in aquatic environments. *Environmental Science and Technology* 9, 1061–1066.

Hooper, F.H. (1973) Origin and fate of organic phosphorus compounds in aquatic systems. In: Griffith, E.J., Beeton, A., Spencer, J.M. and Mitchell, D.T. (eds) *Environmental Phosphorus Handbook*. John Wiley & Sons, New York, pp. 179–202.

Inman, M.P., Beattie, J.K., Jones, D.R. and Baldwin, D.S. (2001) Abiotic hydrolysis of the detergent builder tripolyphosphate by hydrous manganese dioxide. *Water Research* 35, 1987–1993.

Jones, D., Lindoy, L.F. and Sargeson, A.M. (1983) Hydrolysis of phosphate esters bound to cobalt (III). Kinetics and mechanism of intramolecular attack of hydroxide on coordinated 4-nitrophenyl phosphate. *Journal of the American Chemical Society* 105, 7327–7336.

Johnson, D., Chiswell, B. and O'Halloran, K. (1995) Micro-organisms and manganese cycling in a seasonally stratified freshwater dam. *Water Research* 29, 2739–2745.

Kamiya, M. and Kameyama, K. (1998) Photochemical effects of humic substances on the degradation of organophosphorus pesticides. *Chemosphere* 36, 2337–2344.

Kamiya, M. and Kameyama, K. (2001) Effects of selected metal ions on photodegradation of organophosphorus pesticides sensitised by humic acids. *Chemosphere* 45, 231–235.

Kamiya, M., Kameyama, K. and Ishiwata, S. (2001) Effects of cyclodextrins on photodegradation of organophosphorus pesticides in humic water. *Chemosphere* 42, 251–255.

Mansour, M., Feicht, E.A., Behechti, A., Schramm, K.W. and Kettrup, A. (1999) Determination of photostability of selected agrochemicals in water and soil. *Chemosphere* 39, 575–585.

McKenzie, R.M. (1981) The surface charge on manganese dioxides. *Australian Journal of Soil Research* 19, 41–50.

McKnight, D.M. and Aiken, G.R. (1998) Sources and age of aquatic humus. In: Hessen, D.O. and Tranvik, L.J. (eds) *Aquatic Humic Substances: Ecology and Biogeochemistry.* Springer, Berlin, pp. 9–39.

Moore, W.J. (1983) *Basic Physical Chemistry*. Prentice-Hall International, London, 711 pp.

Mortland, M.M. and Gieseking, J.E. (1952) The influence of clay minerals on the enzymatic hydrolysis of organic phosphorus compounds. *Soil Science Society of America Proceedings* 16, 10–13.

Murray, J.W. (1974) The surface chemistry of hydrous manganese dioxide. *Journal of Colloid and Interface Science* 46, 357–371.

Naman, S.A., Khammas, Z.A.A. and Hussein, F.M. (2002) Photo-oxidative degradation of insecticide dichlorovos by a combined semiconductors and organic sensitizers in aqueous media. *Journal of Photochemistry and Photobiology A: Chemistry* 153, 229–236.

Ouvrard, S., Simonnot, M.-O. and Sardin, M. (2002) Reactive behaviour of natural manganese oxides toward the adsorption of phosphate and arsenate. *Industrial and Engineering Chemistry Research* 41, 2785–2791.

Paciolla, M.D., Davies, G. and Jansen, S.A. (1999) Generation of hydroxyl radicals from metal-loaded humic acids. *Environmental Science and Technology* 33, 1814–1818.

Pehkonen, S.O. and Zhang, Q. (2002) The degradation of organophosphorus pesticides in natural waters: a critical review. *Critical Reviews in Environmental Science and Technology* 31, 17–72.

Rietra, R.P.J.J., Hiemstra, T. and Van Riemsdijk, W.H. (2001) Interaction between calcium and phosphate adsorption on goethite. *Environmental Science and Technology* 35, 3369–3374.

Schynowski, F. and Schwack, W. (1996)

Photochemistry of parathion on plant surfaces: relationship between photodecomposition and iodine number of the plant cuticle. *Chemosphere* 33, 2255–2262.

Sega, G.A., Tomkins, B.A., Griest, W.H. and Bayne, C.K. (1998) The hydrolysis of di-isopropyl methyl phosphonate in groundwater. *Journal of Environmental Science and Health: Part A, Toxic Hazardous Substances and Environmental Engineering* 33, 213–236.

Shen, C.Y. and Morgan, F.W. (1973) Hydrolysis of phosphorus compounds. In: Griffith, E.J., Beeton, A., Spencer, J.M. and Mitchell, D.T. (eds) *Environmental Phosphorus Handbook.* John Wiley & Sons, New York, pp. 241–264.

Skoog, D.A. and Leary, J.J. (1992) *Principles of Instrumental Analysis* (4th edn). Saunders College Publishing, Fort Worth, Texas, 829 pp.

Smolen, J. and Stone, A.T. (1997) Divalent metal ion-catalyzed hydrolysis of phosphorothionate ester pesticides and their corresponding oxonates. *Environmental Science and Technology* 31, 1664–1673.

Stone, A. (1987) Reductive dissolution of manganese (III/IV) oxides by substituted phenols. *Environmental Science and Technology* 1987, 979–988.

Tetas, M. and Lowenstein, J.M. (1963) The effect of bivalent metal ions on the hydrolysis of adenosine di- and triphosphate. *Biochemistry* 2, 350–357.

Torrent, J., Barrón, V. and Schwertmann, U. (1990) Phosphate adsorption and desorption by goethites differing in crystal morphology. *Soil Science Society of America Journal* 54, 1007–1012.

Torrents, A. and Stone, A.T. (1994) Oxide surface-catalysed hydrolysis of carboxylate esters and phosphorothioate esters. *Soil Science Society of America Journal* 58, 738–745.

Wijesekera, R. (1992) Coordinated phosphate reactivity. PhD thesis, Australian National University, Canberra, Australia.

Wolterman, G.H., Scott, R.A. and Haight, G.P. (1974) Coupling of adenosine triphosphate hydrolysis to a simple inorganic redox system Vanadyl ion(2+) + hydrogen peroxide. *Journal of the American Chemical Society* 96, 7569–7570.

Yao, W. and Millero, F.J. (1996) Adsorption of phosphate on manganese dioxide in seawater. *Environmental Science and Technology* 30, 536–541.

5 Enzymatic Hydrolysis of Organic Phosphorus

Hervé Quiquampoix and Daniel Mousain

Unité Mixte de Recherche Rhizosphère et Symbiose, INRA-ENSAM, 2 Place Pierre Viala, 34060 Montpellier, France

Introduction

Limiting aspects of phosphorus for plant nutrition

Among the elements essential for plant growth and nutrition, phosphorus is one of the less well represented in the lithosphere (0.1%). Phosphorus is strongly adsorbed to soil colloids and does not form volatile compounds, so its cycle takes place exclusively in the biosphere and losses of phosphorus by leaching are generally small (i.e. between 1 and 12 g P/ha per year; Bieleski and Ferguson, 1983). One major characteristic of this cycle is that only 1% of the soil phosphorus is incorporated in plants during a growing season (Borie and Barea, 1981).

The productivity of many ecosystems is determined by the chemical stability of phosphate, its relative immobility in soil, and its low concentration in the soil solution (approximately 1 μM). Phosphorus uptake by plant roots occurs mainly as phosphate ions ($H_2PO_4^-$ and HPO_4^{2-}) from the soil solution, which depends on the solubilization of mineral phosphates and the degradation or mineralization of organic phosphorus. During phosphate uptake, a phosphorus-depleted area is formed close to the root, since phosphate uptake is faster than its diffusion in soil (Bhat and Nye, 1973). Having an extra-matrical network of mycorrhizae is a way of increasing the volume of soil from which phosphorus can be mobilized (Owusu-Bennoah and Wild, 1979). Phosphorus is intimately involved in the growth and metabolism of plants, so phosphorus deficiency leads to a general reduction in most metabolic processes, including cell division and expansion, respiration and photosynthesis (Terry and Ulrich, 1973; Jacob and Lawlor, 1992).

Organic phosphorus as a widely available phosphorus source in soil

In the upper soil horizons, organic phosphorus represents between 20% and 80% of the total phosphorus (Dalal, 1977). Extreme values of 4% and 90% of the total phosphorus were observed in a podsolic soil and in an alpine humus, respectively (Williams and Steinbergs, 1958). Organic phosphorus in chemically and physically protected forms often represents at least 90% of the total organic phosphorus (Hedley and Stewart, 1982; Hedley *et al.*, 1982). These forms are slowly degraded, contributing to the labile organic phosphorus pool. Phosphorus availability in soils with poor phosphate solubility, but containing significant quantities of organic phosphorus, may be controlled by organic phosphorus mineralization (Tiessen *et al.*, 1984). Therefore, organic phosphorus is significant for plant nutrition.

The main identified organic phosphorus compounds are inositol phosphates, phospholipids and nucleid acids (Anderson, 1967). The content of inositol phosphates is highly variable, both in terms of quantity and as a proportion of the total organic phosphorus. However, they are frequently the dominant component, and may represent 80% or more of the total organic phosphorus (Dalal, 1977). They exist in soil in several stereoisomeric forms, namely *myo*-, *scyllo*-, D-*chiro* and *neo*-inositol phosphates (Dalal, 1977; Cosgrove, 1980). The *myo*-inositol hexakisphosphate is the only compound that is simultaneously found in soils, plants (mainly in mature seeds, roots, tubers, leaves), animals and microorganisms (Lott and Ockenden, 1986; Campbell *et al.*, 1991; Pointillart, 1994; Ravindran *et al.*, 1994; Harland and Morris, 1995). The presence of D-*chiro*-inositol phosphates was reported in soils, seeds and leaves (L'Annunziata and Fuller, 1971; L'Annunziata *et al.*, 1972), while the *scyllo*- and *neo*-inositol phosphates are thought to be exclusively of microbial origin (Laird *et al.*, 1976).

The inositol hexa- and pentakisphosphates are prevalent in soils compared to lower-order esters, probably because stability in the soil is linked to the number of phosphate groups. The most widespread stereoisomer is *myo*-inositol (Dalal, 1977). Inositol phosphates are more resistant to mineralization than the other fractions of the soil organic phosphorus and, therefore, are probably poorly available to plants (Williams and Anderson, 1968). They are present in soils in highly complex forms associated with clay minerals, fulvic and humic acids (Anderson and Arlidge, 1962), proteins and some metallic ions (Rojo *et al.*, 1990). The various forms of inositol phosphates are often imprecisely referred to as phytic acid, which is reserved exclusively for the free acid form of *myo*-inositol hexakisphosphate. Salt forms of *myo*-inositol hexakisphosphate, also known as phytates, are very stable and consequently accumulate in soil. Their behaviour is similar to that of phosphate ions, because they react with iron, aluminium and calcium (Ognalaga *et al.*, 1994). Inositol phosphates form insoluble salts in acid soils, which are unavailable for plants, and precipitate with calcium in alkaline soils (Anderson, 1963; Celi and Barberis, Chapter 6, this volume).

Phospholipid concentrations vary from 0.2 to 14 mg P/kg soil (Kowalenko and McKercher, 1971a). These represent between 0.5% and 7% of the soil organic phosphorus, with an average of 1% (Anderson and Malcolm, 1974). The soil phospholipids may be of microbial, plant or animal origin. Their synthesis and degradation in soil may be quite rapid. Phosphoglycerides (phosphatidylcholine, phosphatidylethanolamine) seem to be prevalent and may represent up to 40% of the phospholipids (Dalal, 1977).

Less than 3% of soil organic phosphorus is present as nucleic acids and derivatives derived from the decomposition of living organisms (Dalal, 1977). The four bases of DNA have been identified in humic acids (Anderson, 1961). The presence of nucleic acids and derivatives in the soil was confirmed by the isolation of two pyrimidine nucleoside diphosphates (Anderson, 1970). Nucleic acids are rapidly mineralized, re-synthesized and combined with other soil constituents, or incorporated into microbial biomass (Anderson and Malcolm, 1974). Nevertheless, the interaction of nucleases with soil constituents can inhibit DNA hydrolysis, with important environmental consequences related to extracellular gene uptake by bacteria (Demanèche *et al.*, 2001).

Sugar phosphates are present in soils (Anderson and Malcolm, 1974; Sanyal and De Datta, 1991). Some organic phosphates isolated from organic matter display the properties characteristic of phosphoproteins (Anderson, 1961). Several monophosphorylated carboxylic acids have also been detected in soil extracts (Anderson and Malcolm, 1974). Much organic phosphorus exists in highly combined forms and remains unidentified.

Compartmental Analysis of Organic Phosphorus in Relation to its Enzymatic Hydrolysis

Throughout this chapter, the enzymatic hydrolysis of organic phosphorus is discussed with reference to the classical inorganic and organic compartmental analysis of phosphorus compounds in ecosystems. However, this is not an ideal way to distinguish classes of phosphorus compounds, since the availability of phosphorus to organisms is largely dependent on biological mobilization processes.

Classical chemical approach: organic versus inorganic phosphorus

One fundamental approach to modelling biogeochemical cycles is the simplification or compartmental representation of the observed phenomena. In particular, due to the large number of different chemical forms of phosphorus in soil and the poorly understood nature of some of these forms, it is usual to group them into generic compartments to understand the phosphorus cycle in ecosystems. As progress in chemistry has historically preceded that in biology (if not in phylogeny then certainly in the fundamental aspects of molecular and cellular biology), agronomists have defined phosphorus availability based on chemical characterization. Organic chemistry is based on carbon chemistry. Thus, organic phosphorus comprises all forms of phosphorus that are directly included in carbon-based molecules. On the other hand, inorganic phosphorus is a compartment comprising crystalline and amorphous forms of phosphorus in which carbon-based molecular structures are not involved. It also includes phosphate in free, adsorbed and complexed forms. This approach to the phosphorus biogeochemical cycle benefits from well-established methods of extraction of organic and inorganic phosphorus (see McKelvie, Chapter 1, this volume). It is also a compartmental approach that is well-placed to benefit from the progress of nuclear magnetic resonance spectroscopy, a powerful technique for characterizing phosphorus in ecosystems. Indeed, the ^{31}P nucleus has a sensitive spin and allows easy determination of the nature of the chemical bonds in which phosphorus atoms are involved (see Cade-Menun, Chapter 2, this volume). Nevertheless, in describing the dynamics of the phosphorus biogeochemical cycle, these organic and inorganic phosphorus compartments must be separated into subcompartments to account for adsorption and precipitation phenomena that regulate the availability or lability of phosphorus.

Despite all the advantages that organic and inorganic phosphorus compartments have in terms of practical concepts for the study of phosphorus cycling (large amounts of data related to its early conceptual acceptance, a simple basis for utilizing data from advanced spectroscopic techniques), when studying the biological use of phosphorus by plants and microorganisms, simple separation into organic and inorganic phosphorus pools does not allow optimal compartment analysis when the real nature of the biological processes are taken into account.

A better biological approach: P_{enz}, P_{memb} and P_{diss}

Abiotic mechanisms leading to a mobilization of freely available phosphorus for plants, fungi, bacteria and invertebrates are several orders of magnitude less efficient than mechanisms of biological origin (see Baldwin *et al.*, Chapter 4, this volume), at least in the soil zone of influence of these biota, including the rhizosphere, hyphosphere, bacterial microhabitats and drillosphere (i.e. the fraction of soil which has gone through the digestive tract of earthworms). Biological mechanisms can be deduced from the main physiological processes governing exchanges of cells with the external medium, i.e. (i) liberation of extracellular enzymes, (ii) liberation of complexing or mineral dissolving ions or molecules, and (iii) cellular membrane transport involving either specific permeases or passive diffusion directly through the membrane lipidic bilayer.

Enzymatically hydrolysable phosphorus (P_{enz})

Fungi, bacteria, plant roots and the digestive tract of invertebrates all produce extracellular enzymes, which can hydrolyse many organic (and some inorganic) compounds in soil. All organic phosphorus forms and some inorganic forms (linear polyphosphate and pyrophosphate of biological origin, tripolyphosphate from the detergent industry) can be hydrolysed by a class of enzymes broadly defined as phosphohydrolases. The efficiency of this hydrolysis is far in excess of any abiotic mechanism that could mobilize this class of phosphorus in the limited soil spatial zone where these biologically released enzymes remain active. This compartment can be defined as P_{enz}.

Dissolvable phosphorus (P_{diss})

Soil biota can also mobilize phosphorus by the active secretion of low-molecular-weight complexing molecules, such as organic acids, or by a modification of the pH around them by proton excretion. These compounds act as solubilizing agents of the network of phosphorus-containing crystalline or amorphous minerals or as desorbing agents for adsorbed phosphate on other soil minerals.

Membrane-permeant phosphorus (P_{memb})

Finally, some low-molecular-weight phosphorus products of enzymatic hydrolysis or dissolution processes, produced respectively from the compartments P_{enz} and P_{diss}, have the ability to cross biological membranes. These compounds represent the P_{memb} compartment. By far the most important is phosphate (included in the inorganic phosphorus pool in the classical chemical approach), for which specific membrane carriers, also known as permeases, permit an active transport towards the intracellular compartment of these organisms. However, other low molecular forms of organic phosphorus, such as glycerophosphate, nucleotide phosphates, and sugar phosphates, have specific permeases, at least on membrane systems of some soil microorganisms or plant roots. If we extend the analysis beyond phosphorus-containing natural molecules, some organophosphorylated pesticides, designed to have hydrophobic properties, can passively cross the lipidic bilayer of membranes and can also be considered as part of the P_{memb} compartment.

Potential Use of Soil Organic Phosphorus by Plants and Microorganisms

The principal components of soil organic phosphorus are inositol phosphates, nucleic acids and phospholipids. Inositol phosphates are the main components of the phosphate monoesters, which can represent 80% or more of the soil organic phosphorus. Thus, the ecological significance of phosphohydrolase activities in plants and microorganisms cannot be actually established, unless enzymes are able to immediately break down inositol phosphates in soils. Attention will be paid to ectomycorrhizal symbiosis, which reinforces the interest in the role of phosphatases in the phosphorus cycle. The two sets of experimental results below may be used to assess the capacity of microorganisms and plants growing under controlled conditions to use *myo*-inositol hexakisphosphate or other significant organic phosphates.

Growth of organisms on organic phosphates

Glucose 1-phosphate and glucose 6-phosphate only acted as adequate sources of phosphorus with small inocula (20,000 cells/ml) of *Ochromonas danica*, a model alga. With larger inocula, all the organic phosphates examined (β-glycerophosphate, glucose phosphates, fructose 1,6-diphosphate) were used as phosphorus sources (Aaronson and Patni, 1976). Soil yeasts of the genus *Cryptococcus*, incubated for 24 h with insoluble phytate salts (1 mg/ml) showed no phosphate release from aluminium or iron phytate, but considerable hydrolysis of calcium phytate. However, *Crytococcus albidus*, *C. macerans* and *C. infirmo-miniatus* hydrolysed sodium phytate at the same rate (*C. albidus*) or at faster rates (*C. macerans*, *C. infirmo-miniatus*) than calcium phytate. The growth of *C. infirmo-miniatus* on agar plates with insoluble *myo*-inositol hexakisphosphate salts also suggested that calcium, but not aluminium or iron phytate, could be utilized (Greenwood and Lewis, 1977).

Certain mycorrhizal fungi cultured *in vitro* are able to partially or totally utilize soluble sodium and calcium phytates, but can rarely utilize the almost insoluble iron phytate (Theodorou, 1968; Mitchell and Read, 1981; Dighton, 1983). For example, after growth either on phosphate-deficient or phosphate-rich medium, growth and phosphorus accumulation in the mycelium of *Pisolithus tinctorius*, cultured on sodium phytate, were similar to those observed on medium rich in phosphate. The hydrolysis of phytate was the result of enzymatic activity which increased gradually with increasing phosphate deficiency (Mousain and Salsac, 1986). The association *Pinus rigida/Pisolithus tinctorius*, growing on sand, was able to utilize phosphorus from phytate (Cumming, 1993). However, other studies have reported contradictory or non-conclusive observations concerning the use of inositol phosphate by ectomycorrhizal associations (Mousain, 1989; Matumoto-Pintro, 1996; Antibus *et al.*, 1997; Colpaert *et al.*, 1997).

Hydrolysis activity and substrate specificity under standard conditions

Examples in plants

ALEPPO PINE (*PINUS HALEPENSIS* MILLER) The three phosphatases purified from roots of *Pinus halepensis* showed very low specificity and low activity against phosphorylated sugars (fructose 6-phosphate, glucose 1-phosphate, glucose 6-phosphate, fructose 1,6-diphosphate). The phenyl-phtaleine phosphates are readily hydrolysed by the phosphatase *1a*, but much less by the phosphatases *3a* and *3b*. Phosphorus deficiency enhanced the utilization of tripolyphosphate and pyrophosphate, mainly by the phosphatases *1a* and *3a* from phosphorus-deficient roots. The phosphatase *1a* hydrolysed phosphoserine, phosphoethanolamine, phosphothreonine and *myo*-inositol hexakisphosphate; phytase activity represented 28% of the hydrolytic activity using the artificial substrate *para*-nitrophenyl phosphate (Doumas *et al.*, 1983).

TOBACCO (*NICOTIANA TABACUM* L.) Fractions *M-Ia* and *M-Ib* from acid phosphatase of tobacco cells did not hydrolyse phosphate monoesters except *para*-nitrophenyl phosphate, but hydrolysed phosphoric anhydrides (inorganic pyrophosphate, thiamine pyrophosphate, nucleoside di- and triphosphates). In addition, both fractions hydrolysed bis-*para*-nitrophenyl phosphate. Fraction *M-II*, a non-specific acid phosphatase, had broad substrate specificity: it hydrolysed pyrophosphate, thiamine pyrophosphate, uridine diphosphate, adenosine triphosphate, uridine triphosphate, and cytidine triphosphate, at rates similar to *para*-nitrophenyl phosphate (Ninomiya *et al.*, 1977).

RICE (*ORYZA SATIVA* L.) The activity of the acid phosphatase purified from the aleurone particle of rice was assessed using various substrates: *para*-nitrophenyl phosphate was readily hydrolysed, as were adenosine triphosphate and pyrophosphate. Among the *myo*-inositol phosphates, inositol trisphosphate was hydrolysed at the highest rate (29% of *para*-nitrophenyl phosphate). The enzyme was also characterized by low or little activity towards β-glycerophosphate, glucose 6-phosphate and nucleotide phosphates (Yamagata *et al.*, 1980).

Examples in microorganisms

Phytase activity measured in mycelial homogenates of the fungus *Pisolithus tinctorius* were strongly enhanced (almost tenfold) by phosphorus deficiency. Phytase activity represented only 1–2% of activity towards *para*-nitrophenyl phosphate (Mousain and Salsac, 1986). The four acid phosphatases purified from mycelium of *Aspergillus niger* were found to be non-specific, hydrolysing synthetic substrates, sugar phosphates, alcohol phosphates, nucleotides and pyrophosphate (Komano, 1975).

In another study (Wyss *et al.*, 1999), some phytases were described as having a broad substrate specificity (*Aspergillus terreus, Emericella nidulans, Myceliophthora thermophila*) and others as being rather specific for *myo*-inositol hexakisphosphate (*Aspergillus niger, A. terreus* 9A1, *A. terreus* CBS). The specificity of these phytases, which are in the group of histidine acid phosphatases (Mullaney and Ullah, 2003), is mainly dependent on the amino acid residue 300 (Mullaney *et al.*, 2002). A high specific activity for *myo*-inositol hexakisphosphate is associated with a basic or acidic amino acid in position 300 and, conversely a low specific activity is associated with a neutral amino acid.

Among the bacterial phosphatases, the alkaline phosphatase of *Lysobacter enzymogenes* is non-specific with respect to the substrate used (von Tigerstrom and Stelmaschuk, 1986). In general, all ribo- and deoxyribonucleoside 5′-phosphates were hydrolysed faster than *para*-nitrophenyl phosphate, in contrast to ribonucleoside 2′- or 3′-phosphates. Glucose 6-phosphate and α- and β-glycerophosphates were hydrolysed at rates similar to *para*-nitrophenyl phosphate. *Bacillus subtilis* phytase is reported to be specific for *myo*-inositol hexakisphosphate with no detectable liberation of phosphate with a number of organic phosphates and pyrophosphate (Powar and Jagannathan, 1982). The specific phytases are often associated with non-specific phosphomonoesterases, from which separation and purification is difficult. Thus, such a complex can hydrolyse *myo*-inositol hexakisphosphate as well as β-glycerophosphate and other phosphate esters, in addition to pyrophosphate (Nayini and Markakis, 1986).

Properties of Phosphohydrolases

Catalytic and kinetic parameters

Phosphomonoesterase

Many living cells contain at least two types of phosphomonoesterase enzymes (EC 3.1.3), the ‘acid’ and the ‘alkaline’ phosphatases, which possess a wide range of specificity (Lehninger, 1981). The *para*-nitrophenyl phosphate assay was developed to test a number of plant cells and organs and microorganisms for phosphomonoester hydrolases (Tibbett, 2002). Below, data relative to the catalytic and kinetic parameters of these enzymes in some plants and microorganisms are given. Their pH optimum varies from 2.5 in one of the four acid phosphatases purified from mycelia of *Aspergillus niger* (Komano, 1975), to 8.5 in the outer membrane of *Lysobacter enzymogenes* (von Tigerstrom and Stelmaschuk, 1986).

The pH optimum for hydrolysis by the major acid phosphatase (EC 3.1.3.2) associated with the aleurone particle of rice grains (*Oryza sativa* Japonica cv. Koshihikari) was 4.8 (Yamagata *et al.*, 1980). The K_m values were respectively 1.74 mM for *para*-nitrophenyl phosphate and 5.26 mM for adenosine triphosphate. They decreased slightly with an increase in the number of phosphate groups of various inositol phosphates, being 0.43 and 11.76 mM for *myo*-inositol hexakisphosphate and *myo*-inositol monophosphate, respectively. In addition, the V_{max} values for *myo*-inositol bis- and trisphosphate were high (564–611 μmol phosphate/min/mg protein).

In contrast to the fractions *M-Ia* and *M-Ib* isolated from the extracellular acid phosphatase of tobacco XD-6 cells, which had maximum activity at pH 6.8, the fraction *M-II* exhibited a pH optimum at 5.8, indicating an acid phosphatase. When a temperature of 55°C (instead of 30°C) was used for the *para*-nitrophenyl phosphatase activity of the three enzyme fractions from phosphate-supplied culture, the activity of *M-Ia* was rather stable, whereas fractions *M-Ib* and *M-II* were inactivated by about 35% and 90%, respectively, in 30 min. The Lineweaver–Burk plot of the rate of *para*-nitrophenyl phosphate hydrolysis by enzyme fractions from a phosphate-supplied culture showed that the apparent Michaelis constant of fractions *M-Ia* and *M-Ib* was 0.9 mM, whereas that of *M-II* was 0.3 mM (Ninomiya *et al.*, 1977).

Partially purified *1a*, *3a* and *3b* phosphatases from the soluble fraction of *Pinus halepensis* roots, assayed between pH 3 and 8, showed their maximum activity at pH 4.3–4.6. At pH 4.5, the values of K_m of the phosphatases *1a*, *3a* and *3b* of *P. halepensis* roots were respectively 1.5, 0.5 and 0.25 mM for *para*-nitrophenyl phosphate, indicating that *1a* possessed a weak affinity for this substrate (Doumas *et al.*, 1983).

Two main acid phosphatases were purified from each extract of phosphate-supplied (*Ia, IIa*) or phosphate-starved (*Ib, IIb*) cultures of the ectomycorrhizal fungus *Pisolithus tinctorius* (Berjaud and d'Auzac, 1986). These soluble phosphatases could not be distinguished either by their pH optimum (pH 5.0–5.5) or by their optimal temperature (60°C). In contrast, the activation energy was lower in the case of the two phosphatases from starved mycelium (50.2–58.2 kJ/mol) than those from phosphate-supplied mycelium (70.5–112.5 kJ/mol). Thus, their reaction rate was higher.

Regarding bacteria, the purified alkaline phosphatase (EC 3.1.3.1) associated with the outer membrane of *Lysobacter enzymogenes* UASM 495 (ATCC 29487) was most active at pH 8.5 and two K_m values (0.056 and 0.34 mM) were estimated from kinetic studies using *para*-nitrophenyl phosphate (von Tigerstrom and Stelmaschuk, 1986).

Phytase

Two classes of enzymes that degrade *myo*-inositol hexakisphosphate are recognized: 3-phytase (EC 3.1.3.8) initially cleaves the phosphate at the 3-position on the *myo*-inositol ring, leaving *myo*-inositol 1,2,4,5,6 pentakisphosphate (this phytase belongs to the family of histidine acid phosphatases; Mitchell *et al.*, 1997), whereas 6-phytase (EC 3.1.3.26) preferentially initiates dephosphorylation at the 6-position, leaving *myo*-inositol 1,2,3,4,5 pentakisphosphate.

The pH optima of a number of phytases ranged between 2.2 (in the yeast *Pichia farinosa*) and 7.5 (in *Bacillus subtilis* and germinated *Phaseolus aureus*), whereas the phytase of *Rhizopus oligosporus* NRRL 2710 exhibited an optimum at pH 5.6 (Nayini and Markakis, 1986). The phytases of *Aspergillus fumigatus* and *A. niger* T213 were both denatured at temperatures between 50 and 70°C. In contrast, *A. niger* pH 2.5 acid phosphatase displayed considerably higher thermostability. Denaturation, conformational changes and irreversible inactivation of this enzyme were observed only at temperatures >80°C. However, if *A. fumigatus* phytase, like *A. niger* T213 phytase, is not thermostable, it is able to refold completely into a native-like, fully active, conformation after heat denaturation (Wyss *et al.*, 1998).

Activation energy values for the hydrolysis of *myo*-inositol hexakisphosphate and of esters with lower phosphate content were between 35.6 kJ/mol in germinated *Phaseolus aureus* (Mandal *et al.*, 1972) and 50.2 kJ/mol in wheat bran (Nagai and Funahashi, 1962). The optimum temperature for the enzymatic hydrolysis of *myo*-inositol hexakisphosphate varies among phytases between 45 and 57°C (Nayini and Markakis, 1986), and even up to 60°C for the phytase of *Bacillus subtilis* (Powar and Jagannathan, 1982).

The lowest and the highest K_m values for *myo*-inositol hexakisphosphate hydrolysis were reported in phytases of *Aspergillus ficuum* (pH 5.3, 0.01 mM) and in that of germinated *Phaseolus aureus* (0.65 mM), respectively. The highest K_m values reported were of wheat bran phytase towards *myo*-inositol tetrakis- and trisphosphate (5 mM; see Nayini and Markakis, 1986). K_m and K_{cat} values for the enzymatic hydrolysis of *myo*-inositol hexakisphosphate by different *Bacillus* spp. were determined to be approximately 0.44 mM and 18.6/s, respectively. The affinity of *myo*-inositol pentakisphosphate for the phytase enzymes and their maximal rates of hydrolysis were lower (K_m = 0.50–0.76 mM; K_{cat} = 7.4–16/s) than that of *myo*-inositol hexakisphosphate (Greiner *et al.*, 2002).

Inhibitors and Activators

Inhibitors

Most acid phosphatases and phytases are inhibited by fluoride, which is a strong non-competitive inhibitor (K_i 1.29 mM) for the major acid phosphatase associated with aleurone particles of rice grains (Yamagata *et al.*, 1980), although it only moderately inhibits the acid phosphatase *M-II* of tobacco XD-6 cells (Ninomiya *et al.*, 1977). Fluoride concentrations determining 50% inhibition for *1a*, *3a* and *3b* soluble phosphatases of *Pinus halepensis* roots were respectively 3.1, 12 and 15 mM (Doumas *et al.*, 1983). Sodium fluoride (0.01 M) had no effect on the alkaline phosphatase activity of *Escherichia coli*, but inhibited the acid phosphatase activity by 85% (Garen and Levinthal, 1960). Phytases of dwarf bean, maize endosperm-scutellar tissue, wheat meal, wheat bran, *Pseudomonas bacterium* and *Aspergillus ficuum* NRRL 3135 were all inhibited by fluoride (Nayini and Markakis, 1986).

Polyvalent anions (phosphate, molybdate, arsenate, etc.), metal ions (silver, zinc, mercury (II), copper (II), iron (II), manganese (II)) and chelating agents (ethylenediaminetetraacetic acid (EDTA), 8-hydroxyquinoline, tartrate, oxalate, etc.) are also inhibitors for phosphatases and phytases. L-tartrate, phosphate, molybdate and arsenate are competitive inhibitors for the acid phosphatase of rice grains (see above); in particular, molybdate is a strong inhibitor (K_i = 0.336 μM) of this phosphatase, which was also inactivated by preincubation with EDTA (Yamagata *et al.*, 1980). The phosphatase *1a* from *Pinus halepensis* roots (see above) was strongly inhibited by molybdate ($I_{50\%}$ = 5 μM) whereas phosphatases *3a* and *3b*, were insensitive to molybdate. The $I_{50\%}$ values for the three phosphates were respectively 1.5, 35 and 37 mM in the case of phosphate (Doumas *et al.*, 1983).

The neutral phosphatase of tobacco XD-6 cells (*M-Ia*) was strongly inhibited by sodium molybdate, and moderately by copper sulphate, zinc sulphate and potassium dihydrogen phosphate (Ninomiya *et al.*, 1977). The alkaline phosphatase of *Lysobacter enzymogenes* was relatively insensitive to phosphate and arsenate, but was 97% inhibited by 0.05 mM EDTA and 99% by 0.5 mM 8-hydroxyquinoline (von Tigerstrom and Stelmaschuk, 1986). Yeast phytase (*Saccharomyces* spp.) was greatly inhibited by EDTA, oxalate and citrate (Nayini and Markakis, 1984), whereas the phytases of wheat bran (Nagai and Funahashi, 1962) and a soil *Pseudomonas* (Irving and Cosgrove, 1971) were not significantly inhibited by these chelating agents. In addition, a *Phaseolus vulgaris* phytase was inhibited by *p*-chloromercuribenzoate (Gibbins and Norris, 1963).

Activators

A number of acid and alkaline phosphatases and specific phytases are activated by divalent metal ions. After its inactivation by *o*-phenanthrolin and α,α′-dipyridil at 50°C, the manganese-containing acid phosphatase from sweet potato was reactivated by the addition of metal ions (Uehara *et al.*, 1974). The EDTA-inhibited alkaline phosphatase of *Lysobacter enzymogenes* was reactivated by divalent metal ions at 0.05 mM (von Tigerstrom and Stelmaschuk, 1986); specifically, reactivation was caused by magnesium chloride (4.6%), calcium chloride (7%), manganese chloride (14.5%), cobalt chloride (24%) and zinc chloride (66%).

Phytase from culture filtrates of *Bacillus subtilis* specifically required calcium ions for its activity, but was not activated by barium, zinc, magnesium or manganese ions (Powar and Jagannathan, 1982). More generally, phytases from plants and microorganisms are activated by several divalent cations, summarized by Nayini and Markakis (1986): magnesium and calcium ions for enzymes of wheat; 10 mM calcium, 1.5 mM magnesium and cobalt (II) for phytases of maize endosperm, germinated lettuce seed and *Phaseolus vulgaris*, respectively, and iron (II) and copper (II) for phytase of *Saccharomyces* spp. An unusual activator, lysolecithin, accelerates the functioning of the fraction F_1 of wheat bran phytase (Lim and Tate, 1973).

Phosphohydrolase Activity in Soil

Hydrolases are a class of enzymes that play an important role in biogeochemical nutrient cycles when they are released in soil, since most of the soil organic matter is in polymerized forms and, as a consequence, is either mainly insoluble or adsorbed on mineral constituents or, even if soluble, cannot cross the transport systems (permeases) of biological membranes of plants or microorganisms. The hydrolytic action of extracellular enzymes allows reduction to smaller molecules (oligomers, monomers, small ions) that are more soluble and can diffuse in the soil pore network, and which can also be recognized by the membrane permeases and taken up by soil biota (Quiquampoix, 2000).

Phosphohydrolases are the extracellular enzymes that play this role in the particular case of the biogeochemical phosphorus cycle. In the biological approach of phosphorus compartments, they catalyse the transfer of phosphorus from P_{enz} to P_{memb}. These extracellular phosphohydrolases can be actively secreted or be intracellular enzymes that are passively released after cell membrane damage. In the definition of extracellular phosphohydrolases *sensu lato*, we can distinguish ectoenzymes that are adsorbed on cell walls of roots, fungi or Gram-positive bacteria, or that are present in the periplasmic space of Gram-negative bacteria, and enzymes that are able to diffuse in the surrounding soil. We reserve the term extracellular phosphohydrolases for the latter class, thus adopting a narrower definition. It is a pervasive problem in soil biology terminology that no unequivocal definition can distinguish both classes of extracellular enzymes: ectoenzymes (extracellular but remaining bound to the organisms secreting them) and enzymes that are free to diffuse in the soil and for which no specific term is widely accepted. It should be noted that the terms *endoenzyme* and *exoenzyme* are not appropriate for indicating the location of enzymes, since they are now widely understood to refer to the site of hydrolysis in a polymer. For example, endocellulases cleave cellulose polymeric chains at random points, whereas exocellulases hydrolyse monomers or dimers of glucose sequentially from the ends of the polymeric cellulose chain.

Thermodynamic aspects of protein interactions with soil constituents

Extracellular phosphohydrolases are proteins. As a general rule, proteins have a strong affinity for solid surfaces. If extended to phosphohydrolases, this phenomenon has two consequences from an ecological point of view. First, phosphohydrolases can be adsorbed, which restricts their mobility by diffusion in the soil pore network and reduces the volume of soil surrounding the producing organism where hydrolysis of P_{enz} can occur. Secondly, adsorption can either modify the conformation of the whole proteic structure (Quiquampoix and Ratcliffe, 1992; Servagent-Noinville *et al.*, 2000) or lead to an unfavourable orientation of its catalytic site relative to the surface of adsorption (Baron *et al.*, 1999).

Second law of thermodynamics

The application of the second law of thermodynamics is useful for understanding complex protein sorption phenomena (Haynes and Norde, 1994; Quiquampoix *et al.*, 2002). It assumes that the spontaneous adsorption of a protein at constant temperature and pressure leads to a decrease in the Gibbs energy of the system. The Gibbs energy (G) depends on enthalpy (H), which is a measure of the potential energy (energy that has to be supplied to separate the molecular constituents from one another), and entropy (S), which is related to the disorder of the system.

$$\Delta_{ads}G = \Delta_{ads}H - T\Delta_{ads}S < 0$$

where T is the absolute temperature and Δ_{ads} is the change in the thermodynamic functions resulting from adsorption.

Enthalpic effects

Two types of intermolecular forces must be taken into account in $\Delta_{ads}H$: coulombic (or electrostatic) interactions and the Lifshitz–van der Waals interactions. Coulombic interactions are long-range and strong intermolecular forces originating from the overlap of the electrical diffuse double layers of the protein and the surface with which it is interacting. For proteins, the electrical charge originates from the ionization of carboxylic, tyrosyl, imidazole and amine groups of the side chains of amino acids. For mineral surfaces, the electrical charge originates from pH-independent isomorphic substitutions in the crystal lattice or from pH-dependent ionization of surface hydroxyls. The strong influence of the pH and the ionic strength of the soil solution make the electrostatic interactions an important factor for understanding the interaction of phosphohydrolases with soil constituents, as illustrated below (Quiquampoix, 1987a,b; Quiquampoix *et al.*, 1989, 1993, 1995). The Lifshitz–van der Waals interactions are short-range and weaker than coulombic interactions, but they act on all molecules, even if they do not bear electrical charge. Since they are independent of the pH and the ionic strength of the soil solution, the influence of this type of force is more difficult to demonstrate experimentally for the interaction of phosphohydrolases with soil constituents.

Entropic effects

Two types of changes in the order of the system must be taken into account in $\Delta_{ads}S$: hydrophobic interactions (Chassin *et al.*, 1986; Staunton and Quiquampoix, 1994) and modifications in the protein molecular structure (Servagent-Noinville *et al.*, 2000). Hydrophobic interactions, when applied to proteins, are related to two main effects. The first is the unshielding of non-polar amino acids in the core of proteins if an unfolding of the protein occurs in the adsorbed state. The second is the displacement of water molecules from the surface with which the protein interacts. Both effects originate from the fact that water molecules establish more hydrogen bounds around a non-polar group than around a polar group. Any perturbation of this state results in an increased disorder of surrounding water, which increases the entropy of the system and thus decreases the Gibbs energy.

Changes in the protein molecular structure in the adsorbed state also contribute to an increase in entropy. The main effect is related to changes in the secondary structure of the proteins from ordered structures such as α-helices or β-sheets to more disordered secondary structures. This phenomenon that can be demonstrated by spectroscopic studies (e.g. Fourier-transform infrared or circular dichroism spectroscopy) on adsorbed proteins (Kondo and Higashitami, 1992; Kondo *et al.*, 1992, 1993; Baron *et al.*, 1999; Servagent-Noinville *et al.*, 2000). Since we can assume four different possible conformations for peptide units in random structures as compared with only one in α-helices or β-sheets, the conformational entropic contribution to adsorption of a protein on a soil constituent surface is:

$$\Delta_{ads} S_{conf} = R \ln 4^n$$

where R is the molar gas constant and n is the number of peptide units involved in the transfer from an ordered secondary structure to a random secondary structure (Norde and Lyklema, 1991; Haynes and Norde, 1994).

Modelling the interaction of enzymes with soil constituents

As discussed above, the adsorption of enzymes on soil mineral surfaces is a complex phenomenon involving enthalpic and entropic effects (Quiquampoix, 2000; Quiquampoix *et al.*, 2002). The determination of conformational changes of the adsorbed enzyme can be followed by nuclear magnetic resonance or Fourier-transform infrared spectroscopy, giving respectively information on the interfacial area of the surface of contact of protein-clay (Quiquampoix and Ratcliffe, 1992; Quiquampoix *et al.*, 1993, 1995) and on the secondary structure of adsorbed proteins (Quiquampoix *et al.*, 1995; Baron *et al.*, 1999; Servagent-Noinville *et al.*, 2000). Such studies have shown the following general mechanisms to be involved in the interaction of enzymes with soil constituents.

Both pH-dependent modification of conformation and pH-dependent orientation of the catalytic site of the enzyme can explain the alkaline pH shift of the enzyme activity on electronegative soil mineral surfaces. Enzymes with a low structural stability are more prone to the first mechanism (Servagent-Noinville *et al.*, 2000), whereas enzymes with a high structural stability are more prone to the second (Baron *et al.*, 1999). It has also been shown, either from experimental observations or from theoretical reasoning, that a surface pH effect cannot explain a shift of the optimal enzyme activity pH when adsorbed on clays (Quiquampoix, 1987a).

In addition to electrostatic forces, hydrophobic interactions are also implied in the interaction of proteins with soil constituents and this results in an interplay between different driving forces in adsorption. For example, hydrophobic interactions with clays can result from an electrostatic exchange of the hydrophilic counter-ions on the clay surface, leaving a hydrophobic siloxane surface (Staunton and Quiquampoix, 1994). The rearrangement of the enzyme structure on the surface can be facilitated when hydrophobic amino acids come into contact with the clay hydrophobic siloxane layer and remain shielded from the water molecules of the solution. If this structural modification is accompanied by a decrease in ordered secondary structures, it will result in an additional increase in conformational entropy. This will lower the Gibbs energy of the system. The combination of all these different sub-processes gives rise to an irreversibility of the modification of conformation of enzymes on clay surfaces.

Phosphohydrolase adsorption and activity in soil

There have been numerous studies on the natural phosphohydrolase activity in soils (Tabatabai and Bremner, 1971; Cervelli *et al.*, 1973; Brams and McLaren, 1974; Pettit *et al.*, 1977; Batistic *et al.*, 1980; Nannipieri *et al.*, 1982, 1988; Harrison, 1983; Malcolm, 1983; Trasar-Cepeda and Gil-Sotres, 1988). Most of these studies dealt with the kinetic constants of phosphatase activity in soils, but it is difficult to draw conclusions from these studies on the effect of interaction with the soil solid phase on phosphatase activity, since it is not possible to separately determine the K_m and V_{max} of the free and adsorbed phosphatases under natural conditions. When model phosphatase–clay systems are studied, allowing the comparison of free and adsorbed phosphatases (Dick and Tabatabai, 1987), the results are diverse, with a global loss of activity due either to a decrease in V_{max} or an increase in K_m. This underlines the strength of the interaction of phosphatases with soil constituents. A study of model systems (Pant and Warman, 2000) showed a decrease of K_m, but the pH of reference, which is in this case the optimum pH, is different for free and adsorbed phosphohydrolases. This complicates any discussion of the effect of adsorption on the kinetic constant, since both the pH effect and the adsorption effect must be considered.

Only a limited number of studies have been devoted to the interaction of phosphohydrolases with soil constituents with the aim of understanding the physicochemical basis of the perturbation in the behaviour of these enzymes when they encounter the solid phase of soil. We emphasize these studies here, since they are better suited to elucidating the mechanisms by which abiotic factors influence the activity of extracellular phosphohydrolases in soil.

Physicochemical interactions of phosphohydrolases with clay surfaces

Due to the number of intermolecular forces and entropic effects that can be implied in the interaction of phosphohydrolases with soil constituents, studies on homogeneous and simplified systems are best suited for understanding the basic phenomena. Montmorillonite is a clay mineral with well-defined surface properties. It is a 2:1 phyllosilicate, in which each clay platelet consists of one layer of octahedral alumina between two layers of tetrahedral silica. It has a large specific surface area (800 m^2/g) mainly represented by basal surfaces whose electrical charge originates from isomorphic substitutions, mainly of aluminium ions by magnesium ions, in the octahedral layer. Considering an average montmorillonite platelet of 200×200×1 nm, the surface area of the edges represents <1% of the total surface area. Thus, the pH-dependent electrical charge due to the dissociation of silica and aluminium hydroxyls from edge surfaces are minimized, which means that only the pH-independent electronegative basal surface requires consideration in discussion of the role of the electrostatic forces in the interaction of phosphohydrolases with this clay surface.

Leprince and Quiquampoix (1996) studied the interaction between montmorillonite and two purified acid phosphatases, P1 and P2, secreted by the ectomycorrhizal fungus *Hebeloma cylindrosporum*. The phosphatase P1 had a molecular weight of 17,000 and an isoelectric point of 6.6, whilst the phosphatase P2, had a molecular weight of 51,000 and an isoelectric point of 7.1. The main results of this work were:

- There was an alkaline shift in the optimal pH of the catalytic activity of both P1 and P2 phosphatases adsorbed on montmorillonite. This effect is often observed when an enzyme interacts with an electronegative surface. It is classically interpreted as a micro-environmental effect caused by the influence of the lower pH at the surface of the electronegative clay due to the Gouy and Chapman diffuse double layer, despite some strong arguments against this theory based on thermodynamic concepts. However, even with a large range of pH (2–9) and ionic strength (20–500 mM) of the solutions used in this study, it was impossible to find conditions where the adsorbed phosphatases had a higher catalytic activity than the free phosphatases in solution on the alkaline side of the pH domain. This observation contradicts the interfacial pH explanation, since the loss of catalytic activity on the acid side is not compensated for by an increase on the alkaline side, and since the alkaline shift was insensitive to ionic strength modifications.
- The effect of the interaction of these acid phosphatases with the clay surfaces is better explained by a pH-dependent modification of the enzymes. This can explain the irreversible effects that are observed when the pH of adsorption is different from the pH at which the phosphohydrolase activity is measured.
- Both adsorption and modification of conformation of P1 and P2 were mainly determined by electrostatic forces. At high pH, electrostatic interactions between the negatively charged phosphohydrolases and the negatively charged clay surfaces are repulsive. This prevents adsorption and the phosphohydrolases are free to diffuse in the water-filled fraction of the soil pore network.
- The pK_a (near 4.5) of amino acids with a carboxyl group on the side chain (aspartic and glutamic acids) gives a better threshold value between stability and unfolding domains of these two enzymes than their respective isoelectric points.
- A modification of the liquid phase from a low pH or ionic strength to higher values induces an irreversible effect on adsorption and modification of conformation of these phosphohydrolases.

Effect of more complex soil constituents

As explained above, pure clay minerals, such as montmorillonite, have a strong effect on the modification of conformation of enzymes. It is thus important to study complex forms of clays ('dirty' clays), which are more representative of natural soil conditions. Rao *et al.* (1996, 2000) studied the effect of aluminium and iron oxide coatings on montmorillonite on the activity of an acid phosphatase. Oxide coatings increased the residual activity of the adsorbed phosphatase as compared to that on pure montmorillonite, which can be explained by a protective effect, since pure aluminium hydrous oxides have the lowest inhibitory influence on this phosphatase. The presence of organic matter also has a protective effect (Quiquampoix *et al.*, 1995; Rao *et al.*, 1996,

2000), a finding that is similar to those of a study on β-glucosidases (Quiquampoix, 1987b). Finally, the large amount of phosphohydrolases from diverse biological sources has a 'buffer' effect, which in natural conditions leads to a less marked effect of pH on inhibition by surfaces than with purified phosphatases (Chapuis-Lardy *et al.*, 2001).

Rhizosphere and Mycorrhizal Aspects of Phosphohydrolase Activity

Controversial interpretations

The phosphorus nutrition of plants is favoured by phosphorus mobilization mechanisms taking place in the rhizosphere. Among these mechanisms, the role of mycorrhizas is of prime importance (Mousain *et al.*, 1997). There are nevertheless some controversial interpretations of the nature of these positive effects regarding the role of phosphohydrolases (also see Richardson *et al.*, Chapter 8, this volume).

Several authors have suggested that the positive effect of mycorrhizas and other rhizosphere microorganisms on plant phosphorus nutrition is related to their phosphohydrolase activity (Doumas *et al.*, 1986; Mousain and Salsac, 1986; Tarafdar and Jungk, 1987; Häussling and Marschner, 1989; MacFall *et al.*, 1991; Antibus *et al.*, 1992; Fox and Comerford, 1992; Tarafdar and Marschner, 1994; Koide and Zabir, 2000; Chen *et al.*, 2002; Feng *et al.*, 2003). For other authors, the presence of phosphohydrolase activity appeared to be of limited benefit to plants (Martin, 1973; Joner and Jakobsen, 1994, 1995a,b; Joner *et al.*, 1995, 2000; Joner and Johansen, 2000; Fransson *et al.*, 2003). An additional group of authors have expressed a more balanced interpretation (e.g. George *et al.*, 2002a, b).

Experiments indicating a central role for phosphohydrolases in the rhizosphere of plants

The experiments can be classified in two groups: (i) a less conclusive set of experiments showing that phosphohydrolases are secreted or liberated in the rhizosphere of plants and can hydrolyse organic phosphorus compounds that may be present in soil, but not based on an integrated approach on what really happens in the rhizosphere (Doumas *et al.*, 1986; Mousain and Salsac, 1986; Häussling and Marschner, 1989; MacFall *et al.*, 1991; Antibus *et al.*, 1992; Fox and Comerford, 1992; Koide and Zabir, 2000; Feng *et al.*, 2003) and (ii) strongly conclusive experiments that, using devices that allow fine spatial determination in the rhizosphere of the relationship between phosphohydrolase activity and organic phosphorus concentration, show an inverse correlation between these two parameters (Tarafdar and Jungk, 1987; Tarafdar and Marschner, 1994; Chen *et al.*, 2002; George *et al.*, 2002b).

In the latter group, Tarafdar and Jungk (1987) devised one of the earlier experimental studies showing unequivocally the essential role of the phosphohydrolases of microorganisms in the rhizosphere in the phosphorus nutrition of plants. Onion, rape, wheat and clover were studied on three types of soil cultivated in pots where a fine-meshed screen separated the roots from the soil, allowing only root hairs to penetrate. The phosphohydrolase activity and the microbial populations with respect to the distance to the root mat were determined in thin slices (approximately 200 μm thick) obtained with a microtome. For all plants and soils studied, the rhizosphere contained increased levels of phosphohydrolase activity compared with control bulk soil without plants. Acid phosphatase activity was enhanced up to 3 mm from the boundary screen and alkaline phosphatase up to 1.5 mm. Bacterial and fungal populations also increased in this rhizospheric volume of soil. However, a more important observation was that organic phosphorus was significantly depleted in the same distance, resulting in a strong correlation between the depletion of organic phosphorus and the phosphohydrolase activity ($r=0.99$ for wheat and $r=0.97$ for clover; $P<0.01$). These results clearly demonstrate the relationship between phosphohydrolase activity in the rhizosphere and the use of organic phosphorus by plants. The same type of correlation between organic phosphorus depletion and enhanced phosphohydrolase activity in the rhizosphere was obtained on ryegrass and *Pinus radiata* (Chen *et al.*, 2002) and on the agroforestry species *Tithonia diversifolia*, *Tephrosia vogelii* and *Crotalaria grahamiana* (George *et al.*, 2002b).

Inconclusive experiments on a central role of phosphohydrolases in the rhizosphere of plants

Joner and collaborators (Joner and Jakobsen, 1994, 1995a,b; Joner *et al.*, 1995, 2000; Joner and Johansen, 2000) also performed experiments using a compartmental separation between plant roots and their associated mycorrhizal hyphae. Their studies on cucumber and subterranean clover clearly showed the positive effect of endomycorrhizal fungi on phosphorus nutrition of the plants, but also showed that: (i) arbuscular mycorrhizal *Glomus* spp. (*G. invermaium*, *G. caledonium*, *G. intraradices* and *G. claroideum*) apparently did not excrete extracellular phosphatases, and (ii) phosphatase activity was not correlated with the concentration of organic phosphorus in soil extracts. By comparing a mycorrhizal plant (*Plantago lanceolata*) and a non-mycorrhizal plant (*Rumex acetosella*), Fransson *et al.* (2003) concluded that, for both plants, soil phosphatase activity decreased relative to the control soil without plants, with no correlation of the remaining activity with any extractable phosphorus fraction. George *et al.* (2002b) demonstrated the importance of phosphohydrolase activity from agroforestry species (see above), but found no such relationship for maize, since the acid phosphatase activity enhancement was low in comparison to agroforestry species, and in fact an increased concentration of organic phosphorus was observed in the rhizosphere of maize (George *et al.*, 2002a).

Possible explanations for the discrepancies

Joner and collaborators proposed a set of explanations for the discrepancies between their results and those favouring an important role for phosphohydrolases:

- The phosphatase activity recorded in fine slices of rhizospheric soil could be an artefact of the liberation of intracellular enzymes after microtome cutting (Joner *et al.*, 1995).
- In the experiments where added organic phosphorus substrates are present, their low adsorption on soil constituents makes the comparison with natural organic phosphorus compounds unrealistic (Findenegg and Nelemans, 1993; Joner *et al.*, 2000).
- The classical method of measuring phosphatase activity by the hydrolysis of *para*-nitrophenyl phosphate could be irrelevant for estimating the potential hydrolysis of natural organic phosphorus compounds such as *myo*-inositol hexakisphosphate, since the relative activity of phytase compared to *para*-nitrophenyl phosphatase activity may be several orders of magnitude lower (Beck *et al.*, 1989; Joner *et al.*, 2000; Tibbett, 2002).
- The liberation of phosphate from organic phosphorus for plant nutrition could simply originate from the large pool of the abundant and ubiquitous immobilized phosphatases on soil constituents (Joner *et al.*, 2000; Fransson *et al.*, 2003).

However, if we exclude the possible influence of microtome cutting on intracellular phosphatase liberation in rhizospheric soil thin slices (which could be demonstrated by analysing other intracellular markers on the same soil slices), the other explanations cannot always explain the lack of apparent involvement of phosphohydrolase activity that is sometimes observed. Indeed, the discrepancies may alternatively be logically explained by the natural diversity of phosphatase properties, making some plant–microorganism associations more efficient in organic phosphorus hydrolysis in a given soil environment than others. Among the phosphatase properties that could explain this variability, the most important are their location (bound to cell walls or freely diffusible in the soil) and their structural stability when adsorbed on soil constituents. The latter affects their relative catalytic activity and the adsorption process itself, thus having an important effect on the diffusivity of phosphohydrolases in soil, and thus on the possibility of reaching insoluble forms of organic phosphorus.

Diversity of phosphohydrolases from mycorrhizal fungi

Matumoto-Pintro (1996) studied the location of phytase and *para*-nitrophenyl phosphatase activity in ten strains of Basidiomycetes (*Laccaria laccata*, *Suillus collinitus* (three strains), *Suillus granulatus*, *Suillus luteus*, *Hebeloma cylindrosporum* (two strains), *Paxillus involutus*, *Rhizopogon rubescens*) after culture of the mycelia in a low-phosphate medium. Four isolates had free extracellular phytase activity (one strain of *Suillus collinitus*, *Suillus granulatus*, and both strains of *Hebeloma cylindrosporum*), two strains had no phytase activity at all (*Suillus luteus* and *Rhizopogon rubescens*), while six had free extracellular *para*-nitrophenyl phosphatase activity (one strain of *Suillus collinitus*, *S. granulatus*, *S. luteus*, both strains of *Hebeloma cylindrosporum* and *Rhizopogon rubescens*). *Hebeloma cylindrosporum* strains were exceptional for the high level of phosphohydrases (phytases as well as acid phosphatases) they secreted into the external medium. Since their enzymes also have a relatively high structural stability, they retain a high catalytic activity when adsorbed on soil clay minerals (Leprince and Quiquampoix, 1996). All these characteristics make *H. cylindrosporum* a potentially efficient ectomycorrhizal fungus for soil organic phosphorus hydrolysis. This behaviour was experimentally demonstrated for the association between *H. cylindrosporum* and the host plant *Pinus nigra* subsp. *laricio* var. *corsicana*. Contrary to two other mycorrhizal associations (with *Suillus luteus* and *S. collinitus*), *H. cylindrosporum* allowed its host plant a significantly higher phosphorus uptake (Matumoto-Pintro, 1996). This shows that some knowledge of the molecular properties of the phosphohydrolases of a given species can allow prediction of the use of soil organic phosphorus for plant nutrition.

A study of the adsorption of several acid phosphatases of ectomycorrhizal fungi on montmorillonite, and its consequences on their catalytic activity, revealed large interspecific as well as intraspecific variability in the properties of the enzymes (J. Abadie and H. Quiquampoix, unpublished results). For example, one strain of *Pisolithus tinctorius* showed no adsorption of its phosphatases on montmorillonite between pH 2 and 8. This is a favourable property, allowing both a large diffusion in soil and a non-perturbed catalytic activity in the soil environment. However, phosphatases of other strains, including one of *Cenococcum geophilum* and one of *Hebeloma cylindrosporum*, behave differently, with adsorption varying between pH 4 and 6, but the adsorbed fraction being completely inhibited and only the free fraction being active. One strain of *Suillus bellini* showed important variation in the adsorption of its phosphatases, but with no effect on the activity of the adsorbed enzymes that have the same catalytic activity as those in solution. Phosphatases from one strain of *Suillus mediterraneensis* adsorbed completely across the whole range of pH, but again without perturbation of their catalytic activity. A strain of *Rhizopogon rubescens* and a strain of *Hebeloma cylindrosporum* have phosphatases with a more complex behaviour: both adsorption properties and catalytic activity in the adsorbed state vary with pH.

These results show that the phosphorus nutrition of a host plant can be heavily dependent on the molecular and physicochemical properties of the extracellular phosphohydrolases of its associated mycorrhizal fungi. This relationship could perhaps be generalized to other rhizospheric microorganisms, but experimental studies are still lacking.

Evolutionary trends

The diversity of physicochemical properties of phosphohydrolases from ectomycorrhizal fungi relative to adsorption on soil constituents makes it difficult to generalize about their ecological role. Only the introduction of other parameters can provide clues. Matumoto-Pintro and Quiquampoix (1997) selected results on the interaction of phytases with clays for which the location of these enzymes in natural conditions can be assessed unequivocally. Two phytases (from wheat and *Suillus collinitus*) are not naturally secreted in soil, being either intracellular (wheat) or cell-wall-bound (*Suillus collinitus*). The adsorption of these enzymes on clay minerals was complete over the large pH range studied and was followed by a complete inhibition of their catalytic activity. In contrast, two other phytases that are naturally secreted in soil (from the saprophytic fungus *Aspergillus ficuum* and the ectomycorrhizal fungus *Hebeloma cylindrosporum*) retain a significant catalytic activity in the presence of clays. This observation suggests that soil solid surfaces have exerted a selection pressure by allowing secreted enzymes to retain higher structural stability when adsorbed on soil constituents.

Regulation of the Phosphohydrolase Activity in the Rhizosphere

Regulation of the expression of phosphohydrolases by phosphate concentration in soil: induction or repression?

The secretion of phosphohydrolases by plants or microorganisms has an important energetic cost for their whole metabolism. It is thus better for these organisms to have evolved a mechanism of regulation of the expression of these enzymes based on the real benefit of phosphate molecules taken up per phosphohydrolase molecule liberated. The classical methods of regulation of enzyme expression (induction or repression) will be discussed and compared with experimental results.

The Burns hypothesis: induction by low levels of enzymatic products

An interesting hypothesis on the possible role of extracellular enzymes that are immobilized on mineral surfaces or on humic substances was proposed by Burns (1982). Since most of the substrates available in soil for the nutrition of organisms are either adsorbed or insoluble, the presence of adsorbed enzymes in their vicinity could be a valuable way to signal their presence by a small but continuous liberation of assimilable products. These soluble products could then induce the synthesis of new hydrolytic extracellular enzymes by plants or microorganisms. This hypothesis, which represents an 'economical' way of managing organic phosphorus for soil organisms (as inducible rather than constitutive phosphohydrolases, i.e. the cost of synthesis of extracellular enzymes is limited to situations where organic phosphorus is available in the vicinity), has never been studied in representative soil conditions, despite the existence of simplified experimental approaches well adapted to overcoming the technical difficulties (Hope and Burns, 1985; Vetter and Deming, 1999). Nevertheless, as it will be shown below, this hypothesis seems not to be relevant to the regulation of the secretion of phosphohydrolases, at least at first sight.

Repression of the expression of extracellular phosphohydrolase synthesis by high levels of phosphate

All the studies devoted to the regulation of secreted phosphatases point to a mechanism of repression by high phosphate concentrations, whatever the organisms being considered: plants, fungi or bacteria (Shieh *et al.*, 1969; Ferminan and Dominguez, 1998; Aleksieva and Micheva-Viteva, 2000; Antelmann *et al.*, 2000; Haran *et al.*, 2000). There are nevertheless some exceptions, such as an acid phosphatase from the bacterium *Morganella morganii*, which is not repressed by high levels of phosphate (Thaller *et al.*, 1994). The molecular mechanism of phosphate repression has been fully characterized in more recent studies. For example, the *Arabidopsis* acid phosphatase promoter region was cloned and fused to the GUS and GFP reporter genes (Haran *et al.*, 2000). In roots starved of phosphate, the expression of GUS was initiated in the meristems of lateral roots. Later, the expression was generalized throughout the root. The addition of phosphate decreased the expression. In the GFP reporter gene experiment, the expression was also initiated in lateral root meristems in low-phosphorus conditions and the recombinant GFP with the acid phosphatase signal peptide was secreted in the external medium.

Why regulation of phosphohydrolase expression does not follow the Burns hypothesis

Competition of phosphate produced by inorganic phosphorus mobilization processes

A first possible explanation for the lack of observed inductive effect of phosphate on secreted phosphohydrolases is based on the peculiarity of the biogeochemical phosphorus cycle. In contrast to the soil carbon cycle, in which the assimilable carbon compartment is in organic form and, therefore, almost completely dependent on enzymes for its hydrolysis, and for which the Burns hypothesis is relevant, phosphorus in soil is distributed in two compartments, organic and inorganic phosphorus, which are of more or less equal importance. Thus, the repression of phosphohydrolase expression by increasing the concentration of phosphate could be simply the consequence of a more efficient mobilization of phosphate by the excretion of organic acids and protons than the liberation of phosphate by enzymatic hydrolysis of organic phosphorus. In this interpretation, plants and microorganisms have a better strategy if they prioritize phosphate, and consequently repress the synthesis of extracellular phosphohydrolases resources, in conditions of sufficient phosphate concentration in the soil solution.

The effect of low phosphate concentrations has not been studied

The above explanation has a logical basis, but it is unclear whether it is supported by the available data. Most studies on the regulation of the expression of extracellular phosphohydrolases are based on phosphorus-starving media that contain between 10 and 100 μM phosphate, or even more (Ferminan and Dominguez, 1998; Aleksieva and Micheva-Viteva, 2000; Haran *et al.*, 2000). This threshold experimental value is justifiable, since the lowest values in a batch experiment in the presence of actively phosphorus-consuming organisms would result in uncertainties in the evolution of the phosphate concentration during the time course of the experiment. However, the concentration of phosphate that can be expected in most terrestrial and aquatic ecosystems is nearer 1 μM. Buffer systems based on resins could eventually provide a phosphate-releasing device to maintain a constant phosphate concentration for such studies, but to our knowledge they have not yet been used to study the regulation of phosphatases in the low range of phosphate.

Studies of the regulation of phosphohydrolases in the presence of complex phosphorus forms and without phosphate

Investigation by Aleksieva and Micheva-Viteva (2000) into the regulation of extracellular acid phosphatase biosynthesis by the fungus *Humicola lutea* suggested that the Burns hypothesis may be applicable to the enzymatic hydrolysis of organic phosphorus. The phosphorus source in this study was a phosphoprotein (casein). Phosphohydrolase secretion increased in an experiment where the concentration of casein was increased up to 4 g/l in the total absence of added phosphate, demonstrating an induction of expression, contrary to the repression effect in the presence of increasing phosphate. A few enzymes liberated from senescing fungal cells may have played a similar role to the soil-adsorbed enzymes of the Burns hypothesis, allowing the sub-micromolar concentration of liberated phosphate to trigger an activation of the casein-hydrolysing extracellular phosphatases.

Effect of phosphate concentration on the location of phosphohydrolases

The increased quantity of extracellular phosphohydrolases synthesized following derepression by low concentrations of phosphate in the external medium are not the only aspect that we need to take into account in order to appreciate the ecological role of these enzymes in phosphorus nutrition. There is also a qualitative aspect in the distribution of the phosphohydrolases. Tibbett and collaborators (Tibbett *et al.*, 1998; Tibbett, 2002) showed on 12 different strains of the fungus *Hebeloma* that a decrease from 6 mM to 20 μM phosphate resulted in a dramatic change in the ratio of free extracellular *para*-nitrophenyl phosphatase activity to activity associated with the cell wall. At high phosphate concentrations, the cell-bound phosphatase fraction exceeded the free extracellular fraction in nine of the 12 strains. However, at low phosphate concentrations, the free extracellular fraction was higher than the cell-bound fraction for 11 of the 12 strains. This phenomenon may be interpreted as an adaptation to access sparingly soluble sources of soil organic phosphorus under conditions of phosphorus starvation.

Concluding Remarks and Perspectives

Enzymatic hydrolysis of organic phosphorus is an essential step in the biogeochemical phosphorus cycle, including the phosphorus nutrition of plants and microorganisms (see Oberson and Joner, Chapter 7; Richardson *et al.*, Chapter 8; Heath, Chapter 9, this volume) and the transfer of organic phosphorus from soils to water bodies (see Turner, Chapter 12, this volume). It also plays a role in gene fluxes in the environment by its effect on the residual extracellular DNA in soil, and in the degradation of organophosphate pesticides in the environment. The regulation of phosphohydrolase secretion by the availability of phosphate and organic phosphorus in soil is insufficiently studied. The physicochemical aspects of the interaction of phosphohydrolases with soil constituents, in particular the consequences on their proteic structure and diffusion properties, also warrants further investigation. Such studies would help to design, on a rational basis, genetically modified plants or root symbiotic microorganisms with improved phosphatase activity in the rhizosphere (see Richardson *et al.*, Chapter 8, this volume). Finally, a better integration of the biological mechanisms responsible for phosphorus mobilization in the rhizosphere could lead to a re-definition of phosphorus fractions (P_{enz}, P_{diss} and P_{memb}) that are more suited to modelling the fate of phosphorus in ecosystems than the classical organic and inorganic phosphorus compartments. Solution and solid-state nuclear magnetic resonance spectroscopy is well suited for the identification of the P_{enz} and P_{memb} compartments, and the P_{diss} compartment, respectively (see Cade-Menun, Chapter 2, this volume).

References

Aaronson, S. and Patni, N.J. (1976) The role of surface and extracellular phosphatases in the phosphorus requirement of *Ochromonas*. *Limnology and Oceanography* 21, 838–845.

Aleksieva, P. and Micheva-Viteva, S. (2000) Regulation of extracellular acid phosphatase biosynthesis by phosphates in proteinase producing fungus *Humicola lutea* 120-5. *Enzyme and Microbial Technology* 27, 570–575.

Anderson, G. (1961) Estimation of purines and pyrimidines in soil humic acid. *Soil Science* 91, 156–161.

Anderson, G. (1963) Effect of iron/phosphorus ratio and acid concentration on the precipitation of ferric inositol hexaphosphate. *Journal of the Science of Food and Agriculture* 14, 352–359.

Anderson, G. (1967) Nucleic acids, derivatives, and organic phosphates. In: McLaren, A.D. and Peterson, G.H. (eds) *Soil Biochemistry*, Vol. 1. Marcel Dekker, New York, pp. 67–90.

Anderson, G. (1970) The isolation of nucleoside

diphosphates from alkaline extracts of soil. *Journal of Soil Science* 21, 96–104.

Anderson, G. and Arlidge, E.Z. (1962) The adsorption of inositol phosphates and glycerophosphate by soil clays, clay minerals and hydrated sesquioxides in acid media. *Journal of Soil Science* 13, 216–224.

Anderson, G. and Malcom, R.E. (1974) The nature of alkali-soluble soil organic phosphates. *Journal of Soil Science* 25, 282–297.

Antelmann, H., Scharf, C. and Hecker, M. (2000) Phosphate starvation-inducible proteins of *Bacillus subtilis*: proteomics and transcriptional analysis. *Journal of Bacteriology* 182, 4478–4490.

Antibus, R.K., Sinsabaugh, R.L. and Linkins, A.E. (1992) Phosphatase activities and phosphorus uptake from inositol phosphate by ectomycorrhizal fungi. *Canadian Journal of Botany* 70, 794–801.

Antibus, R.K., Bower, D. and Dighton, J. (1997) Root surface phosphatase activities and uptake of ^{32}P-labelled inositol phosphate in field-collected gray birch and red maple roots. *Mycorrhiza* 7, 39–46.

Baron, M.H., Revault, M., Servagent-Noinville, S., Abadie, J. and Quiquampoix, H. (1999) Chymotrypsin adsorption on montmorillonite: enzymatic activity and kinetics FTIR structural analysis. *Journal of Colloid and Interface Science* 214, 319–332.

Batistic, L., Sarkar, J.M. and Mayaudon, J. (1980) Extraction, purification and properties of soil hydrolases. *Soil Biology and Biochemistry* 12, 59–63.

Beck, E., Fusseder, A. and Kraus, M. (1989) The maize root system *in situ*: evaluation of structure and capability of utilization of phytate and inorganic soil phosphates. *Zeitschrift für Pflanzenernärhrung und Bodenkunde* 152, 159–167.

Berjaud, C. and d'Auzac, J. (1986) Isolement et caractérisation des phosphatases d'un champignon ectomycorhizogène typique, *Pisolithus tinctorius*. Effets de la carence en phosphate. *Physiologie Végétale* 24, 163–172.

Bhat, K.K.S. and Nye, P.H. (1973) Diffusion of phosphate to plant roots. I. Quantitative autoradiography of the depletion zone. *Plant and Soil* 38, 161–175.

Bieleski, R.L. and Ferguson, I.B. (1983) Physiology and metabolism of phosphate and its compounds. In: Läuchli, A. and Bieleski, R.L. (eds) *Inorganic Plant Nutrition*. Springer, Berlin, pp. 422–449.

Borie, F. and Barea, J.M. (1981) Ciclo del fosforo. I. Formas del elemento en los suelos y su disponibilidad para plantas y microorganismos. *Annales de Edafologia y Agrobiologia* 40, 2351–2364.

Brams, W.H. and McLaren, A.D. (1974) Phosphatase reactions in soil columns. *Soil Biology and Biochemistry* 6, 183–189.

Burns, R.G. (1982) Enzyme activity in soil: location and a possible role in microbial ecology. *Soil Biology and Biochemistry* 14, 423–427.

Campbell, M., Dunn, R., Ditterline, R., Pickett, S. and Raboy, V. (1991) Phytic acid represents 10 to 15% of total phosphorus in alfalfa root and crown. *Journal of Plant Nutrition* 14, 925–937.

Cervelli, S., Nannipieri, P., Ceccanti, B. and Sequi, P. (1973) Michaelis constant of soil acid phosphatase. *Soil Biology and Biochemistry* 5, 841–845.

Chapuis-Lardy, L., Brossard, M. and Quiquampoix, H. (2001) Assessing organic phosphorus status of Cerrado oxisols (Brazil) using ^{31}P-NMR spectroscopy and phosphomonoesterase activity measurement. *Canadian Journal of Soil Science* 81, 591–601.

Chassin, P., Jouany, C. and Quiquampoix, H. (1986) Measurement of the surface free energy of calcium-montmorillonite. *Clay Minerals* 21, 899–907.

Chen, C.R., Condron, L.M., Davis, M.R. and Sherlock, R.R. (2002) Phosphorus dynamics in the rhizosphere of perennial ryegrass (*Lolium perenne* L.) and radiata pine (*Pinus radiata* D. Don.). *Soil Biology and Biochemistry* 34, 487–499.

Colpaert, J.V., Van Laere, A., Van Tichelen, K.K. and Van Assche, J.A. (1997) The use of inositol hexaphosphate as a phosphorus source by mycorrhizal and non-mycorrhizal Scots pine (*Pinus sylvestris*). *Functional Ecology* 11, 407–415.

Cosgrove, D.J. (1980) *Inositol Phosphates: Their Chemistry, Biochemistry and Physiology*. Elsevier Science, Amsterdam, 197 pp.

Cumming, J.R. (1993) Growth and nutrition of non-mycorrhizal and mycorrhizal pitch pine (*Pinus rigida*) seedlings under phosphorus limitation. *Tree Physiology* 13, 173–187.

Dalal, R.C. (1977) Soil organic phosphorus. *Advances in Agronomy* 29, 83–117.

Demanèche, S., Jocteur-Monrozier, L., Quiquampoix, H. and Simonet, P. (2001) Evaluation of biological and physical protection against nuclease degradation of clay bound plasmid DNA. *Applied and Environmental Microbiology* 67, 293–299.

Dick, W.A. and Tabatabai, M.A. (1987) Kinetics and activities of phosphatase–clay complexes. *Soil Science* 143, 5–15.

Dighton, J. (1983) Phosphatase production by mycorrhizal fungi. *Plant and Soil* 71, 455–462.

Doumas, P., Coupé, M. and d'Auzac, J. (1983) Effet

de la carence en phosphate sur les activités des phosphatases racinaires du Pin d'Alep. *Physiologie Végétale* 21, 651–663.

Doumas, P., Berjaud, C., Calléja, M., Coupé, M., Espiau, C. and d'Auzac, J. (1986) Phosphatases extracellulaires et nutrition phosphatée chez les champignons ectomycorhiziens et les plantes hôtes. *Physiologie Végétale* 24, 173–184.

Feng, G., Song, Y.C., Li, X.L. and Christie, P. (2003) Contribution of arbuscular mycorrhizal fungi to utilization of organic sources of phosphorus by red clover in a calcareous soil. *Applied Soil Ecology* 22, 139–148.

Ferminan, E. and Dominguez, A. (1998) Heterologous protein secretion directed by a repressible acid phosphatase system of *Kluyveromyces lactis*: characterization of upstream region-activating sequences in the *KIPHO5* gene. *Applied and Environmental Microbiology* 64, 2403–2408.

Findenegg, G.R. and Nelemans, J.A. (1993) The effect of phytase on the availability of P from myoinositol hexaphosphate (phytate) for maize roots. *Plant and Soil* 154, 189–196.

Fox, T.R. and Comerford, N.B. (1992) Rhizosphere phosphatase activity and phosphatase hydrolyzable organic phosphorus in two forested spodosols. *Soil Biology and Biochemistry* 24, 579–583.

Fransson, A.M., van Aarle, I.M., Olsson, P.A. and Tyler, G. (2003) *Plantago lanceolata* L. and *Rumex acetosella* L. differ in their utilisation of soil phosphorus fractions. *Plant and Soil* 248, 285–295.

Garen, A. and Levinthal, C. (1960). A fine-structure genetic and chemical study of the enzyme alkaline phosphatase of *E. coli*. 1. Purification and characterization of alkaline phosphatase. *Biochimica et Biophysica Acta* 38, 470–483.

George, T.S., Gregory, P.J., Robinson, J.S. and Buresh, R.J. (2002a) Changes in phosphorus concentrations and pH in the rhizosphere of some agroforestry and crop species. *Plant and Soil* 246, 65–73.

George, T.S., Gregory, P.J., Wood, M., Read, D. and Buresh, R.J. (2002b) Phosphatase activity and organic acids in the rhizosphere of potential agroforestry species and maize. *Soil Biology and Biochemistry* 34, 1487–1494.

Gibbins, L.N. and Norris, F.W. (1963) Phytase and acid phosphatase in the dwarf bean, *Phaseolus vulgaris*. *Biochemical Journal* 86, 67–71.

Greenwood, A.J. and Lewis, D.H. (1977) Phosphatases and the utilisation of inositol hexaphosphate by soil yeasts of the genus *Cryptococcus*. *Soil Biology and Biochemistry* 9, 161–166.

Greiner, R., Farouk, A., Alminger, M.L. and Carlsson, N.-G. (2002) The pathway of dephosphorylation of *myo*-inositol hexakisphosphate by phytate-degrading enzymes of different *Bacillus* spp. *Canadian Journal of Microbiology* 48, 986–994.

Haran, S., Logendra, S., Seskar, M., Bratanova, M. and Raskin, I. (2000) Characterization of *Arabidopsis* acid phosphatase promoter and regulation of acid phosphatase expression. *Plant Physiology* 124, 615–626.

Harland, B.F. and Morris, E.R. (1995) Phytate: a good or bad food component? *Nutrition Research* 15, 733–754.

Harrison, A.F. (1983) Relationship between intensity of phosphatase activity and physical–chemical properties of woodland soils. *Soil Biology and Biochemistry* 15, 93–99.

Häussling, M. and Marschner, H. (1989) Organic and inorganic soil phosphates and acid phosphatase activity in the rhizosphere of 80-year-old Norway spruce. *Biology and Fertility of Soils* 8, 125–133.

Haynes, C.A. and Norde, W. (1994) Globular proteins at solid/liquid interfaces. *Colloids and Surfaces B: Biointerfaces* 2, 517–566.

Hedley, M.J. and Stewart, J.W.B. (1982) A method to measure microbial phosphorus in soils. *Soil Biology and Biochemistry* 14, 377–385.

Hedley, M.J., Stewart, J.W.B. and Chauhan, B.S. (1982) Changes in inorganic and organic soil phosphorus fractions induced by cultivation practices and by laboratory incubations. *Soil Science Society of America Journal* 46, 970–976.

Hope, C.F.A. and Burns, R.G. (1985) The barrier-ring plate technique for studying extracellular enzyme diffusion and microbial growth in model soil environments. *Journal of General Microbiology* 131, 1237–1243.

Irving, G.C.J. and Cosgrove, D.J. (1971) Inositol phosphate phosphatases of microbiological origin: observations on the nature of this active centre of a bacterial (*Pseudomonas* sp.) phytase. *Australian Journal of Biological Science* 24, 559–564.

Jacob, J. and Lawlor, D.W. (1992) Dependence of photosynthesis of sunflower and maize leaves on phosphate supply, ribulose-1,5-bisphosphate pool size. *Plant Physiology* 98, 801–807.

Joner, E.J. and Jakobsen, I. (1994) Contribution by two arbuscular mycorrhizal fungi to P uptake by cucumber (*Cucumis sativus* L.) from ^{32}P-labelled organic matter during mineralization in soil. *Plant and Soil* 163, 203–209.

Joner, E.J. and Jakobsen, I. (1995a) Uptake of ^{32}P from labelled organic matter by mycorrhizal and non-mycorrhizal subterranean clover (*Trifolium subterraneum* L.). *Plant and Soil* 172, 221–227.

Joner, E.J. and Jakobsen, I. (1995b) Growth and extracellular phosphatase activity of arbuscular mycorrhizal hyphae as influenced by soil organic matter. *Soil Biology and Biochemistry* 27, 1153–1159.

Joner, E.J. and Johansen, A. (2000) Phosphatase activity of external hyphae of two arbuscular mycorrhizal fungi. *Mycological Research* 104, 81–86.

Joner, E.J., Magid, J., Gahoonia, T.S. and Jakobsen, I. (1995) P depletion and activity of phosphatase in the rhizosphere of mycorrhizal and non-mycorrhizal cucumber (*Cucumis sativus* L.). *Soil Biology and Biochemistry* 9, 1145–1151.

Joner, E.J., van Aarle, I.M. and Vosatka, M. (2000) Phosphatase activity of extra-radical arbuscular mycorrhizal hyphae: a review. *Plant and Soil* 226, 199–210.

Koide, R.T. and Kabir, Z. (2000) Extraradical hyphae of the mycorrhizal fungus *Glomus intraradices* can hydrolyse organic phosphate. *New Phytologist* 148, 511–517.

Komano, T. (1975) Formation of multiple forms of acid and alkaline phosphatases in relation to the culture age of *Aspergillus niger*. *Plant and Cell Physiology* 16, 643–658.

Kondo, A. and Higashitani, K. (1992) Adsorption of model proteins with wide variation in molecular properties on colloidal particles. *Journal of Colloid and Interface Science* 150, 344–351.

Kondo, A., Murakami, F. and Higashitani, K. (1992) Circular dichroism studies on conformational changes in protein molecules upon adsorption on ultrafine polystyrene particles. *Biotechnology and Bioengineering* 40, 889–894.

Kondo, A., Murakami, F., Kawagoe, M. and Higashitani, K. (1993) Kinetic and circular dichroism of enzymes adsorbed on ultrafine silica particles. *Applied Microbiology and Biotechnology* 39, 726–731.

Kowalenko, G.C. and Mc Kercher, R.B. (1971) Phospholipid P content of Saskatchewan soils. *Soil Biology and Biochemistry* 3, 243–247.

Laird, M.H., Allen, H.J., Danielli, J.F. and Winzler, R.J. (1976) The identification of phosphorylated *neo*-inositol as an anionic component of the *Amoeba* cell surface. *Archives of Biochemistry and Biophysics* 175, 384–391.

L'Annunziata, M.F. and Fuller, W.H. (1971) Soil and plant relationship of inositol phosphate stereoisomers; the identification of D-*chiro*- and plant system. *Soil Science Society of America Proceedings* 35, 587–595.

L'Annunziata, M.F., Fuller, W.H. and Brantley, D.S. (1972) D-*chiro*-inositol phosphate in a forest soil. *Soil Science Society of America Proceedings* 36, 183–184.

Lehninger, A.L. (1981) *Biochemistry: the Molecular Basis of Cell Structure and Function*, 2nd edn. Worth Publishers, New York, 833 pp.

Leprince, F. and Quiquampoix, H. (1996) Extracellular enzyme activity in soil: effect of pH and ionic strength on the interaction with montmorillonite of two acid phosphatases secreted by the ectomycorrhizal fungus *Hebeloma cylindrosporum*. *European Journal of Soil Science* 47, 511–522.

Lim, P.E. and Tate, M.E. (1973) The phytases. III. Properties of phytase fractions F_1 and F_2 from wheat bran and the *myo*-inositol phosphates produced by fraction F_2. *Biochimica et Biophysica Acta* 302, 316–328.

Lott, J.N.A. and Ockenden, I. (1986) The fine structure of phytate-rich particles in plants. In: Graf, E. (ed.) *Phytic Acid: Chemistry and Applications*. Pilatus Press, Minneapolis, pp. 43–55.

MacFall, J., Slack, S.A. and Iyer, J. (1991) Effects of *Hebeloma arenosa* and phosphorus fertility on root acid phosphatase activity of red pine (*Pinus resinosa*) seedlings. *Canadian Journal of Botany* 69, 380–383.

Malcolm, R.E. (1983) Assessment of phosphatase activity in soils. *Soil Biology and Biochemistry* 15, 403–408.

Mandal, N.C., Burman, S. and Biswas, B.B. (1972) Isolation, purification and characterization of phytase from germinating mung beans. *Phytochemistry* 11, 495–502.

Martin, J.K. (1973) The influence of rhizosphere microflora on the availability of ^{32}P-*myo*inositol hexaphosphate phosphorus to wheat. *Soil Biology and Biochemistry* 5, 473–483.

Matumoto-Pintro, P.T. (1996) Rôle des phosphatases dans l'utilisation du phosphore organique par les champignons mycorhiziens et leurs associations avec le Pin laricio de Corse. PhD thesis, Ecole Nationale Agronomique de Montpellier, Montpellier, France.

Matumoto-Pintro, P.T. and Quiquampoix, H. (1997) La phase solide des sols comme contrainte au fonctionnement des enzymes sécrétées par les microorganismes: comparaison de phytases intra et extracellulaires. In: Baleux, B., Desmazeaud, M., Divies, C., Gendre, F. and Moletta, R. (eds) *Microbiologie Industrielle et Environnement*. Société Française de Microbiologie, Paris, pp. 195–204.

Mitchell, D.T. and Read, D.J. (1981) Utilisation of inorganic and organic phosphates by the mycorrhizal endophytes of *Vaccinium macrocarpon* and *Rhododendron ponticum*. *Transactions of the British Mycological Society* 76, 255–260.

Mitchell, D.B., Vogel, K., Weimann, B., Pasamontes, L. and van Loon, A.P.G.M. (1997) The phytase subfamily of histidine acid phosphatase: isolation of genes for two novel phytases from the fungi *Aspergillus terreus* and *Myceliophthora thermophila*. *Microbiology* 143, 245–252.

Mousain, D. (1989) Etude de la nutrition phosphatée de symbiotes ectomycorhiziens. Thèse de Doctorat d'Etat, Université des Sciences et Techniques du Languedoc, Montpellier, France.

Mousain, D. and Salsac, L. (1986) Utilisation du phytate et activités phosphatases acides chez *Pisolithus tinctorius*, basidiomycète mycorhizien. *Physiologie Végétale* 24, 193–200.

Mousain, D., Matumoto-Pintro, P. and Quiquampoix, H. (1997) Le rôle des mycorhizes dans la nutrition phosphatée des arbres forestiers. *Revue Forestière Française* 49, 67–81.

Mullaney, E.J. and Ullah, A.H.J. (2003) The term phytase comprises several different classes of enzymes. *Biochemical and Biophysical Research Communications* 312, 179–184.

Mullaney, E.J., Daly, C.B., Kim, T., Porres, J.M., Lei, X.G., Sethumadhavan, K. and Ullah, A.H.J. (2002) Site-directed mutagenesis of *Aspergillus niger* NRRL 3135 phytase at residue 300 to enhance catalysis at pH 4.0. *Biochemical and Biophysical Research Communications* 297, 1016–1020.

Nagai, Y. and Funahashi, S. (1962) Phytase from wheat bran. I. Purification and substrate specificity. *Agricultural and Biological Chemistry* 26, 794–803.

Nannipieri, P., Ceccanti, B., Cervelli, S. and Conti, C. (1982) Hydrolases extracted from soil: kinetic parameters of several enzymes catalysing the same reaction. *Soil Biology and Biochemistry* 14, 429–432.

Nannipieri, P., Ceccanti, B. and Bianchi, D. (1988) Characterization of humus–phosphatase complexes extracted from soil. *Soil Biology and Biochemistry* 20, 683–691.

Nayini, N.R. and Markakis, P. (1984) The phytase of yeast. *Lebensmittel Wissenschaft und Technologie* 17, 24–26.

Nayini, N.R. and Markakis, P. (1986) Phytases. In: Graf, E. (ed.) *Phytic Acid: Chemistry and Applications.* Pilatus Press, Minneapolis, pp. 101–118.

Ninomiya, Y., Ueki, K. and Sato, S. (1977) Chromatographic separation of extracellular acid phosphatase of tobacco cells cultured under Pi-supplied and omitted conditions. *Plant and Cell Physiology* 18, 413–420.

Norde, W. and Lyklema, J. (1991) Why proteins prefer surfaces. *Journal of Biomaterial Science, Polymer Edition* 2, 183–202.

Ognalaga, M., Frossard, E. and Thomas, F. (1994) Glucose-1-phosphate and *myo*-inositol hexaphosphate adsorption mechanisms on goethite. *Soil Science Society of America Journal* 58, 332–337.

Owusu-Bennoah, E. and Wild, A. (1979) Autoradiography of the depleted zone of phosphate around onion roots in the presence of vesicular-arbuscular mycorrhiza. *New Phytologist* 82, 133–140.

Pant, H.K. and Warman, P.R. (2000) Enzymatic hydrolysis of soil organic phosphorus by immobilized phosphatases. *Biology and Fertility of Soils* 30, 306–311.

Pettit, N.M., Lindsay, J., Gregory, R.B., Freedman, R.B., and Burns, R.G. (1977) Differential stabilities of soil enzymes. Assay and properties of phosphatase and arylsulphatase. *Biochimica et Biophysica Acta* 485, 357–366.

Pointillart, A. (1994) Phytates, phytases; leur importance dans l'alimentation des monogastriques. *INRA Production Animales* 7, 29–39.

Powar, V.K. and Jagannathan, V. (1982) Purification and properties of phytate-specific phosphatase from *Bacillus subtilis*. *Journal of Bacteriology* 151, 1102–1108.

Quiquampoix, H. (1987a) A stepwise approach to the understanding of extracellular enzyme activity in soil. I. Effect of electrostatic interactions on the conformation of a β-D-glucosidase on different mineral surfaces. *Biochimie* 69, 753–763.

Quiquampoix, H. (1987b) A stepwise approach to the understanding of extracellular enzyme activity in soil. II. Competitive effects on the adsorption of a β-D-glucosidase in mixed mineral or organo-mineral systems. *Biochimie* 69, 765–771.

Quiquampoix, H. (2000) Mechanisms of protein adsorption on surfaces and consequences for extracellular enzyme activity in soil. In: Bollag, J.M. and Stotzky, G. (eds) *Soil Biochemistry,* Vol. 10. Marcel Dekker, New York, pp. 171–206.

Quiquampoix, H. and Ratcliffe, R.G. (1992) A ^{31}P NMR study of the adsorption of bovine serum albumin on montmorillonite using phosphate and the paramagnetic cation Mn^{2+}: modification of conformation with pH. *Journal of Colloid and Interface Science* 148, 343–352.

Quiquampoix, H., Chassin, P. and Ratcliffe, R.G. (1989) Enzyme activity and cation exchange as tools for the study of the conformation of proteins adsorbed on mineral surfaces. *Progress in Colloid and Polymer Science* 79, 59–63.

Quiquampoix, H., Staunton, S., Baron, M.H. and Ratcliffe, R.G. (1993) Interpretation of the pH dependence of protein adsorption on clay mineral surfaces and its relevance to the understanding of extracellular enzyme activity in soil.

Colloids and Surfaces, A: Physicochemical and Engineering Aspects 75, 85–93.
Quiquampoix, H., Abadie, J., Baron, M.H., Leprince, F., Matumoto-Pintro, P.T., Ratcliffe, R.G. and Staunton, S. (1995) Mechanisms and consequences of protein adsorption on soil mineral surfaces. *ACS Symposium Series* 602, 321–333.
Quiquampoix, H., Servagent-Noinville, S. and Baron, M.H. (2002) Enzyme adsorption on soil mineral surfaces and consequences for the catalytic activity. In: Burns, R. and Dick, R. (eds) *Enzymes in the Environment: Activity, Ecology and Applications.* Marcel Dekker, New York, pp. 285–306.
Rao, M.A., Gianfreda, L., Palmiero, F. and Violante, A. (1996) Interactions of acid phosphatase with clays, organic molecules and organo-mineral complexes. *Soil Science* 161, 751–760.
Rao, M.A., and Violante, A. and Gianfreda, L. (2000) Interaction of acid phosphatase with clays, organic molecules and organo-mineral complexes: kinetics and stability. *Soil Biology and Biochemistry* 32, 1007–1014.
Ravindran, V., Ravindran, G. and Sivalogan, S. (1994) Total and phytate phosphorus contents of various foods and feedstuffs of plant origin. *Food Chemistry* 50, 133–136.
Rojo, M.J., Carcedo, S.G. and Mateos, M.P. (1990) Distribution and characterisation of phosphatase and organic phosphorus in soil fractions. *Soil Biology and Biochemistry* 22, 169–174.
Sanyal, S.K. and De Datta, S.K. (1991) Chemistry of phosphorus transformations in soil. *Advances in Soil Science* 16, 1–120.
Servagent-Noinville, S., Revault, M., Quiquampoix, H. and Baron, M.H. (2000) Conformational changes of bovine serum albumin induced by adsorption on different clay surfaces: FTIR analysis. *Journal of Colloid and Interface Science* 221, 273–283.
Shieh, T.R., Wodzinski, R.J. and Ware, J.H. (1969) Regulation of the formation of acid phosphatases by inorganic phosphate in *Aspergillus ficuum. Journal of Bacteriology* 100, 1161–1165.
Staunton, S. and Quiquampoix, H. (1994). Adsorption and conformation of bovine serum albumin on montmorillonite: modification of the balance between hydrophobic and electrostatic interactions by protein methylation and pH variation. *Journal of Colloid and Interface Science* 166, 89–94.
Tabatabai, M.A. and Bremner, J.M. (1971) Michaelis constants of soil enzymes. *Soil Biology and Biochemistry* 3, 317–323.
Tarafdar, J.C. and Jungk, A. (1987) Phosphatase activity in the rhizosphere and its relation to the depletion of soil organic phosphorus. *Biology and Fertility of Soils* 3, 199–204.
Tarafdar, J.C. and Marschner, H. (1994) Phosphatase activity in the rhizosphere and hyphosphere of VA mycorrhizal wheat supplied with inorganic and organic phosphorus. *Soil Biology and Biochemistry* 26, 387–395.
Terry, N. and Ulrich, A. (1973) Effects of phosphorus deficiency on the photosynthesis and respiration of leaves in sugar beet. *Plant Physiology* 51, 43–47.
Thaller, M.C., Berlutti, F., Schippa, S., Lombardi, G. and Rossolini, G.M. (1994) Characterization and sequence of PhoC, the principal phosphate-irrepressible acid phosphatase of *Morganella morganii. Microbiology* 140, 1341–1350.
Theodorou, C. (1968) Inositol phosphates in needles of *Pinus radiata* D. Don and the phytase activity of mycorrhizal fungi. *Transactions of the 9th International Congress of Soil Science* (Adelaide) 3, 483–490.
Tibbett, M. (2002) Considerations on the use of the *p*-nitrophenyl phosphomonoesterase assay in the study of the phosphorus nutrition of soil-borne fungi. *Microbiological Research* 157, 221–231.
Tibbett, M., Sanders, F.E. and Cairney, J.W.G. (1998) The effect of temperature and inorganic phosphorus supply on growth and acid phosphatase production in arctic and temperate strains of ectomycorrhizal *Hebeloma* spp. *Mycological Research* 102, 129–135.
Tiessen, H., Stewart, J.W.B. and Cole, C.V. (1984) Pathways of phosphorus transformations in soil of differing pedogenesis. *Soil Science Society of America Journal* 48, 853–858.
Trasar-Cepeda, M.C. and Gil-Sotres, F. (1988) Kinetics of acid phosphatase activity in various soils of Galicia (NW Spain). *Soil Biology and Biochemistry* 20, 275–280.
Uehara, K., Fujimoto, S., Taniguchi, T. and Nakai, K. (1974) Studies on violet-colored acid phosphatase of sweet potato. II. Enzymatic properties and amino acid composition. *Journal of Biochemistry* 75, 639–649.
Vetter, Y.A. and Deming, J.W. (1999) Growth rates of marine bacterial isolates on particulate organic substrates solubilized by freely released extracellular enzymes. *Microbial Ecology* 37, 86–94.
von Tigerstrom, R.G. and Stelmaschuk, S. (1986) Purification and characterization of the outer membrane associated alkaline phosphatase of *Lysobacter enzymogenes. Journal of General Microbiology* 132, 1379–1387.
Williams, C.H. and Anderson, G. (1968) Inositol phosphates in some Australian soils. *Australian Journal of Soil Research* 6, 121–130.
Williams, C.H. and Steinbergs, A. (1958) Sulphur and

phosphorus in some eastern Australian soils. *Australian Journal of Agricultural Research* 9, 483–491.

Wyss, M., Pasamontes, L., Rémy, R., Kohler, J., Kusznir, E., Gadient, M., Müller, F. and van Loon, A.P.G.M. (1998) Comparison of the thermostability properties of three acid phosphatases from molds: *Aspergillus fumigatus* phytase, *A. niger* phytase, and *A. niger* pH 2.5 acid phosphatase. *Applied and Environmental Microbiology* 64, 4446–4451.

Wyss, M., Brugger, R., Kronenberger, A., Rémy, R., Fimbel, R., Oesterhelt, G., Lehmann, M. and van Loon, A.P.G.N. (1999) Biochemical characterization of fungal phytases (*myo*-inositol hexakisphosphate phosphohydrolases): catalytic properties. *Applied and Environmental Microbiology* 65, 367–373.

Yamagata, H., Yanaka, K. and Kasai, Z. (1980) Purification and characterization of acid phosphatase in aleurone particle of rice grains. *Plant and Cell Physiology* 21, 1449–1460.

6 Abiotic Stabilization of Organic Phosphorus in the Environment

Luisella Celi and Elisabetta Barberis

Università di Torino, DIVAPRA-Chimica Agraria, via Leonardo da Vinci 44, Grugliasco, 10095 Torino, Italy

Introduction

Most soils contain between 50 and 1000 mg/kg of phosphorus, although reported concentrations vary between 20 and 1800 mg/kg (Harrison, 1987). Conversely, phosphorus concentrations in water are substantially lower, often less than 10 μg/l in uncontaminated systems (Correll, 1998). Organic phosphorus constitutes between 20% and 80% of the total phosphorus in soils (Anderson, 1980). While less than 50% of the organic phosphorus in many soils has been properly identified, there appear to be substantially more phosphate monoesters in soils than would be expected from the distribution of organic phosphorus species in the biosphere. This has been attributed to selective stabilization of phosphate monoesters by a series of processes that hamper their biodegradation (Anderson, 1980; Stewart and Tiessen, 1987; Condron *et al.*, 1990). Abiotic stabilization processes, such as adsorption to soil minerals and/or precipitation reactions, can limit the loss of phosphate monoesters from soils (Fig. 6.1), as can their incorporation into humic substances. In contrast, phosphate diesters and other organic phosphorus compounds are less strongly stabilized and are more likely to be found in soil solution, where they can be readily degraded by biological processes. All these processes affect the accumulation of organic phosphorus in soils, its availability to plants, and its leaching to waters. In aquatic environments, a large proportion of the organic phosphorus, especially inositol phosphates, probably originates from soils in both dissolved and particulate forms (Föllmi, 1996; Turner, Chapter 2, this volume).

This chapter reviews the abiotic processes that can lead to the stabilization of organic phosphorus in soils and the aquatic environment. In particular, we examine the role of adsorption to soil minerals, complexation reactions, precipitation with polyvalent cations and the incorporation of organic phosphorus into humic substances in stabilizing organic phosphorus. We then discuss the effects of soil solution chemistry on these reactions, as well as the effects of these reactions on the environment.

Forms of Organic Phosphorus in Terrestrial and Aquatic Environments

Details of the forms of organic phosphorus that occur in the environment can be found in several published reviews (e.g. Dalal, 1977; Anderson, 1980; Harrison, 1987) and elsewhere in this volume, but a brief overview is provided here. Advances in solution ^{31}P nuclear magnetic resonance spectroscopy

 Organic Phosphorus in the Environment (eds B.L. Turner, E. Frossard and D.S. Baldwin)

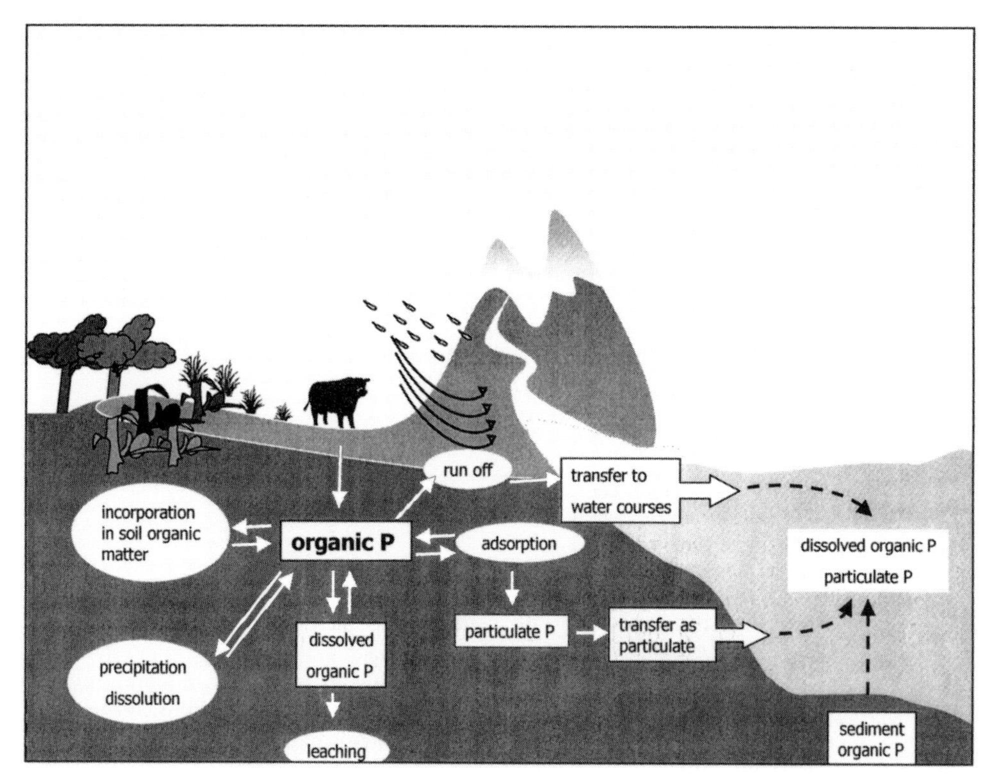

Fig. 6.1. Abiotic reactions involved in the biogeochemical cycle of organic phosphorus in terrestrial and aquatic environments.

have been influential in determining the structure of organic phosphorus compounds in environmental samples (Newman and Tate, 1980; Tate and Newman, 1982; Hawkes *et al.*, 1984; Zech *et al.*, 1985; Condron *et al.*, 1990; Bedrock *et al.*, 1994; Rubæk *et al.*, 1999; Turner *et al.*, 2003; Cade-Menun, Chapter 2, this volume), although characterization has been limited to the fraction that can be chemically extracted.

Inositol phosphates are often the most abundant organic phosphorus compounds in soils, although they are less common in aquatic samples. They include a sequence of phosphate monoesters, from inositol hexakisphosphate to inositol monophosphate (Mandal and Islam, 1979; Harrison, 1987), in various stereoisomeric configurations (*myo*, *scyllo*, *neo*, D-*chiro*) of the sugar inositol (Anderson and Malcolm, 1974; Anderson *et al.*, 1974; Cosgrove, 1980; Harrison, 1987; Turner *et al.*, 2002). These compounds are characterized by high acidity (Table 6.1) and, hence, high charge density in the normal pH range of soils and sediments. The most common inositol phosphate in the environment, especially in soils, is *myo*-inositol hexakisphosphate (phytic acid). This compound has been widely studied in part due to its role in promoting mineral deficiency in humans and animals (Cosgrove, 1980; Frossard *et al.*, 2000). Inositol phosphates are often found in soils as components of polymers or insoluble complexes with proteins and lipids (Harrison, 1987).

Other phosphorus-containing organic compounds found in soils and waters include phospholipids, nucleic acids (DNA and RNA) and their nucleotide residues (Newman and Tate, 1980; Condron *et al.*, 1990), phosphorylated polymers of mannose produced by yeasts (Harrison, 1987), and teichoic acids that form the cell walls of many Gram-positive bacteria (Zhang *et al.*, 1998;

Table 6.1. Dissociation acid constants of *myo*-inositol hexakisphosphate (Costello *et al.*, 1976) and phosphate (Corbridge, 1985).

Molecule	pK_1	$pK_{2,3}$	$pK_{4\text{–}6}$	pK_7	pK_8	pK_9	$pK_{10,11}$	pK_{12}
Myo-inositol hexakisphosphate	1.1	1.5	1.8	5.7	6.9	7.6	10.0	12.0
Phosphate	2.0	6.8	12.3					

Table 6.2. Distribution of organic phosphorus fractions (% of total phosphorus) in soils and growing organisms (Magid *et al.*, 1996).

	Escherichia coli	Fungi	*Nicotiana*	Soils
Nucleic acids	65	58	52	2
Phospholipids	15	20	23	5
Phosphate monoesters	20	22	25	50

Rubæk *et al.*, 1999). Phospholipids are often present in soils (0.5–7.0% of total organic phosphorus; Dalal, 1977) and sediments (3–14% total organic phosphorus; Suzumura and Ingall, 2001). In soils, phosphoglycerides are the predominant phospholipid, although little information on the other forms (e.g. phosphoglycolipids, phosphosphingolipids) is available. The molecules responsible for biochemical energy transfer, notably adenosine triphosphate, are also detected in soils and leachate (Espinosa *et al.*, 1999).

Most of the organic phosphorus compounds listed above can be found in living plants, animals and microorganisms. Magid *et al.* (1996) compared the organic phosphorus composition in soils with that found in living bacteria, fungi and plants. In the biosphere, nucleic acids account for more than half of the organic phosphorus, but only 2% of the organic phosphorus in soil (Table 6.2). Moreover, phosphate monoesters account for only around 20% of biotic organic phosphorus, but more than 50% of soil organic phosphorus, mostly as inositol phosphates. It should be noted that the values for soil correspond to the amounts identified by chemical methods, while the remaining 40–50% has eluded extraction and identification. It was suggested that part of the organic phosphorus occurs in association with humic acids, fulvic acids and humin (Stevenson, 1994). Goh and Williams (1982) found that carbon-to-phosphorus ratios were higher in the low-molecular-weight components of soil organic matter than in the high-molecular-weight components, indicating a greater phosphorus association with humin or humic acids. Makarov *et al.* (1997) found that the phosphorus species present in humic and fulvic acids comprised phosphate monoesters, phosphate diesters, phosphonates, sugar diesters, pyrophosphate and polyphosphate. However, other authors have discounted the presence of phosphorus in humic and fulvic acid structures (Barrow, 1961; Kowalenko, 1978).

In aquatic environments, substantial amounts of inositol phosphates have been reported in lake waters (McKelvie *et al.*, 1995) and estuarine sediments (Suzumura and Kamatani, 1995). Other more labile organic phosphorus compounds, such as phosphate diesters and adenosine triphosphate, which account for only a small proportion of the total organic phosphorus in soils, are particularly important in lake and marine sediments (Carman *et al.*, 2000), as are labile polyphosphates (Hupfer *et al.*, 1995).

Besides naturally occurring organic phosphorus compounds, the soil organic phosphorus concentration can be increased by manure and slurry application and, to a lesser extent, by the use of phosphorus-containing pesticides. Many soils receive animal manure, in which typical phosphorus concentrations are many times

Table 6.3. Chemical designations of the main phosphorus-containing pesticides (Sánchez-Camazano and Sánchez-Martín, 1983).

Common name	Chemical name
Azinphosmethyl	0,0-dimethyl, *S*-(4-oxobenzotriazino-3-methyl) phosphorothiolothionate
Bidrin	0,0-dimethyl, 0-(*N*,*N*-dimethylcarbamoyl-1-methylvinyl) phosphate
Carbophenothion	0,0-diethyl, *S*-(4-chlorophenylthiomethyl) phosphorothiolothionate
Ciodrin	0,0-dimethyl, 0-(1-methyl, 2-(1-phenylethoxycarbonyl) vinyl) phosphate
Diazinon	0,0-diethyl, 0-(2-isopropyl-4-methylpyrimidyl-6) phosphorothionate
Dichlorvos	0,0-dimethyl, 0-2,2-dichlorovinyl phosphate
Dimethoate	0,0-dimethyl, *S*-(*N*-methylcarbamoylmethyl) phosphorothiolothionate
Dursban	0,0-diethyl, 0-(3,5,6-trichloropyridyl) phosphorothionate
Malathion	0,0-dimethyl, *S*-(1,2-dicarboethoxyethyl) phosphorothiolothionate
Metasystox	0,0-dimethyl, *S*-(2-ethylsulfinyl) ethyl phosphorothiolate
Methyltrithion	0,0-dimethyl, *S*-(4-chlorophenylthiomethyl) phosphorothiolothionate
Monocrotophos	0,0-dimethyl, 0-(*N*-methylcarbamoyl-1-methylvinyl) phosphate
Phosidrin	0,0-dimethyl, 0-(1-methyl-2-carbomethoxyvinyl) phosphate
Phosmet	0,0-dimethyl, *S*-(*N*-phthalimidomethyl) phosphorothiolothionate
Phosphamidion	0,0-dimethyl, 0-(2-chloro-2-*N*,*N*-diethylcarbamoyl-1-methylvinyl) phosphate
Sumithion	0,0-dimethyl, 0-(4-nitro-3-methylphenyl) phosphorothionate
Zinophos	0,0-diethyl, 0-pyrazinyl phosphorothionate

greater than those of soils. On a dry-weight basis, dairy manure may contains between 4 and 7 g P/kg dry weight, while poultry manure can contain more than 20 g P/kg depending on animal species and diet (Barnett, 1994). The proportion of organic phosphorus in manure can vary from 10% to 80% of the total, and decreases with manure ageing (Peperzak *et al.*, 1959; Gerritse and Zugec, 1977; Sharpley and Moyer, 2000). The main forms of organic phosphorus are inositol phosphates and more labile compounds, such as nucleic acids and phospholipids (Turner, 2004). A number of synthetic phosphorus products (Table 6.3) are used in agriculture for weed and insect control and can accumulate in soils or be readily transferred to waters depending on their degradability and on soil properties (Sánchez-Martin and Sánchez-Camazano, 1991).

Abiotic Processes in the Environment

Organic phosphorus generally accumulates in the finest soil fractions, with the clay fraction often containing more organic phosphorus than silt (Williams and Saunders, 1956a, 1956b; Bates and Baker, 1960; Syers *et al.*, 1969; Hanley and Murphy, 1970; Tiessen *et al.*, 1983; Zhu *et al.*, 1983; Choudhry, 1988; Guzel and Ibrikci, 1994). Similarly, organic phosphorus in aquatic ecosystems is often associated with fine sediment (Föllmi, 1996). The high affinity of organic phosphorus compounds for colloids has been attributed to the high surface area and anion retention capacity exhibited by many colloidal particles (Tiessen *et al.*, 1983; Guzel and Ibrikci, 1994). Therefore, sorption on to soil minerals (particularly colloids) seems to be the main process affecting the distribution of organic phosphorus and the selective accumulation of some compounds in soils. However, complexation, precipitation with polyvalent cations, and/or physical incorporation into soil organic matter structures may also control the fate of organic phosphorus in the environment.

Sorption and desorption

Sorption models

Phosphorus sorption is one of the most widely studied reactions in soils. Iyamuremye and Dick (1996) traced an interesting history of the literature regarding phosphate sorption, attributing the first studies to Way (1850). However, the study of organic phosphorus sorption to soil is more recent, with the earliest studies being undertaken in the mid-1930s (e.g. Spencer and Stewart, 1934).

From a theoretical point of view, the term sorption is used instead of adsorption to cover any process that removes a reactant from the solution, including adsorption and precipitation (Barrow, 1993). In the adsorption process for phosphate, the concentration of phosphate in solution controls the extent of adsorption, while the extent of precipitation is determined by the solubility product of the least soluble phosphate compound, which in turn controls the phosphate concentration in solution. As with phosphate, the adsorption of organic phosphorus compounds on minerals can be described using the Langmuir equation:

$$Q_a = \frac{K_L X_{max} C_e}{1 + K_L C_e} \quad (6.1)$$

where Q_a is the amount of adsorbed phosphorus, C_e the solution concentration, X_{max} is the maximum amount that can be adsorbed on the substrate, and K_L is a constant that can give information on the affinity of adsorbate for the substrate surface.

The assumptions underlying the Langmuir isotherm (e.g. monolayer cover, all adsorption sites of equal energy, and adsorption at one site will not affect adsorption at other sites) are rarely met when dealing with real samples (Detenbeck and Brezonik, 1991), but data from adsorption studies at higher phosphorus loadings ($>10^{-3}$ M) nevertheless often give a good fit to the Langmuir equation. For example, the Langmuir model has been successfully used for describing *myo*-inositol hexakisphosphate adsorption on pure iron oxides and phyllosilicates (Celi *et al.*, 1999, 2003). At lower equilibrium phosphorus concentrations ($<10^{-6}$–10^{-4} M), the Freundlich equation (equation (6.2)) often models the data better (Barrow, 1983; Barrón *et al.*, 1988; Colombo *et al.*, 1994).

$$Q_a = K C_e^{\,n} \quad (6.2)$$

An extension of the Langmuir equation (equation (6.3)) has been proposed for phosphate adsorption to soils and metal oxides in order to relate adsorption to the electrostatic potential of mineral surfaces and to the bulk solution characteristics, such as pH, electrolyte concentration, temperature and competing ions (Barrow *et al.*, 1980; Barrow, 1983, 1985, 1993; Bolan *et al.*, 1986):

$$Q_a = \frac{X_{max} K_i C_e \alpha \gamma \exp(-z_i \psi F/RT)}{1 + K_i C_e \alpha \gamma \exp(-z_i \psi F/RT)} \quad (6.3)$$

where K_i is the affinity constant for the ion species i, α the proportion of ion i, γ the activity coefficient, z_i the valency of the ion i, ψ is the electrostatic potential at the plane of adsorption, F the Faraday constant, R the gas constant, and T the temperature.

Barrow (1993) successfully applied this model to phosphate adsorption by taking into account pH and electrolyte concentration. The use of this equation to describe organic phosphorus adsorption can be important as well, due to the high charge density of molecules such as *myo*-inositol hexakisphosphate, but the complexity of organic phosphorus compounds and the different mechanism of adsorption can make the model difficult to apply.

Organic phosphorus adsorption

Adsorption of organic phosphorus on soil minerals initially proceeds by a fast reaction that is often considered to reach an equilibrium value. However, a true equilibrium is not found for phosphate ions within a very long period. For example, one study showed that phosphate adsorption on to soils had not reached equilibrium after 1000 days at 25°C (Barrow and Shaw, 1975). The long-term adsorption processes for organic phosphorus compounds on to soils and oxides have not been well studied. Shang *et al.* (1990) compared the kinetics of adsorption of phosphate and selected organic phosphates on aluminium precipitates. They found that adsorption obeyed first-order

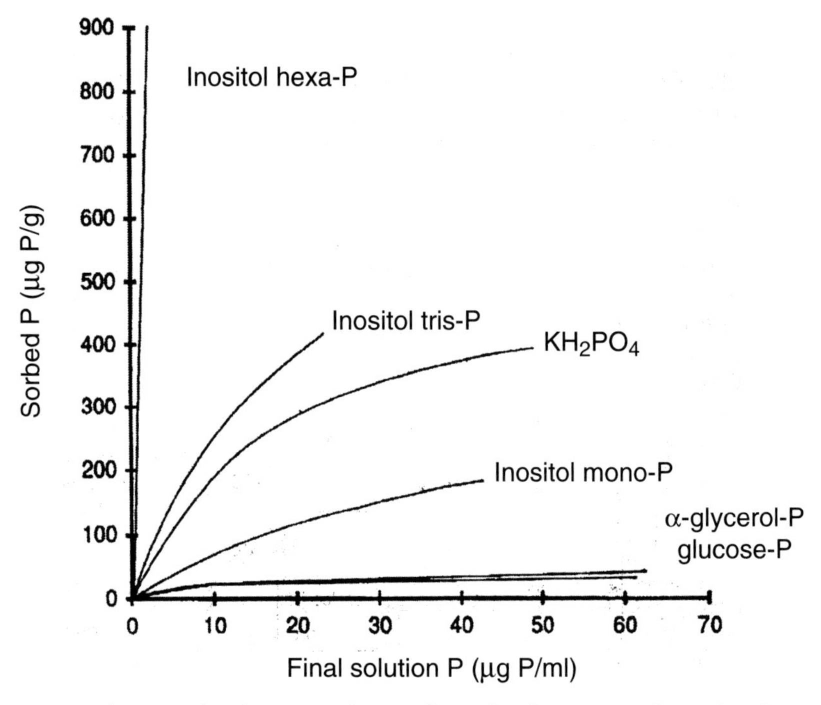

Fig. 6.2. Sorption of organic phosphates on a clayey soil (P = phosphate) (McKercher and Anderson, 1989).

kinetics and occurred in two stages, a fast adsorption before 1 h and a slow adsorption between 1 and 24 h. However, Anderson *et al.* (1974) found that true equilibrium had not been reached after 72 h of interaction of *myo*-inositol hexakisphosphate with soils.

As for phosphate, the adsorption reaction of organic phosphorus is not readily reversible, although some phosphorus can pass to the solution depending on the time of desorption, the solution-to-soil ratio, and temperature (Barrow, 1983). As solution characteristics play an important role in desorption, different approaches using free water, dilute electrolyte solutions, or chemical extractants with variable pH and ionic composition have been used to quantify the amount of phosphorus desorbable from minerals and then available to plants (Frossard *et al.*, 1995). Desorption of inorganic and organic phosphorus was found to increase with pH (Cabrera *et al.*, 1981; Celi *et al.*, 2003; Martin *et al.*, 2003), with the percentage of phosphorus saturation (Parfitt, 1979; He *et al.*, 1991, 1994; Martin *et al.*, 2002), and in the presence of competing ligands such as citrate, oxalate or carbonate (Nagarajah *et al.*, 1968; He *et al.*, 1991; Martin *et al.*, 2003).

The extent and the rate of adsorption of organic phosphorus in soils, sediments, and on to their minerals depend on the structure of the organic phosphorus compound, notably in terms of the number of phosphorus groups and molecular size. Among the different compounds, inositol phosphates are sorbed to clay minerals to a greater extent than nucleic acids, phospholipids and simple sugar phosphates (Fig. 6.2; Goring and Bartholomew, 1950; Anderson and Arlidge, 1962; Anderson *et al.*, 1974; McKercher and Anderson, 1989; Leytem *et al.*, 2002). Moreover, sorption can be very different depending on soil type. In acid soils the sorption of inositol phosphates was reported to be dependent on the content of amorphous aluminium and iron oxides (Anderson *et al.*, 1974), while in neutral and basic soils it was governed by clays and organic matter (McKercher and Anderson,

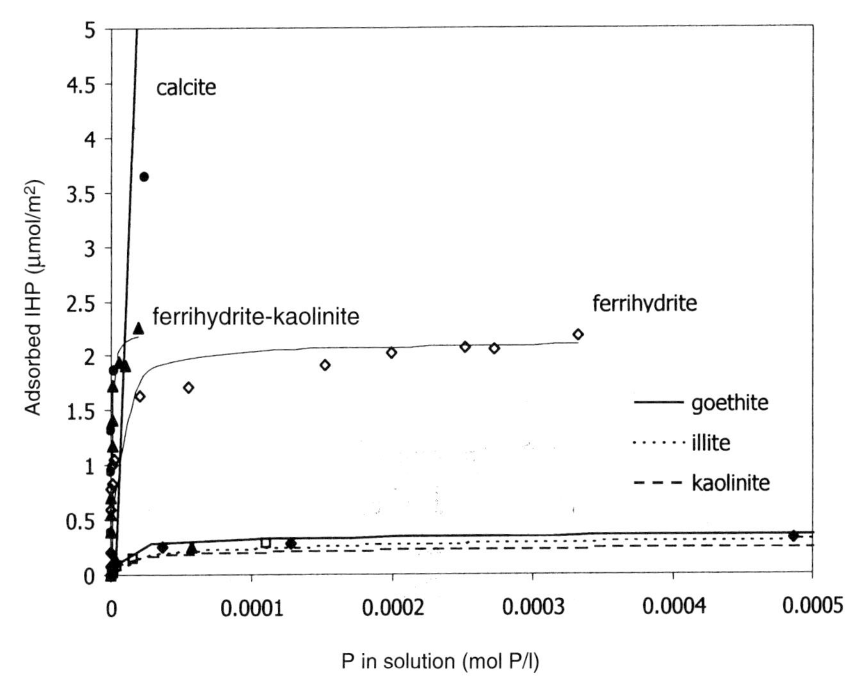

Fig. 6.3. Isotherms of adsorption of *myo*-inositol hexakisphosphate (IHP) on goethite, ferrihydrite, illite, kaolinite, calcite and ferrihydrite–kaolinite mixed systems at pH 4.5 and in 0.01 M KCl (adapted from Celi *et al.*, 1999, 2000, 2003).

1989). Sorption of organic phosphorus by neutral and basic soils increases or reaches a maximum with increasing organic phosphorus concentration (McKercher and Anderson, 1989), while acid soils can show similar or contrasting behaviour (Leytem *et al.*, 2002). The extent and rate of sorption can also be regulated by mineral properties (i.e. surface porosity, specific area and degree of crystallinity) and soil solution chemistry.

Iron and aluminium oxides

Organic phosphorus can be strongly adsorbed on ferric oxide surfaces in soils (Anderson and Arlidge, 1962) and sediments (de Groot and Golterman, 1993). Indeed, in most acidic soils and sediments the amount of iron oxide in the substrate governs the extent of adsorption of organic phosphorus (Anderson *et al.*, 1974; Harrison, 1987; Pant *et al.*, 1994). Phosphate monoesters such as *myo*-inositol hexakisphosphate, glucose 1-phosphate and glucose 6-phosphate probably adsorb on the same sites of iron oxides as phosphate ions (Ognalaga *et al.*, 1994; Celi *et al.*, 1999).

The maximum amount of *myo*-inositol hexakisphosphate adsorbed on iron oxides at pH 4.5 and after 24 h of interaction (Fig. 6.3) was found to be 2.12 $\mu mol/m^2$ for ferrihydrite (Celi *et al.*, 2003), 0.64 $\mu mol/m^2$ for goethite (Ognalaga *et al.*, 1994; Celi *et al.*, 1999), and 0.40 $\mu mol/m^2$ for haematite (L. Celi, unpublished data). Such adsorption is adequately described by the Langmuir equation, although K values are always higher than those reported for phosphate adsorbed on the same iron oxides (Table 6.4), indicating a greater affinity of *myo*-inositol hexakisphosphate for these minerals compared to phosphate. Adsorption occurs through the phosphate groups, which react with the oxide in the same way as the free phosphate ion via a ligand exchange with the water and hydroxide ion groups of the surfaces (Parfitt *et al.*, 1976; Goldberg and Sposito, 1985).

Table 6.4. Langmuir coefficients of adsorption isotherms of *myo*-inositol hexakisphosphate (*myo*-IP_6) and phosphate on goethite, ferrihydrite, haematite, illite, kaolinite and ferrihydrite–kaolinite systems (Fh–KGa2) at pH 4.5 in 0.01 M KCl (Celi *et al.*, 1999, 2003).

		X_{max} (μmol/m)	K (l/mol)	r^2 ($n = 10$)
Goethite	*myo*-IP_6	0.64 (3.8)[a]	$8.0 \cdot 10^3$	0.978
	Phosphate	2.4	$5.6 \cdot 10^3$	0.996
Ferrihydrite	*myo*-IP_6	2.12 (12.7)[a]	$8.4 \cdot 10^4$	0.994
	Phosphate	4.57	$0.3 \cdot 10^3$	0.996
Haematite	*myo*-IP_6	0.40 (2.4)[a]	$6.2 \cdot 10^4$	0.993
	Phosphate	0,82	$4.0 \cdot 10^4$	0.994
Illite	*myo*-IP_6	0.38 (2.3)[a]	$1.0 \cdot 10^5$	0.999
	Phosphate	1.0	$7.5 \cdot 10^3$	0.986
Kaolinite	*myo*-IP_6	0.27 (1.6)[a]	$1.0 \cdot 10^5$	0.999
	Phosphate	0.79	$1.6 \cdot 10^3$	0.923
(Fh–KGa2)	*myo*-IP_6	2.24 (13.4)[a]	$2.2 \cdot 10^6$	0.993
	Phosphate	2.96	$9.1 \cdot 10^4$	0.991

[a] The values reported in parentheses indicate the amount of adsorbed *myo*-inositol hexakisphosphate expressed as moles of phosphorus.

Evidence of this has been obtained by Fourier-transform infrared spectroscopy, which shows changes in the P=O and P–O bands of *myo*-inositol hexakisphosphate after adsorption on to goethite, due to the formation of Fe–O–P bonds (Celi *et al.*, 1999). Ognalaga *et al.* (1994) and Celi *et al.* (1999) suggested that adsorption occurred on goethite through four of the six phosphate groups (see Fig. 6.4), with the remaining two groups being free. This explains the 3:2 sorption ratio observed between *myo*-inositol hexakisphosphate and phosphate. The involvement of such a large number of phosphate groups in *myo*-inositol hexakisphosphate adsorption leads to a very stable complex with goethite. In fact, desorption of this compound from goethite was shown to be limited, especially at low pH, even in the presence of citrate and bicarbonate probably due to the inability of these ions to simultaneously detach the four phosphate groups involved in the bonding (Martin *et al.*, 2003). Adsorption of *myo*-inositol hexakisphosphate on another metal oxide, ferrihydrite, is believed to occur through two of the six phosphate groups (Celi *et al.*, 2003).

If iron oxides are associated with other soil components, then surface properties are modified and organic phosphorus interaction can be affected. For instance, the association of ferrihydrite with kaolinite did not cause any variation in the amount of adsorbed *myo*-inositol hexakisphosphate, but resulted in the involvement of only one phosphate group in the adsorption mechanism (Celi *et al.*, 2003). This results in lower desorption of *myo*-inositol hexakisphosphate from ferrihydrite than from the ferrihydrite–kaolinite system. As for phosphate, desorption of *myo*-inositol hexakisphosphate from iron oxides increases with increasing pH, reaching maximum values at pH ≥ 7 (Celi *et al.*, 2003).

Adsorption of glucose 1-phosphate on goethite (Ognalaga *et al.*, 1994) occurs through the phosphate group, leading to the complete coverage of the surface (2.5 $\mu mol/m^2$) if a spatial area of 0.341 nm^2 per singly coordinated hydroxyl (OH_A) on the {110} face of goethite is assumed (Torrent *et al.*, 1990). However, it could be hypothesized that this phosphate monoester was easily hydrolysed and glucose was released, since the new surface obtained after adsorption was observed to be negative (Ognalaga *et al.*, 1994; see also Baldwin *et al.*, Chapter 4, this volume), indicating that the goethite remained covered only by the phosphate group.

Organophosphorus pesticides can also interact with iron oxides. Glyphosate (*N*-

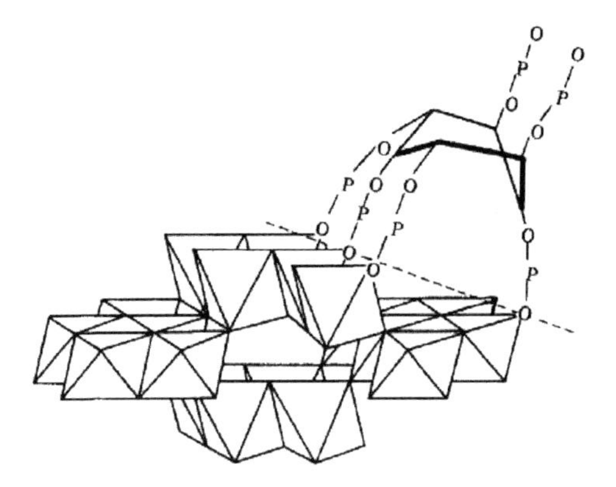

Fig. 6.4. Mechanism of adsorption of *myo*-inositol hexakisphosphate on the goethite surface at pH 4.5. For simplicity, the hydrogen and hydroxide groups of the phosphate molecule are omitted (Ognalaga *et al.*, 1994).

[phosphonomethyl] glycine) is reported to adsorb on goethite through a ligand exchange between the phosphonate moiety and the hydroxide ion or water groups of the iron oxide surface (McBride and Kung, 1989). The adsorption extent is enhanced in the presence of copper (II) ions, due to formation of copper–glyphosate complexes with a higher tendency to be adsorbed compared with free pesticide (Maqueda *et al.*, 2002).

Besides iron oxides, aluminium oxides can control the retention of organic phosphorus compounds in soils (Anderson *et al.*, 1974) depending on the type of oxide. Finely divided boehmite was reported to be effective in removing *myo*-inositol hexakisphosphate from solution, while gibbsite was ineffective (Anderson and Arlidge, 1962). Among different phosphate monoesters such as *myo*-inositol hexakisphosphate, *myo*-inositol monophosphate, and glucose 6-phosphate, *myo*-inositol hexakisphosphate has the highest chemical affinity to the aluminium oxide surface, confirming that the interaction is regulated by the functionality of the phosphate groups (Shang *et al.*, 1990, 1992). This results in a higher thermodynamic stability of *myo*-inositol hexakisphosphate–aluminium oxide complexes and a lower required activation energy with a higher rate constant. From a kinetic point of view, the fast adsorption before 1 h proceeds easily because of the high concentration of adsorbate in the surrounding particle, but as adsorption proceeds the adsorbed *myo*-inositol hexakisphosphate can hinder the approach of other molecules to the surface by imposing electrical and steric effects (Shang *et al.*, 1990). Once sorbed on aluminium precipitates, organic phosphorus compounds are poorly available to plants and microorganisms (Shang *et al.*, 1996) and their bioavailability is determined primarily by the stability of the phosphate surface complexes rather than by the total amount of phosphate adsorbed.

Clays

Clay minerals adsorb less organic phosphates than pure iron and aluminium oxides, but with a higher affinity. Montmorillonite, and to a much smaller extent bentonite, illite and kaolinite, can adsorb *myo*-inositol hexakisphosphate and glycerophosphate (Goring and Bartholomew, 1950). The adsorption is related to their active aluminium content and to the solution pH (Anderson and Arlidge, 1962) and limits microbial hydrolysis of organic phosphate (Greaves and Webley, 1969). In contrast to phosphate, the adsorption isotherm curve of *myo*-inositol hexakisphosphate on clays is regular (Fig. 6.3) and equilibrium adsorption models such as the Langmuir equation can be applied (Table 6.4; Celi *et al.*, 1999). At pH 4.5, the maximum amount of *myo*-inositol hexakisphosphate adsorbed by illite and kaolinite has been reported to be 0.38 and 0.27 $\mu mol/m^2$, respectively, with a higher affinity ($>K_L$) than phosphate. The adsorption mechanism, as hypothesized from the *myo*-inositol hexakisphosphate-to-phosphate ratio, suggests that three phosphate groups interact with the clay surface, although it is possible that the occupation of adsorption sites is partly hindered by the organic moiety of *myo*-inositol hexakisphosphate.

The sorption of nucleic acids in soils is governed by the presence of montmorillonite and by the molecular weight of DNA (Ogram *et al.*, 1988). DNA is weakly

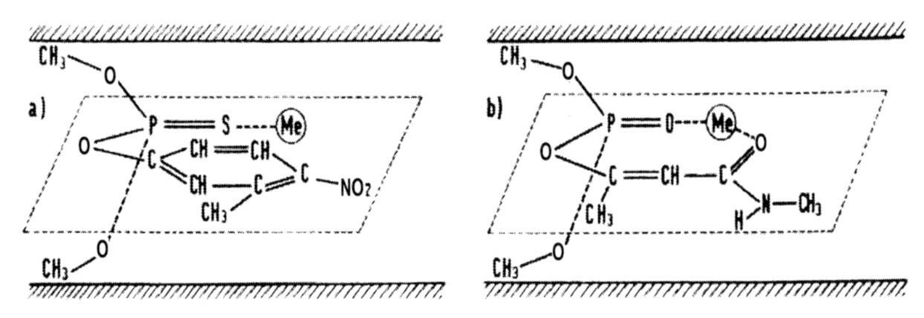

Fig. 6.5. Proposed arrangements of pesticide molecules in the interlayer space of montmorillonite: (a) sumithion; (b) monocrotophos (Sánchez-Camazano and Sánchez-Martín, 1983).

adsorbed on pure montmorillonite, since only one hydroxide group of the phosphate diester is free for ligand exchange (Greaves and Wilson, 1969), leaving the compound vulnerable to microbial degradation.

A summary of the results obtained from the study of adsorption of many organophosphorus pesticides on montmorillonite was reported by Sánchez-Camazano and Sánchez-Martín (1983). The authors showed by X-ray diffraction and Fourier-transform infrared spectroscopy that many organophosphorus compounds are adsorbed into the interlayer space of montmorillonite, causing a different degree of expansion as a function of the pesticide chemical structure (Fig. 6.5). In particular, in the presence of P = O or P = S groups, the length and composition of the radicals determine the interaction with the montmorillonite surface, affecting the number of molecular layers of pesticides. The interaction can be enhanced by the presence of interlamellar calcium in montmorillonite, as observed with dimethyl methylphosphonate (Bowen *et al.*, 1988) and other organophosphorus pesticides (Barba *et al.*, 1991; Baiares-Munoz *et al.*, 1995). Such sorption can modify the pesticide properties (Mingelgrin and Tsvetkov, 1985).

Complexation, precipitation and mineral dissolution

The ability to form complexes or precipitates depends on the type of organic phosphorus, in particular the number of phosphate groups that are present in the molecule. For instance, inositol phosphates are characterized by a high ability to complex metal cations (Cosgrove, 1980; Nolan and Duffin, 1987). Studies made on *myo*-inositol hexakisphosphate using potentiometric titration methods listed the order of stability as copper (II) > zinc > nickel (II) > cobalt (II) > manganese (II) > iron (III) > calcium (Fig. 6.6; Martin and Evans, 1987; Nolan and Duffin, 1987). These complexes are soluble at low pH, while insoluble complexes are formed at mid-range pH values (Martin and Evans, 1987). The formation of insoluble complexes with the above-cited cations has important consequences from a nutritional point of view, since *myo*-inositol hexakisphosphate can act as an anti-nutritional agent by interfering with mineral bioavailability (Maga, 1982).

In soils and sediments, complexation can increase organic phosphorus stabilization, especially with iron (III) and calcium ions and their minerals (Harrison, 1987; House and Denison, 2002). The interaction with iron (III) was reported to transform a large part of the labile and moderately labile organic phosphorus forms supplied with manure to paddy soils into more resistant organic phosphorus, possibly because inositol phosphates initially bound to calcium or magnesium were transformed into iron-bound compounds (Zhang *et al.*, 1994). In the presence of calcium, *myo*-inositol hexakisphosphate can form two soluble calcium complexes with one or two calcium ions (Ca_1- or Ca_2-phytate), but when three calcium ions are involved (Ca_3-phytate), the complex precipitates at all pH values (Graf, 1983). This enhances the interaction of *myo*-

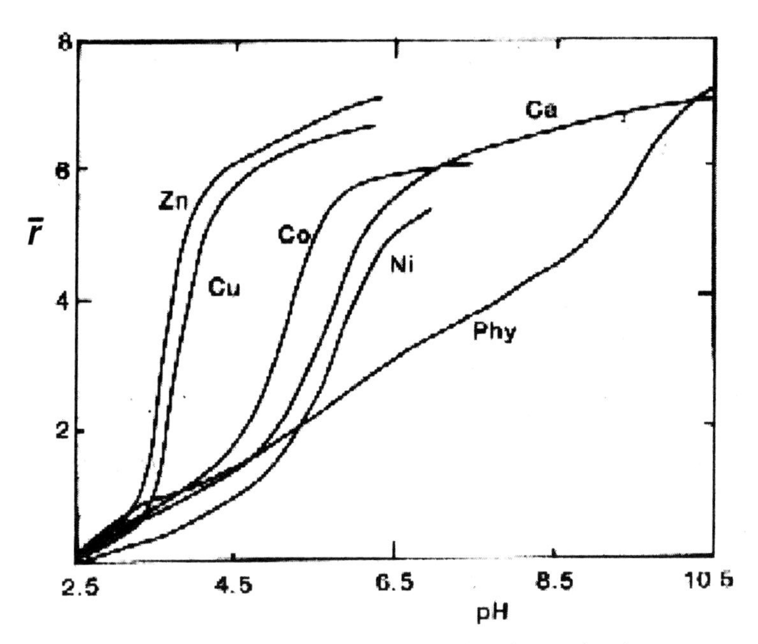

Fig. 6.6. Titration curves of *myo*-inositol hexakisphosphate (Phy) alone and in the presence, at 6:1 molar ratio, of zinc (II), copper (II), calcium (II), cobalt (II) and nickel (II) (Martin and Evans, 1987).

inositol hexakisphosphate with calcium-containing minerals, since phosphate can be adsorbed on the mineral surface at low concentration and precipitate as calcium phosphate at higher concentrations, depending on the initial metal ion-to-ligand molar ratio (Cole *et al.*, 1953; Griffin and Jurinak, 1973; Freeman and Rowell, 1981). Precipitation of calcium salts with calcite occurs at even very low concentrations of *myo*-inositol hexakisphosphate (Celi *et al.*, 2000) and overlays the adsorption process (Fig. 6.3). This explains why in some calcareous soils the organic phosphorus content is positively correlated to the calcium content (Harrison, 1987).

The high complexing ability of inositol phosphates and other phosphate monoesters can favour dissolution of minerals containing bi- and trivalent ions, due to the fact that, once adsorbed, these compounds can weaken and break the metal–oxygen bonds in the minerals, with the consequent release of the metal to solution. The formation of insoluble metal salts further enhances the dissolution process by removing metal ions from the reaction equilibrium. In fact, precipitation of calcium salts of *myo*-inositol hexakisphosphate can cause dissolution of calcite (Celi *et al.*, 2000). Unlike phosphate (Zinder *et al.*, 1986), the addition of *myo*-inositol hexakisphosphate can lead to the dissolution of ferrihydrite and, to a lesser extent, of goethite (Celi *et al.*, 2003; Martin *et al.*, 2003).

Under anaerobic conditions, iron (III) is reduced to iron (II), causing dissolution of iron oxides (Schwertmann, 1991). The adsorption of organic phosphates can then either increase through oxidation of iron (II) back to iron (III) leading to the formation of highly reactive amorphous iron oxyhydroxides (Sah *et al.*, 1989), or decrease due to the release of the adsorbed anions that become more labile (Zhang *et al.*, 1994). The release of organic phosphorus due to redox changes is frequently observed during the transport of particulate phosphorus in rivers (Föllmi, 1996).

Incorporation of organic phosphorus in organic matter

Pant *et al.* (1994) reported that organic phosphorus in soil solution is associated with a wide molecular-size range of organic material. In soils, inositol phosphates are bound to both humic and non-humic organic compounds and tend to accumulate in the humic acid fraction. Conversely, fulvic acids often contain a range of organic phosphorus compounds besides inositol phosphates (Hong and Yamane, 1980, 1981; Borie *et al.*, 1989). A similar situation occurs in lake waters (Nanny and Minear, 1994).

The interaction of organic phosphorus compounds with organic matter can occur through various abiotic processes, including physical or chemical incorporation in the organic matter fraction, direct adsorption on the organic surfaces, or indirect adsorption through bi- or trivalent cations that act as bridges and form ternary organic matter–metal–phosphorus complexes. Physical incorporation of phosphorus compounds into the core structure of organic matter (Schulten and Schnitzer, 1993) can occur either through weak van der Waals interactions, or alternatively through hydrophobic forces that aggregate organic molecules into a supramolecular structure (Piccolo, 2001). Organic phosphorus species can also react with functional groups on the organic matter to form covalent bonds (Brannon and Sommers, 1985).

The affinity of the organic moiety of phosphorus-containing pesticides with humic substances can lead to their retention in soils and sediments. For example, Sánchez-Martin and Sánchez-Camazano (1991) reported that adsorption of the thiophosphates methyl parathion and ethyl parathion by soils is controlled by the organic matter content, while the adsorption of methyl paraoxon and ethyl paraoxon is related to the clay–organic matter complexes.

Effects of the bulk solution composition on the abiotic processes

As sorption of organic phosphates occurs on surfaces with variable charges, the process is affected by the characteristics of the bulk solution, including pH, ionic strength, the nature and concentration of electrolytes, and the presence of competing anions. Solution pH affects phosphorus adsorption by influencing the charge of the reacting surfaces (Barrow *et al.*, 1980; Bolan *et al.*, 1986; Barrow, 1993) and the relative concentrations of the anionic forms of adsorbates, leading to variations in the amounts adsorbed at equilibrium rather than in the kinetics of adsorption (Shang *et al.*, 1992). In general, adsorption decreases with increasing pH, due to the formation of repulsive forces between the highly negative phosphorus compound and the progressively more negative surfaces. The pH at which a maximum sorption occurs varies with the type of organic phosphorus (Goring and Bartholomew, 1950). The sorption of DNA on montmorillonite increases below pH 5 (Greaves and Wilson, 1969), while above this value it decreases to a minimum and is confined to the external surfaces of the phyllosilicate. Adsorption of *myo*-inositol hexakisphosphate in acid soils, clays, and iron or aluminium oxides decreases with increasing pH (Fig. 6.7; Anderson and Arlidge, 1962; Anderson *et al.*, 1974; Shang *et al.*, 1992; Celi *et al.*, 2001), although different results have been reported in some acid soils (Leytem *et al.*, 2002). The decrease in adsorption capacity for *myo*-inositol hexakisphosphate is more pronounced than for phosphate (Anderson *et al.*, 1974; Shang *et al.*, 1992; Celi *et al.*, 2001), due to the fact that, at the same pH, *myo*-inositol hexakisphosphate has a lower buffer capacity than phosphate to neutralize the hydroxide ions released from the surface during adsorption (Table 6.1). Adsorption is enhanced at pH levels lower than the point of zero charge of the mineral, but in some cases formation of insoluble iron- or aluminium-phytates can occur if the pH is sufficiently low to cause mineral dissolution (Anderson and Arlidge, 1962; Anderson *et al.*, 1974). Other organic phosphorus compounds undergo a lesser pH effect during adsorption. For instance, the adsorption kinetics of glucose 6-phosphate on iron and aluminium oxides were not substantially affected by the suspension pH (Shang *et al.*, 1992).

Several studies (van Olphen, 1977; Barrow *et al.*, 1980; Bowden *et al.*, 1980;

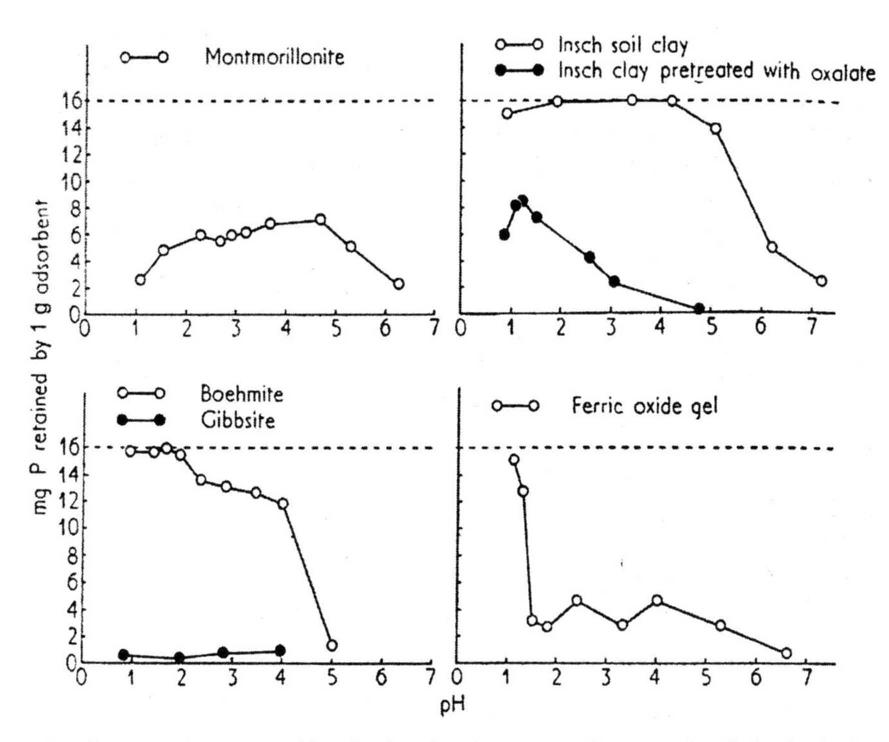

Fig. 6.7. The adsorption of *myo*-inositol hexakisphosphate by montmorillonite, Insch soil clay, boehmite, gibbsite and ferric oxide gel as affected by pH (Anderson and Arlidge, 1962).

Barrow, 1993) showed that the electric potential of the reacting surfaces changes with pH in different ways depending on the nature and concentration of the electrolyte present in the bulk solution. On increasing the concentration of monovalent ions, the adsorption of phosphate monoesters should decrease at pH values lower than the point of zero charge of the mineral and increase at pH values higher than the point of zero charge. This is a consequence of the compression of the double layer, which in turn causes a reduction of the absolute value of the surface charge. The pH at which there is no effect of the electrolyte concentration is called the 'point of zero salt effect'.

In the presence of polyvalent cations, the electric potential can be affected by the formation of bridges between cations and the adsorbents or by the formation of organic phosphate salts that precipitate on the reacting surfaces. For instance, with calcium ions the adsorption of *myo*-inositol hexakisphosphate increases above pH 5, well beyond its capacity to form a monolayer on goethite, due to the simultaneous occurrence of adsorption and precipitation of insoluble calcium-phytate (Celi *et al.*, 2001). However, even at low concentrations of calcium, the adsorption is higher than in the presence of monovalent cations (Presta *et al.*, 2000), since calcium ions cause a surface-excess of positive charge over a large pH range.

Adsorption of organic phosphates is affected by the presence of other ligands in the bulk solution, which can compete for the same sites of adsorption. For example, the presence of phosphate slightly decreases the adsorption of *myo*-inositol hexakisphosphate in soils (Anderson *et al.*, 1974) and on goethite (Presta *et al.*, 2000). Moreover, once adsorbed, *myo*-inositol hexakisphosphate is not readily displaced by other ligands, such as phosphate, citrate or carbonate (Presta *et al.*, 2000; Martin *et al.*, 2003). Conversely, *myo*-inositol hexakisphosphate can displace phosphate from mineral surfaces, either

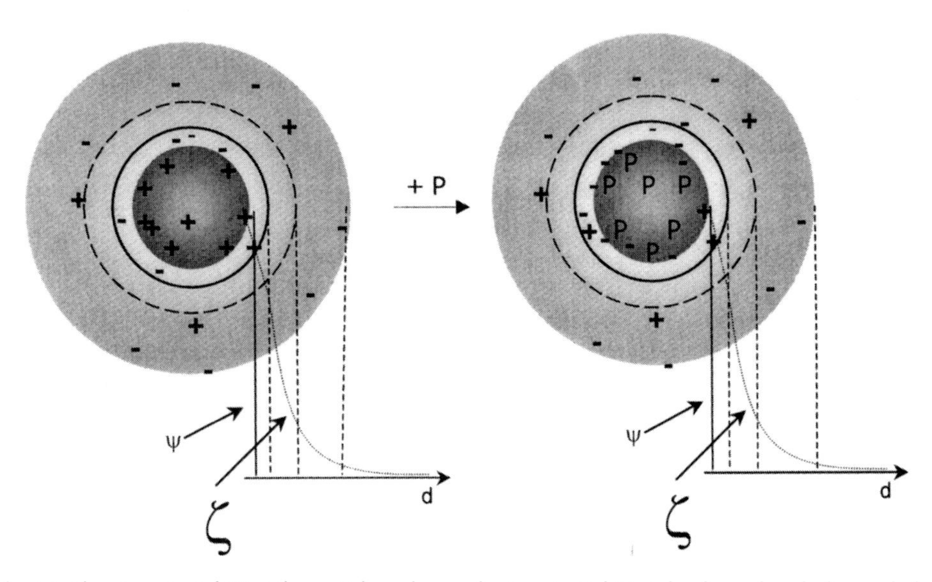

Fig. 6.8. Electric potential (ψ) at the particle surface and zeta potential (ζ) at the shear plane before and after phosphorus adsorption on oxides at initial pH lower than the point of zero charge.

before or during treatment with phosphate (Anderson *et al.*, 1974). The release of phosphate and organic matter into solution upon *myo*-inositol hexakisphosphate addition and the inhibition of their re-sorption have also been observed in soils and sediments (Anderson *et al.*, 1974; De Groot and Golterman, 1993).

Another effect caused by the presence of ligands is related to their ability to chelate metal ions by preventing the reaction with organic phosphates, although this has been shown only with phosphate (Earl *et al.*, 1979). Furthermore, the presence of organic compounds can hamper the crystallization of amorphous oxides, increasing their surface area and thus enhancing the sorption of organic phosphates.

Effects of organic phosphorus sorption on soil properties

Organic phosphorus can affect soil properties. Adsorption of organic phosphates affects the charge and electric potential of colloidal particles and, therefore, their dispersion/flocculation behaviour (Fig. 6.8). The adsorption of *myo*-inositol hexakisphosphate on goethite, illite and kaolinite can cause a significant change in the surface properties of these minerals (Celi *et al.*, 1999). The phosphate groups of *myo*-inositol hexakisphosphate that do not bind to the mineral surface have hydroxide groups that dissociate at pH>2.0 (Table 6.1), which leads to an overall net negative charge of the surface. This effect is more pronounced for phyllosilicate surfaces than for iron oxide surfaces. The specific adsorption of inositol phosphates and the charge reversal of the initially net positively charged surface cause an increase in inter-particle separation due to development of particle–particle repulsive forces. After adsorption of *myo*-inositol hexakisphosphate, minerals are dispersed in the normal soil or water pH range (Fig. 6.9), while after adsorption of phosphate dispersion starts only at pH >5 and with a high percentage of phosphorus coverage.

Colloidal behaviour is affected by the type of electrolytes present in the bulk solution, since dispersion occurs in the presence of monovalent cations, while with calcium ions the *myo*-inositol hexakisphosphate–goethite complexes flocculate (Celi *et al.*, 2001). These findings have important environmental implications, since the charge-

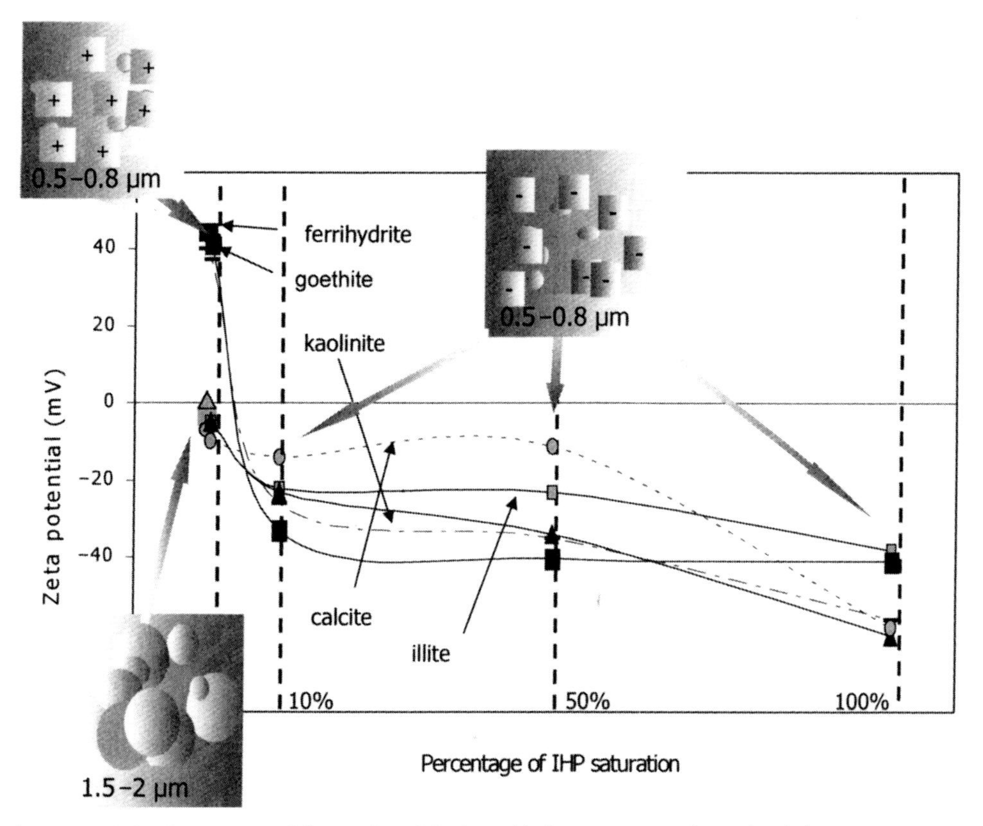

Fig. 6.9. Variation in zeta potential (ζ) and particle size with the percentage of *myo*-inositol hexakisphosphate (IHP) adsorbed on goethite, ferrihydrite, calcite, illite and kaolinite at pH 4.5 in 0.01 M KCl, except for the calcite system, for which the pH was 8.0 (adapted from Celi *et al.*, 1999, 2000, 2003).

reversal effect caused by the presence of organic phosphates can dramatically affect the transport of colloidal particles in soils and may explain the high amounts of *myo*-inositol hexakisphosphate in particulate form found in lakes and oceans (McKelvie *et al.*, 1995; Suzumura and Kamatani, 1995), contributing to the eutrophication that is currently threatening global water quality (Correll, 1998; Turner *et al.*, 2002).

Concluding Remarks

Abiotic reactions affect the amount of organic phosphorus that can be fixed or released in soils and sediments, depending on environmental properties. The high affinity of phosphate monoesters for iron and aluminium oxide surfaces and clays, their ability to form insoluble complexes with polyvalent cations, and their interaction with organic matter, account for the high accumulation of organic phosphorus in many soils and sediments. Phosphate diesters and simple organic phosphorus compounds are less readily involved in these reactions and are more likely to participate in the biological cycle, accounting for their low concentrations in soil. In aquatic environments, reducing conditions guarantee a greater resistance to degradation.

Recent significant advances have involved the expansion of studies on the interaction of organic phosphorus compounds with soils to those involving pure or more complex minerals. Particular attention has been devoted to the effects of adsorption on changes in mineral and soil properties, which have important environmental, rather than agronomic, implications. Integration of these aspects with biological processes has allowed the evaluation of the effects of abiotic reactions on phosphorus availability to plants and transport to water bodies, although further investigation is needed to better understand these processes.

In light of the studies reviewed, some directions for future research may be proposed. Eventually adsorption studies must address a greater variety of organic phosphorus compounds, including those that may be identified in the unextractable pool of soil organic phosphorus. In particular, a large proportion of the soil organic phosphorus occurs as stereoisomeric forms of inositol phosphates other than *myo* (Cosgrove, 1980; Turner *et al.*, 2002), yet possible differences in their behaviour in soil are unknown. The interaction between the different forms of phosphorus-containing compounds with organic matter should also be investigated. Attention should be paid to the influence of reaction kinetics on the fate of sorbed organic phosphorus, especially for long time periods. Finally, the major results described in this chapter should be integrated with those concerning phosphate, to obtain a more comprehensive conceptualization of phosphorus dynamics in the environment through modelling initiatives. This will allow prediction of the influence of environmental and anthropogenic actions on the phosphorus cycle, and contribute to the development of strategies for limiting transfer of phosphorus to waterbodies.

References

Anderson, G. (1980) Assessing organic phosphorus in soils. In: Khasawneh, F.E., Sample, E.C. and Kamprath, E.J. (eds) *The Role of Phosphorus in Agriculture.* American Society of Agronomy, Madison, Wisconsin, pp. 411–431.

Anderson, G. and Arlidge, E.Z. (1962) The adsorption of inositol phosphates and glycerophosphate by soil clays, clay minerals and hydrated sesquioxides in acid media. *Journal of Soil Science* 13, 216–224.

Anderson, G. and Malcolm, R.E. (1974) The nature of alkali-soluble soil organic phosphates. *Journal of Soil Science* 25, 282–297.

Anderson, G., Williams, E.G. and Moir, J.O. (1974) A comparison of the sorption of inorganic phosphate and inositol-hexaphosphate by six acid soils. *Journal of Soil Science* 25, 51–62.

Baiares-Munoz, M.A., Vivar-Cerrato, M.A. and Prieto-Garcia, M.C. (1995) Interlaminar adsorption of the organophosphorus pesticide NALED by Gador montmorillonite. *Agrochimica* 39, 290–298.

Barba, A., Navarro, M., Garcia, S.N., Camara, M.A. and Coste, C.M. (1991) Adsorption of chlorofenvinphos and methidathion on saturated clays using different cations. *Journal of Environmental Science and Health B: Pesticides, Food Contaminants and Agricultural Wastes* 26, 547–556.

Barnett, G.M. (1994) Phosphorus forms in animal manures. *Bioresource Technology* 49, 139–147.

Barrón, V., Herruzo, M. and Torrent, J. (1988) Phosphate adsorption by aluminous hematites of different shapes. *Soil Science Society of America Journal* 52, 647–651.

Barrow, N.J. (1961) Phosphorus in soil organic matter. *Soils and Fertilizers* 24, 69–73.

Barrow, N.J. (1983) A mechanistic model for describing the sorption and desorption of phosphate by soil. *Journal of Soil Science* 34, 733–750.

Barrow, N.J. (1985) Reaction of anions and cations with variable-charge soils. *Advances in Agronomy* 38, 183–230.

Barrow, N.J. (1993) Effects of surface heterogeneity on ion adsorption by metal oxides and by soils. *Langmuir* 9, 2606–2611.

Barrow, N.J. and Shaw, T.C. (1975) The slow reactions between soil and anions. 2. Effect of time and temperature on the decrease in phosphate concentration in the soil solution. *Soil Science* 119, 167–177.

Barrow, N.J., Bowden, J.W., Posner, A.M. and Quirk, J.P. (1980) Describing the effects of electrolyte on adsorption of phosphate by variable charge surface. *Australian Journal of Soil Resources* 18, 395–404.

Bates, J.A.R. and Baker, T.C.N. (1960) Studies on a Nigerian forest soil. II. The distribution of phosphorus in the profile and in various soil fractions. *Journal of Soil Science* 11, 257–265.

Bedrock, C.N., Cheshire, M.V., Chudek, J.A., Goodman, B.A. and Shand, C.A. (1994) Use of ^{31}P-NMR to study the forms of phosphorus in peat. *Science of the Total Environment* 152, 1–8.

Bolan, N.S., Syers, J.K. and Tillman, R.W. (1986) Ionic strength effects on surface charge and adsorption of phosphate and sulphate by soils. *Journal of Soil Science* 37, 379–388.

Borie, F., Zunino, H. and Martinez, L. (1989) Macromolecular-P associations and inositol phosphates in some Chilean volcanic soils of temperate regions. *Communications in Soil Science and Plant Analysis* 20, 1881–1894.

Bowden, J.W., Posner, A.M. and Quirk, J.P. (1980) Adsorption and charging phenomena in variable charge soils. In: Theng, B.K.G. (ed.) *Soils with Variable Charge*. New Zealand Soil Science Society, Lower Hutt, New Zealand, pp. 147–166.

Bowen, J.M., Powers, C.R., Ratcliffe, A.E., Rockley, M.G. and Hounslow, A.W. (1988) Fourier transform infrared and Raman spectra of dimethyl phosphonate adsorbed on montmorillonite. *Environmental Science and Technology* 22, 1178–1181.

Brannon, C.A. and Sommers, L.B. (1985) Preparation and characterization of model humic polymers containing organic phosphorus. *Soil Biology and Biochemistry* 17, 213–219.

Cabrera, F., De Arambarri, P., Madrid, L. and Toca, C.G. (1981) Desorption of phosphate from iron oxides in relation to equilibrium pH and porosity. *Geoderma* 26, 203–216.

Carman, R., Edlund, G. and Damberg, C. (2000) Distribution of organic and inorganic phosphorus compounds in marine and lacustrine sediments: a ^{31}P NMR study. *Chemical Geology* 163, 101–114.

Celi, L., Lamacchia, S., Ajmone-Marsan, F. and Barberis, E. (1999) Interaction of inositol hexaphosphate on clays: adsorption and charging phenomena. *Soil Science* 164, 574–585.

Celi, L., Lamacchia, S. and Barberis, E. (2000) Interaction of inositol phosphate with calcite. *Nutrient Cycling in Agroecosystems* 57, 271–277.

Celi, L., Presta, M., Ajmone-Marsan, F. and Barberis, E. (2001) Effects of pH and electrolyte on inositol hexaphosphate interaction with goethite. *Soil Science Society of America Journal* 65, 753–760.

Celi, L., De Luca, G. and Barberis, E. (2003) Effects of interaction of organic and inorganic P with ferrihydrite and kaolinite-iron oxide systems on iron release. *Soil Science* 168, 479–488.

Choudhry, F.A. (1988) Distribution of total inorganic and organic phosphate with respect to soil particle size. *Thailand Journal of Agriculture Science* 21,149–155.

Cole, C.V., Olsen, S.R. and Scott, C.O. (1953) The nature of phosphate sorption by calcium carbonate. *Soil Science Society of America Proceedings* 17, 352–356.

Colombo, C., Barrón, V. and Torrent, J. (1994) Phosphate adsorption and desorption in relation to morphology and crystal properties of synthetic hematites. *Geochimica et Cosmochimica Acta* 58, 1261–1269.

Condron, L.M., Frossard, E., Tiessen, H., Newman, R.H. and Stewart, J.W.B. (1990) Chemical nature of organic phosphorus in cultivated and uncultivated soils under different environmental conditions. *Journal of Soil Science* 41, 41–50.

Corbridge, D.E.C. (1985) *Phosphorus: an Outline of its Chemistry, Biochemistry and Technology*. Elsevier, Amsterdam, 761 pp.

Correll, D.L. (1998) The role of phosphorus in the eutrophication of receiving waters: a review. *Journal of Environmental Quality* 27, 261–266.

Cosgrove, D.J. (1980) *Inositol Phosphates: Their Chemistry, Biochemistry and Physiology*. Elsevier, Amsterdam, 197 pp.

Costello, A.J., Glonek, T. and Myers, T.C. (1976) ^{31}P nuclear magnetic resonance-pH titrations of myo-inositol hexaphosphate. *Carbohydrate Research* 46, 159–171.

Dalal, R.C. (1977) Soil organic phosphorus. *Advances in Agronomy* 29, 83–117.

de Groot, C.J. and Golterman, H.L. (1993) On the presence of organic phosphate in some Camargue sediments: evidence for the importance of phytate. *Hydrobiologia* 252, 117–126.

Detenbeck, N.E. and Brezonick, P.L. (1991) Phosphorus sorption by sediments from a soft water seepage lake. 1. An evaluation of kinetic and equilibrium models. *Environmental Science and Technology* 25, 395–403.

Earl, K.D., Syers, J.K. and McLaughlin, J.R. (1979) Origin of the effects of citrate, tartrate and acetate on phosphate sorption by soils and synthetic gels. *Soil Science Society of America Journal* 43, 674–678.

Espinosa, M., Turner, B.L. and Haygarth, P.M. (1999) Preconcentration and separation of trace phosphorus compounds in soil leachate. *Journal of Environmental Quality* 28, 1497–1504.

Föllmi, K.B. (1996) The phosphorus cycle, phosphogenesis and marine phosphate-rich deposits. *Earth-Science Review* 40, 55–124.

Freeman, J.S. and Rowell, D.L. (1981) The adsorption and precipitation of phosphate on to calcite. *Journal of Soil Science* 32, 75–84.

Frossard, E., Brossard, M., Hedley, M.J. and Metherell, A. (1995) Reactions controlling the cycling of P in soils. In: Tiessen, H. (ed.) *Phosphorus in the Global Environment*. John Wiley & Sons, Chichester, UK, pp. 107–135.

Frossard, E., Bucher, M., Mächler, F., Mozafar, A. and Hurrell, R. (2000) Potential for increasing the content and bioavailability of Fe, Zn and Ca in plants for human nutrition. *Journal of the Science of Food and Agriculture* 80, 861–879.

Gerritse, R.G. and Zugec, I. (1977) The phosphorus

cycle in pig slurry measured from $^{32}PO_4$ distribution rates. *Journal of Agricultural Science* 88, 101–109.
Goh, K.M. and Williams, M.R. (1982). Distributions of carbon, nitrogen, phosphorus, sulphur, and acidity in two molecular weight fractions of organic matter in soil chronosequences. *Journal of Soil Science* 33, 73–87.
Goldberg, S. and Sposito, G. (1985) On the mechanism of specific phosphate adsorption by hydroxylated mineral surfaces: a review. *Communications in Soil Science and Plant Analysis* 16, 801–821.
Goring, C.A.I. and Bartholomew, W.V. (1950) Microbial products and soil organic matter. III Adsorption of carbohydrate phosphates by clays. *Soil Science Society of America Proceedings* 15, 189–194.
Graf, E. (1983) Calcium binding to phytic acid. *Journal of Agricultural and Food Chemistry* 31, 851–855.
Greaves, M.P. and Webley, D.M. (1969) The hydrolysis of myoinositol hexaphosphate by soil microorganisms. *Soil Biology and Biochemistry* 1, 37–43.
Greaves, M.P. and Wilson, M.D. (1969) The adsorption of nucleic acids by montmorillonite. *Soil Biology and Biochemistry* 1, 317–323.
Griffin, R.A. and Jurinak, J.J. (1973) The interaction of phosphate with calcite. *Soil Science Society of America Proceedings* 37, 847–850.
Guzel, N. and Ibrikci, H. (1994) Distribution and fractionation of soil phosphorus in particle-size separates in soils of western Turkey. *Communications in Soil Science and Plant Analysis* 25, 2945–2958.
Hanley, P.K. and Murphy, M.D. (1970) Phosphate forms in particle size separates of Irish soils in relation to drainage and parent materials. *Soil Science Society of America Proceedings* 34, 587–590.
Harrison, A.F. (1987) *Soil Organic Phosphorus: a Review of World Literature.* CAB International, Wallingford, UK, 257 pp.
Hawkes, G.E., Powlson, D.S., Randall, E.W. and Tate, K.R. (1984) A ^{31}P nuclear magnetic resonance study of the phosphorus species in alkali extracts of soils from long-term field experiments. *Journal of Soil Science* 35, 35–45.
He, Z.L., Yuan, K.N., Zhu, X.Z. and Zhang, Q.Z. (1991) Assessing the fixation and availability of sorbed phosphate in soil using an isotopic exchange method. *Journal of Soil Science* 42, 661–669.
He, Z.L., Yang, X., Yuan, K.N. and Zhu, X.Z. (1994) Desorption and plant-availability of phosphate sorbed by some important minerals. *Plant and Soil* 162, 89–97.
Hong, J.K. and Yamane, I. (1980) Inositol phosphate and inositol in humic acid and fulvic acid fractions extracted by three methods. *Soil Science and Plant Nutrition* 26, 491–496.
Hong, J.K. and Yamane, I. (1981) Distribution of inositol phosphate in the molecular size fractions of humic and fulvic acid fractions. *Soil Science and Plant Nutrition* 27, 295–303.
House, W.A. and Denison, F.H. (2002) Total phosphorus content of river sediments in relationship to calcium, iron and organic matter concentrations. *Science of the Total Environment* 282/283, 341–351.
Hupfer, M., Gächter, R. and Rüegger, H. (1995) Polyphosphate in lake-sediments: ^{31}P NMR-spectroscopy as a tool for its identification. *Limnology and Oceanography* 40, 610–617.
Iyamuremye, F. and Dick, R.P. (1996) Organic amendments and phosphorus sorption by soils. *Advances in Agronomy* 56, 139–185.
Kowalenko, C.G. (1978) Organic nitrogen, phosphorus and sulfur in soils. In: Schnitzer, M. and Khan, S.U. (eds) *Soil Organic Matter.* Elsevier, New York, pp. 95–136.
Leytem, A.B., Mikkelsen, R.L. and Gilliam, J.W. (2002) Sorption of organic phosphorus compounds in Atlantic coastal plain soils. *Soil Science* 167, 652–658.
Maga, J.A. (1982) Phytate: its chemistry, occurrence, food interaction, nutritional significance, and methods of analysis. *Journal of Agricultural and Food Chemistry* 30, 1–9.
Magid, J., Tiessen, H. and Condron, L.M. (1996) Dynamics of organic phosphorus in soils under natural and agricultural ecosystems. In: Piccolo, A. (ed.) *Humic Substances in Terrestrial Ecosystems.* Elsevier Science, Amsterdam, pp. 429–466.
Makarov, M.I., Malysheva, T.I., Haumaier, L., Alt, H.G. and Zech, W. (1997) The forms of phosphorus in humic and fulvic acids of a toposequence of alpine soils in the northern Caucasus. *Geoderma* 80, 61–73.
Mandal, R. and Islam, A. (1979) Inositol phosphate esters in some surface soils of Bangladesh. *Geoderma* 22, 315–321.
Maqueda, C., Morillo, E. and Undabeytia, T. (2002) Cosorption of glyphosate and copper (II) on goethite. *Soil Science* 167, 659–665.
Martin, C.J. and Evans, W.J. (1987) Phytic acid: divalent cation interactions. V. Titrimetric, calorimetric, and binding studies with cobalt (II) and nickel (II) and their comparison with other metal ions. *Journal of Inorganic Biochemistry* 30, 101–119.
Martin, M., Celi, L. and Barberis, E. (2002) The influence of the phosphatic saturation of goethite on

phosphorus extractability and availability to plants. *Communications in Soil Science and Plant Analysis* 33, 143–153.
Martin, M., Celi, L. and Barberis, E. (2003) Desorption and plant availability of myo-inositol hexaphosphate adsorbed on goethite. *Soil Science* 169, 115–124.
McBride, M. and Kung, K.H. (1989) Complexation of glyphosate and related ligands with iron (III). *Soil Science of Society America Journal* 53, 1668–1673.
McKelvie, I.D., Hart, B.T., Cardwell, T.J. and Cattrall, R.W. (1995) Use of immobilized 3-phytase and flow-injection for the determination of phosphorus species in natural waters. *Analytica Chimica Acta* 316, 277–289.
McKercher, R.B. and Anderson, G. (1989) Organic phosphate sorption by neutral and basic soils. *Communications in Soil Science and Plant Analysis* 20, 723–732.
Mingelgrin, U. and Tsvetkov, F. (1985) Surface condensation of organophosphate esters on smectites. *Clays and Clay Minerals* 3, 62–70.
Nagarajah, S., Posner, A.M. and Quirk, J.P. (1968) Desorption of phosphate from kaolinite by citrate and bicarbonate. *Soil Science Society of America Proceedings* 32, 507–510.
Nanny, M.A. and Minear, R.A. (1994) Organic phosphorus in the hydrosphere: characterization via ^{31}P Fourier-transform nuclear magnetic resonance spectroscopy. In: Baker, L.A. (ed.) *Environmental Chemistry of Lakes and Reservoirs.* American Chemical Society, Washington, DC, pp. 161–191.
Newman, R.H. and Tate, K.R. (1980) Soil phosphorus characterization by ^{31}P nuclear magnetic resonance. *Communications in Soil Science and Plant Analysis* 11, 835–842.
Nolan, K.B. and Duffin, P.A. (1987) Effects of phytate on mineral availability: in vitro studies on Mg^{2+}, Ca^{2+}, Fe^{3+}, Cu^{2+} and Zn^{2+} (also Cd^{2+}) solubilities in the presence of phytate. *Journal of the Science of Food and Agriculture* 40, 79–85.
Ognalaga, M., Frossard, E. and Thomas, F. (1994) Glucose-1-phosphate and myo-inositol hexaphosphate adsorption mechanisms on goethite. *Soil Science Society of America Journal* 58, 332–337.
Ogram, A., Sayler, G.S., Gustin, D. and Lewis, R.J. (1988) DNA adsorption to soils and sediments. *Environmental Science and Technology* 22, 982–984.
Pant, H.K., Edwards, A.C. and Vaughan, D. (1994) Extraction, molecular fractionation and enzyme degradation of organically associated phosphorus in soil solutions. *Biology and Fertility of Soils* 17, 196–200.
Parfitt, L.R. (1979) The availability of P from phosphate–goethite bridging complexes: desorption and uptake by ryegrass. *Plant and Soil* 53, 55–65.
Parfitt, R.L., Russell, J.D. and Farmer, V.C. (1976) Confirmation of the surface structure of goethite and phosphated goethite. *Journal of Chemical Society: Faraday Transactions* 72, 1082–1087.
Peperzak, P.A., Caldwell, G., Hunziker, R.R. and Black, C.A. (1959) Phosphorus fractions in manures. *Soil Science* 87, 293–302.
Piccolo, A. (2001) The supramolecular structure of humic substances. *Soil Science* 166, 810–832.
Presta, M., Celi, L. and Barberis, E. (2000) Competizione tra fosfato e inositol fosfato su goethite. In: *Proceedings of the XVIII SICA Congress,* Catania, Italy, 22–24 September, pp. 100–108.
Rubæk, G.H., Guggenberger, G., Zech, W. and Christensen, B.T. (1999) Organic phosphorus in soil size separates characterized by phosphorus-31 nuclear magnetic resonance and resin extraction. *Soil Science Society of America Journal* 63, 1123–1132.
Sah, R.N., Mikkelsen, D.S. and Hafez, A.A. (1989) Phosphorus behavior in flooded-drained soils. II. Iron transformation and phosphorus sorption. *Soil Science Society of America Journal* 53, 1723–1729.
Sánchez-Camizano, M. and Sánchez-Martin, M.J. (1983) Factors influencing interactions of organophosphorus pesticides with montmorillonite. *Geoderma* 29, 107–118.
Sánchez-Martin, M.J. and Sánchez-Camizano, M. (1991) Relationship between the structure of organophosphorus pesticides and adsorption by soil components. *Soil Science* 152, 283–288.
Schulten, H.R. and Schnitzer, M. (1993). A state of the art structural concept for humic substances. *Naturwissenschaften* 80, 29–30.
Schwertmann, U. (1991) Solubility and dissolution of iron oxides. *Plant and Soil* 130, 1–25.
Shang, C., Huang, P.M. and Stewart, J.W.B. (1990) Kinetics of adsorption of organic and inorganic phosphates by short-range ordered precipitate of aluminium. *Canadian Journal of Soil Science* 70, 461–470.
Shang, C., Stewart, J.W.B. and Huang, P.M. (1992) pH effects on kinetics of adsorption of organic and inorganic phosphates by short-range ordered aluminum and iron precipitates. *Geoderma* 53, 1–14.
Shang, C., Caldwell, D.E., Stewart, J.W.B., Tiessen, H. and Huang, P.M. (1996) Bioavailability of organic and inorganic phosphates adsorbed on short-range ordered aluminium precipitate. *Microbial Ecology* 31, 29–39.
Sharpley, A.N. (1985) The selective erosion of plant

nutrients in runoff. *Soil Science Society of America Journal* 49, 1527–1534.

Sharpley, A.N. and Moyer, B. (2000) Phosphorus forms in manure and compost and their release during simulated rainfall. *Journal of Environmental Quality* 29, 1462–1469.

Spencer, J. and Stewart, K. (1934) Phosphate studies. I. Soil penetration of some organic and inorganic phosphates. *Soil Science* 38, 65–79.

Stevenson, F.J. (1994) *Humus Chemistry: Genesis, Composition and Reactions*. John Wiley & Sons, Chichester, UK, 496 pp.

Stewart, J.W.B. and Tiessen, T. (1987) Dynamics of soil organic phosphorus. *Biogeochemistry* 4, 1–60.

Suzumura, M. and Kamatani, A. (1995) Origin and distribution of inositol hexaphosphate in estuarine and coastal sediments. *Limnology and Oceanography* 40, 1254–1261.

Suzumura, M. and Ingall, E.D. (2001) Concentrations of lipid phosphorus and its abundance in dissolved and particulate organic phosphorus in coastal seawater. *Marine Chemistry* 75, 141–149.

Syers, J.K., Shah, R. and Walker, T.W. (1969) Fractionation of phosphorus in two alluvial soils and particle-size separates. *Soil Science* 108, 283–289.

Tate, K.R. and Newman, R.H. (1982) Phosphorus fractions of a climosequence of soils in New Zealand tussock grassland. *Soil Biology and Biochemistry* 14, 191–196.

Tiessen, H., Stewart, J.W.B. and Moir, J.O. (1983) Changes in organic and inorganic phosphorus composition of two grassland soils and their particle size fractions during 60–90 years of cultivation. *Journal of Soil Science* 34, 815–823.

Torrent, J., Barron, V. and Schwertmann, U. (1990) Phosphate adsorption and desorption by goethites differing in crystal morphology. *Soil Science Society of America Proceedings* 36, 587–593.

Turner, B.L. (2004) Optimizing phosphorus characterization in animal manures by solution phosphorus-31 nuclear magnetic resonance spectroscopy. *Journal of Environmental Quality* 33, 757–766.

Turner, B.L., Papházy, M.G., Haygarth, P.M. and McKelvie, I.D. (2002) Inositol phosphates in the environment. *Philosophical Transactions of the Royal Society of London, Series B* 357, 449–469.

Turner, B.L., Mahieu, N. and Condron, L.M. (2003) The phosphorus composition of temperate pasture soils determined by NaOH–EDTA extraction and solution ^{31}P NMR spectroscopy. *Organic Geochemistry* 34, 1199–1210.

van Olphen, H. (1977) *An Introduction to Clay Colloid Chemistry*. Wiley Interscience, London, 318 pp.

Way, J.T. (1850) On the power of soils to absorb manure. *Journal of the Royal Agriculture Society* 11, 313–379.

Williams, E.G. and Saunders, W.M.H. (1956a) Distribution of phosphorus in profiles and particle-size fractions of some Scottish soils. *Journal of Soil Science* 7, 90–108.

Williams, E.G. and Saunders, W.M.H. (1956b) Significance of particle-size fractions in readily soluble phosphorus extractions by the acetic, Truog, and lactate methods. *Journal of Soil Science* 7, 189–202.

Zech, W., Alt, H.G., Zucker, A. and Kögel, I. (1985) ^{31}P-NMR spectroscopic investigations of NaOH extracts from soils with different land use in Yucatan, Mexico. *Zeitschrift Fur Pflanzenernahrung und Bodenkunde* 148, 626–632.

Zhang, X., Amelung, W.Y.Y. and Zech, W. (1998) Amino sugar signature of particle-size fractions in soils of the native prairie as affected by climate. *Soil Science* 163, 220–229.

Zhang, Y.S., Werner, W., Scherer, H.W. and Sun, X. (1994) Effect of organic manure on organic phosphorus fractions in two paddy soils. *Biology and Fertility of Soils* 17, 64–68.

Zhu, Y., Pardini, G., Poggio, G. and Sequi, P. (1983) Distribution of phosphorus in particle-size fractions of soils treated with organic wastes. *Agrochimica* 27, 105–111.

Zinder, B., Furrer, G. and Stumm, W. (1986) The coordination chemistry of weathering. II. Dissolution of Fe (III) oxides. *Geochimica et Cosmochimica Acta* 50, 1861–1869.

7 Microbial Turnover of Phosphorus in Soil

Astrid Oberson[1] and Erik J. Joner[2]

[1]*Institute of Plant Sciences, Swiss Federal Institute of Technology (ETH), Eschikon 33, PO Box 185, CH-8315 Lindau, Switzerland;* [2]*Norwegian Forest Research Institute, Hogskoleveien 12, N-1432 Aas, Norway*

Introduction

Microorganisms constitute a large pool of phosphorus in the soil and mediate several key processes in the biogeochemical phosphorus cycle. Microbial uptake of phosphorus and its subsequent release and redistribution strongly affect the availability of phosphorus to plants in natural and managed ecosystems, especially when the latter receive organic amendments (Walbridge, 1991; Attiwill and Adams, 1993; Seeling and Zasoski, 1993; Oehl *et al.*, 2004). Phosphorus dynamics in the rhizosphere are further influenced by extracellular root and microbial phosphatase enzymes that hydrolyse organic phosphorus, microbial exudates that solubilize and/or desorb phosphate, and symbiotic mycorrhizal fungi that absorb phosphorus and transport it to plant roots (Smith and Read, 1997; Richardson, 2001).

The microbial phosphorus pool is in close contact with the soil solution (Fig. 7.1). Direct uptake of organophosphates has been reported in *Escherichia coli*, but most organic phosphorus compounds are hydrolysed in the periplasm and taken up as phosphate (Wanner, 1996; Heath, Chapter 9, this volume). Microorganisms possess a range of mechanisms to solubilize or desorb poorly available phosphate (Kucey *et al.*, 1989; Richardson, 2001) and may also absorb phosphate from the soil solution. Solubilized or desorbed phosphate that is not taken up by microorganisms may replenish soil solution phosphate. Microorganisms synthesize extracellular phosphatase to hydrolyse organic phosphorus, or access phosphate protected in cell structures of organic material, either present in native soil organic matter or added to the soil in the form of organic amendments (Gressel and McColl, 1997; Richardson *et al.*, 2001). Microbial phosphorus uptake may be less than, equal to, or greater than organic phosphorus mineralization. Therefore, phosphorus in the soil solution can be replenished or depleted during organic phosphorus mineralization (Fig. 7.1; the by-passing arrows denote solubilized/mineralized phosphorus that is not taken up by the agent-exuding microorganism).

To assign processes and substrates that determine net mineralization, we adopt the concept of soil nitrogen mineralization proposed by Mary and Recous (1994). Net mineralization is the difference between gross mineralization and immobilization. Gross mineralization consists of different processes: flush effects, basal mineralization and re-mineralization. Flush effects are caused by sequences of drying and wetting or freezing and thawing and are partly due to death and subsequent decomposition of

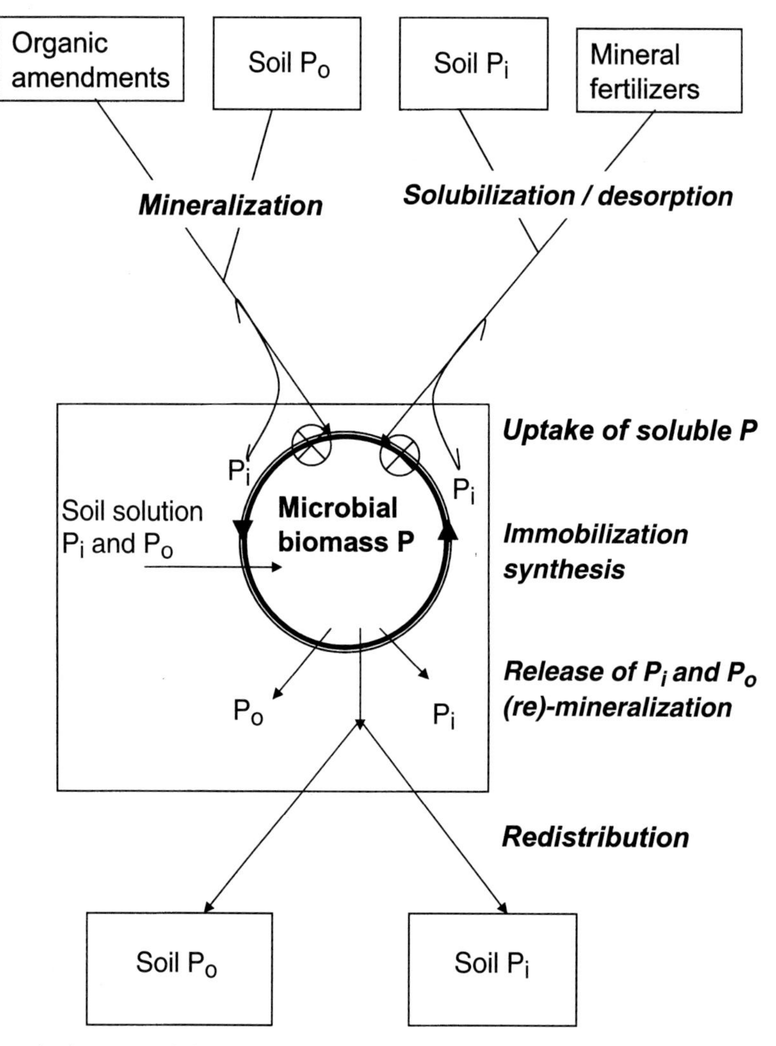

Fig. 7.1. Microbial turnover of phosphorus in soils: processes (*italic*) and related pools (rectangle). The dark arrows on the microbial phosphorus pool in the figure indicate phosphorus recycling between different generations or groups of microbes. The symbol ⊗ denotes extracellular processes in the periplasm mediated by phosphatase exoenzymes or inorganic phosphorus (P_i) solubilizing agents (P_o, organic phosphorus).

microbial cells. They probably also result from physical disturbance of the soil, which exposes physically protected soil organic matter. Basal phosphorus mineralization can be defined as the gross mineralization of organic matter in a soil that has not received fresh organic matter inputs recently (i.e. at basal soil respiration with constant respiration rates). It presents the basal potential of a soil to deliver phosphate to the soil solution from soil organic phosphorus. Re-mineralization is the release of microbial phosphorus after the main decomposition phase of organic materials (e.g. crop residues) due to microbial death and predation, and represents mineralization of recently synthesized organic phosphorus and release of inorganic phosphorus held in soil micro-organisms. Similarly to nitrogen, some phosphorus may be released immediately during fresh organic matter decomposition. Phosphorus release from microorganisms occurs when cells die, for example in response to environmental changes, starvation or predation (Bloem *et al.*, 1997; Grierson *et al.*, 1998). In addition, inorganic and organic phosphorus efflux (Maloney, 1996) and diffusion (Gächter and Meyer, 1993) from microorganisms have been reported. Microbial phosphorus immobilization signifies the incorporation of phosphorus into the living microbial biomass, reducing phosphate in the soil solution, and is accompanied by the synthesis of microbial phosphorus compounds. The composition of microbial phosphorus has implications for the fate of phosphorus released from disrupted microbial cells (Macklon *et al.*, 1997) and determines which soil phosphorus pool it will subsequently enter.

We define microbial turnover of phosphorus in soil as the sum of all microbially mediated transformations and the related fluxes of phosphorus in the soil. The quantitative description found in the literature only encompasses the phosphorus flux (Ψ) through the soil microbial biomass (i.e. turnover of microbial phosphorus), and thus excludes phosphorus fluxes to the cell externally caused by microbial activity, indicated by the by-passing arrows in Fig. 7.1:

$$\Psi = \frac{P_{mic}}{T} = r \times P_{mic}$$

where P_{mic} is the microbial phosphorus pool, T is its turnover time, and r is its turnover rate (Coe, 1968; McGill *et al.*, 1986; Chen *et al.*, 2003). The turnover time of a compartment corresponds to the mean residence time of a nutrient in the compartment and to the time required to completely renew the nutrients held in a given compartment (Coe, 1968; Jenkinson and Ladd, 1981). Its reciprocal value is the turnover rate.

This chapter reviews information on the composition of the microbial phosphorus pool and current approaches used to quantify the amount of phosphorus held in the soil microbial biomass. We compare data on microbial phosphorus concentrations in soils under different land-use systems and agricultural practices, and discuss the sources and extent of variation in the reported values. Finally, information is presented on phosphorus flux and turnover through the soil microbial biomass, including the processes regulating these values, with emphasis on immobilization and mineralization.

Forms and Quantities of Phosphorus in Soil Microbial Biomass

Phosphorus concentrations and forms in microbial cells

Soil microorganisms involved in the turnover of phosphorus include all cellular members of the soil microflora and microfauna smaller than 100 μm in diameter. These comprise bacteria, fungi, chromista (i.e. algae and oomycetes), protozoa and some nematodes (Stewart and Tiessen, 1987; Paul and Clark, 1996). Even acellular microorganisms (viruses, prions and viroids) are included, but their mass is so small and their detection so uncertain that they are rarely included in nutrient turnover studies. Fungi and bacteria together form the bulk of soil microorganisms (Killham, 1994), although other trophic groups influence their turnover (Bloem *et al.*, 1997). Bacteria are the most numerous microorganisms in soil, although their biomass is generally less than that of the fungi (Killham, 1994). For example, in 17 agricultural and forest soils with a wide range of carbon contents, fungal-to-bacterial-biomass ratios were between 2 and 9, with an average value of 3 (Anderson and Domsch, 1980). However, based on phospholipid fatty acid membrane markers, the fungal-to-bacterial-biomass ratio can be as low as 0.03–0.06 in tropical soils (Waldrop *et al.*, 2000) and 0.07–0.20 in temperate grassland soils (Grayston *et al.*, 2001), whereas it is commonly around 0.5 in coniferous forest soils (Bååth *et al.*, 1995; Pennanen *et al.*, 1998). In 25 young volcanic ash soils of Nicaragua under arable cropping, fungal biomass calculated from ergosterol measurements represented on average 26% of total biomass carbon (Joergensen and Castillo, 2001).

Table 7.1. Phosphorus concentrations of cultured soil microorganisms.[a]

Group and organism	Phosphorus (mg/g dry wt)	Reference
Fungi		
Candida utilis (yeast)	6.4	Brookes *et al.* (1982)
Cylindrocarpon sp.	3.8	Brookes *et al.* (1982)
Fusarium oxysporum	2.2–4.2[b]	Hedley and Stewart (1982)
Penicillum chryosogenum	6.1	Brookes *et al.* (1982)
Saccharomyces cerevisiae (Type 1)	15.0	Myers *et al.* (1999)
Saccharomyces cerevisiae (Type 2)	8.4	Myers *et al.* (1999)
Trichoderma harzianum	3.0–14[b]	Hedley and Stewart (1982)
Bacteria		
Aerobacter aerogenes	20.0	Brookes *et al.* (1982)
Aerobacter aerogenes	22.0	Myers *et al.* (1999)
Arthrobacter globiformus	1.7–7.0[b]	Hedley and Stewart (1982)
Azotobacter vinelandii	18.0	Brookes *et al.* (1982)
Azotobacter vinelandii	25.0	Myers *et al.* (1999)
Bacillus subtilis	14.6	Brookes *et al.* (1982)
Bacillus subtilis	48.0	Myers *et al.* (1999)
Pseudomonas fluorescens	27.4	Brookes *et al.* (1982)
Pseudomonas cepaeca	2.7–8.6[b]	Hedley and Stewart (1982)
Streptomyces clavuligerus	13.9	Brookes *et al.* (1982)

[a] Phosphorus concentrations of microorganisms used for K_p determination (Table 7.3).
[b] Grown in media of different phosphorus concentrations.

The calculation of biomass from phospholipid fatty acids is based on the percentage of the marker in total extracted phospholipid fatty acids, and from average ergosterol contents in fungi. However, fungal-to-bacterial ratios obtained from specific markers should be used with caution, because it is difficult to convert fatty acid yield into microbial biomass appropriately for individual genera, and inherent taxonomic variations in fatty acid yield affect whether signature fatty acids are seen at all in community level profiles (Haack *et al.*, 1994). Similarly, Joergensen and Castillo (2001) could not exclude the possibility that the fungi in the Nicaraguan soils they studied contained low ergosterol contents. Nevertheless, fungal-to-bacterial-biomass ratios are of interest because the two groups contribute differently to specific functions in the phosphorus cycle, react differently to environmental and nutritional factors, and in turn affect microbial phosphorus turnover (see below).

Information on the concentration and forms of phosphorus in soil microorganisms is usually obtained from cultures. Microbial phosphorus concentrations vary widely (Table 7.1) and are affected by cell age and the phosphorus concentration of the growth medium (Hedley and Stewart, 1982). Information on carbon-to-phosphorus ratios in cultured soil microorganisms is scarce. Anderson and Domsch (1980) grew ten representative species of soil bacteria and 14 soil fungi in cultures supplied with the same concentrations of glucose and phosphate. The average phosphorus concentration (mg P/g dry wt ± standard deviation) was 24.2 ± 11.4 and 31.0 ± 11.0 in bacteria and fungi, respectively. The average biomass carbon-to-phosphorus ratio was 18 in bacteria and 12 in fungi, although a tenfold reduction of glucose in the media decreased the ratio in bacteria to 7. More information exists on the nutrient composition of microorganisms growing in aquatic systems. For natural assemblages of freshwater bacteria grown with different ratios of organic carbon, nitrogen and phosphorus, Tezuka (1990) found that carbon-to-phosphorus ratios varied from 30 to 500. For *E. coli* grown in chemostat (i.e. at constant population size and in a culture media with carbon and nitrogen concentrations held constant), Makino *et al.* (2003) found only a twofold variation in the biomass carbon-to-phosphorus ratio, despite two orders of magnitude variation between carbon and phosphorus supply. This suggested that *E. coli* was strongly homoeostatic in its elemental composition. From this result, as well as a literature survey on heterotrophic prokaryotes, the authors concluded that each bacterial species regulates its elemental composition homoeostatically within a relatively narrow range of characteristic biomass carbon-to-phosphorus ratios and that shifts in the dominant bacterial species in the environment would be responsible for the large variation in bacterial carbon-to-phosphorus ratio. To our knowledge, no similar studies exist for soils, although microbial carbon-to-phosphorus ratios in soil do vary greatly between soils and during the year. For example, in a selection of arable, grassland and woodland soils in England, Brookes *et al.* (1982) reported an average microbial carbon-to-phosphorus ratio of 14 (minimum 11, maximum 36), while in 25 young volcanic ash soils in Nicaragua, the mean ratio ranged between 34 and 214 (mean 107; Joergensen and Castillo, 2001). Chen *et al.* (2003) reported microbial carbon-to-phosphorus ratios to vary seasonally from 20 to 80 in a grassland and mineral forest soil, with only minor differences between the two sites. In an incubation experiment using a chernozemic black soil, microbial carbon-to-phosphorus ratios ranged from 12, when phosphorus was readily available, to 45 when phosphorus was in limited supply (Chauhan *et al.*, 1981).

The forms of phosphorus in microorganisms are almost as diverse as the microorganisms themselves. Phosphorus is a major building block of numerous biomolecules, being especially important in energy metabolism because of the biological role of high-energy phosphoanhydride bonds (Wanner, 1996). According to Alexander (1977), phosphorus-containing compounds in bacteria and fungi include 30–50% RNA, 15–20% acid-soluble inorganic and organic phosphorus compounds (including sugar and nucleotide esters and various phosphorylated coenzymes and polyphosphates), <10% phospholipids, 5–10% DNA, and small amounts of inositol phosphates. Phosphorus is incorporated into many proteins post-translationally. The phosphorus content in growing *E. coli* was 3.2% of dry weight, of which 65% was in nucleic acids, 15% in phospholipids and 20% in the acid-soluble fraction, mainly as phosphate esters (Webley and Jones, 1971). The contribution of RNA to total bacterial cellular phosphorus increases with increasing growth rate, and there is evidence that increased growth rates also reduce bacterial carbon-to-phosphorus ratios (Makino *et al.*, 2003). In a review of the role of microorganisms in phosphorus dynamics in lake sediments, Gächter and Meyer (1993) reported that many bacteria common in aquifers, surface waters and soils are able to take up phosphate when it is readily available and store it as polyphosphate. Indeed, polyphosphate can constitute up to 20% of the dry weight in *Acinetobacter* species. In addition, Gram-positive bacteria can accumulate phosphorus in their cell walls as teichoic acids, which can account for 30% of the phosphorus in *Bacillus subtilis* cells growing exponentially (Grant, 1979). Teichoic acids are phosphorylated polymers held together by phosphate diester linkages (Archibald, 1974) and can be used as a phosphate reserve in times of phosphorus limitation (Grant, 1979).

These examples suggest that microbial phosphorus compounds are affected by the composition of the microbial community, the phosphorus content of the culture medium, and cell age (which in turn is affected by the overall living conditions of the soil microorganisms). Microorganisms directly extracted from the soil (Smith and Stribley, 1994; Bakken, 1997) should thus be analysed in order to improve our knowledge of the composition of microbial phosphorus in different soils.

With respect to the function of the microbial biomass as a source of phosphorus for plants or for transfer from soils to waterbodies, it is the reactivity and ease of hydrolysis that is of interest. Filtered extracts of a culture of the soil bacterium *Serratia liquefasciens* labelled with ^{32}P contained 87% organic phosphorus, 7% molybdate-reactive phosphorus and 6% condensed phosphorus, assumed to be polyphosphate (Macklon *et al.*, 1997). The organic phosphorus, but not the condensed phosphorus, was quickly (<10 h) hydrolysed and made available for phosphorus uptake by seedlings of *Agrostis capillaris* growing in a nutrient solution containing these extracts. In an incubation study, van Veen *et al.* (1987) showed that ^{32}P in cultured bacterial cells, presumably present in organic compounds, was mineralized within a few days after addition to soil. The incorporation of this labelled bacterial phosphorus into soil microorganisms was not any different from that observed when ^{32}P-phosphate was added in a nutrient solution. Considerable amounts of water-soluble phosphorus are released from lysed bacterial cells following re-wetting of dry soils, suggesting that microbial cells may be an important source of phosphorus transferred to watercourses following rainfall on to dry soils (Turner *et al.*, 2003b). On the other hand, organic phosphorus compounds of microbial origin, such as DNA, can accumulate in the soil due to abiotic retention (Makarov *et al.*, 2002; Turner *et al.*, 2003a; Celi and Barberis, Chapter 6, this volume).

Concentrations of soil microbial phosphorus

Methods of quantifying soil microbial phosphorus

Methods used to quantify microbial phosphorus have so far been based on fumigation and extraction. The basic methods were published early in the 1980s (Brookes *et al.*, 1982; Hedley and Stewart, 1982) and, of the methods presented in Table 7.2, that of Brookes *et al.* (1982) has been used the most frequently. Phosphorus released by fumigation is calculated as the difference between inorganic or total phosphorus extracted from fumigated and non-fumigated soil. Most phosphorus in microbial cells is organic, but is recovered as phosphate due to rapid enzymatic hydrolysis during fumigation and extraction (Brookes *et al.*, 1982). The difference between the fumigated and non-fumigated sample is then corrected for the adsorption of phosphate in the soil solid phase during fumigation–extraction by determining the recovery of a phosphate spike added to a parallel sample of unfumigated soil. The resulting value is the concentration of microbial phosphorus released by fumigation. Because not all phosphorus in the biomass is rendered extractable by fumigation, total microbial phosphorus is calculated from microbial phosphorus using a conversion factor, K_p (Brookes *et al.*, 1982; Hedley and Stewart, 1982).

The original methods have been modified to increase the accuracy of microbial phosphorus estimates on specific soils, and different protocols now exist for each step of the method (Table 7.2). Chloroform is most commonly used as fumigant, either gaseous, liquid or both. The non-carcinogenic hexanol is also effective as a fumigation biocide (McLaughlin *et al.*, 1986) and must replace chloroform when certain brands of resin strips are used for extraction, since these are damaged by chloroform (Bünemann *et al.*, 2004). The time of fumigation varies between 1 and 24 h, which depends partly on the extraction procedure. For example, Brookes *et al.* (1982) showed that sodium bicarbonate-extractable phosphorus concentrations were greatest after 2–4 h of chloroform fumigation, and that 6 h fumigation was optimal for maximum conversion of organic phosphorus into inorganic phosphorus. In contrast, 24 h seemed optimal in the case of simultaneous fumigation–resin extraction (Kouno *et al.*, 1995).

The choice of extractant depends on soil properties, and in addition to the extractants shown in Table 7.2, 0.1 M H_2SO_4 (van Veen *et al.*, 1987) and 0.5 M $KHCO_3$ (Ross *et al.*, 1999) have been used to recover phosphorus released by fumigation. Bicarbonate is the appropriate extractant for neutral and alkaline soils (Brookes *et al.*, 1982), while for strongly phosphorus-sorbing soils with relatively low microbial phosphorus concentrations, the extractant must overcome

Table 7.2. Overview of fumigation extraction methods for microbial phosphorus determination.

Reference and method number	Fumigation	Extraction method[a]	Sorption correction	Phosphorus fraction[b]	Conversion factor (K_p)
Brookes *et al.* (1982) [1]	Chloroform vapour, 24 h	0.5 M $NaHCO_3$, 30 min, 1:20	Spike of 25 mg P/kg during extraction	Phosphate	0.4
Hedley and Stewart (1982) [2]	Chloroform liquid, 1 h	0.5 M $NaHCO_3$, 16 h, 1:60	None	Total P	0.37
McLaughlin *et al.* (1986) [3]	Hexanol, 36 h	0.5 M $NaHCO_3$, 30 min, 1:20	None	Total P	Separate for each soil
Kouno *et al.* (1995) [4]	Chloroform liquid	HCO_3^- resin strips, 24 h	2.5 mg P/l	Phosphate	0.4
Morel *et al.* (1996) [5]	Chloroform liquid and vapour, 75 min; thereafter 40 h equilibration prior to extraction	Deionized water, 16 h,1:10	0–50 mg P/kg soil during 24 h equilibration and 16 h extraction	Phosphate	None
Oberson *et al.* (1997) [6]	Chloroform liquid and vapour, 75 min; thereafter 24 h equilibration prior to extraction	Bray (0.03 M NH_4F–0.025 M HCl) 1 min, 1:10	0–50 mg P/kg soil, during 24 h equilibration and extraction	Phosphate	None
Myers *et al.* (1999) [7]	Hexanol liquid, 24 h	Fe-coated paper or HCO_3^--resin strips, 24 h	25 mg P/kg during extraction	Phosphate	None
Wu *et al.* (2000) [8]	Chloroform vapour, 24 h	Bray-1 (0.03 M NH_4F–0.025 M HCl), 30 min	None	Phosphate	Separate for each soil

[a] Extractant, extraction time, soil-to-extractant ratio.
[b] Extracted phosphorus fraction (phosphate or total phosphorus) recommended for microbial phosphorus determination.

problems of low phosphorus concentrations and be sensitive to variations in concentration caused by phosphorus additions of <5 mg/P kg soil (see Tables 7.5 and 7.6). For acid tropical Oxisols with low available phosphorus contents and high phosphate sorption capacity, Bray extractant (NH_4F+ HCl) is adequate (Oberson *et al.*, 1997) and was also appropriate for temperate acid soils (Wu *et al.*, 2000). Anion exchange membranes (resin strips) were originally used on granitic soils and Andosols of high phosphate sorption capacity (Kouno *et al.*, 1995), but work equally well on strongly phosphate-sorbing acid and calcareous tropical soils (Weisbach *et al.*, 2002; Bünemann *et al.*, 2004) and excessively fertilized temperate grassland soils (Schärer, 2003). The coloration of extracts of high-organic-matter soils does not affect phosphorus determination when using resin strips. For isotope work using carrier-free radioactive phosphate (Oehl *et al.*, 2001b), deionized water should be used for phosphorus extraction (Frossard and Sinaj, 1997), although this is unsuitable for soils with low available phosphorus concentrations and high phosphate sorption capacity (Oberson *et al.*, 1997).

Correction for phosphate sorption accounts for sorption occurring during extraction (Brookes *et al.*, 1982) or during both extraction and fumigation (Morel *et al.*, 1996). According to Morel *et al.* (1996), significant underestimation of microbial phosphorus can occur if phosphorus sorption during fumigation is not accounted for. The method of Kouno *et al.* (1995) overcomes this, because the resin strips used for extraction are added at the same time as the liquid fumigant. However, since microbial phosphorus is mostly organic, the addition of a phosphate spike at the onset of fumigation cannot mimic sorption of microbial phosphate derived from hydrolysed microbial organic phosphorus compounds. To overcome this problem, Morel *et al.* (1996) proposed a short fumigation time followed by a long equilibration period. A single phosphate spike is often sufficient to determine sorption if this is close to the expected amount of phosphorus released by fumigation, and if the sorption curve is linear in the relevant range (Oberson *et al.*, 1997). Not all methods include a correction for phosphate sorption (Table 7.2). This is acceptable for soils where sorption is so low that it can be ignored (Grierson *et al.*, 1998), but may be problematic if microbial phosphorus is compared for different soils or for treatments that affect the phosphorus sorption capacity of the soil (Buchanan and King, 1992; He *et al.*, 1997).

Finally, K_p factors have been determined from the recovery of a known amount of phosphorus added to soils in lyophilized or fresh microbial cells (radiolabelled or not), but these vary considerably. Several factors may explain this variability (Table 7.3), including differences in species, age and phosphorus content of the added cells, the soil, the biocide, the extractant and the time of extraction (Brookes *et al.*, 1982; Hedley and Stewart, 1982; McLaughlin *et al.*, 1986; Ross *et al.*, 1987; Myers *et al.*, 1999). By studying changes in phospholipid fatty acid fractions after chloroform fumigation, Zelles *et al.* (1997) discovered that Gram-positive bacteria were only slightly affected by fumigation (around 30%), whereas Gram-negative bacteria and fungi decreased by around 70%. Phosphorus recovery from fungal cells may be more (Myers *et al.*, 1999) or less (McLaughlin *et al.*, 1986) than from bacterial cells. The efficiency of chloroform in lysing microbial cells is also affected by soil physical properties (Sparling and Zhu, 1993; Badalucco *et al.*, 1997), so K_p factors should be determined for each soil. However, freshly added microorganisms may react differently upon fumigation, because they are less protected physically than organisms naturally present in the soil, and the selected species may not be representative of the natural population. The K_p approach is additionally complicated because Brookes *et al.* (1982) proposed the conversion using K_p in addition to correction for phosphate sorption, while Hedley and Stewart (1982) proposed to correct the difference between phosphorus extracted from a fumigated and a non-fumigated sample with a K_p factor only (i.e. using K_p to account for both incomplete

Table 7.3. Experimental factors (K_p) to convert phosphate or total phosphorus extracted after biocide treatment of soils into the total amount of phosphorus held in the soil microbial biomass. The factors were obtained from the recovery of bacterial and fungal cells added to the soils prior to fumigation and extraction.

Reference and method number[a]	Microorganisms added for K_p determination[b]	Number of soils	Fumigant	Extractant	K_p (for phosphate)[c]	K_p (for total phosphorus)[c]
Brookes *et al.* (1982) [1][d]	Lyophilized bacteria (5) and fungi (3)	3	Chloroform	0.5 M $NaHCO_3$	0.34 (0.10–0.58)[g]	0.47 (0.17–0.76)
Hedley and Stewart (1982) [2][def]	Fresh bacterial (2) and fungal (2) cells	1	Chloroform	0.5 M $NaHCO_3$	0.32 (0.18–0.54)[h] 0.27 (0.17–0.38)[h]	0.48 (0.35–0.61)[h] 0.37 (0.35–0.38)[h]
Hedley and Stewart (1982) [2][def]	^{33}P-labelled fresh bacterial (2) and fungal (2) cells	1	Chloroform	0.5 M $NaHCO_3$	nd	0.38
McLaughlin *et al.* (1986) [3][d]	Fresh cultured soil bacterial and fungal populations	2	Hexanol	0.5 M $NaHCO_3$ HCO_3^- resin Bray (0.03 M NH_4F–0.1 M HCl)	0.24 (0.09–0.36) 0.18 (0.04–0.31) 0.39 (0.20–0.77)	0.54 (0.19–0.85) nd 0.44 (0.26–0.84)
Myers *et al.* (1999) [7][de]	Lyophilized bacteria (3) and dried fungi (2)	4	Hexanol	0.5 M $NaHCO_3$ FeO-coated paper HCO_3^- resin	0.36 (0.03–0.75) 0.49 (0.08–0.99) 0.52 (0.01–1.0)	nd
Wu *et al.* (2000) [1]	^{32}P-labelled cultured soil bacterial and fungal populations	8	Chloroform	0.5 M $NaHCO_3$ Bray-1 (0.03 M NH_4F–0.025 M HCl)	0.10 (0.02–0.29) 0.23 (0.09–0.37)	nd

nd, not determined.
[a] For fumigation and extraction the method with the corresponding number described in Table 7.2 was followed.
[b] Number of species indicated in brackets.
[c] Mean value for all soils and microorganisms included in the study, with minimum and maximum values in parentheses.
[d] Study contained separate K_p values for bacterial and fungal cells.
[e] Study contained information about the recovery of microbial P from the same species differing in phosphorus content.
[f] Effect of extraction time tested.
[g] K_p value corrected for sorption.
[h] The two experiments differed in extraction time and microorganisms; in second experiment resin extraction previous to chloroform treatment.

release of microbial phosphorus by fumigation and subsequent sorption). Because phosphorus released from added microbial cells inevitably interacts with the surrounding soil, one may in fact consider K_p capable of accounting for both processes.

Due to these uncertainties, several recent studies have avoided K_p factors altogether (Table 7.2, Tables 7.4–7.6) and reported the difference between fumigated and non-fumigated samples corrected for sorption. Some studies simply report the measured difference between fumigated and non-fumigated samples (Selles *et al.*, 1995; Grierson *et al.*, 1998; Myers *et al.*, 1999; Daroub *et al.*, 2000). Many studies, however, have applied a correction for phosphate sorption in addition to the K_p factor of 0.4 proposed by Brookes *et al.* (1982), without considering the specificity of the studied soils and their microbial community. This may partly explain the wide variations in biomass carbon-to-phosphorus ratios between soils and ecotypes (Smith and Paul, 1990; Joergensen *et al.*, 1995). Likewise, the finding of Lukito *et al.* (1998) that a phosphorus concentration as high as 64 mg/P g biomass was required to achieve 80% of the maximum biomass carbon may be explained by non-adapted K_p and K_c factors. This concentration is markedly greater than the critical phosphorus concentration in leaves or shoots of plants, and greater than most concentrations measured in cultivated soil microorganisms (Table 7.1). Thus, fumigation–extraction coupled with inappropriate conversion factors is an unsuitable tool for studying physiological traits of the biomass. Furthermore, the size of the microbial phosphorus pool, and in turn the phosphorus flux through this pool, cannot be accurately quantified with the current procedures.

Microbial phosphorus concentrations in different soils and ecosystems

The most recent reviews on soil microbial phosphorus summarized data from agricultural soils published up until 1992 (Smith and Paul, 1990; Richardson, 1994). Therefore, we review here the recent literature on forest (Table 7.4), grassland (Table 7.5) and cropped soils (Table 7.6). We include data obtained in field studies of the tropics and the temperate zone dedicated to the comparison of land-use or agricultural systems, and as far as possible, only consider studies where the collected data are complete (see Tables 7.4, 7.5 and 7.6). The compiled data illustrate (i) the quantity of microbial phosphorus and its percentage of total phosphorus; (ii) the extent of modifications in the size of microbial phosphorus in response to different land-use and agricultural practices; and (iii) linkages between changes in microbial phosphorus and the presented soil characteristics. We present all microbial phosphorus concentrations as phosphorus extractable after biocide treatment (P_{mic}) and corrected for sorption except where data were obtained using methods that include no sorption correction (i.e. methods 2, 3 and 8 in Table 7.2) and hence the uncorrected difference between fumigated and non-fumigated samples is presented. Taking into account the variation in K_p factors presented in Table 7.3, the total quantity of microbial phosphorus and its percentage of total soil phosphorus may be equal to or several times greater than microbial phosphorus. When comparing the different studies, it is worthwhile considering the implications of the different methods used to determine total, organic (see Nziguheba and Bünemann, Chapter 11, this volume), and microbial phosphorus.

Concentrations of microbial phosphorus in soils range between 0.75 for the mineral layer of a sandy Spodosol planted with loblolly pine (Grierson *et al.*, 1998) and 184 mg/P kg for the litter layer of a planted pine forest (Ross *et al.*, 1999). Microbial phosphorus contributes between 0.5% of the total phosphorus in grassland soils of New Zealand (Chen *et al.*, 2000) and 26% of the total phosphorus in the litter layer of an indigenous broadleaved–podocarp forest, also in New Zealand (Ross *et al.*, 1999). In mineral topsoil layers, microbial phosphorus increases in the order arable < forest ≤ grassland (Tables 7.4, 7.5 and 7.6), but there is extensive overlap. Similarly to microbial carbon (Smith and Paul, 1990), soil type, texture and related soil organic

Table 7.4. Microbial phosphorus concentration in soils of indigenous and planted forests.

Location, forest type and management	Soil type and texture	Layer (cm)	pH	Organic carbon (g/kg)	Total phosphorus (mg/kg)	Organic phosphorus (mg/kg)	Microbial phosphorus[a] (mg/kg)	Microbial phosphorus[a] (% of total P)	Method[b]	Reference
Central Sweden, Norway spruce forest										
Unfertilized	Haplic Podzol,	Humus layer	nd	370	650	nd	~90–120[cd]	~20	[1] [2]	Clarholm (1993)
Fertilized	sandy						~40–60[cd]	~10		
Central Germany, forests dominated by beech trees										
	No indication	0–10 mineral	3.5–8.3	18.2–180.5	180–1773	nd	24 (7–70)[c]	5.2 (2–10.5)	[1]	Joergensen *et al.* (1995)
North central Florida, loblolly pine plantation										
Unfertilized	Spodosol,	0–5 mineral	3.9	16.7	36.7	nd	1.0	2.7	[3] modified	Grierson *et al.*
Fertilized	sandy		3.8	18.3	51.1	nd	0.75	1.5		(1998)
New Zealand: comparison of forest and grassland soils (see Table 7.5)										
Indigenous forest	Typic	Litter	5.2	520	436	364	114[c]	26.1	[1] modified	Ross *et al.* (1999)
	Udivitrand,	Litter: FH	4.4	430	748	643	169[c]	22.6		
	sandy loam	0–10 mineral	4.2	116	287	221	38[c]	13.2		
Planted pine		Litter	4.9	550	902	729	184[c]	20.4		
(*Pinus radiata*)		Litter: FH	4.3	480	909	753	92[c]	10.1		
forest		0–10 mineral	4.7	82	999	666	20[c]	2.0		
New Zealand: comparison of forest and grassland soils (see Table 7.5)										
Planted pine	Dystrochrept	L layer	4.0	516	639	nd	15[c]	2.3	[1]	Chen *et al.* (2000)
(*Pinus ponderosa*		F layer	4.4	454	803	nd	201[c]	25		
and *P. nigra*)		0–5	4.6	63	948	598	5[c]	0.5		

nd, not determined.
[a] All values present microbial phosphorus extracted after soil fumigation (i.e. no conversion factor (K_p) applied).
[b] See Table 7.2.
[c] Original data were published after conversion using a K_p value of 0.4.
[d] Data read from graph in original publication.

Table 7.5. Microbial phosphorus concentration in the topsoil layer of permanent grassland and pasture soils.

Location and grassland type	Soil type and texture	Layer (cm)	pH	Organic carbon (g/kg)	Total phosphorus (mg/kg)	Organic phosphorus (mg/kg)	Microbial phosphorus[a] (mg/kg)	Microbial phosphorus[a] (% of total P)	Method[b]	Reference
Colombia, introduced pastures replacing native savanna										
Native savanna	Oxisol, 39% clay,	0–10	4.8	23.5	179	64	5.2	2.9	[6]	Oberson *et al.* (1999)
Grass-only	19% sand	0–10	4.85	23.2	228	68	5.9	2.6		
Grass–legume		0–10	4.96	24.9	226	77	7.3	3.2		
New Zealand, comparison of forest and grassland soils (see Table 7.4)										
Grass–legume pasture	Typic Udivitrand sandy loam	0–10	5.2	110	1390	872	58[c]	4.2	[1] modified	Ross *et al.* (1999)
		10–20	5.3	62	885	470	18[c]	2.0		
New Zealand, comparison of forest and grassland soils (see Table 7.4)										
Permanent grassland (grass pasture)	Dystrochrept (total 4 layers)	0–5	5.2	76.5	1035	724	6.4[c]	0.6	[1]	Chen *et al.* (2000)
		5–10	5.1	57.3	960	675	4.8[c]	0.5		
Switzerland, permanent grassland										
Mixed swards dominated by grasses	Calcaric regosol, 34% clay, 43% sand	0–4	6.9	90.3	1421	944	106	7.5	[4]	Schärer (2003)
	Dystric cambisol, 40% clay, 54% sand		4.7	50.7	1194	822	87	7.3		
UK, permanent grassland soils										
	Range of soil types and texture; 22–68% clay	0–10	4.4–6.8	29–80	400–2000	nd	41 (12–96)[c]	4.6 (1.6–8.1)	[1]	Turner *et al.* (2001)
UK, Northumberland, meadow hay trial with different fertilization since 1897										
FYM+NPK	Wigton Moor	0–15	5.3	53	2200[d]	nd	71.2	3.2	[8]	He *et al.* (1997)
FYM	Clay loam		5.4	53	1900[d]		77.2	4.1		
Non-fertilized			5.2	33	350[d]		14.8	4.2		
N			3.7	63	750[d]		4.4	0.6		
P			5.7	43	1200[d]		61.6	5.1		
NPK			5.5	37	820[d]		35.2	4.3		

nd, not determined.

[a] All values represent microbial phosphorus extracted after soil fumigation (i.e. no conversion factor (K_p) applied).

[b] See Table 7.2.

[c] Original data was published after conversion using a K_p value of 0.4.

[d] Original data reported in percentages.

matter content are the major factors producing this broad range in microbial phosphorus concentrations.

The linkage between organic matter and microbial phosphorus content is illustrated by the different layers in forest soils (Table 7.4). In the forest floor, absolute and relative concentrations of microbial phosphorus are much greater than in the underlying mineral soil. Where the humus layer was divided into a litter and fermentation layer, higher microbial phosphorus concentrations were clearly found in the fermentation layer (Chen *et al.*, 2000). The mineral layers of forest soils also contain significant concentrations of microbial phosphorus (Joergensen *et al.*, 1995). In contrast, the sandy Spodosol studied by Grierson *et al.* (1998) contained a low microbial phosphorus concentration, but also a low organic carbon concentration. Fertilization with phosphorus decreased the absolute and relative microbial phosphorus concentration in the humus layer in the studies of Clarholm (1993) and Grierson *et al.* (1998), while Joergensen and Scheu (1999) found an increase in microbial phosphorus concentration in the litter layer of forest soils and no change in the underlying mineral layer. The different response to phosphorus fertilization may be due to differences in the availability of soil phosphorus for microorganisms, because several studies suggest that nutrient cycling is limited by phosphorus in forests growing on highly weathered soils (Attiwill and Adams, 1993; Cleveland *et al.*, 2002).

Microbial phosphorus concentrations between 4 and >100 mg P/kg have been determined in grassland soils (Table 7.5). Concentrations decrease with increasing soil depth in parallel with decreasing soil organic matter concentrations. Tropical pasture soils containing little available and total phosphorus have a similar proportion of total phosphorus in the soil microbial biomass as temperate grassland soils, except for the excessively fertilized sites sampled by Schärer (2003). In a fertilization experiment, farmyard manure and phosphorus fertilization resulted in the largest microbial phosphorus concentrations, whereas ammonium sulphate decreased microbial phosphorus in spite of this treatment having a higher organic matter content than the phosphorus treatment (He *et al.*, 1997). As no correction for sorption was made in this study, lower total and available phosphorus contents in the nitrogen treatment were probably related to stronger sorption and may thus have lowered the recovery of microbial phosphorus. Significantly higher microbial phosphorus in grass–legume than in grass-only swards and savanna showed that the botanical composition of grasslands affects the size of the microbial phosphorus pool through changes in the overall biological activity in the soil–plant system caused by legumes (Oberson *et al.*, 1999). Organic carbon and organic phosphorus were also greater in grass–legume than in other pasture types. A strong positive correlation between soil organic matter and microbial phosphorus in grassland topsoils (0–4 cm) was found by Schärer (2003), whose values for extractable microbial phosphorus ranged from 30 to about 115 mg P/kg soil, with soil organic matter varying from 9% to about 18%.

The linkage between concentrations of soil organic carbon and microbial phosphorus may be the major reason why cropped soils usually contain less microbial phosphorus than forest and grassland soils. The repeated addition of organic materials to cropped soils tends to increase microbial phosphorus concentrations (Table 7.6). Such additions can be plant biomass from fallow plants in tropical agroforestry systems (Nziguheba *et al.*, 1998; Maroko *et al.*, 1999; Lehmann *et al.*, 2001; Smestad *et al.*, 2002; Bünemann *et al.*, 2004), or a ley phase included in the crop rotation (Gijsman *et al.*, 1997; Oberson *et al.*, 2001). Likewise, cropped soils in organic farming systems, which rely on organic fertilizers, contain almost twice as much microbial phosphorus as conventionally cultivated soils receiving only mineral fertilizers (Table 7.6) (Oberson *et al.*, 1996; Oehl *et al.*, 2001b). However, the absence of pesticides may also contribute to higher microbial phosphorus concentrations in organically cultivated soil (Oberson *et al.*, 1993). Higher microbial phosphorus concentrations in the topsoil layer of no-tillage than ploughed soils also coincide with higher soil

Table 7.6. Microbial phosphorus concentration in the topsoil layer of cropped soils.

Location and cropping or farming system	Soil type and texture	Layer (cm)	pH	Organic carbon (g/kg)	Total phosphorus (mg/kg)	Organic phosphorus (mg/kg)	Microbial phosphorus[a] (mg/kg)	Microbial phosphorus[a] (% of total P)	Method[b]	Reference
Colombia, introduced pastures and rice replacing native savanna										
Savanna	Oxisol, silty clay	0–10	4.9	26.1	216	82	5.4	2.4	[6]	Oberson *et al.* (2001)
Grass–legume			4.8	28.4	272	98	6.6	2.4		
Continuous rice			4.7	24.7	35	98	2.6	0.7		
Kenya, maize cropping systems, average of 0P and P fertilized treatments										
Continuous maize	Oxisol, 49% clay,	0–15	5.5	19.2	694	227	2.6	0.4	[4]	Smestad *et al.* (2002)
Weed fallow	30% sand			19.1	703	228	2.6	0.4		
Tithonia fallow				20.7	716	233	3.1	0.4		
Crotalaria fallow				20.8	718	245	3.5	0.5		
Kenya, maize cropping systems, average of 0P and P fertilization treatments										
Continuous maize	Oxisol, 39% clay,	0–15	4.9	23.6	779	264	3.5	0.4	[4]	Bünemann *et al.* (2004)
Natural fallow	37% sand		5.1	25.5	775	272	5.3	0.7		
Crotalaria fallow			5.0	26.8	770	286	6.4	0.8		
Saskatchewan, Canada, crop rotations										
Continuous wheat	Orthic brown	0–15	6.5	nd	868	270	19	2.2	[2]	Selles *et al.* (1995)
Fallow–wheat–wheat	Chernozemic soil				943	305	21	2.2		
Michigan, USA, tillage systems (sites Kalamazoo and Misteguay)										
No-tillage	Mesic typic	0–2	5.5	10.2	367	108	3.9	1.1	[2]	Daroub *et al.* (2000)
Conventional tillage	Hapludalf, 14% clay, 42% sand		6.3	7.0	360	57	2.4	0.7		
No-tillage	Mesic aeric									
Conventional tillage	Endoaquept;	0–2	7.9	17.3	955	159	11.5	1.2		
	54% clay, 5% sand		8.1	12.5	871	160	7.9	0.9		
Switzerland, arable soils under different farming systems										
Conventional-mineral	Haplic Luvisol on	0–20	5.9	13.3	727	359	9.0	1.2	[5]	Oehl *et al.* (2001a)
Biodynamic	loess; loamy silt		6.8	15	696	398	17.6	2.5		
Bio-organic			6.4	13.7	694	375	16.5	2.4		

nd, not determined.

[a] All values present microbial phosphorus extracted after soil fumigation (i.e. no conversion factor (K_p) applied).

[b] See Table 7.2.

organic matter concentrations (Aslam *et al.*, 1999; Daroub *et al.*, 2000). As in forest soils, soil texture affects microbial phosphorus concentrations through its relationship with soil organic carbon dynamics. For example, microbial phosphorus decreased in the order loam > sandy loam > sandy soil in a crop residue decomposition incubation study (Huffman *et al.*, 1996). Microbial phosphorus and total organic carbon concentrations were almost three times higher in loamy than in sandy soil.

In contrast to the forest soils, a recent study (Bünemann *et al.*, 2004) provided evidence that microbial phosphorus in cropped tropical soils containing little available phosphate was limited by the availability of carbon and nitrogen, rather than phosphorus, because microbial phosphorus concentrations increased substantially when supplied with carbon and nitrogen in a laboratory experiment. This confirmed field observations in a soil where phosphorus availability is usually the primary factor limiting plant growth (Jama *et al.*, 1997). Thus, the soil microbial population seems to be adapted to the severe chemical constraints of highly weathered tropical soils (Oberson *et al.*, 2001). The key question of whether microbial phosphorus in low-phosphorus tropical soils protects phosphorus from being sorbed and improves phosphorus availability for plants, however, requires a knowledge of fluxes through the microbial pool, plus the sources and sinks for microbial phosphorus, and cannot be answered using simple estimates of microbial phosphorus alone.

Phosphorus Flux through the Soil Microbial Biomass

The turnover of microbial phosphorus has two components. One is regulated by temporal net changes in microbial phosphorus, the other is determined by microbial activity and occurs even if biomass remains unchanged (Wardle, 1998). Because turnover does not necessarily manifest itself as a change in net pool size, the fluxes resulting from net pool size changes may be considered as additional to the fluxes at steady state. The rate of turnover regulated by net changes can be estimated from the fluctuations in the microbial phosphorus pool, while the detection of the fluxes at steady state requires a tracer.

Net changes in soil microbial biomass

Temporal fluctuations in microbial phosphorus concentrations under field conditions have been monitored to study the role of microbial phosphorus as a sink and source of plant-available phosphorus (Tate *et al.*, 1991; Ross *et al.*, 1995). An increase in microbial phosphorus indicates immobilization in microorganisms, while a decrease indicates that phosphorus has been released from them (Fig. 7.2). Significant net changes in the microbial biomass result from sudden changes in environmental conditions. The flush of phosphorus from microbial cells that lyse during a freeze–thaw cycle constitutes the main phosphorus source at the beginning of the growing season in Alaskan coastal tundra (Chapin *et al.*, 1978). Similarly, soil drying and re-wetting causes flushes of phosphorus that are at least partly derived from lysed microbial cells (Campo *et al.*, 1998; Chepkwony *et al.*, 2001; Turner *et al.*, 2003b). Subsequent increases in extractable phosphorus may contribute to the phosphorus uptake of growing plants (Chepkwony *et al.*, 2001) as well as increasing the risk of phosphorus losses from soils (Turner and Haygarth, 2001; see Turner, Chapter 12, this volume). Water potential increases lower than 2.0 MPa, however, did not cause significant cell lysis of bacteria cultured at a low solute water potential (−3.0 MPa) (Halverson *et al.*, 2000), although intracellular organic solutes such as amino acids and low-molecular-weight neutral sugars were released into the extracellular environment during desiccation. Because phosphorus is a building block of some of these compounds, microbial phosphorus release may also occur through this process. Related phosphorus fluxes may be low, but according to our knowledge have

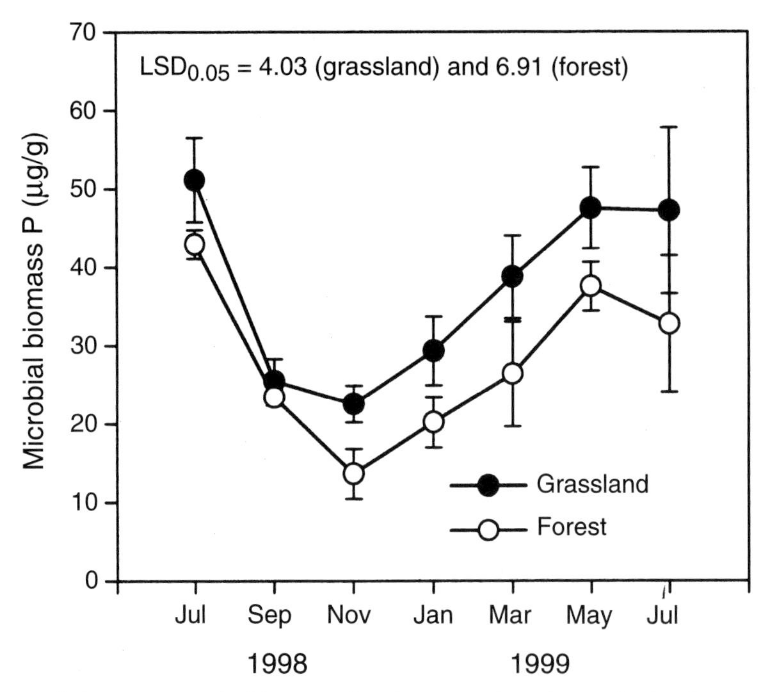

Fig. 7.2. Seasonal changes in microbial biomass phosphorus in soils under grassland or forest. Reprinted from Chen *et al.* (2003) with permission from Elsevier Science.

not been quantified. Because the various groups of soil microorganisms differ widely in their susceptibility to water stress (Killham, 1994), phosphorus fluxes related to the osmoregulatory strategy, as well as the extent of phosphorus flushes due to cell death upon wetting–drying cycles, may be determined partly by the composition of the microbial community.

Seasonal changes in soil microbial phosphorus seem to be mostly related to gravimetric water content in the soil (McLaughlin *et al.*, 1988; Chen *et al.*, 2003). Microbial phosphorus increased markedly (sixfold) and rapidly during the first 7 days after wheat sowing and after wetting of an initially dry soil (McLaughlin *et al.*, 1988). At high moisture content, only small and mainly non-significant temporal fluctuations in microbial carbon and phosphorus occurred in pasture soils, although these were related to variability in total carbon and nitrogen concentrations (Ross *et al.*, 1995). Likewise, temporal fluctuations in microbial biomass in cropped soils were mainly attributed to differences in climatic conditions, with smaller effects of soil tillage, crop rotation and the growth cycle of the crop (Buchanan and King, 1992; Oberson *et al.*, 1996). Seasonal fluctuations in microbial phosphorus concentrations usually counteract similar fluctuations in available phosphorus. While Chen *et al.* (2003) reported decreasing microbial phosphorus concentrations and increased concentrations of readily available phosphorus coinciding with increased phosphorus demand by growing plants in grassland and forest soils, Oberson *et al.* (1999) observed increased microbial phosphorus and decreased available phosphorus when tropical pastures started growing at the onset of the rainy season. None the less, grass–legume pasture soils in the latter study, which contained the highest microbial phosphorus concentrations, maintained higher organic and available phosphate levels with less variation than the other pasture types.

Under optimal soil moisture and temperature, a marked immobilization–

re-mineralization sequence follows the addition of organic material to soils. The pattern, duration and quantity of phosphorus turned over during such a sequence depend on the substrate, the microbial biomass (size, activity, composition), soil properties and the community structure of soil flora and fauna. For example, phosphorus immobilization in microorganisms was reported to increase in parallel with the proportion of soluble carbon in the added substrate (glucose, cellulose and various plant materials) and with the initial size of the microbial biomass (Bünemann, 2003). Rapid growth of the soil microbial biomass after glucose addition resulted in a marked (two- to sixfold) increase in microbial phosphorus and a marked decrease in concentrations of phosphorus extractable by water (Oehl *et al.*, 2001b) or resin (Bünemann, 2003). The depletion of the water- or resin-extractable phosphorus amounted to less than the increase in microbial phosphorus, suggesting that the microbial biomass took up phosphorus from other pools as well, or that extractable phosphorus was rapidly replenished. From a range of glucose and nitrogen additions to Oxisols having little plant-available phosphorus, Bünemann *et al.* (2004) suggested that microorganisms acquired phosphorus from poorly-available mineral phosphates, or through mineralization of organic phosphorus. Thus, immobilization in microorganisms may retain phosphorus in a potentially bioavailable form. The decrease in easily extractable phosphorus fractions, however, illustrates the potential of microbial phosphorus immobilization to decrease phosphorus availability to plants. The immobilization of soil-derived phosphorus in microorganisms has also been repeatedly observed following incorporation of plant residues, and subsequent increases in microbial phosphorus have been larger than the amount of phosphorus added in the litter (Ofori-Frimpong and Rowell, 1999; Schomberg and Steiner, 1999). In contrast, the addition of plant residues from *Tithonia diversifolia* alone or with triple superphosphate increased microbial phosphorus, resin-, bicarbonate- and hydroxide-extractable phosphate (Nziguheba *et al.*, 1998). This concomitant increase in microbial phosphorus and labile phosphate fractions contrasts with most field and incubation studies, in which these phosphorus pools fluctuated in opposite directions. When both these parameters react similarly, as observed in the study of Nziguheba *et al.* (1998), this may be partly explained by drying of the samples prior to analysis of labile phosphate.

Upon release, phosphorus from microorganisms may increase concentrations of organic phosphorus (Chauhan *et al.*, 1979) or soluble phosphate (Oehl *et al.*, 2001b) in solution, or be absorbed by roots if release is synchronized with the demands of growing plants. The time elapsed between phosphorus immobilization and subsequent re-mineralization is affected by substrate quality and soil properties. Readily available carbon substrates result in immobilization maxima a few days after substrate addition, followed by a sharp decrease in microbial phosphorus. Immobilization and re-mineralization patterns are much less pronounced after the addition of substrates that are less easily degradable (Bünemann, 2003). During an incubation experiment with plant residues varying in degradability and mineral nutrient content, microbial phosphorus did not return to the initial level after 180 days with any of the plant residue additions (Bünemann, 2003). The addition of glucose and nitrogen to a ^{33}P-labelled soil resulted in almost complete re-mineralization of ^{31}P and ^{33}P after 70 days of incubation (Oehl *et al.*, 2001b). The mineralization and fate of phosphorus from bacteria are also affected by soil texture. Mineralization of phosphorus from ^{32}P-labelled bacteria was more rapid in a sandy loam than in a clay, and a lower proportion of the applied label was accounted for in the soil biota of the sandy loam (van Veen *et al.*, 1987).

Besides cell death due to environmental changes or a declining population after substrate depletion, mineral nutrients immobilized in microbial biomass are released when microorganisms are grazed by microbivores such as protozoa and nematodes. Their nutrient ratios are similar or even wider than those of their prey and their growth efficiencies are generally 40% or less. Thus, predation will result in excretion of surplus nitrogen and phosphorus in the form of ammonium and phosphate (Bloem *et al.*, 1997). Microcosm experiments that include bacterial grazers have demonstrated that net phosphorus mineralization can occur less than 1 week after carbon supply is increased, whereas microcosms containing bacteria but no grazers maintained high microbial phosphorus immobilization and showed no net phosphorus mineralization over >3 weeks (Cole *et al.*, 1978). Soil fauna (e.g. earthworms) also stimulate microbial phosphorus turnover (Tiunov *et al.*, 2001; Jiménez *et al.*, 2003), and an increase in species and trophic diversity of mesofauna affects fungal biomass and organic matter decomposition processes (Cortet *et al.*, 2003). Reported carbon-to-phosphorus ratios of the microbial biomass vary widely, and we have discussed the uncertainty of these ratios earlier in this chapter. Therefore, to better understand phosphorus mineralization in the food web of soil organisms, there is a need to accurately determine carbon-to-phosphorus ratios of microorganisms and to study the effects of different ratios on phosphorus release following predation.

Phosphorus flux at constant size of the microbial biomass

Recent studies using isotope techniques suggest that significant phosphorus fluxes occur through the soil microbial biomass in the absence of net changes in the size of the microbial phosphorus pool. Thus, a few days after labelling with isotopically exchangeable phosphate, between 2% and 25% of the label was recovered in microbes (Oberson *et al.*, 2001; Oehl *et al.*, 2001b; Kouno *et al.*, 2002). This high incorporation may be partly explained by increased microbial activity following stirring and mixing of soils at labelling (Bünemann, 2003), but phosphorus cycling in the absence of net changes in microbial phosphorus suggests a dynamic equilibrium between phosphorus uptake and release by the soil microbial biomass (i.e. that immobilization and mineralization of microbial phosphorus occur simultaneously). The respective evolution of the specific activity in the microbial biomass and the soil solution (i.e. their convergence; see Fig. 7.3), also underlines the close linkage between microbial phosphorus turnover and the soil solution phosphate concentration (Oehl *et al.*, 2001b). By comparing sterile and non-sterile grassland soils, Seeling and Zasoski (1993) demonstrated that microbial phosphorus is a major factor controlling organic and inorganic phosphorus concentrations in soil solution.

Processes underlying phosphorus release from microorganisms include microbial decay and predation (Bloem *et al.*, 1997). In the absence of net changes in the soil microbial biomass, these processes are balanced by growth and related phosphorus uptake by another generation of microorganisms. Furthermore, the exchange of phosphate and sugar phosphates between microbial cells and the external cell medium (Maloney, 1996) and phosphorus diffusion (Gächter and Meyer, 1993) from microbial cells may also explain phosphorus fluxes in the absence of net changes. The two latter processes have been studied in detail in cultured bacteria and aquatic systems, respectively, but there is a lack of information about their significance in the soil phosphorus cycle, notably concerning their role in regulating phosphorus concentrations in soil solution.

Turnover time of microbial phosphorus

Temporal fluctuations in microbial phosphorus (Fig. 7.2) have been used to estimate the annual phosphorus flux through the soil microbial biomass and the turnover time of microbial phosphorus under field conditions (Chen *et al.*, 2003). This approach assumes that the sum of the fluctuations representing a decrease in microbial phosphorus is equivalent to the mean annual biomass turnover (McGill *et al.*, 1986). Results depend on the frequency and time of sampling, but allow a qualitative compari-

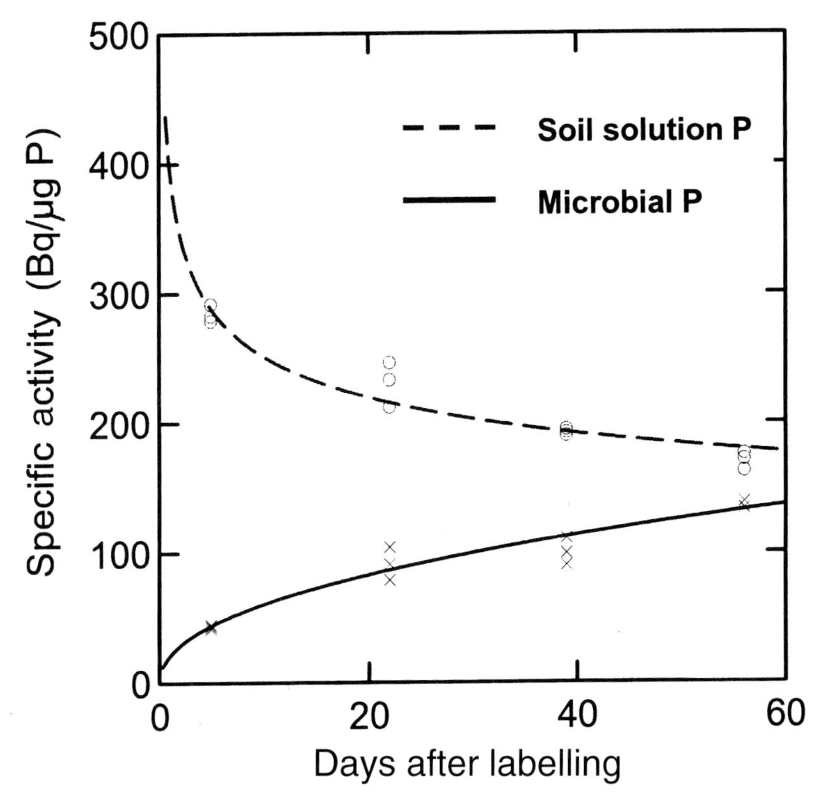

Fig. 7.3. Evolution of the specific activities in the soil solution and in microbial phosphorus in the absence of significant changes in the phosphorus concentration in the soil solution or in microbial phosphorus. Reprinted from Oehl *et al.* (2001b) with permission from Springer-Verlag.

son of phosphorus fluxes in different soils, given a carefully designed sampling regime. For instance, Chen *et al.* (2003) estimated the annual phosphorus release from the soil microbial biomass in New Zealand soils to be approximately 30 mg P/kg (equivalent to 14 kg P/ha) under grassland and 35 mg P/kg (equivalent to 15 kg P/ha) under forest. The estimated turnover rate was higher in the forest soil (1.3/year) than in the grassland soil (0.8/year), resulting in turnover times of 0.8 and 1.3 years for soil microbial biomass in forest and grassland soils, respectively (Table 7.7).

Kouno *et al.* (2002) estimated the turnover times of soil microbial biomass carbon and phosphorus by following decreases in biomass ^{14}C and ^{32}P following the addition to soil of ^{14}C-labelled glucose and $KH_2{}^{32}PO_4$, or of ryegrass which had been double labelled with $^{14}CO_2$ and $KH_2{}^{32}PO_4$. The turnover time of biomass phosphorus was about 37 days in the case of glucose addition and 42 days in the case of ryegrass addition (Table 7.7). Biomass carbon had a much longer turnover time of about 82 days for glucose amendment and 95 days for ryegrass amendment. These shorter turnover times for microbial phosphorus than for carbon may indicate that phosphorus enters a part of the biomass that is continuously degraded and re-synthesized within living cells, or that is more easily degradable upon cell death. In contrast, microbial carbon may be incorporated to a greater extent into structural materials that remain unaltered during the lifetime of a cell, or that are subsequently degraded more slowly. However, the conjunct application of isotopes and fumigation–extraction is difficult (van Veen

Table 7.7. Turnover times of microbial phosphorus in soil.

Reference	Climate	Method	Turnover time
Brookes *et al.* (1982)	Temperate soils, field measurement of microbial carbon in England	Calculated for biomass carbon by (Jenkinson and Ladd, 1981)	2.5 years
Oehl *et al.* (2001a)	Temperate soils, incubation at 20°C	^{33}P isotope tracer at constant biomass	70–160 days
Kouno *et al.* (2002)	Temperate soil, incubation at 25°C	Addition of glucose with $KH_2{}^{32}PO_4$ or $^{32}PO_4$ labelled ryegrass	Around 40 days
Chen *et al.* (2003)	Temperate soil, field measurement of microbial phosphorus in New Zealand	Temporal fluctuations of microbial phosphorus in the field	0.8 and 1.25 years for forest and grassland soils

et al., 1987) and requires correction for the incomplete recovery of labelled phosphorus derived from lysed microbial cells due to sorption and/or exchange of the label (McLaughlin *et al.*, 1988; Oberson *et al.*, 2001; Oehl *et al.*, 2001b). If these corrections are not applied, then over- or underestimation of the ^{33}P released from microbial cells lysed upon fumigation cannot be excluded.

More studies have been conducted on the turnover of microbial carbon and nitrogen than phosphorus. Turnover times range from around 70 days to 2.5 years (Prosser, 1997), although values are much shorter for specific groups in specific habitats (e.g. the rhizosphere; Breland and Bakken, 1991). In turn, spatial variation of related microbial nutrient turnover in the soil may be large. The experiment of Oehl *et al.* (2001b) provided some information on the turnover time of microbial phosphorus in the absence of net changes in size. The specific activity of microbial phosphorus and water-extractable phosphorus converged with increasing incubation time (Fig. 7.3). This convergence is used to indicate the time required for renewal of microbial phosphorus (i.e. the turnover time), which in this case was 70 days. Because at constant pool size the phosphorus flux is obtained by dividing the pool size by the turnover time (Coe, 1968), the related annual phosphorus flux at a microbial phosphorus content of 17.6 mg/kg soil would be approximately 92 mg P/kg soil.

Microbially Mediated Processes in the Soil Phosphorus Cycle

Mineralization of soil organic phosphorus

Mineralization roughly concerns three distinct organic phosphorus sources: microorganisms, soil organic matter and freshly added organic material (e.g. plant material, animal manure, compost). This distinction is useful in assigning factors that determine mineralization processes and in analysing different mineralization patterns. Grierson *et al.* (1998) quantified net phosphorus mineralization from net changes in potassium-chloride-extractable phosphate during drying and re-wetting cycles because, in the studied Florida Spodosol, sorption of added phosphate was considered negligible. Net mineralization in rewetted soils progressed in three stages: (i) an initial flush where phosphate was brought into solution, the source most probably being dead microbial biomass from the previous drying period and mineralization of organic substrates; (ii) a lag of a few days during which no net release of phosphorus occurred; and (iii) a period that followed similar kinetics to the undried soil, where the microbial biomass had recovered sufficiently to mineralize phosphorus from soil organic matter. Net mineralization during phase (i) was between 4 and 5 mg P/kg soil, of which around 99% was released during the first 24 h after re-wetting. During phase (iii), which is the phase of basal soil organic phosphorus mineralization, net daily mineralization rates were between 0.07 and 0.1 mg P/kg soil per day.

Phosphorus mineralization includes biological and biochemical processes (McGill and Cole, 1981). Biological mineralization is defined as the release of inorganic phosphorus from organic materials during oxidation of carbon by soil organisms, and is driven by the need for energy (McGill and Cole, 1981), being closely linked to carbon mineralization (Gressel *et al.*, 1996). Biochemical mineralization is defined as the release of inorganic phosphorus from organic compounds through enzymatic hydrolysis outside the cell. A negative feedback exists between phosphorus availability to the organism and phosphatase biosynthesis in the cell and its excretion into the extracellular medium (Wanner, 1996; Wykoff *et al.*, 1999; Moura *et al.*, 2001), so biochemical mineralization can be considered to be controlled by the need of the microorganism for phosphate (McGill and Cole, 1981). Phosphatase remains functional in soils sterilized by γ-irradiation (Zou *et al.*, 1992; Oehl *et al.*, 2001a), indicating that biochemical mineralization can continue independently of the presence of living microorganisms. Richardson *et al.* (2001) demonstrated that the presence of soil microorganisms enhanced plant growth with *myo*-inositol hexakisphosphate as the sole phosphorus source. Plant roots also contribute to the phosphatase activity in soil, notably in the rhizosphere (Joner *et al.*, 1995). For further information on phosphatases see Quiquampoix and Mousain (Chapter 5, this volume).

The mineralization of microbial phosphorus is treated in detail above, and organic phosphorus mineralization is also addressed elsewhere in this volume (Nziguheba and Bünemann, Chapter 11; Condron and Tiessen, Chapter 13). Therefore, we focus here on some quantitative aspects of soil organic phosphorus mineralization and summarize studies where the role of microorganisms in the decomposition of organic amendments was studied in detail.

Except for soils in which phosphate sorption can be ignored (e.g. the Spodosols studied by Grierson *et al.*, 1998), soil organic phosphorus mineralization cannot be estimated from changes in extractable phosphate, because some of the mineralized phosphorus will be rapidly sorbed. Estimates of organic phosphorus mineralization are therefore essentially based on two approaches: comparison of native and cultivated soils at sites with known cultivation history (Tiessen *et al.*, 1994), and determination of gross phosphorus mineralization rates by isotopic dilution techniques (Walbridge and Vitousek, 1987; López-Hernandez *et al.*, 1998; Oehl *et al.*, 2001a). Long-term field trials have shown that cultivation and cropping without fertilizer inputs deplete not only soil organic matter but also mineralizable nitrogen and organic phosphorus. For instance, the rate of net organic matter mineralization during the first 40–60 years of cultivation of soils of the North American Great Plains liberated sufficient phosphorus to prevent phosphorus limitation of crops (Tiessen and Stewart, 1983).

The principle of the most recent isotope dilution method is the comparison of the isotopic composition of soil solution phosphate resulting only from physicochemical processes with the isotopic composition obtained under the influence of physicochemical, biological and biochemical processes (Oehl *et al.*, 2001a). Basal mineralization rates obtained using this approach were between 1.4 and 2.5 mg P/kg per day for soils under different farming systems (Oehl *et al.*, 2004). Using a similar method, López-Hernandez *et al.* (1998) obtained rates between 0.2 and 0.9 mg P/kg day for several northern American Mollisols. The rates obtained by Oehl *et al.* (2004) were less than 10% of the amount of phosphorus isotopically exchangeable during one day in the same soils, showing that physical and chemical processes are more important than organic phosphorus mineralization in releasing plant-available phosphate in the studied soils. Higher rates occurred in soils with higher soil organic matter concentrations and higher soil microbial biomass and activity, which makes sense if low microbial biomass is a result of low availability of carbon, energy, nutrients, or other factors that render a large part of the biomass inactive. It should be noted that the isotopic dilution technique of Oehl *et al.* (2001a) cannot be applied to strongly phosphate-sorbing and phosphorus-deficient soils, which presents a shortcoming of the method, especially since in such soils, which mostly prevail in the tropics, organic phosphorus turnover may play a more important role than in soils well supplied with phosphorus (Gijsman *et al.*, 1996).

Phosphorus immobilization by microorganisms, turnover of microbial phosphorus and mineralization of microbial by-products, appear to be major processes regulating phosphorus cycling and availability from organic material. For example, organic phosphorus mineralization, mostly in the litter layer, is generally recognized as the major process providing available phosphate in forests (Yanai, 1992; Attiwill and Adams, 1993). If decomposing organic material contains less phosphorus than microorganisms, no net phosphorus mineralization occurs initially. Phosphorus mineralization by bacteria appears to cease above a substrate carbon-to-phosphorus ratio of 60 (Tezuka, 1990), confirming the suggestion that the critical substrate carbon-to-phosphorus ratio for phosphorus mineralization from microbial growth is between 50 and 70 (White, 1981). Other authors have concluded that phosphorus mineralization may occur once the carbon-to-phosphorus ratio is below 200 (see Nziguheba and Bünemann, Chapter 11, this volume), but such ratios have been used with much less success to predict phosphorus mineralization than carbon-to-nitrogen ratios for prediction of nitrogen mineralization, probably due to the high reactivity of mineralized phosphorus with the soil solid phase. The carbon-to-phosphorus ratios of plant residues are often above these values, for example, between 400 and 2700 for whole litter of tropical forage plants, and between 380 and 550 for rice leaves before harvest (Gijsman *et al.*, 1996). Thus, in spite of a high proportion (40–70%) of plant residue phosphorus being resin-extractable phosphate (Daroub *et al.*, 2000; Bünemann, 2003), most of this is taken up by the microbial biomass when a simultaneous application of carbon and nitrogen induces microbial growth. In a pot experiment, for example, about 66% of the ^{33}P that had been applied in plant residues was recovered in the microbial biomass, as compared with 23–29% from ^{32}P-labelled mineral fertilizer (McLaughlin and Alston, 1986). In addition, microorganisms cover their need for phosphorus by concomitant phosphorus immobilization from the soil, resulting in an up to fourfold phosphorus enrichment in decomposing agricultural residues (Schomberg and Steiner, 1999) and up to twofold phosphorus enrichment in decomposing pine and oak litter (Conn and Dighton, 2000). Fungi may be responsible for this temporal immobilization (Salas *et al.*, 2003).

Radiolabelled phosphorus studies provide information about subsequent re-mineralization and the crop availability of phosphorus added in organic material. Under field conditions, between 22% and 28% of the ^{33}P applied in plant residues was recovered in the microbial biomass, compared with <5% of ^{32}P in banded ^{32}P-labelled fertilizer (McLaughlin *et al.*, 1988). The amount of microbial phosphorus and the proportion of microbial phosphorus derived from plant residues remained constant between 7 and 95 days after residue incorporation. The microbial biomass therefore appeared to compete successfully for phosphorus from plant residues, leading to only 5% recovery of ^{33}P in plants, compared with 12% from banded fertilizer. Phosphorus applied with plant residues may continue to be re-mineralized after longer periods.

Factors that affect the uptake of phosphorus from added plant residues by a crop include the carbon-to-phosphorus ratio (Umrit and Friesen, 1994), soil phosphorus availability (Thibaud *et al.*, 1988), nitrogen availability (Umrit and Friesen, 1994) and the amount of added residues (Joseph *et al.*, 1995). Armstrong *et al.* (1993) reported that the net release of phosphorus from ^{32}P-

labelled residues increased significantly in the presence of plants, but was not affected by the species present. Besides the carbon-to-phosphorus ratio, the same quality attributes that determine the decomposition of the residues, including the carbon-to-nitrogen ratio, the content of lignin, and secondary metabolites such as polyphenols, may affect P release (Handayanto *et al.*, 1997). In addition, the ratio of water-soluble carbon to total phosphorus has been proposed as a predictor of phosphorus mineralization from organic amendments (Nziguheba *et al.*, 2000). Finally, the initial size, activity and composition of the soil microbial biomass can affect the decomposition of organic material (Fliessbach *et al.*, 2000; Bünemann, 2003). In the presence of mycorrhiza, plants compete efficiently for mineralized phosphorus and mycorrhizal phosphorus uptake can reduce the amount of mineralized phosphorus that is adsorbed in soil (Joner and Jakobsen, 1994).

Information on the immobilization and re-mineralization pattern of phosphorus derived from animal manure is scarce, partly because manure cannot be homogeneously labelled with phosphorus isotopes. The application of diammonium phosphate and animal manure to soils in which isotopically exchangeable phosphorus was previously labelled, resulted in a lower use efficiency of manure phosphorus than of diammonium phosphate by Italian ryegrass (Tagmann, 2000), which may be partly explained by an increase in microbial phosphorus after manure application. However, the hypothesis that this phosphorus may become available at a later stage was rejected, because the residual fertilizer value of animal manure was found to be lower than that of soluble mineral fertilizers (Tagmann, 2000). Similarly, phosphorus applied with compost has a lower availability than phosphorus applied as soluble mineral fertilizer, despite most of the phosphorus in composts being in mineral form. Apart from the fact that a significant proportion of the inorganic phosphorus contained in compost is of limited availability to plants (Frossard *et al.*, 2002; Sinaj *et al.*, 2002), it is likely that the simultaneous addition of organic matter stimulates microbial phosphorus immobilization within the microbial biomass.

Solubilization of phosphate and mycorrhizal phosphorus transport

A wide range of soil microorganisms (*Bacillus*, *Pseudomonas*, *Penicillium* and *Aspergillus* spp.) can solubilize precipitated phosphates by releasing organic acids and/or by decreasing pH in their vicinity. For the involvement of specific groups of microorganisms and processes underlying microbially mediated phosphorus solubilization, the reader is referred to the reviews of Kucey *et al.* (1989), Whitelaw (2000) and Richardson (2001). The term phosphate-solubilizing microorganisms has been used to designate these as a functional group within the soil. In their reviews, Richardson (2001) and Gyaneshwar *et al.* (2002) discuss why solubilization of phosphorus compounds by microorganisms is common under laboratory conditions, but results in the field are highly variable. Shang *et al.* (1996) point to the importance of direct microbial colonization of particle surfaces for microbially-mediated phosphorus solubilization. Within the micro-environment of these colonies or bio-aggregates, the concentration of organic ligands is much greater than in the bulk phase and may be high enough to permit ligand exchange and phosphate release. Similarly, as microbial activity is generally limited by energy, the most important niche for their function is in the rhizosphere of plants where carbon supply is abundant (Laheurte and Berthelin, 1988; Toro *et al.*, 1997).

Our review would not be complete without a note on mycorrhizal phosphorus transport, which is another key microbial process in the phosphorus cycle. It consists of three distinct parts: (i) fungal absorption of phosphorus from the soil solution; (ii) translocation of phosphorus within extraradical hyphae from the site of absorption to the site of exchange with the host plant; and (iii) transfer from the fungus to the host plant (*sensu* Cooper and Tinker, 1981). This process is as ancient as the colonization of Earth by terrestrial plants, and is the initial basis for all phototrophic land plants (Simon *et al.*, 1993). The importance of this process partly resides in the ubiquitous distribution of mycorrhiza among plant species, and partly in the extent of exploitation of soil by the extra-radical part of the involved fungi. Thus, arbuscular mycorrhizal fungi establish a symbiotic relationship with most terrestrial plant species (estimated by Fitter and Moyersoen (1996) to reach up to 80%), and the fungal biomass in soil is often dominated by mycorrhizal mycelium. However, their function in symbiotic nutrient transport to plants, whereby the host plant acts as a sink, may leave their biomass rather poor in phosphorus compared with the amount of phosphorus that has been absorbed by the mycelium. In ectomycorrhiza, the dominant type of mycorrhiza formed in forests from sub-Arctic to Mediterranean regions, the phosphorus concentration of mycelium in soil was reported to range between 0.5 and 1.7 mg P/g (Wallander *et al.*, 2003). Ectomycorrhizal fungi may constitute a biomass of up to 35 mg/g and may contain as much as 20% of the total phosphorus in organic topsoil (Bååth and Söderström, 1979). In arbuscular mycorrhiza, which are formed globally by most herbaceous plants, the phosphorus concentration in living hyphae may be two orders of magnitude greater (Nielsen *et al.*, 2002). Most of this phosphorus occurs in polyphosphate granules, which are streaming rapidly towards the host root where they are hydrolysed to phosphate and transferred to the plant. Dying hyphae are typically drained of cytoplasm as the cell content is withdrawn into active parts of the mycelium, so dead hyphae constitute a rather modest pool of microbial phosphorus. An exception can be observed when hyphae are excised by grazing arthropods, thus cutting an active mycelium from its host plant, although this does not seem to be a phenomenon of any quantitative importance in terms of phosphorus cycling (Larsen and Jakobsen, 1996).

When the phosphorus source is recently mineralized phosphorus, mycorrhizal transport becomes increasingly efficient. This is partly due to local proliferation of mycorrhizal hyphae in response to enhanced organic matter content (e.g. around plant debris) where mineralization is likely to occur (Bending and Read, 1995; Joner and Jakobsen, 1995), and the fact that mycorrhizal fungi produce extra-cellular phosphatases that may enhance the mineralization process (Joner *et al.*, 2000; Perez-Moreno and Read, 2000). Since mycorrhizal fungi are fed with carbon and energy from their host plant (together with nutrient transport, this is the basis of the symbiosis), they have a strong competitive advantage over other microorganisms when they exploit soil for mineral nutrients (Jakobsen *et al.*, 1994). For further information about the function of mycorrhiza in mineral plant nutrition, see the reviews of Jakobsen (1999) and Jakobsen *et al.* (2002).

Conclusions

The importance of soil microorganisms as a pool of phosphorus and mediator of major phosphorus transformation processes means that they have been included in many studies related to phosphorus dynamics during the last decade. In particular, phosphorus contained in the soil microbial biomass has often been estimated as a function of land use, farming system, agronomic practice and environmental factors. The results have demonstrated the impact of soil organic matter management on microbial phosphorus, and shown that the soil microbial biomass is limited by phosphorus availability only in situations where carbon is very abundant and phosphorus availability is low, such as in the forest floor of highly weathered tropical soils. Studies with radioisotopes have revealed the highly dynamic nature of the microbial phosphorus pool. The close linkage between soil solution and soil microorganisms has been highlighted, both under steady state conditions and after stimulation of the microorganisms by organic amendments. Much progress has been made in our understanding and quantification of organic phosphorus mineralization (either native soil organic phosphorus, re-mineralization of microbial phosphorus, or mineralization of organic amendments). Changes in climatic factors, including freezing/thawing and drying/rewetting of soils, cause flush effects, while organic amendments cause an immobilization–re-mineralization sequence with the pattern, duration and quantity of phosphorus turnover depending on properties of the added substrate, the microbial biomass and the soil. Seasonal fluctuations in microbial phosphorus (i.e. cycles of net phosphorus immobilization and net re-mineralization), mostly occur in response to changes in soil moisture and usually counteract plant-available phosphorus, although they do not necessarily hamper plant growth. Microbial phosphorus mineralization and immobilization also occur in the absence of net changes in the soil microbial biomass. Therefore, gross mineralization may far exceed net mineralization.

We anticipate the following future research needs. Since the concentrations and forms of phosphorus contained in bacterial and fungal cells vary widely, and different phosphorus compounds undergo different reactions once they are released from disrupted microbial cells, more information is clearly required on the composition of microbial cells as a function of land use and environmental factors. For this purpose, studies should go beyond cultured microorganisms and analyse microorganisms extracted directly from the soil (e.g. by density gradient centrifugation; Bakken, 1997). Identification and quantification of labelled phosphorus species within microbial cells as a function of phosphorus source and degree of phosphorus starvation is another possible approach. In combination with molecular tools, this may reveal whether shifts in the composition of the microbial phosphorus pool result from different management of the phosphorus resource by a given microbial population, or from a shift in the composition of the microbial population. It is important that such measurements are made in oligotrophic as well as eutrophic systems.

Determination of the microbial phosphorus pool in soil requires suitable correction factors to account for incomplete nutrient release upon fumigation (K_p) and sorption of released phosphorus in the soil. These are currently imprecise and almost certainly account for some of the variation in microbial phosphorus concentrations among different organisms and soils. Therefore, available data on the size of microbial phosphorus and, in turn, on the phosphorus fluxes through this pool, are currently only estimates. Specific marker molecules seem unlikely to be applicable for the quantification of microbial phosphorus, because the concentrations of phosphorus-containing compounds vary across taxa (e.g. phospholipids; Haack *et al.*, 1994) and even within a single strain, depending on nutritional status (e.g. polyphosphate and teichoic acids).

Our understanding of rates of net and gross immobilization and mineralization under different situations and at different scales (e.g. in the rhizosphere) is far from complete. The distinction between microsites in soil and their respective phosphorus pools is difficult to resolve, so spatial heterogeneity will probably remain a hurdle for some time. This has implications for deciphering such aspects as the importance of substrate affinity versus the advantages of localization. Furthermore, the significance of mineralization driven by extracellular phosphatases will continue to be clarified using novel phosphatase assays (van Aarle *et al.*, 2001).

Microbial phosphorus turnover also takes place in the absence of net changes in the microbial biomass. Isotope studies suggest that related phosphorus fluxes are important, but the underlying processes in soils are not well understood. Studies should test whether diffusion and efflux from microbial cells are important. Models would help to integrate different processes and factors related to gross phosphorus mineralization and microbial phosphorus turnover in general.

Acknowledgements

The senior author gratefully acknowledges stimulating discussions with Else Bünemann and Emmanuel Frossard (Institute of Plant Sciences, ETH Zurich), which contributed greatly to the improvement of this chapter.

References

Alexander, M. (1977) *Introduction to Soil Microbiology*, 2nd edn. John Wiley & Sons, New York, 467 pp.

Anderson, J.P.E. and Domsch, K.H. (1980) Quantities of plant nutrients in the microbial biomass of selected soils. *Soil Science* 130, 211–216.

Archibald, A.R. (1974) The structure, biosynthesis and function of teichoic acid. *Advances in Microbiology and Physiology* 11, 53–95.

Armstrong, R.D., Helyar, K.R. and Prangnell, R. (1993) Direct assessment of mineral phosphorus availability to tropical crops using ^{32}P-labelled compounds. *Plant and Soil* 150, 279–287.

Aslam, T., Choudhary, M.A. and Saggar, S. (1999) Tillage impacts on soil microbial biomass C, N and P, earthworms and agronomy after two years of cropping following permanent pasture in New Zealand. *Soil and Tillage Research* 51, 103–111.

Attiwill, P.M. and Adams, M.A. (1993) Nutrient cycling in forests. *New Phytologist* 124, 561–582.

Bååth, E. and Söderström, B. (1979) Fungal biomass and fungal immobilization of plant nutrients in Swedish coniferous forest soils. *Revue d'Ecologie et de Biologie du Sol* 16, 477–489.

Bååth, E., Frostegård, Å., Pennanen, T. and Fritze, H. (1995) Microbial community structure and pH response in relation to soil organic matter quality in wood-ash fertilized, clear-cut or burned coniferous forest soils. *Soil Biology and Biochemistry* 27, 229–240.

Badalucco, L., De Cesare, F., Grego, S., Landi, L. and Nannipieri, P. (1997) Do physical properties of soil affect chloroform efficiency in lysing microbial biomass? *Soil Biology and Biochemistry* 29, 1135–1142.

Bakken, L. (1997) Culturable and nonculturable bacteria in soil. In: van Elsas, L., Trevors, J. and Wellington, E. (eds) *Modern Soil Microbiology*. Marcel Dekker, New York, pp. 47–61.

Bending, G.D. and Read, D.J. (1995) The structure and function of the vegetative mycelium of ectomycorrhizal plants. 5. Foraging behaviour and translocation of nutrients from exploited litter. *New Phytologist* 130, 401–409.

Bloem, J., De Ruiter, P. and Bouwman, L. (1997) Soil food webs and nutrient cycling in agroecosystems. In: Wellington, E.M.H. (ed.) *Modern Soil Microbiology*. Marcel Dekker, New York, pp. 245–278.

Breland, T.A. and Bakken, L.R. (1991) Microbial growth and nitrogen immobilization in the root zone of barley (*Hordeum vulgare* L.), Italian ryegrass (*Lolium multiflorum* Lam.), and white clover (*Trifolium repens* L.). *Biology and Fertility of Soils* 12, 154–160.

Brookes, P.C., Powlson, D.S. and Jenkinson, D.S. (1982) Measurement of microbial biomass phosphorus in soil. *Soil Biology and Biochemistry* 14, 319–329.

Buchanan, M. and King, L.D. (1992) Seasonal fluctuations in soil microbial biomass carbon, phosphorus, and activity in no-till and reduced-chemical-input maize agroecosystems. *Biology and Fertility of Soils* 13, 211–217.

Bünemann, E. (2003) Phosphorus dynamics in a Ferralsol under maize–fallow rotations: the role of the soil microbial biomass. PhD thesis, Swiss Federal Institute of Technology, Zurich.

Bünemann, E., Smithson, P.C., Jama, B., Frossard, E. and Oberson, A. (2004) Maize productivity and nutrient dynamics in maize–fallow rotations in western Kenya. *Plant and Soil* 260.

Campo, J., Jaramillo, V.J. and Maass, J.M. (1998) Pulses of soil phosphorus availability in a Mexican tropical dry forest: effects of seasonality and level of wetting. *Oecologia* 115, 167–172.

Chapin, F.S., Barsdate, R.J. and Barel, D. (1978) Phosphorus cycling in Alaskan coastal tundra: a hypothesis for the regulation of nutrient cycling. *Oikos* 31, 189–199.

Chauhan, B.S., Stewart, J.W.B. and Paul, E.A. (1979) Effect of carbon additions on soil labile inorganic, organic and microbially held phosphate. *Canadian Journal of Soil Science* 59, 387–396.

Chauhan, B.S., Stewart, J.W.B. and Paul, E.A. (1981) Effect of labile inorganic phosphate status and organic carbon additions on the microbial uptake of phosphorus in soils. *Canadian Journal of Soil Science* 61, 373–385.

Chen, C.R., Condron, L.M., Davis, M.R. and Sherlock, R.R. (2000) Effects of afforestation on phosphorus dynamics and biological properties in a New Zealand grassland soil. *Plant and Soil* 220, 151–163.

Chen, C.R., Condron, L.M., Davis, M.R. and Sherlock, R.R. (2003) Seasonal changes in soil phosphorus and associated microbial properties under adjacent grassland and forest in New Zealand. *Forest Ecology and Management* 177, 539–557.

Chepkwony, C.K., Haynes, R.J., Swift, R.S. and Harrison, R. (2001) Mineralization of soil organic P induced by drying and rewetting as a source of plant-available P in limed and unlimed samples of an acid soil. *Plant and Soil* 234, 83–90.

Clarholm, M. (1993) Microbial biomass-P, labile-P, and acid-phosphatase-activity in the humus layer of a spruce forest, after repeated additions of fertilizers. *Biology and Fertility of Soils* 16, 287–292.

Cleveland, C.C., Townsend, A.R. and Schmidt, S.K. (2002) Phosphorus limitation of microbial processes in moist tropical forests: evidence from short-term laboratory incubations and field studies. *Ecosystems* 5, 680–691.

Coe, F.L. (1968) Mean life in steady-state populations. *Journal of Theoretical Biology* 18, 171–180.

Cole, C.V., Elliott, E.T., Hunt, H.W. and Coleman, D.C. (1978) Trophic interactions in soils as they affect energy and nutrient dynamics. V. Phosphorus transformations. *Microbial Ecology* 4, 381–387.

Conn, C. and Dighton, J. (2000) Litter quality influences on decomposition, ectomycorrhizal community structure and mycorrhizal root surface acid phosphatase activity. *Soil Biology and Biochemistry* 32, 489–496.

Cooper, K.M. and Tinker, P.B. (1981) Translocation and transfer of nutrients in vesicular-arbuscular mycorrhizas. IV. Effect of environmental variables on movement of phosphorus. *New Phytologist* 88, 327–339.

Cortet, J., Joffre, R., Elmholt, S. and Krogh, P.H. (2003) Increasing species and trophic diversity of mesofauna affects fungal biomass, mesofauna community structure and organic matter decomposition processes. *Biology and Fertility of Soils* 37, 302–312.

Daroub, S.H., Pierce, F.J. and Ellis, B.G. (2000) Phosphorus fractions and fate of phosphorus-33 in soils under plowing and no-tillage. *Soil Science Society of America Journal* 64, 170–176.

Fitter, A.H. and Moyersoen, B. (1996) Evolutionary trends in root–microbe symbioses. *Philosophical Transactions of the Royal Society, London, Series B* 351, 1367–1375.

Fliessbach, A., Mäder, P. and Niggli, U. (2000) Mineralization and microbial assimilation of ^{14}C-labeled straw in soils of organic and conventional agricultural systems. *Soil Biology and Biochemistry* 32, 1131–1139.

Frossard, E. and Sinaj, S. (1997) The isotopic exchange kinetic technique: a method to describe the availability of inorganic nutrients. Applications to K, P, S and Zn. *Isotopes in Environmental and Health Studies* 33, 61–77.

Frossard, E., Skrabal, P., Sinaj, S., Bangerter, F. and Traoré, O. (2002) Form and exchangeability of inorganic phosphate in composted solid organic wastes. *Nutrient Cycling in Agroecosystems* 62, 103–113.

Gächter, R. and Meyer, J.S. (1993) The role of microorganisms in mobilization and fixation of phosphorus in sediments. *Hydrobiologia* 253, 103–121.

Gijsman, A.J., Oberson, A., Tiessen, H. and Friesen, D.K. (1996) Limited applicability of the Century model to highly weathered tropical soils. *Agronomy Journal* 88, 894–903.

Gijsman, A.J., Oberson, A., Friesen, D.K., Sanz, J.I. and Thomas, R.J. (1997) Nutrient cycling through microbial biomass under rice–pasture rotations replacing native savanna. *Soil Biology and Biochemistry* 29, 1433–1441.

Grant, W.D. (1979) Cell wall teichoic acid as a reserve phosphate source in *Bacillus subtilis*. *Journal of Bacteriology* 137, 35–68.

Grayston, S.J., Griffith, G.S., Mawdsley, J.L., Campbell, C.D. and Bardgett, R.D. (2001) Accounting for variability in soil microbial communities of temperate upland grassland ecosystems. *Soil Biology and Biochemistry* 33, 533–551.

Gressel, N. and McColl, J.G. (1997) Phosphorus mineralization and organic matter decomposition: a critical review. In: Cadisch, G. and Giller, K.E. (eds) *Driven by Nature: Plant Litter Quality and Decomposition.* CAB International, Wallingford, UK, pp. 297–309.

Gressel, N., McColl, J.G., Preston, C.M., Newman, R.H. and Powers, R. (1996) Linkages between phosphorus transformations and carbon decomposition in a forest soil. *Biogeochemistry* 33, 97–123.

Grierson, P.F., Comerford, N.B. and Jokela, E.J. (1998) Phosphorus mineralization kinetics and response of microbial phosphorus to drying and rewetting in a Florida Spodosol. *Soil Biology and Biochemistry* 30, 1323–1331.

Gyaneshwar, P., Kumar, G.N., Parekh, L.J. and Poole, P.S. (2002) Role of soil microorganisms in improving P nutrition of plants. *Plant and Soil* 245, 83–93.

Haack, S.K., Garchow, H., Odelson, D.A., Forney, L.J. and Klug, M.J. (1994) Accuracy, reproducibility, and interpretation of fatty-acid methyl-ester profiles of model bacterial communities. *Applied and Environmental Microbiology* 60, 2483–2493.

Halverson, L.J., Jones, T.M. and Firestone, M.K. (2000) Release of intercellular solutes by four soil bacteria exposed to dilution stress. *Soil Science Society of America Journal* 64, 1630–1637.

Handayanto, E., Cadisch, G. and Giller, K.E. (1997) Regulating N mineralization from plant residues by manipulation of quality. In: Cadisch, G. and

Giller, K.E. (eds) *Driven by Nature: Plant Litter Quality and Decomposition*. CAB International, Wallingford, UK, pp. 175–186.

He, Z.L., Wu, J., O'Donnell, A.G. and Syers, J.K. (1997) Seasonal responses in microbial biomass carbon, phosphorus and sulphur in soils under pasture. *Biology and Fertility of Soils* 24, 421–428.

Hedley, M.J. and Stewart, J.W.B. (1982) Method to measure microbial phosphate in soils. *Soil Biology and Biochemistry* 14, 377–385.

Huffman, S.A., Cole, C.V. and Scott, N.A. (1996) Soil texture and residue addition effects on soil phosphorus transformations. *Soil Science Society of America Journal* 60, 1095–1101.

Jakobsen, I. (1999) Transport of phosphorus and carbon in arbuscular mycorrhizas. In: Varma, A. and Hock, B. (eds) *Mycorrhiza: Structure, Function, Molecular Biology and Biotechnology*, 2nd edn. Springer, Berlin, pp. 306–332.

Jakobsen, I., Joner, E.J. and Larsen, J. (1994) Hyphal phosphorus transport, a keystone to mycorrhizal enhancement of plant growth. In: Schüepp, H. (ed.) *Impact of Arbuscular Mycorrhizas on Sustainable Agriculture and Natural Ecosystems*. Birkhäuser, Basel, pp. 133–146.

Jakobsen, I., Smith, S.E. and Smith, F.A. (2002) Function and diversity of arbuscular mycorrhizae in carbon and mineral nutrition. In: Van der Heijden, M.G.A. and Sanders, I. (eds) *Mycorrhizal Ecology*. Springer, Berlin, pp. 75–92.

Jama, B., Swinkels, R.A. and Buresh, R. (1997) Agronomic and economic evaluation of organic and inorganic sources of phosphorus in Western Kenya. *Agronomy Journal* 89, 597–604.

Jenkinson, D.S. and Ladd, J.N. (1981) Microbial biomass in soil: measurement and turnover. In: Paul, E.A. and Ladd, J.N. (eds) *Soil Biochemistry*. Marcel Dekker, New York, pp. 415–471.

Jiménez, J.J., Cepeda, A., Decaëns, T., Oberson, A. and Friesen, D. (2003) Phosphorus fractions and dynamics in surface earthworm casts under native and improved grasslands in a Colombian savanna Oxisol. *Soil Biology and Biochemistry* 35, 715–727.

Joergensen, R.G. and Castillo, X. (2001) Interrelationships between microbial and soil properties in young volcanic ash soils of Nicaragua. *Soil Biology and Biochemistry* 33, 1581–1589.

Joergensen, R.G. and Scheu, S. (1999) Response of soil microorganisms to the addition of carbon, nitrogen and phosphorus in a forest Rendzina. *Soil Biology and Biochemistry* 31, 859–866.

Joergensen, R.G., Kubler, H., Meyer, B. and Wolters, V. (1995) Microbial biomass phosphorus in soils of beech (*Fagus sylvatica* L.) forests. *Biology and Fertility of Soils* 19, 215–219.

Joner, E.J. and Jakobsen, I. (1994) Contribution by two arbuscular mycorrhizal fungi to P-uptake by cucumber (*Cucumis sativus* L.) from ^{32}P-labeled organic matter during mineralization in soil. *Plant and Soil* 163, 203–209.

Joner, E.J. and Jakobsen, I. (1995) Growth and extracellular phosphatase-activity of arbuscular mycorrhizal hyphae as influenced by soil organic matter. *Soil Biology and Biochemistry* 27, 1153–1159.

Joner, E.J., Magid, J., Gahoonia, T.S. and Jakobsen, I. (1995) P depletion and activity of phosphatases in the rhizosphere of mycorrhizal and non-mycorrhizal cucumber (*Cucumis sativus* L.). *Soil Biology and Biochemistry* 27, 1145–1151.

Joner, E.J., van Aarle, I. and Vosatka, M. (2000) Phosphatase activity of extra-radical arbuscular mycorrhizal hyphae: a review. *Plant and Soil* 226, 199–210.

Joseph, P., George, M., Wahid, P.A., John, P.S. and Kamalm, N.V. (1995) Dynamics of phosphorus mineralization from ^{32}P-labelled green manure. *Journal of Nuclear Agriculture and Biology* 24, 158–162.

Killham, K. (1994) *Soil Ecology*, 1st edn. Cambridge University Press, Cambridge, UK, 242 pp.

Kouno, K., Tuchiya, Y. and Ando, T. (1995) Measurement of soil microbial biomass phosphorus by an anion exchange membrane method. *Soil Biology and Biochemistry* 10, 1353–1357.

Kouno, K., Wu, J. and Brookes, P.C. (2002) Turnover of biomass C and P in soil following incorporation of glucose or ryegrass. *Soil Biology and Biochemistry* 34, 617–622.

Kucey, R.M.N., Janzen, H.H. and Leggett, M.E. (1989) Microbially mediated increases in plant-available phosphorus. *Advances in Agronomy* 42, 199–228.

Laheurte, F. and Berthelin, J. (1988) Effect of phosphate-solubilizing bacteria on maize growth and root exudation over four levels of labile phosphorus. *Plant and Soil* 105, 11–17.

Larsen, J. and Jakobsen, I. (1996) Interactions between a mycophagous collembola, dry yeast and the external mycelium of an arbuscular mycorrhizal fungus. *Mycorrhiza* 6, 259–264.

Lehmann, J., Cravo, M.D., de Macedo, J.L.V., Moreira, A. and Schroth, G. (2001) Phosphorus management for perennial crops in central Amazonian upland soils. *Plant and Soil* 237, 309–319.

López-Hernandez, D., Brossard, M. and Frossard, E. (1998) P-isotopic exchangeable values in relation to P_o mineralization in soils with very low P-sorbing capacities. *Soil Biology and Biochemistry* 30, 1663–1670.

Lukito, H.P., Kouno, K. and Ando, T. (1998) Phos-

phorus requirements of microbial biomass in a regosol and an andosol. *Soil Biology and Biochemistry* 30, 865–872.
Macklon, A.E.S., Grayston, S.J., Shand, C.A., Sim, A., Sellars, S. and Ord, B.G. (1997) Uptake and transport of phosphorus by *Agrostis capillaris* seedlings from rapidly hydrolysed organic sources extracted from ^{32}P-labelled bacterial cultures. *Plant and Soil* 190, 163–167.
Makarov, M.I., Haumaier, L. and Zech, W. (2002) The nature and origins of diester phosphates in soils: a ^{31}P- NMR study. *Biology and Fertility of Soils* 35, 136–146.
Makino, W., Cotner, J.B., Sterner, R.W. and Elser, J.J. (2003) Are bacteria more like plants or animals? Growth rate and resource dependence of bacterial C:N:P stoichiometry. *Functional Ecology* 17, 121–130.
Maloney, P.C. (1996) Pi-linked anion exchange in bacteria: biochemical and molecular studies of UhpT and related porters. In: Konings, W.N., Kaback, H.R. and Lolkeman, J.S. (eds) *Transport Processes in Eukaryotic and Prokaryotic Organisms.* Elsevier, Amsterdam, pp. 261–280.
Maroko, J.B., Buresh, R.J. and Smithson, P.C. (1999) Soil phosphorus fractions in unfertilized fallow–maize systems on two tropical soils. *Soil Science Society of America Journal* 63, 320–326.
Mary, B. and Recous, S. (1994) Measurement of nitrogen mineralization and immobilization fluxes in soil as a means of predicting net mineralization. *European Journal of Agronomy* 3, 291–300.
McGill, W.B. and Cole, C.V. (1981) Comparative aspects of cycling of organic C, N, S and P through soil organic matter. *Geoderma* 26, 267–286.
McGill, W.B., Cannon, K.R., Robertson, J.A. and Cook, F.D. (1986) Dynamics of soil microbial biomass and water-soluble organic C in Breton L after 50 years of cropping to two rotations. *Canadian Journal of Soil Science* 66, 1–19.
McLaughlin, M.J. and Alston, A.M. (1986) The relative contribution of plant residues and fertilizer to the P nutrition of wheat in a pasture/cereal system. *Australian Journal of Soil Research* 24, 517–526.
McLaughlin, M.J., Alston, A.M. and Martin, J.K. (1986) Measurement of phosphorus in the soil microbial biomass: a modified procedure for field soils. *Soil Biology and Biochemistry* 18, 437–443.
McLaughlin, M.J., Alston, A.M. and Martin, J.K. (1988) Phosphorus cycling in wheat–pasture rotations. II. The role of the microbial biomass in phosphorus cycling. *Australian Journal of Soil Research* 26, 333–342.
Morel, C., Tiessen, H. and Stewart, J.W.B. (1996) Correction for P-sorption in the measurement of soil microbial biomass P by $CHCl_3$ fumigation. *Soil Biology and Biochemistry* 28, 1699–1706.
Moura, R.S., Martin, J.F., Martin, A. and Liras, P. (2001) Substrate analysis and molecular cloning of the extracellular alkaline phosphatase of *Streptomyces griseus. Microbiology* 147, 1525–1533.
Myers, R.G., Thien, S.J. and Pierzynski, G.M. (1999) Using an ion sink to extract microbial phosphorus from soil. *Soil Science Society of America Journal* 63, 1229–1237.
Nielsen, J.S., Joner, E.J., Declerck, S., Olsson, S. and Jakobsen, I. (2002) Phospho-imaging as a tool for visualisation and non-invasive measurement of P transport dynamics in arbuscular mycorrhiza. *New Phytologist* 154, 809–819.
Nziguheba, G., Palm, C.A., Buresh, R.J. and Smithson, P.C. (1998) Soil phosphorus fractions and adsorption as affected by organic and inorganic sources. *Plant and Soil* 198, 159–168.
Nziguheba, G., Merckx, R., Palm, C.A. and Rao, M.R. (2000) Organic residues affect phosphorus availability and maize yields in a Nitisol of western Kenya. *Biology and Fertility of Soils* 32, 328–339.
Oberson, A., Fardeau, J.C., Besson, J.M. and Sticher, H. (1993) Soil-phosphorus dynamics in cropping systems managed according to conventional and biological agricultural methods. *Biology and Fertility of Soils* 16, 111–117.
Oberson, A., Besson, J.M., Maire, N. and Sticher, H. (1996) Microbiological processes in soil organic phosphorus transformations in conventional and biological cropping systems. *Biology and Fertility of Soils* 21, 138–148.
Oberson, A., Friesen, D.K., Morel, C. and Tiessen, H. (1997) Determination of phosphorus released by chloroform fumigation from microbial biomass in high P sorbing tropical soils. *Soil Biology and Biochemistry* 29, 1579–1583.
Oberson, A., Friesen, D.K., Tiessen, H., Morel, C. and Stahel, W. (1999) Phosphorus status and cycling in native savanna and improved pastures on an acid low-P Colombian Oxisol. *Nutrient Cycling in Agroecosystems* 55, 77–88.
Oberson, A., Friesen, D.K., Rao, I.M., Buhler, S. and Frossard, E. (2001) Phosphorus transformations in an Oxisol under contrasting land-use systems: the role of the soil microbial biomass. *Plant and Soil* 237, 197–210.
Oehl, F., Oberson, A., Sinaj, S. and Frossard, E. (2001a) Organic phosphorus mineralization studies using isotopic dilution techniques. *Soil Science Society of America Journal* 65, 780–787.

Oehl, F., Oberson, A., Probst, M., Fliessbach, A., Roth, H.R. and Frossard, E. (2001b) Kinetics of microbial phosphorus uptake in cultivated soils. *Biology and Fertility of Soils* 34, 31–41.

Oehl, F., Frossard, E., Fliessbach, A., Dubois, D. and Oberson, A. (2004) Basal organic phosphorus mineralization in soils under different farming systems. *Soil Biology and Biochemistry* 36, 667–675.

Ofori-Frimpong, K. and Rowell, D.L. (1999) The decomposition of cocoa leaves and their effects on phosphorus dynamics in tropical soil. *European Journal of Soil Science* 50, 165–172.

Paul, E.A. and Clark, F.E. (1996) *Soil Microbiology and Biochemistry*, 2nd edn. Academic Press, San Diego, 340 pp.

Pennanen, T., Perkiomaki, J., Kiikkila, O., Vanhala, P., Neuvonen, S. and Fritze, H. (1998) Prolonged, simulated acid rain and heavy metal deposition: separated and combined effects on forest soil microbial community structure. *FEMS Microbial Ecology* 27, 291–300.

Perez-Moreno, J. and Read, D.J. (2000) Mobilization and transfer of nutrients from litter to tree seedlings via the vegetative mycelium of ectomycorrhizal plants. *New Phytologist* 145, 301–309.

Prosser, J.I. (1997) Microbial processes within the soil. In: van Elsas, J.D., Trevors, J.T. and Wellington, E.M.H. (eds) *Modern Soil Microbiology*. Marcel Dekker, New York, pp. 183–213.

Richardson, A.E. (1994) Soil microorganisms and phosphorus availability. In: Pankhurst, C.E., Doube, E.M., Gupta, V.V.S.R. and Grace, P.R. (eds) *Soil Biota Management in Sustainable Farming Systems*. CSIRO Publishing, East Melbourne, pp. 50–62.

Richardson, A.E. (2001) Prospects for using soil microorganisms to improve the acquisition of phosphorus by plants. *Australian Journal of Plant Physiology* 28, 897–906.

Richardson, A.E., Hadobas, P.A., Hayes, J.E., O'Hara, C.P. and Simpson, R.J. (2001) Utilization of phosphorus by pasture plants supplied with *myo*-inositol hexaphosphate is enhanced by the presence of soil micro-organisms. *Plant and Soil* 229, 47–56.

Ross, D.J., Sparling, G.P. and West, A.W. (1987) Influence of *Fusarium oxysporum* age on proportions of C, N and P mineralized after chloroform fumigation in soil. *Australian Journal of Soil Research* 25, 563–566.

Ross, D.J., Speir, T.W., Kettles, H.A. and Mackay, A.D. (1995) Soil microbial biomass, C and N mineralization and enzyme activities in a hill pasture: influence of season and slow-release P and S fertilizer. *Soil Biology and Biochemistry* 27, 1431–1443.

Ross, D.J., Tate, K.R., Scott, N.A. and Feltham, C.W. (1999) Land-use change: effects on soil carbon, nitrogen and phosphorus pools and fluxes in three adjacent ecosystems. *Soil Biology and Biochemistry* 31, 803–813.

Salas, A.M., Elliott, E.T., Westfall, D.G., Cole, C.V. and Six, J. (2003) The role of particulate organic matter in phosphorus cycling. *Soil Science Society of America Journal* 67, 181–189.

Schärer, M. (2003) Processes controlling phosphorus availability in two grassland soils. PhD thesis, Swiss Federal Institute of Technology, Zurich.

Schomberg, H. and Steiner, J. (1999) Nutrient dynamics of crop residues decomposing on a fallow no-till soil surface. *Soil Science Society of America Journal* 63, 607–613.

Seeling, B. and Zasoski, R.J. (1993) Microbial effects in maintaining organic and inorganic solution phosphorus concentrations in a grassland topsoil. *Plant and Soil* 148, 277–284.

Selles, F., Campbell, C.A. and Zentner, R.P. (1995) Effect of cropping and fertilization on plant and soil phosphorus. *Soil Science Society of America Journal* 59, 140–144.

Shang, C., Caldwell, D.E., Stewart, J.W.B., Tiessen, H. and Huang, P.M. (1996) Bioavailability of organic and inorganic phosphates adsorbed on short-range ordered aluminum precipitate. *Microbial Ecology* 31, 29–39.

Simon, L., Bousquet, J., Lévesque, R.C. and Lalonde, M. (1993) Origin and diversification of endomycorrhizal fungi and coincidence with vascular plants. *Nature* 363, 67–69.

Sinaj, S., Traoré, O. and Frossard, E. (2002) Effect of compost and soil properties on the availability of compost phosphate for white clover (*Trifolium repens* L.). *Nutrient Cycling in Agroecosystems* 62, 89–102.

Smestad, T.B., Tiessen, H. and Buresh, R.J. (2002) Short fallows of *Tithonia diversifolia* and *Crotalaria grahamiana* for soil fertility improvement in western Kenya. *Agroforestry Systems* 55, 181–194.

Smith, J.L. and Paul, E.A. (1990) The significance of soil microbial biomass estimations. In: Bollag, J.M. and Stotzky, G. (eds) *Soil Biochemistry*. Marcel Dekker, New York, pp. 357–396.

Smith, S.E. and Read, D.J. (1997) *Mycorrhizal Symbiosis*, 2nd edn. Academic Press, San Diego, 605 pp.

Smith, N.C. and Stribley, D.P. (1994) A new approach to direct extraction of microorganisms from soil. In: Ritz, K., Dighton, J. and Giller, K.E. (eds) *Beyond the Biomass*. John Wiley & Sons, Chichester, pp. 49–55.

Sparling, G. and Zhu, C. (1993) Evaluation and calibration of biochemical methods to measure microbial biomass C and N in soils from Western Australia. *Soil Biology and Biochemistry* 25, 1793–1801.

Stewart, J.W.B. and Tiessen, H. (1987) Dynamics of soil organic phosphorus. *Biogeochemistry* 4, 41–60.

Tagmann, H.U. (2000) *Nach- und aktuelle Wirkung von Hof- und Mineraldüngerphosphat in langjährig konventionell und biologisch bewirtschafteten Böden.* Diploma thesis, Swiss Federal Institute of Technology, Zurich.

Tate, K.R., Speir, T.W., Ross, D.J., Parfitt, R.L., Whale, K.N. and Cowling, J.C. (1991) Temporal variations in some plant and soil P pools in two pasture soils of widely different P fertility status. *Plant and Soil* 132, 219–232.

Tezuka, Y. (1990) Bacterial regeneration of ammonium and phosphate as affected by the carbon:nitrogen:phosphorus ratio of organic substrates. *Microbial Ecology* 19, 227–238.

Thibaud, M.C., Morel, C. and Fardeau, J.C. (1988) Contribution of phosphorus issued from crop residues to plant nutrition. *Soil Science and Plant Nutrition* 34, 481–491.

Tiessen, H. and Stewart, J.W.B. (1983) Particle-size fractions and their use in studies of soil organic matter. II. Cultivation effects of organic matter composition in size fractions. *Soil Science Society of America Journal* 47, 509–514.

Tiessen, H., Stewart, J.W.B. and Oberson, A. (1994) Innovative phosphorus availability indices: assessing organic phosphorus. In: Havlin, J.L., Jacobsen, J., Fixen, J. and Hergert, G. (eds) *Soil Testing: Prospects for Improving Nutrient Recommendations,* SSSA Special Publication 40. Soil Science Society of America, Madison, Wisconsin, pp. 141–162.

Tiunov, A.V., Bonkowski, M., Alphei, J. and Scheu, S. (2001) Microflora, protozoa and nematoda in *Lumbricus terrestris* burrow walls: a laboratory experiment. *Pedobiologia* 45, 46–60.

Toro, M., Azcon, R. and Barea, J.M. (1997) Improvement of arbuscular mycorrhizae development by inoculation of soil with phosphate-solubilizing rhizobacteria to improve rock phosphate bioavailability (^{32}P) and nutrient cycling. *Applied and Environmental Microbiology* 63, 4408–4412.

Turner, B.L. and Haygarth, P.M. (2001) Phosphorus solubilization in rewetted soils. *Nature* 411, 258.

Turner, B.L., Bristow, A.W. and Haygarth, P.M. (2001) Rapid estimation of microbial biomass in grassland soils by ultra-violet absorbance. *Soil Biology and Biochemistry* 33, 913–919.

Turner, B.L., Mahieu, N. and Condron, L.M. (2003a) The phosphorus composition of temperate pasture soils determined by NaOH–EDTA extraction and solution ^{31}P NMR spectroscopy. *Organic Geochemistry* 34, 1199–1210.

Turner, B.L., Driessen, J.P., Haygarth, P.M. and McKelvie, I.D. (2003b) Potential contribution of lysed bacterial cells to phosphorus solubilisation in two rewetted Australian pasture soils. *Soil Biology and Biochemistry* 35, 187–189.

Umrit, G. and Friesen, D.K. (1994) The effect of C:P ratio of plant residues added to soils of contrasting phosphate sorption capacities on P uptake by *Panicum maximum* (Jacq.). *Plant and Soil* 158, 275–285.

van Aarle, I.M., Olsson, P.A. and Söderström, B. (2001) Microscopic detection of phosphatase activity of saprophytic and arbuscular mycorrhizal fungi using a fluorogenic substrate. *Mycologia* 93, 17–24.

van Veen, J.A., Ladd, J.N., Martin, J.K. and Amato, M. (1987) Turnover of carbon, nitrogen and phosphorus through the microbial biomass in soils incubated with ^{14}C-, ^{15}N- and ^{32}P-labelled bacterial cells. *Soil Biology and Biochemistry* 19, 559–565.

Walbridge, M.R. (1991) Phosphorus availability in acid organic soils of the lower North-Carolina coastal-plain. *Ecology* 72, 2083–2100.

Walbridge, M.R. and Vitousek, P.M. (1987) Phosphorus mineralization potentials in acid organic soils: processes affecting $^{32}PO_4^-$ isotope dilution measurements. *Soil Biology and Biochemistry* 19, 709–717.

Waldrop, M.P., Balser, T.C. and Firestone, M.K. (2000) Linking microbial community composition to function in a tropical soil. *Soil Biology and Biochemistry* 32, 1837–1846.

Wallander, H., Mahmood, S., Hagerberg, D., Johansson, L. and Pallon, J. (2003) Elemental composition of ectomycorrhizal mycelia identified by PCR-RFLP analysis and grown in contact with apatite or wood ash in forest soil. *FEMS Microbiology Ecology* 44, 57–65.

Wanner, B.L. (1996) Phosphorus assimilation and control of the phosphate regulon. In: Neidhardt, F.C. (ed.) Escherichia coli *and* Salmonella*: Cellular and Molecular Biology.* ASM, Washington, DC, pp. 1357–1381.

Wardle, D.A. (1998) Controls of temporal variability of the soil microbial biomass: a global-scale synthesis. *Soil Biology and Biochemistry* 30, 1627–1637.

Webley, D.M. and Jones, D. (1971) Biological transformations of microbial residues in soils. In: Skujins, J. (ed.) *Soil Biochemistry.* Marcel Dekker, New York, pp. 446–485.

Weisbach, C., Tiessen, H. and Jimenez-Osornio, J.J.J. (2002) Soil fertility during shifting cultivation in the tropical Karst soils of Yucatan. *Agronomie* 22, 253–263.

White, R.E. (1981) Pathways of phosphorus in soil. In: Hucker, T.W.G. and Catroux, G. (eds) *Proceedings of a Symposium on Phosphorus in Sewage Sludge and Animal Manure Slurries.* D. Reidel, Dortrecht, pp. 21–44.

Whitelaw, M.A. (2000) Growth promotion of plants inoculated with phosphate-solubilizing fungi. *Advances in Agronomy* 69, 99–151.

Wu, J., He, Z.L., Wei, W.X., O'Donnell, A.G. and Syers, J.K. (2000) Quantifying microbial biomass phosphorus in acid soils. *Biology and Fertility of Soils* 32, 500–507.

Wykoff, D.D., Grossman, A.R., Weeks, D.P., Usuda, H. and Shimogawara, K. (1999) Psr1, a nuclear localized protein that regulates phosphorus metabolism in *Chlamydomonas. Proceedings of the National Academy of Sciences of the United States of America* 96, 15336–15341.

Yanai, R.D. (1992) Phosphorus budget of a 70-year-old northern hardwood forest. *Biogeochemistry* 17, 1–22.

Zelles, L., Palojärvi, A., Kandeler, E., von Lützow, M., Winter, K. and Bai, Q.Y. (1997) Changes in soil microbial properties and phospholipid fatty acid fractions after chloroform fumigation. *Soil Biology and Biochemistry* 29, 1325–1336.

Zou, X., Binkley, D. and Doxtader, K.G. (1992) A new method for estimating gross phosphorus mineralization and immobilization rates in soils. *Plant and Soil* 147, 243–250.

8 Utilization of Soil Organic Phosphorus by Higher Plants

Alan E. Richardson, Tim S. George, Maarten Hens and Richard J. Simpson

CSIRO Plant Industry, PO Box 1600, Canberra, ACT 2601, Australia

Introduction

Phosphorus is a critical macronutrient for plant growth and cellular function. However, plant growth is commonly limited by poor availability of phosphorus in soil, and consequently phosphorus deficiency represents a major constraint to crop production globally (Runge-Metzger, 1995). Limited availability of soil phosphorus is also a key factor that constrains the accumulation and turnover of plant biomass in a range of natural ecosystems (McGill and Cole, 1981; Attiwill and Adams, 1993). Plants therefore possess a number of adaptive features that enhance phosphorus acquisition from soil and these are postulated to be important mechanisms by which the efficiency of phosphorus uptake by plants may be improved. The ability of plants to obtain phosphate from soil organic phosphorus is of particular significance because, in most soils, organic phosphorus is a substantial component of the total soil phosphorus.

Plants are a key component of the phosphorus cycle in terrestrial ecosystems, because they contain a significant pool of phosphorus. Plants are also an important route for the incorporation of phosphate into organic forms, and through death, decay and herbivory facilitate the return of both organic and inorganic phosphorus to soil. Inputs of phosphorus to soil through these processes may directly contribute organic phosphorus to soil or, following decomposition and microbial utilization of inorganic and organic phosphorus, indirectly through the accumulation of stabilized soil organic phosphorus. Indeed the contribution of plants to the soil phosphorus cycle and the importance of organic phosphorus within this cycle has been the subject of a number of reviews (e.g. McGill and Cole, 1981; Stewart and Tiessen, 1987; Tate and Salcedo, 1988; Magid *et al.*, 1996). In this chapter, we specifically focus on the importance of organic phosphorus as a source of phosphorus for plant nutrition. In particular, the importance of rhizosphere phosphatase activity and its role in hydrolysing organic phosphorus in soil is considered. Increased knowledge of the processes that govern the availability and utilization of organic phosphorus by plants is critical for understanding ecosystem function and for the development of sustainable agricultural systems with improved efficiency in phosphorus fertilizer use and lesser environmental impacts.

Phosphorus as a Plant Nutrient

Organic phosphorus in soil

Soils generally contain a significant amount of phosphorus (i.e. commonly in the order of 200–3000 mg P kg/soil; Harrison, 1987) but only a small proportion of this (generally less than 1%) is immediately available to plants. Soil phosphorus occurs in organic and inorganic forms; detailed reviews concerning its chemistry and factors that influence its composition have been published elsewhere (Sample *et al.*, 1980; Sanyal and De Datta, 1991). Organic phosphorus generally accounts for at least 30%, and occasionally up to 80%, of the total soil phosphorus (Dalal, 1977; Anderson, 1980), of which a large proportion is associated with high-molecular-weight complexes. However, various techniques based on chemical extraction and fractionation, combined with a range of analytical procedures (e.g. solution ^{31}P nuclear magnetic resonance spectroscopy; see Cade-Menun, Chapter 2, this volume) have shown that organic phosphorus in soil largely comprises phosphate monoesters (up to 90%), with lesser concentrations of phosphate diesters and phosphonates (Newman and Tate, 1980; Hawkes *et al.*, 1984; Condron *et al.*, 1990). Most phosphate monoesters in soil are in the form of phytate (i.e. cation derivatives of inositol hexakisphosphate).[1] In a range of Australian soils, Williams and Anderson (1968) showed that phytate accounted for on average 16% (and up to 38%) of soil organic phosphorus, whereas other studies reported phytate to constitute more than 50% (Anderson and Malcolm, 1974; Halstead and McKercher, 1975; Islam and Mandal, 1977; Turner *et al.*, 2002b). In contrast, sugar phosphates and phosphate diesters (mainly nucleic acids and phospholipids) constitute only a small component, typically around 5% of the total soil organic phosphorus (Dalal, 1977; Anderson, 1980). However, assessing the importance of various pools of organic phosphorus based on their overall content in soils may be misleading, as the turnover rates of sugar phosphates and phosphate diesters are likely to be higher than that of phytate, and therefore their contribution to phosphorus cycling in soils may be of greater significance.

A number of competing processes contribute to the accumulation and predominance of different forms of organic phosphorus in soils (Hedley *et al.*, 1982; Tate, 1984; Stewart and Tiessen, 1987; Magid *et al.*, 1996; Frossard *et al.*, 2000). The potential for organic phosphorus substrates to form metal ion complexes and their interaction with physical, chemical and biological properties of soil are important. Likewise, the effects of land management practices such as fertilization, cultivation and the use of different cropping or forest systems have a significant impact on the short- and long-term dynamics of organic phosphorus accumulation and mineralization (McLaughlin *et al.*, 1988b; Condron *et al.*, 1990; Magid *et al.*, 1996; Chen *et al.*, 2002; also see Nziguheba and Bünemann, Chapter 11: Condron and Tiessen, Chapter 13, this volume).

To be available to plants, organic phosphorus must first be mineralized to release phosphate, with hydrolysis of phosphate ester (C–O–P), phosphoanhydride (P–O–P) or phosphonate (C–P) bonds being predominantly mediated by the action of phosphatase enzymes. In soil, the process is largely controlled by microbial activity (see Oberson and Joner, Chapter 7, this volume) and a number of studies have demonstrated significant net rates of organic phosphorus mineralization in relation to soil phosphatase activity (Tate and Salcedo, 1988; Rojo *et al.*, 1990; Trasar-Cepeda and Carballas, 1991; George *et al.*, 2002b). In natural ecosystems, mineralization of soil organic phosphorus is considered to provide a major proportion of

[1] Phytate is used here collectively in reference to metal ion derivatives of inositol hexa- and pentakisphosphates and includes all isomeric forms. As a purified compound, phytate is available as *myo*-inositol hexakis[dihydrogen] phosphate (a salt of phytic acid or inositol hexaphosphoric acid) and is commonly abbreviated as IHP or InsP6. Throughout this chapter the term phytate is used for generic reference and *myo*-inositol hexakisphosphate in relation to the use of purified compound.

plant-available phosphorus (Fox and Comerford, 1992; Polglase *et al.*, 1992). Laboratory studies have shown that gross mineralization rates in soil are substantial in relation to plant uptake (1–4 mg P kg/soil/day) and thus potentially provide an important pool of plant-available phosphorus (Chauhan *et al.*, 1981; Zou *et al.*, 1992; Jungk *et al.*, 1993; Lopez-Hernandez *et al.*, 1998; Oehl *et al.*, 2001). However, the direct contribution of mineralized phosphorus to plant nutrition in soils remains relatively poorly understood. Incubation of soils under laboratory conditions provides limited information regarding spatial and temporal variations, and disregards the significance of modified soil chemistry, biochemistry and biology found in the rhizosphere. To be of immediate benefit to plants, mineralization of phosphorus must occur in close proximity to roots. This is necessary in order to give plants a competitive advantage to acquire the phosphate over microbial immobilization and various edaphic physical and chemical reactions that limit the availability of free phosphate in soil.

Phosphorus uptake by plants and response to phosphorus deficiency

It is generally considered that plants obtain phosphorus exclusively as phosphate anions from soil solution. However, in most soils, phosphorus acquisition by plants is limited by low concentrations of phosphate in soil solution (typically less than 5 μM), low diffusion rates, or a limited capacity to replenish phosphate levels in soil solution (Bieleski, 1973). Soil solution also contains significant amounts of organic (and condensed) phosphorus compounds at concentrations that may be appreciably higher than that of phosphate (Ron Vaz *et al.*, 1993; Shand *et al.*, 1994). Despite this, the contribution of non-phosphate forms of phosphorus to plant nutrition is not well understood and currently there is no evidence to suggest that plants are able to take up such forms of phosphorus directly.

Recent developments in plant phosphorus nutrition have shown that roots specifically take up and transport phosphate anions (primarily as HPO_4^{2-} and $H_2PO_4^-$) via membrane-associated proteins, for which a number of genes have now been cloned and characterized from a wide range of plant species (Rausch and Bucher, 2002). Specific members of the *Pht1* family of phosphate transporters are expressed in root epidermal cells, primarily in root hair cells (trichoblasts), and show high affinity (e.g. K_m of ~3 μM) for the transport of phosphate (Mitsukawa *et al.*, 1997; Mudge *et al.*, 2002). These transporters are specifically induced by phosphorus deficiency and are energized by adenosine triphosphatase-mediated proton extrusion, which allows phosphate anions to be transported over the plasma membrane against steep concentration gradients (Bieleski, 1973; Schachtman *et al.*, 1998). Indeed, the difference between the concentration of phosphate in soil solution and that of internal plant tissues is at least two or three orders of magnitude. Plant roots are effective in depleting soil solution phosphorus from their immediate vicinity and the activity of high-affinity transporters at the root surface is unlikely to be a major limitation to phosphorus acquisition by plants. Rather, the rate of supply of phosphorus to the roots and factors that govern phosphorus availability in soil solution and rates of root growth into new soil are generally considered to be of greater significance (Bieleski, 1973).

Plants possess a number of other physiological adaptations to cope with low phosphorus availability, which are also generally induced under conditions of phosphorus deficiency. These include strategies that conserve phosphorus (e.g. reduced growth rate, remobilization of internal phosphorus, or modified phosphorus metabolism) or strategies that increase the availability of phosphorus through either modified root structure or function (reviewed by Raghothama, 1999; Vance *et al.*, 2003). Root morphological changes, such as rate of root growth, total and specific root length, degree of root branching, depth of rooting and abundance and elongation of root hairs are important mechanisms that allow larger volumes of soil to be explored, better exploitation of nutrient-rich regions and consequently greater access to soil phosphorus (Caradus, 1995; Lynch, 1995; Gahoonia and Nielsen, 1997). Symbiotic association with mycorrhizal fungi is similarly an important means by which plants are able to access soil phosphorus, as the fungi essentially provide an 'extension' to the root system (Robson *et al.*, 1994; Smith and Read, 1997).

Plants also have the capacity to modify the chemical and physical environment of the rhizosphere through the release of exudates, which may directly affect phosphorus availability or have indirect effects through stimulation of soil microorganisms (Richardson, 1994; Randall *et al.*, 2001; Richardson, 2001; Read *et al.*, 2003; Jakobsen *et al.*, 2005). Acidification of the rhizosphere and release of organic acids are widely recognized as key processes by which roots are able to increase the availability of soil phosphorus, particularly of phosphate (reviewed by Jones, 1998; Hinsinger, 2001; Hocking, 2001; Ryan *et al.*, 2001). For example, the development of proteoid roots and the release of large amounts of organic acids by white lupin (*Lupinus albus* L.) in response to phosphorus deficiency is well documented (Gardner *et al.*, 1983; Dinkelaker *et al.*, 1995; Keerthisinghe *et al.*, 1998; Vance *et al.*, 2003), as are similar mechanisms for a range of plants belonging to the Proteaceae that have evolved survival mechanisms on low-phosphorus soils (Grierson, 1992; Roelofs *et al.*, 2001; Pate and Watt, 2002). Increased phosphatase enzyme activity in the rhizosphere of plants also occurs in response to phosphorus deficiency, and consequently phosphatases are widely regarded as being important for increasing the availability of phosphate from soil organic phosphorus.

Phosphatases and Plant Phosphorus Nutrition

Characterization of extracellular phosphatases

Plants possess a variety of phosphatases and these are associated with a wide range of cellular functions, including energy metabolism, nutrient transport, metabolic regulation, protein activation and hydrolysis of organic phosphorus compounds (Duff *et al.*, 1994). Phosphatases are broadly grouped into acid or alkaline (E.C. 3.1.3.2 and 3.1.3.1, respectively) on the basis of their optimal pH for activity, and are further differentiated according to substrate specificity, primarily as phosphomonoesterase and phosphodiesterases (or monoester and diester phosphohydrolases; see Quiquampoix and Mousain, Chapter 5, this volume). Of particular relevance to enhancing plant phosphorus nutrition from soil organic phosphorus are root extracellular monoester and diester acid phosphatases, which are generally considered to be substrate non-specific (Duff *et al.*, 1994). Extracellular phosphatase can either be associated with cell walls (McLachlan, 1980; Dracup *et al.*, 1984; Barrett-Lennard *et al.*, 1993; Hayes *et al.*, 1999; Hunter and McManus, 1999) or are released from roots into the external environment as exoenzymes (Tadano *et al.*, 1993; Ascencio 1997; Li *et al.*, 1997a; Gilbert *et al.*, 1999; Wasaki *et al.*, 2000; Zhang and McManus, 2000; Gaume *et al.*, 2001; Yun and Kaeppler, 2001). Most studies have shown that extracellular phosphatase activities are enhanced at least several-fold under conditions of phosphorus deprivation. Whilst phosphatase activity is distributed throughout the entire root apoplasm, it is particularly evident in the root epidermis (Hübel and Beck, 1996). The significance of whole-root phosphatase activity and its presence at the root surface has been verified by assays of intact roots (McLachlan, 1980; Richardson *et al.*, 2000; Fig. 8.1) and direct visualization techniques (Dinkelaker and Marschner, 1992; Grierson and Comerford, 2000). Moreover, the recent cloning of genes encoding extracellular phosphatase from arabidopsis (*Arabidopsis thaliana* (L.) Heynh.) (Haran *et al.*, 2000) and white lupin (Wasaki *et al.*, 2000; Miller *et al.*, 2001) provides strong evidence for direct secretion and regulation of their expression in a root-specific manner and in response to phosphorus deprivation.

The presence of extracellular phospha-

(a)

Fraction from wheat roots	Total APase activity[a]		Phytase activity[a]	
	P-fed plants	No-P plants	P-fed plants	No-P plants
Whole root (soluble) extract (mU/g root fresh wt)	469.8±94.5	509.7±80.6	4.4±1.1	23.9±1.2
Total intact root (mU/g root fresh wt)	466.7±141.7	1029.7±130.3	n.d.	12.1±4.0
External-root solution (mU/g root fresh wt/h)	21.5±4.7	46.7±19.07	<0.3	<0.3

[a] Activities (±1 standard error) determined using *p*-nitrophenyl phosphate and *myo*-inositol hexakisphosphate (phytate) as substrates for total monoesterase-APase and phytase activities, respectively (n.d.=not determined, <0.3=limit of assay sensitivity).

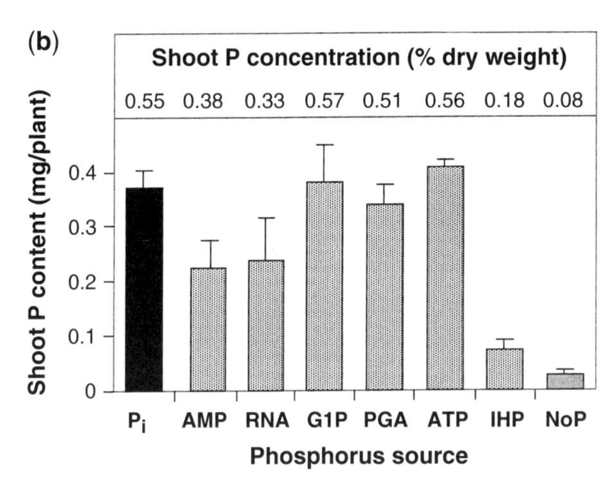

Fig. 8.1. Acid phosphatase (phosphomonoesterase; APase) and phytase activities of wheat (*Triticum aestivum*) roots and utilization of phosphate from organic phosphorus sources. Shown are (a) enzyme activities of different root fractions from 13-day-old wheat seedlings grown in agar with (P-fed; 0.8 mM Na_2HPO_4) or without added phosphorus (No-P) under sterile conditions and (b) total phosphorus uptake and percentage phosphorus content (dry weight basis) of wheat shoots after 19 days' growth in sterile agar supplied with different sources of phosphorus (1.0 mM with respect to phosphorus; total supply of 1.24 mg P/plant) provided as phosphate (P_i, Na_2HPO_4), phosphate diesters as adenosine 3′:5′-cyclic monophosphate (cAMP) and ribonucleic acid (RNA), phosphate monoesters as glucose phosphate (α-D-glucose 1-phosphate; G1P), glycerol phosphate (D(-)3-phosphoglyceric acid; PGA), adenosine-5′-triphosphate (ATP), and *myo*-inositol hexakisphosphate (IHP), or grown without added phosphorus (No-P). Error bars show 1 standard deviation. Data are taken from Richardson *et al.* (2000), with permission from Blackwell Publishing.

tase in plant roots is supported by a number of studies that have directly measured increases in soil phosphatase activity (generally phosphomonoesterase) of up to tenfold in close proximity to roots (Tarafdar and Jungk, 1987; Asmar *et al.*, 1995; Li *et al.*, 1997b; Chen *et al.*, 2002; George *et al.*, 2002c). However, the direct contribution of plants to observed increases in phosphatase activity within the rhizosphere is difficult to ascertain, because in most instances concomitant increases in microbial populations also occur (Tarafdar and Jungk, 1987; Chen *et al.*, 2002).

Microorganisms may make a substantial contribution to the total phosphatase activity of the rhizosphere, and distinction between the relative importance of enzymes from either source for the utilization of soil organic phosphorus remains to be established. Recent evidence suggests that phosphatases derived from fungal sources show higher efficiency for utilization of model organic phosphorus compounds than those derived from plant roots (Tarafdar *et al.*, 2001). The identification of specific plant genes that encode extracellular phosphatase (Haran *et al.*, 2000; Wasaki *et al.*, 2000; Miller *et al.*, 2001) provides new opportunities for investigating the significance of plant-derived enzymes through the development of reporter gene constructs, the use of specific probes, or the generation of modified plants that show enhanced or reduced phosphatase expression.

Extracellular phosphatase activities from plant roots

Despite the evidence for extracellular phosphatase activities around plant roots, their significance for the mineralization of organic phosphorus substrates that occur in soil environments is poorly understood, and the contribution of phosphatase to plant phosphorus nutrition is to a large extent still an assumed benefit. Much of the work reported on root phosphatase has relied on the use of either artificial (e.g. *para*-nitrophenyl phosphate, 4-methylumbelliferyl phosphate or bis-*para*-nitrophenyl phosphate; for phosphomonoesterases and phosphodiesterases, respectively) or specific organic phosphorus substrates. For example, high levels of extracellular phosphatase activity against *para*-nitrophenyl phosphate and various purified substrates (e.g. phosphate monoesters such as β-glycerophosphate and glucose phosphate; phosphate diesters such as phosphatidyl choline, cyclic-adenosine monophosphate, and nucleic acids and derivatives) have been recorded (Tarafdar and Claassen, 1988; Barrett-Lennard *et al.*, 1993; Zhang and McManus, 2000). While such substrates are convenient for assay purposes, they provide limited information regarding either the specificity or the affinity of enzyme for other organic phosphorus substrates and their interactions in the rhizosphere. Moreover, on the basis of high activities, it has been suggested that total phosphatase activity of roots is unlikely to limit the hydrolysis of soil organic phosphorus (Tarafdar and Claassen, 1988). Such assumptions take little account of the importance of substrate specificity, the nature of the substrates that are likely to be encountered in soil, or the rhizosphere environment where pH, ionic strength and substrate concentrations may differ substantially from 'ideal' enzyme assay conditions. For example, phytase (phosphomonoesterase specific for the release of phosphate from *myo*-inositol hexakisphosphate; EC 3.1.3.8 and EC 3.1.3.26 for 3-phytase and 6-phytase, respectively) generally constitutes less than 5% of the total root phosphomonoesterase activity (Barrett-Lennard *et al.*, 1993; Bosse and Köck, 1998; Gilbert *et al.*, 1999; Hayes *et al.*, 1999; Richardson *et al.*, 2000; Fig. 8.1). Furthermore, the extracellular component of the phytase activity may represent only 1% or 2% of the total root activity, or be undetectable (Asmar, 1997; Bosse and Köck, 1998; Gilbert *et al.*, 1999; Hayes *et al.*, 1999; Richardson *et al.*, 2000; Fig. 8.1). In contrast, some studies have indicated that root exudates contain phytase activities that are comparable with total phosphatase (Tarafdar and Claassen, 1988; Li *et al.*, 1997a). The significance of such activities and the relative availability of organic phosphorus substrates, such as phytate, in soil and soil solution require more detailed investigation.

On the basis that root phosphatases generally show high specificity towards defined organic phosphorus compounds that are found in plant cells, extracellular phosphatases may also play a role in the recycling of organic phosphorus lost from roots (Lefebvre *et al.*, 1990; Barrett-Lennard *et al.*, 1993, Duff *et al.*, 1994). Although it is well established that plants lose a significant amount of carbon through root exudates, mucigels and via sloughed-off cells (e.g. up to 50% of the photosynthetic carbon transported to root tissues; Grayston *et al.*, 1996), quantitative estimates of phosphorus loss from roots are not common. It is reasonable to assume that sloughed-off cells and other cellular debris contains organic phosphorus in various monoester and diester forms (e.g. Read *et al.*, 2003), but there is currently little evidence for the direct exudation of organic phosphorus compounds from roots. One study using barley plants grown in water under axenic conditions reported the presence of organic phosphorus in root exudates equating to ~3% of seed phosphorus reserves (Pant *et al.*, 1994). In contrast, the efflux of phosphate from roots can be significant and is dependent on plant species, relative phosphorus status of the plant and other environmental conditions (McLaughlin *et al.*, 1987; Pellet *et al.*, 1997; Dunlop and Phung, 1999; Clark *et al.*, 2000). The significance of phosphorus loss from roots (either as phosphate or as organic phosphorus) and the capacity of plants to recapture any released phosphorus in competition with immobilization by rhizosphere microorganisms, or through soil physical and chemical processes, remains to be determined.

Utilization of Organic Phosphorus by Plants

The ability of plants to obtain phosphorus from organic sources has been investigated either by direct measurement of growth and phosphorus nutrition when provided with organic phosphorus substrates supplied in controlled media or soil, or by measuring the depletion of various fractions of organic phosphorus in rhizosphere soil.

Utilization of organic phosphorus substrates by plants

Extracellular secretion of phosphatases by plant roots is consistent with their ability to obtain phosphorus from a range of organic phosphorus sources when grown under sterile conditions (Tarafdar and Claassen, 1988; Hayes *et al.*, 2000b; Richardson *et al.*, 2000). In the study by Richardson *et al.* (2000), plants grown in sterile agar were able to effectively obtain phosphorus from a range of both phosphate monoester and diester substrates, although there was some evidence for poorer utilization of phosphate diesters relative to phosphate (Fig. 8.1). In contrast, plants (including a range of dicot and monocot species) were unable to utilize phosphorus supplied as soluble *myo*-inositol hexakisphosphate (Fig. 8.1; Hayes *et al.*, 2000b; Richardson *et al.*, 2000, 2001b). The inability of plants to obtain phosphorus from *myo*-inositol hexakisphosphate was associated with insufficient extracellular phytase activity (Fig. 8.1), and growth was improved significantly when either a purified phytase or a phytase-producing microorganism was added to the growth media (Hayes *et al.*, 2000b; Richardson *et al.*, 2001b). This is consistent with work by Findenegg and Nelemans (1993), where the addition of phytase to maize (*Zea mays* L.) roots increased the acquisition of phosphorus from phytate, but only when plants were grown in a non-phosphorus-fixing sand medium. The critical need for extracellular secretion of phytase for utilization of phytate was further demonstrated by the enhanced phosphorus nutrition of transgenic arabidopsis that overexpress an *Aspergillus niger* phytase gene and release extracellular phytase from their roots (Richardson *et al.*, 2001a). Control transgenic plants that express phytase, but without extracellular secretion, were unable to obtain phosphorus from phytate when supplied as soluble *myo*-inositol hexakisphosphate. More recent studies have shown that expression and extracellular secretion of phytase from root hair cells alone is sufficient to confer the ability to utilize phosphorus from phytate in both arabidopsis and potato (*Solanum tuberosum* L.) plants (Mudge *et al.*, 2003; Zimmermann *et al.*, 2003). Subterranean clover (*Trifolium subterraneum* L.) transformed with *Aspergillus* phytase similarly has a significantly enhanced (77-fold) root-secreted phytase activity and an increased capacity for growth and phosphorus nutrition fourfold when grown in agar and supplied with *myo*-inositol hexakisphosphate as the sole phosphorus source (Richardson *et al.*, 2001c; George *et al.*, 2004). When grown in a range of soils, however, the growth and phosphorus nutrition advantage of these transgenic plants was compromised (George *et al.*, 2004). Further work is required in order to understand factors that restrict the efficacy of phytase–phytate interaction in the soil.

The capacity of plants to utilize organic phosphorus has also been examined by the addition of different phosphorus substrates to soil (under non-sterile conditions) with general indications that organic phosphorus is effective for plant nutrition (Tarafdar and Claassen, 1988; Adams and Pate, 1992; Findenegg and Nelemans, 1993; Taranto *et al.*, 2000). However, the effectiveness of different organic phosphorus substrates is dependent on their availability in soils and is likely to be further influenced by the presence of soil microorganisms. For example, Adams and Pate (1992) showed that ribonucleic acid and β-glycerophosphate were approximately equivalent to phosphate for phosphorus nutrition of lupins grown in either sand or soil, but in contrast, phytate was only accessed by plants grown in sand. Phytate was also of limited availability to maize and wheat (*Triticum aestivum* L.) in high phosphate-fixing soils (Martin and Cartwright, 1971; Martin, 1973; Findenegg and Nelemans, 1993; Hübel and Beck, 1993). Analogous to phosphate, phytate is subject to adsorption–desorption reactions in soil and is readily complexed by cations to form precipitates of low solubility under both acidic (in the presence of iron and aluminium) and alkaline (in the presence of calcium and magnesium) conditions (Jackman and Black, 1951; McKercher and Anderson, 1968; Anderson *et al.*, 1974; Shang *et al.*, 1992; Celi and Barberis, Chapter 6, this volume). If phytate is a potential source of phosphorus for plants then its poor availability in soil is further confounded by a lack of extracellular phytase activity in roots. Alternatively, the apparent absence of extracellular phytase activities in plant roots may indicate that such an attribute confers little benefit in soil environments, where plant access to this substrate is essentially governed by its availability.

Depletion of soil organic phosphorus from the rhizosphere

A variety of techniques and sampling procedures have been used to demonstrate that plants are effective in depleting phosphorus around their roots (Jungk, 1987; Tarafdar and Jungk, 1987; Gahoonia and Nielsen, 1991). In particular, 'rhizobox' systems, in which plant roots are separated from the soil by a nylon mesh, have been used to illustrate rhizosphere effects on soil phosphorus dynamics in close proximity to roots. Although rhizoboxes may result in exaggerated effects due to the dense root mats that form at the soil interface, and the non-cylindrical nature of the contact between plant and soil, such studies are useful for understanding the mechanisms of phosphorus depletion in the rhizosphere. The importance of factors such as root hairs and biochemical changes, including pH and root exudates, on the availability of soil phosphorus have been demonstrated (Tarafdar and Jungk, 1987; Armstrong and Helyar, 1992; Gahoonia and Nielsen, 1992; Hedley *et al.*, 1994; Asmar *et al.*, 1995; Gahoonia and Nielsen, 1997; Li *et al.*, 1997b; Chen *et al.*, 2002; George *et al.*, 2002a), as has the contribution of mycorrhizal hyphae for depletion of soil phosphorus at large distances from roots (Tarafdar and Marschner, 1994a,b; Joner and Jakobsen, 1995; Joner *et al.*, 1995). Roots with and without mycorrhizae are able to deplete various pools of soil phosphorus, including those considered to be labile (e.g. resin- and sodium bicarbonate-extractable phosphorus; Bowman and Cole, 1978) and others previously considered to be only of poor availability to plants (e.g. sodium hydroxide- and hydrochloric acid-extractable phosphorus and, in some cases, residual phosphorus). Generally it is evident that mycorrhizal roots can access the same pools of soil phosphorus as non-mycorrhizal roots (reviewed by Bolan, 1991; Joner *et al.*, 2000). Despite this, some studies have shown that mycorrhizal association results in a specific depletion of either soil organic phosphorus or added organic phosphorus substrates (Jayachandran *et al.*, 1992; Tarafdar and Marschner, 1994b), while other studies are contrary to this (Joner and Jakobsen, 1995; Joner *et al.*, 1995; Colpaert *et al.*, 1997). The importance of mycorrhizae for the utilization of soil organic phosphorus remains to be resolved.

Depletion of organic phosphorus from the rhizosphere has been established in a number of studies. For instance, Tarafdar and Jungk (1987) showed a 65–86% depletion of total soil organic phosphorus in the rhizosphere of berseem clover (*Trifolium alexandrinum* L.) and wheat roots that extended 1 and 2 mm, respectively, into soil. Depletion of specific organic phosphorus fractions has similarly been reported, with utilization of phosphorus from sodium bicarbonate-extractable organic phosphorus by barley (*Hordeum vulgare* L.) roots (Asmar *et al.*, 1995) and from sodium hydroxide-extractable organic phosphorus by a range of other plant species (Gahoonia and Nielsen, 1992; Chen *et al.*, 2002; George *et al.*, 2002a; Fig. 8.2). In contrast, some studies have found either no net change of soil organic phosphorus (Armstrong and Helyar, 1992; Hedley *et al.*, 1994) or, in some cases, an increase in specific pools of organic phosphorus (Saleque and Kirk, 1995; Zoysa *et al.*, 1997, 1999; Chen *et al.*, 2002). Although it is difficult to identify reasons for these different findings, differences in plant species and soil characteristics are clearly important, as is the contribution of microorganisms in mediating phosphorus transformations between phosphorus pools.

The capacity of plants to utilize organic phosphorus in the rhizosphere is generally associated with an increase in the activity of phosphatases (Tarafdar and Jungk, 1987). A correlation between the amount of sodium bicarbonate-extractable organic phosphorus withdrawn from the rhizosphere by different barley cultivars and the activity of soluble phosphatases in the soil was observed by Asmar *et al.* (1995). Differential depletion of sodium hydroxide extractable organic phosphorus has also been reported in the rhizosphere of radiata pine (*Pinus radiata* D. Don) as compared to perennial ryegrass (*Lolium perenne* L.) when grown in two different soils (Chen *et al.*, 2002; Fig. 8.2). A significant depletion of organic phosphorus by both species occurred in one of the soils, but was evident for radiata pine only in the other soil, despite ryegrass having at least twice as much root surface-area in the root mat. Notably, the activities of acid and alkaline monoester and total diester phosphatases were increased to a similar extent in the rhizosphere of both species in the two soils and, in all cases, were accompanied by a significant increase in microbial biomass (Fig. 8.2). Correlations between various phosphatase activities, enhanced microbial biomass and depletion of sodium hydroxide extractable organic phosphorus were therefore evident, but were significant for radiata pine only. These results indicate that increased activities of phosphatases in the rhizosphere alone may not necessarily result in a depletion of organic phosphorus (Fig. 8.2). Either differences in the efficacy or substrate specificity of phosphatases from different plant species are important, or the capacity for organic phosphorus depletion is affected by other factors that influence the availability of organic phosphorus. Radiata pine forms ectomycorrhizal associations, and Chen *et al.* (2002) suggested that this may specifically enhance the capacity of this species to deplete the rhizosphere.

The differential ability of plant species to deplete organic phosphorus was also reported by George *et al.* (2002a,c), who compared white lupin, maize and two agroforestry species (*Tithonia diversifolia* (Hemsl.) A. Gray and *Tephrosia vogelii* Hook. f.). Lupin, tithonia and tephrosia depleted the sodium hydroxide-extractable organic phosphorus pool in their rhizospheres, but there was no net depletion of organic phosphorus around maize roots. Depletion of organic phosphorus in field soil was related to phosphatase activity. However, on the basis of the differential depletion of organic phosphorus, George *et al.* (2002a,c) suggested that phosphatase activity alone did not entirely explain differences in organic phosphorus utilization by the different plant species and that rhizosphere-induced changes to the availability of organic phosphorus substrate were important. Preliminary results from NMR analysis of rhizosphere soil suggest that there is specific depletion of phosphate

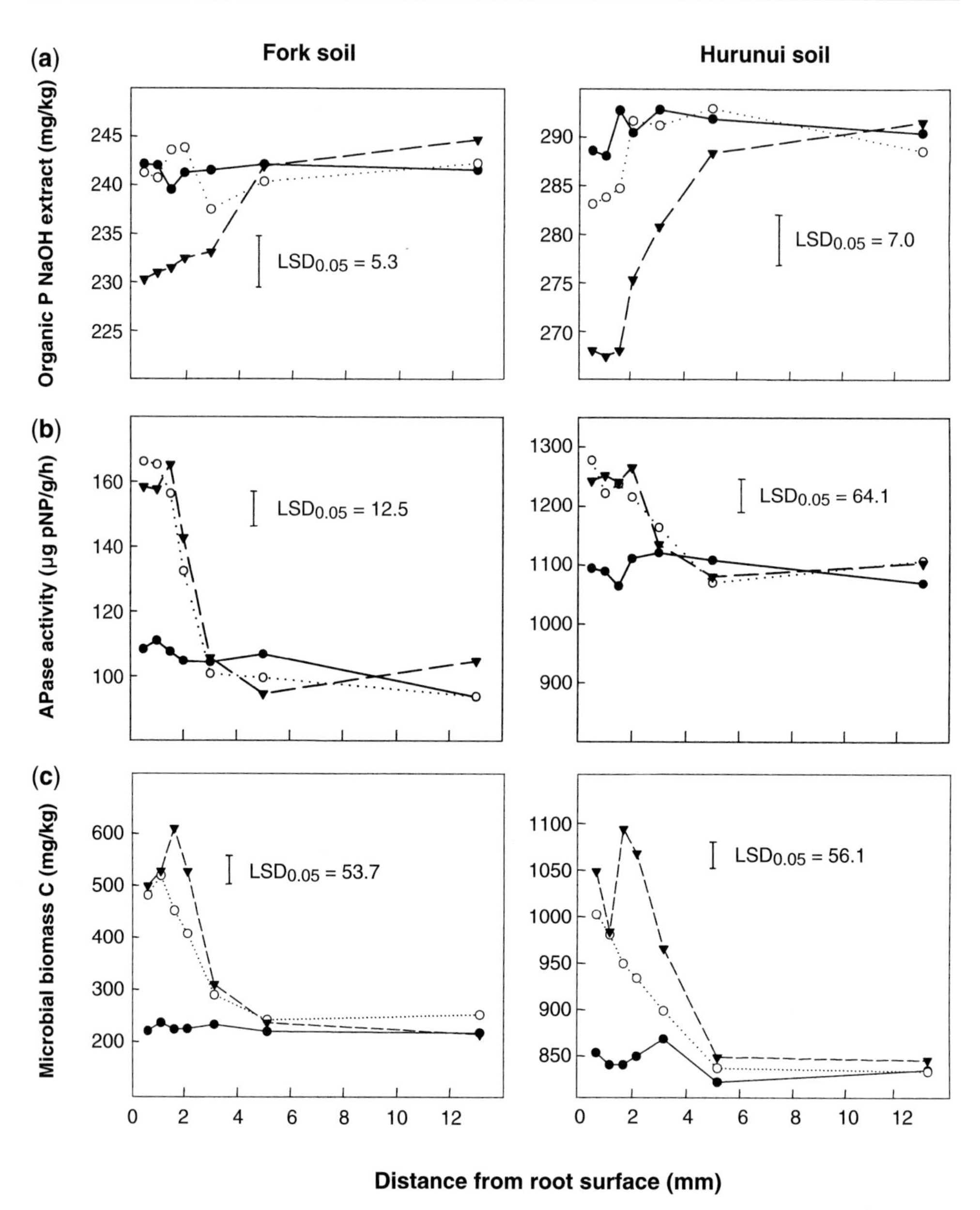

Fig. 8.2. Selected features of the rhizosphere from roots of perennial ryegrass (*Lolium perenne*; ○····○) and radiata pine (*Pinus radiata*; ▼– –▼), as compared to control (unplanted ●—●) soils when grown in a rhizobox system. Shown are changes in (a) sodium hydroxide (NaOH)-extractable organic phosphorus following sequential extraction, (b) total acid phosphomonoesterase activity (APase) and (c) microbial biomass carbon in various rhizosphere sections (up to 13 mm from the root surface) for two different soils. Briefly, the Fork and Hurunui soils were pH 6.0 and 5.6 (soil-water extract), respectively, and contained total phosphorus concentrations of 827 and 958 mg P/kg soil, and extractable organic phosphorus concentrations of 344 and 375 mg P/kg soil (sum of sodium bicarbonate- and sodium hydroxide-extractable fractions). For each panel the error bar shows least significant difference (LSD; $P=0.05$). The data are reproduced from Chen *et al.* (2002), with permission from Elsevier Science.

monoesters around roots of species that exclude phosphate (T.S. George, A.E. Richardson and B.L. Turner, unpublished data), but further work to establish the importance of rhizosphere modification on the solubility and subsequent availability of soil organic phosphorus is required. Studies using bulk soils have previously suggested that mineralization of organic phosphorus is associated predominantly with a decline in phosphate diesters due to the activity of soil microorganisms (Hawkes *et al.*, 1984). There is some evidence to suggest that different plants have differing abilities to deplete specific phosphate monoesters in their rhizospheres. For example, maize was unable to acquire phosphorus from labelled phytate in soil (Hübel and Beck, 1993) and wheat had only a limited capacity to use phytate (Martin, 1973). By contrast, white lupin appears to be more effective at obtaining phosphorus from phytate when added to soil (Adams and Pate, 1992). This ability may be related to the formation of proteoid roots, which release organic acids and show extracellular phytase activity (Dinkelaker *et al.*, 1995; Keerthisinghe *et al.*, 1998; Gilbert *et al.*, 1999). As such, white lupin may have the capacity to both solubilize phytate (by ion-chelation) and to liberate phosphate from this substrate through a secreted phytase. Transgenic plants that overexpress extracellular phytase (Richardson *et al.*, 2001a,c, Mudge *et al.*, 2003; Zimmermann *et al.*, 2003; George *et al.*, 2004) are an important tool for further examining the ability of plants to utilize phytate and to establish the significance of phytase activity and other root exudates within the rhizosphere.

It is evident from a number of studies that organic phosphorus may also accumulate in the root zone of plants. Increases in organic phosphorus extractable by sodium hydroxide (Saleque and Kirk, 1995; Zoysa *et al.*, 1997, 1999; George *et al.*, 2000a) and sodium bicarbonate (Helal and Sauerbeck, 1984; Chen *et al.*, 2002) have been reported and are generally considered to be due to the presence of microorganisms. Rapid assimilation of phosphorus by microorganisms and subsequent deposition into organic phosphorus fractions has been documented in bulk soils (McLaughlin *et al.*, 1988a; Oberson *et al.*, 2001) and transformation of phosphorus is likely to be of considerable significance in the rhizosphere (McLaughlin *et al.*, 1987; Helal and Dressler, 1989; Jakobsen *et al.*, 2005). Microorganisms contain a substantial pool of immobilized phosphorus, which predominantly occurs in organic form (~80%) and, if not specifically accounted for, may contribute to the increase in labile sodium bicarbonate-extractable organic phosphorus observed directly around roots (Helal and Sauerbeck, 1984; Thien and Myers, 1992; Chen *et al.*, 2002; Fig. 8.2). The significance of microbial phosphorus dynamics for either the accumulation or utilization of organic phosphorus within the rhizosphere warrants further investigation (Jakobsen *et al.*, 2005; see Oberson and Joner, Chapter 7, this volume).

Organic Phosphorus in Soil Solution

Although organic phosphorus and/or molybdate-unreactive phosphorus[2] may constitute more than 50% of the total phosphorus in soil solutions (Ron Vaz *et al.*, 1993; Shand *et al.*, 1994), its chemical nature and subsequent availability to plants roots has received only limited attention. Dissolved organic phosphorus is highly mobile in soil (Hannapel *et al.*, 1964; Hoffman and Rolston, 1980; Frossard *et al.*, 1989; Turner and Haygarth, 2000) and is therefore likely to be of particular significance, both for the movement of phosphorus towards plant roots and in the leaching of phosphorus to water-courses (see Turner, Chapter 12, this volume). Water and dilute calcium chloride extracts of soil have been shown to contain as many as 30 distinct organic phosphorus compounds, with initial studies indicating the

[2] The organic phosphorus content of solutions is generally determined by the difference between total phosphorus and molybdate-reactive phosphorus. As this fraction may also contain condensed phosphates (Ron Vaz *et al.*, 1993), the operational term molybdate-unreactive phosphorus is used throughout the chapter, rather than organic phosphorus, which is used in a generic sense.

presence of inositol monophosphates (Wild and Oke, 1966; Martin, 1970). More recently, the presence of sugar phosphates, phosphonates and inositol hexakisphosphate was verified in soil leachate (Espinosa *et al.*, 1999). Using solution ^{31}P nuclear magnetic resonance spectroscopy, Toor *et al.* (2003) showed that the molybdate-unreactive phosphorus in leachate from a grassland soil contained phosphate monoesters (77%) and diesters (23%).

Whilst direct solubilization of soil organic phosphorus would be expected to contribute to organic phosphorus in soil solution, it is evident that a large component of solution molybdate-unreactive phosphorus is derived directly from microbial turnover. Increases in water-extractable molybdate-unreactive phosphorus were observed in soil following drying and rewetting (Turner and Haygarth, 2001) and in response to soil sterilization (Seeling and Jungk, 1996). Such increases are associated with a release of phosphorus from microbial biomass. In controlled incubation studies, Seeling and Zasoski (1993) further demonstrated that microorganisms are critical for maintaining the concentration of molybdate-unreactive phosphorus in soil solution. Organic phosphorus derived from microbial turnover similarly accounted for up to 90% of the soil solution phosphorus within the rhizosphere of maize (Helal and Dressler, 1989), and organic phosphorus compounds identified in water extracts by Martin (1970) were considered to be derived mostly from soil microorganisms. Organic phosphorus contained within microorganisms occurs predominantly as nucleic acids, phospholipids and a range of sugars, and is rapidly mineralized in soil environments (van Veen *et al.*, 1987).

Plant availability and enzyme lability of soil solution organic phosphorus

Using ^{32}P-labelled bacterial extracts, Macklon *et al.* (1997) showed that dissolved organic phosphorus and condensed phosphates accounted for 87% and 6%, respectively, of microbial phosphorus and that these compounds were readily available to plant roots (~75% utilization over 24 h). Other studies showed that molybdate-unreactive phosphorus in aqueous soil extracts was effectively utilized by roots (Wild and Oke, 1966; Seeling and Jungk, 1996; Tarafdar and Claassen, 2003). In the study by Seeling and Jungk (1996), solution-grown barley plants were able to utilize up to 55% of the molybdate-unreactive phosphorus in soil solution extracts over a 24-h period. The capability of plants to utilize phosphorus was similar in extracts prepared from non-sterile and sterilized soil, despite an approximately eightfold difference in the total amount of phosphorus present. It is further evident from these studies that a significant proportion of the molybdate-unreactive phosphorus in solution extracts cannot be hydrolysed by root-exuded phosphatases. Whether or not the plant availability of this organic phosphorus increases in the soil, where it will be amenable to microbial utilization and other physical and chemical reactions, remains to be determined.

The availability of organic and condensed phosphorus in soil solution and other extractable soil phosphorus fractions has also been investigated using commercially available phosphatase enzymes (Fox and Comerford, 1992; Pant *et al.*, 1994; Shand and Smith, 1997; Otani and Ae, 1999; Hayes *et al.*, 2000a; Hens and Merckx, 2001; Turner *et al.*, 2002a; Toor *et al.*, 2003). The use of a range of phosphatases, either substrate-specific or non-specific, allows classification of the hydrolysable molybdate-unreactive phosphorus into functional groups. Collectively, these studies indicate that a significant proportion of organic phosphorus in soil solution and soil water extracts is amenable to enzyme hydrolysis. For example, Fox and Comerford (1992) showed that between 20% and 30% of water-soluble organic phosphorus from two contrasting soils was hydrolysed by a non-specific phosphomonoesterase. Similar results were observed in other studies, although the degree of organic phosphorus hydrolysis showed wide variation, ranging from negligible amounts to almost 100% of the filterable molybdate-unreactive phosphorus being hydrolysed (Pant *et al.*, 1994; Hayes *et al.*, 2000a; Hens and Merckx, 2001; Turner *et al.*, 2002a; Toor *et al.*, 2003). Whilst the susceptibility of organic phosphorus to enzymatic hydrolysis is dependent on soil type and organic phosphorus composition, it is evident that sampling procedures (e.g. extraction, centrifugation, filtration and the use of suction-cup apparatus; Magid *et al.*, 1992; Shand *et al.*, 2000), interaction between the enzyme and solution components, and the specificity of the phosphatase enzyme preparation, all affect the degree of hydrolysis. Using a non-specific commercial preparation of *Aspergillus* phytase (as compared to a highly purified preparation of the same enzyme; Wyss *et al.*, 1999), Hayes *et al.* (2000a) showed that the amount of enzyme-labile organic phosphorus in various soil extracts was around twice as high for the non-specific phytase. This was associated with wider substrate specificity due to the presence of other phosphatases (Shand and Smith, 1997; Hayes *et al.*, 2000a). Likewise, other studies have shown that crude phytase also consistently hydrolyses a greater proportion of organic phosphorus than phosphomonoesterase alone (Pant *et al.*, 1994; Shand and Smith, 1997; Hens and Merckx, 2001; Turner *et al.*, 2002a), thereby confirming the heterogenic nature of dissolved organic phosphorus. The specificity of different phosphatases towards hydrolysis of water-extractable molybdate-unreactive phosphorus has been further characterized by Turner *et al.* (2002a), who found that up to 63% of the phosphorus was hydrolysed by a mixture of phosphodiesterase and alkaline phosphomonoesterase, but only 6% was hydrolysed by the phosphomonoesterase alone. This indicates that, for the soils examined, phosphate diesters were the predominant form of organic phosphorus in soil solution. However, it is also possible that the accessibility of monoester substrates to alkaline phosphatase (as compared to acid phosphatase) was poor, or that phosphate monoesters are readily hydrolysed by endogenous activities during extraction and processing procedures. Such possibilities remain to be tested and may contribute to the variability in the susceptibility of organic phosphorus to hydrolysis that has been observed in different studies. Moreover, there is a need to adopt standardized methodologies for assessing the enzyme lability of soil organic phosphorus fractions.

Complete enzymatic degradation of molybdate-unreactive phosphorus in aqueous soil extracts may not occur despite excess phosphatases being added. Hens and Merckx (2001) demonstrated that a large proportion (up to 95%) of the 'dissolved' molybdate-unreactive phosphorus in soil solutions from sandy soils was colloidal-sized and showed only limited (<30%) susceptibility to enzymatic hydrolysis. Molybdate-unreactive phosphorus may either contain phosphorus bonds that are resistant to hydrolysis by the phosphatase preparations used, or may consist of hydrolysable phosphorus bonds that are protected from enzyme degradation through occlusion within colloidal particles (Hens and Merckx, 2001). The degree of hydrolysis also depends on the extractant used. Enzyme lability of molybdate-unreactive phosphorus was higher, for instance, in citric acid extracts of soil, as compared with water or sodium bicarbonate extracts, even though the latter extracted larger amounts of organic phosphorus (Otani and Ae, 1999; Hayes *et al.*, 2000a). In the study by Hayes *et al.* (2000a), up to 79% of the molybdate-unreactive phosphorus in citric acid extracts was amenable to hydrolysis by non-specific phytase, whereas less than 17% of the phosphorus was hydrolysed in other extracts. Citrate, presumably through its capacity for chelation, either mobilizes different fractions of soil organic phosphorus or modifies the physico-chemical nature of organic phosphorus compounds extracted, thereby increasing their susceptibility to enzymatic hydrolysis. Further work to assess the significance of chemical and physical changes in the rhizosphere of different plant species and their impact on the amenability of soil solution organic phosphorus to hydrolysis by phosphatases is warranted.

Contribution of Organic Phosphorus to Plant Nutrition: Directions for the Future

Organic phosphorus is an important part of the soil phosphorus cycle and potentially makes an important contribution to the phosphorus nutrition of plants (Jungk *et al.*, 1993; Frossard *et al.*, 2000; Tarafdar and Claassen, 2003). Despite this, our understanding of the processes involved in the mineralization of organic phosphorus in close proximity to plant roots and the subsequent availability of released phosphate for plant nutrition remains limited. In particular, the availability of organic phosphorus substrates in soil and soil solution, and factors that influence its availability, need further investigation, as substrate availability is clearly of major importance. Similarly, the significance of specific extracellular phosphatases released from roots and the relative contribution of plant and microbial phosphatases needs to be determined. It is assumed that improved exudation of extracellular phosphatases from plants roots may improve plant phosphorus nutrition, but this remains to be verified. The cloning of extracellular phosphatases from plants and analysis of their expression in roots will assist in determining the role that these enzymes play in the utilization of organic phosphorus. Determining the function of specific phosphatase activities in relation to the utilization of defined organic phosphorus substrates in soils, and better characterization of the nature of the organic phosphorus substrates, is critical. Success in achieving these goals will require a combination of approaches that integrate plant physiology and molecular biology with practical methods to sample rhizosphere phosphorus pools, combined with the development of analytical procedures that allow meaningful chemical and biochemical characterization of the functional organic phosphorus substrates that are present. Finally, the capacity of plants to modify the rhizosphere, and subsequent effects of this on the solubility and availability to hydrolysis of soil organic phosphorus, also warrants further investigation. An understanding of critical features of the rhizosphere that are involved in mobilization of soil organic phosphorus may provide an opportunity for the manipulation of plants for improved phosphorus utilization. Understanding of the importance of microbial-mediated transformation of phosphorus and its interaction with plant roots is paramount. The increased availability of organic phosphorus within the rhizosphere and the greater cycling of phosphorus in soils remain important requirements for improving the sustainability of agriculture and for minimizing the environmental impacts of agricultural production systems.

References

Adams, M.A. and Pate, J.S. (1992) Availability of organic and inorganic forms of phosphorus to lupins (*Lupinus* spp.). *Plant and Soil* 145, 107–113.

Anderson, G. (1980) Assessing organic phosphorus in soils. In: Khasawneh, F.E., Sample, E.C. and Kamprath, E.J. (eds) *The Role of Phosphorus in Agriculture*. American Society of Agronomy, Madison, Wisconsin, pp. 411–432.

Anderson, G. and Malcolm, R.E. (1974) The nature of alkali-soluble soil organic phosphates. *Journal of Soil Science* 25, 282–297.

Anderson, G., Williams, E.G. and Moir, J.O. (1974) A comparison of the sorption of inorganic orthophosphate and inositol hexaphosphate by six acid soils. *Journal of Soil Science* 25, 51–62.

Armstrong, R.D. and Helyar, K.R. (1992) Changes in soil phosphate fractions in the rhizosphere of semi-arid pasture grasses. *Australian Journal of Soil Research* 30, 131–143.

Ascencio, J. (1997) Root-secreted acid phosphatase kinetics as a physiological marker for phosphorus deficiency. *Journal of Plant Nutrition* 20, 9–26.

Asmar, F. (1997) Variation in activity of root extracellular phytase between genotypes of barley. *Plant and Soil* 195, 61–64.

Asmar, F., Gahoonia, T.S. and Nielsen, N.E. (1995) Barley genotypes differ in activity of soluble extracellular phosphatase and depletion of organic phosphorus in the rhizosphere soil. *Plant and Soil* 172, 117–122.

Attiwill, P.M. and Adams, M.A. (1993) Nutrient cycling in forests. *New Phytologist* 124, 561–582.

Barrett-Lennard, E.G., Dracup, M. and Greenway, H.

(1993) Role of extracellular phosphatase in the phosphorus-nutrition of clover. *Journal of Experimental Botany* 44, 1595–1600.

Bieleski, R.L. (1973) Phosphate pools, phosphate transport and phosphate availability. *Annual Reviews of Plant Physiology* 24, 225–252.

Bolan, N.S. (1991) A critical review on the role of mycorrhizal fungi in the uptake of phosphorus by plants. *Plant and Soil* 134, 189–207.

Bosse, D. and Köck, M. (1998) Influence of phosphate starvation on phosphohydrolases during development of tomato seedlings. *Plant, Cell and Environment* 21, 325–332.

Bowman, R.A. and Cole, C.V. (1978) Transformations of organic phosphorus substrates in soils as evaluated by $NaHCO_3$ extraction. *Soil Science* 125, 49–54.

Caradus, J.R. (1995) Genetic control of phosphorus uptake and phosphorus status in plants. In: Johanses, C., Lee, K.K., Sharma, K.K., Subbarao, G.V. and Kueneman, E.A. (eds) *Genetic Manipulation of Crop Plants to Enhance Integrated Nutrient Management in Cropping Systems. 1. Phosphorus*. ICRISAT, India, pp. 55–74.

Chauhan, B.S., Stewart, J.W.B. and Paul, E.A. (1981) Effect of labile inorganic phosphate status and organic carbon additions on the microbial uptake of phosphorus in soils. *Canadian Journal of Soil Science* 61, 373–385.

Chen, C.R., Condron, L.M., Davis, M.R. and Sherlock, R.R. (2002) Phosphorus dynamics in the rhizosphere of perennial ryegrass (*Lolium perenne* L.) and radiata pine (*Pinus radiata* D.Don). *Soil Biology and Biochemistry* 34, 487–499.

Clark, G.T., Dunlop, J. and Phung, H.T. (2000) Phosphate absorption by *Arabidopsis thaliana*: interactions between phosphorus status and inhibition by arsenate. *Australian Journal of Plant Physiology* 27, 959–965.

Colpaert, J.V., van Laere, A., van Tichelen, K.K. and van Assche, J.A. (1997) The use of inositol hexaphosphate as a phosphorus source by mycorrhizal and non-mycorrhizal Scots pine (*Pinus sylvestris*). *Functional Ecology* 11, 407–415.

Condron, L.M., Frossard, E., Tiessen, H., Newman, R.H. and Stewart, J.W.B. (1990) Chemical nature of organic phosphorus in cultivated and uncultivated soils under different environmental conditions. *Journal of Soil Science* 41, 41–50.

Dalal, R.C. (1977) Soil organic phosphorus. *Advances in Agronomy* 29, 85–117.

Dinkelaker, B. and Marschner, H. (1992) *In vivo* demonstration of acid phosphatase activity in the rhizosphere of soil-grown plants. *Plant and Soil* 144, 199–205.

Dinkelaker, B., Hengeler, C. and Marschner, H. (1995) Distribution and function of proteoid roots and other root clusters. *Botanica Acta* 108, 183–200.

Dracup, M.N.H., Barrett-Lennard, E.G., Greenway, H. and Robson, A.D. (1984) Effect of phosphorus deficiency on phosphatase activity of cell walls from roots of subterranean clover. *Journal of Experimental Botany* 35, 466–480.

Duff, S.M.G., Sarath, G. and Plaxton, W.C. (1994) The role of acid phosphatases in plant phosphorus metabolism. *Physiologia Plantarum* 90, 791–800.

Dunlop, J. and Phung, T. (1999) Efflux and influx as factors in the relative abilities of ryegrass and white clover to compete for phosphate. In: Gissel-Nielsen, G. and Jensen, A. (eds) *Plant Nutrition: Molecular Biology and Genetics*. Kluwer Academic, Dordrecht, pp. 105–110.

Espinosa, M., Turner, B.L. and Haygarth, P.M. (1999) Preconcentration and separation of trace phosphorus compounds in soil leachate. *Journal of Environmental Quality* 28, 1497–1504.

Findenegg, G.R. and Nelemans, J.A. (1993) The effect of phytase on the availability of phosphorus from *myo*-inositol hexaphosphate (phytate) for maize roots. *Plant and Soil* 154, 189–196.

Fox, T.R. and Comerford, N.B. (1992) Rhizosphere phosphatase activity and phosphatase hydrolyzable organic phosphorus in two forested spodosols. *Soil Biology and Biochemistry* 24, 579–583.

Frossard, E., Stewart, J.W.B. and St Arnaud, R.J. (1989) Distribution and mobility of phosphorus in grassland and forest soils of Saskatchewen. *Canadian Journal of Soil Science* 69, 401–416.

Frossard, E., Condron, L.M., Oberson, A., Sinaj, S. and Fardeau, J.C. (2000) Processes governing phosphorus availability in temperate soils. *Journal of Environmental Quality* 29, 15–23.

Gahoonia, T.S. and Nielsen, N.E. (1991) A method to study rhizosphere processes in this soil layers of different proximity to roots. *Plant and Soil* 135, 143–146.

Gahoonia, T.S. and Nielsen, N.E. (1992) The effect of root-induced pH changes on the depletion of inorganic and organic phosphorus in the rhizosphere. *Plant and Soil* 143, 185–191.

Gahoonia, T.S. and Nielsen, N.E. (1997) Variation in root hairs of barley cultivars doubled phosphorus uptake from soil. *Euphytica* 98, 177–182.

Gardner, W.K., Barber, D.A. and Parbery, D.G. (1983) The acquisition of phosphorus by *Lupinus albus* L. II. The probable mechanism by which phosphorus movement in the soil/root interface is enhanced. *Plant and Soil* 70, 107–114.

Gaume, A., Machler, F., Deleon, C., Narro, L. and Frossard, E. (2001) Low-P tolerance by maize

(*Zea mays* L.) genotypes: significance of root growth, and organic acids and acid phosphatase root exudation. *Plant and Soil* 228, 253–264.

George, T.S., Gregory, P.J., Robinson, J.S. and Buresh, R.J. (2002a) Changes in phosphorus concentrations and pH in the rhizosphere of some agroforestry and crop species. *Plant and Soil* 246, 65–73.

George, T.S., Gregory, P.J., Robinson, J.S., Buresh, R.J. and Jama, B. (2002b) Utilisation of soil organic P by agroforestry species in the field, western Kenya. *Plant and Soil* 246, 53–63.

George, T.S., Gregory, P.J., Wood, M., Read, D. and Buresh, R.J. (2002c) Phosphatase activity and organic acids in the rhizosphere of potential agroforestry species and maize. *Soil Biology and Biochemistry* 34, 1487–1494.

George, T.S., Richardson, A.E., Hadobas, P.A. and Simpson, R.J. (2004) Characterization of transgenic *Trifolium subterraneum* L. which expresses *phyA* and releases extracellular phytase: growth and P nutrition in laboratory media and soil. *Plant, Cell and Environment* 27 (in press).

Gilbert, G.A., Knight, J.D., Vance, C.P. and Allan, D.L. (1999) Acid phosphatase activity in phosphorus-deficient white lupin roots. *Plant, Cell and Environment* 22, 801–810.

Grayston, S.J., Vaughan, D. and Jones, D. (1996) Rhizosphere carbon flow in trees in comparison with annual plants: the importance of root exudation and its impact on microbial activity and nutrient availability. *Applied Soil Ecology* 5, 29–56.

Grierson, P.F. (1992) Organic acids in the rhizosphere of *Banksia integrifolia* L.f. *Plant and Soil* 144, 259–265.

Grierson, P.F. and Comerford, N.B. (2000) Non-destructive measurement of acid phosphatase activity in the rhizosphere using nitrocellulose membranes and image analysis. *Plant and Soil* 218, 49–57.

Halstead, R.L. and McKercher, R.B. (1975) Biochemistry and cycling of phosphorus. In: Paul, E.A. and McLaren, A.D. (eds) *Soil Biochemistry,* Vol. 4. Marcel Dekker, New York, pp. 31–63.

Hannapel, R.J., Fuller, W.H. and Fox, R.H. (1964) Phosphorus movement in calcareous soil. II. Soil microbial activity and organic phosphorus movement. *Soil Science* 97, 421–427.

Haran, S., Logendra, S., Seskar, M., Bratanova, M. and Raskin, I. (2000) Characterization of *Arabidopsis* acid phosphatase promoter and regulation of acid phosphatase expression. *Plant Physiology* 124, 615–626.

Harrison, A.F. (1987) *Soil Organic Phosphorus: a Review of World Literature.* CAB International, Wallingford, UK, 257 pp.

Hawkes, G.E., Powlson, D.S., Randall, E.W. and Tate, K.R. (1984) A ^{31}P nuclear magnetic resonance study of the phosphorus species in alkali extracts of soils from long-term field experiments. *Journal of Soil Science* 35, 35–45.

Hayes, J.E., Richardson, A.E. and Simpson, R.J. (1999) Phytase and acid phosphatase activities in roots of temperate pasture grasses and legumes. *Australian Journal of Plant Physiology* 26, 801–809.

Hayes, J.E., Richardson, A.E. and Simpson, R.J. (2000a) Components of organic phosphorus in soil extracts that are hydrolysed by phytase and acid phosphatase. *Biology and Fertility of Soils* 32, 279–286.

Hayes, J.E., Simpson, R.J. and Richardson, A.E. (2000b) The growth and utilisation of plants in sterile media when supplied with inositol hexaphosphate, glucose 1-phosphate or inorganic phosphate. *Plant and Soil* 220, 165–174.

Hedley, M.J., Stewart, J.W.B. and Chauhan, B.S. (1982) Changes in inorganic and organic soil phosphorus fractions induced by cultivation practices and by laboratory incubations. *Soil Science Society of America Journal* 46, 970–976.

Hedley, M.J., Kirk, G.J.D. and Santos, M.B. (1994) Phosphorus efficiency and the forms of soil phosphorus utilized by upland rice cultivars. *Plant and Soil* 158, 53–62.

Helal, H.M. and Dressler, A. (1989) Mobilization and turnover of soil phosphorus in the rhizosphere. *Zeitschrift für Pflanzenernährung und Bodenkunde* 152, 175–180.

Helal, H.M. and Sauerbeck, D.R. (1984) Influence of plant roots on C and P metabolism in soil. *Plant and Soil* 76, 175–182.

Hens, M. and Merckx, R. (2001) Functional characterization of colloidal phosphorus species in the soil solution of sandy soils. *Environmental Science and Technology* 35, 493–500.

Hinsinger, P. (2001) Bioavailability of soil inorganic P in the rhizosphere as affected by root-induced chemical changes: a review. *Plant and Soil* 237, 173–195.

Hocking, P.J. (2001) Organic acids exuded from roots in phosphorus uptake and aluminum tolerance of plants in acid soils. *Advances in Agronomy* 74, 63–97.

Hoffman, D.L. and Rolston, D.E. (1980) Transport of organic phosphate in soil as affected by soil type. *Soil Science Society of America Journal* 44, 46–52.

Hübel, F. and Beck, E. (1993) *In-situ* determination of the P-relations around the primary root of maize with respect to inorganic and phytate-P. *Plant and Soil* 157, 1–9.

Hübel, F. and Beck, E. (1996) Maize root phytase: purification, characterization and localization of enzyme activity and its putative substrate. *Plant Physiology* 112, 1429–1436.

Hunter, D.A. and McManus, M.T. (1999). Comparison of acid phosphatase in two genotypes of white clover with different responses to applied phosphate. *Journal of Plant Nutrition* 22, 679–692.

Islam, A. and Mandal, R. (1977) Amounts and mineralization of organic phosphorus compounds and derivatives in some surface soils of Bangladesh. *Geoderma* 17, 57–68.

Jakobsen, I., Leggett, M.E. and Richardson, A.E. (2005) Rhizosphere microorganisms and plant phosphorus uptake. In: Sims, J.T. and Sharpley, A.N. (eds) *Phosphorus, Agriculture and the Environment.* American Society of Agronomy/Soil Science Society of America, Madison, Wisconsin (in press).

Jackman, R.H. and Black, C.A. (1951) Solubility of iron, aluminium, calcium and magnesium inositol phosphates at different pH values. *Soil Science* 72, 179–186.

Jayachandran, K., Schwab, A.P. and Hetrick, B.A.D. (1992) Mineralization of organic phosphorus by vesicular-arbuscular mycorrhizal fungi. *Soil Biology and Biochemistry* 24, 897–903.

Joner, E.J. and Jakobsen, I. (1995) Growth and extracellular phosphatase activity of arbuscular mycorrhizal hyphae as influenced by soil organic matter. *Soil Biology and Biochemistry* 27, 1153–1159.

Joner, E.J., Magid, J., Gahoonia, T.S. and Jakobsen, I. (1995) P depletion and activity of phosphatases in the rhizosphere of mycorrhizal and non-mycorrhizal cucumber (*Cucumis sativus* L.). *Soil Biology and Biochemistry* 27, 1145–1151.

Joner, E.J., van Aarle, I.M. and Vosatka, M. (2000) Phosphatase activity of extra-radical arbuscular mycorrhizal hyphae: a review. *Plant and Soil* 226, 199–210.

Jones, D.L. (1998) Organic acids in the rhizosphere: a critical review. *Plant and Soil* 205, 25–44.

Jungk, A. (1987) Soil–root interactions in the rhizosphere affecting plant availability of phosphorus. *Journal of Plant Nutrition* 10, 1197–1204.

Jungk, A., Seeling, B. and Gerke, J. (1993) Mobilisation of different phosphate fractions in the rhizosphere. *Plant and Soil* 155/156, 91–94.

Keerthisinghe, G., Hocking, P.J., Ryan, P.R. and Delhaize, E. (1998) Effect of phosphorus supply on the formation and function of proteoid roots of white lupin (*Lupinus albus* L.). *Plant, Cell and Environment* 21, 467–478.

Lefebvre, D.D., Duff, S.M.G., Fife, C.A., Julien-Inalsingh, C. and Plaxton, W.C. (1990) Response to phosphate deprivation in *Brassica nigra* suspension cells: enhancement of intracellular, cell surface and secreted phosphatase activities compared to increases in Pi-absorption rate. *Plant Physiology* 93, 504–511.

Li, M., Osaki, M., Rao, I.M. and Tadano, T. (1997a) Secretion of phytase from the roots of several plant species under phosphorus-deficient conditions. *Plant and Soil* 195, 161–169.

Li, M., Shinano, T. and Tadano, T. (1997b) Distribution of exudates of lupin roots in the rhizosphere under phosphorus deficient conditions. *Soil Science and Plant Nutrition* 43, 237–245.

Lopez-Hernandez, D., Brossard, M. and Frossard, E. (1998) P-isotopic exchange values in relation to Po mineralisation in soils with very low P-sorbing capacities. *Soil Biology and Biochemistry* 30, 1663–1670.

Lynch, J. (1995) Root architecture and plant productivity. *Plant Physiology* 109, 7–13.

Macklon, A.E.S., Grayston, S.J., Shand, C.A., Sim, A., Sellars, S. and Ord, B.G. (1997) Uptake and transport of phosphorus by *Agrostis capillaris* seedlings from rapidly hydrolysed organic sources extracted from ^{32}P-labelled bacterial cultures. *Plant and Soil* 190, 163–167.

Magid, J., Christensen, N. and Nielsen, H. (1992) Measuring phosphorus fluxes through the root zone of a layered sandy soil: comparisons between lysimeter and suction cell solution. *Journal of Soil Science* 43, 739–747.

Magid, J., Tiessen, H. and Condron, L.M. (1996) Dynamics of organic phosphorus in soils under natural and agricultural ecosystems. In: Piccolo, A. (ed.) *Humic Substances in Terrestrial Ecosystems.* Elsevier Science, Amsterdam, pp. 429–466.

Martin, J.K. (1970) Organic phosphate compounds in water extracts of soils. *Soil Science* 109, 362–375.

Martin, J.K. (1973) The influence of rhizosphere microflora on the availability of ^{32}P-*myo*inositol hexaphosphate phosphorus to wheat. *Soil Biology and Biochemistry* 5, 473–483.

Martin, J.K. and Cartwright, B. (1971) The comparative plant availability of ^{32}P *myo*-inositol hexaphosphate and $KH_2{}^{32}PO_4$ added to soils. *Communications in Soil Science and Plant Analysis* 2, 375–381.

McGill, W.B. and Cole, C.V. (1981) Comparative aspects of cycling of organic C, N, S and P through soil organic matter. *Geoderma* 26, 267–286.

McKercher, R.B. and Anderson, G. (1968) Content of inositol penta- and hexaphosphates in some Canadian soils. *Journal of Soil Science* 19, 47–55.

McLachlan, K.D. (1980) Acid phosphatase activity of

intact roots and phosphorus nutrition of plants. I. Assay conditions and phosphatase activity. *Australian Journal of Agricultural Research* 31, 429–440.

McLaughlin, M.J., Alston, A.M. and Martin, J.K. (1987) Transformations and movement of P in the rhizosphere. *Plant and Soil* 97, 391–399.

McLaughlin, M.J., Alston, A.M. and Martin, J.K. (1988a) Phosphorus cycling in wheat–pasture rotations. II. The role of the microbial biomass in phosphorus cycling. *Australian Journal of Soil Research* 26, 333–342.

McLaughlin, M.J., Alston, A.M. and Martin, J.K. (1988b) Phosphorus cycling in wheat–pasture rotations. III. Organic phosphorus turnover and phosphorus cycling. *Australian Journal of Soil Research* 26, 343–353.

Miller, S.S., Liu, J.Q., Allan, D.L., Menzhuber, C.J., Fedorova, M. and Vance, C.P. (2001) Molecular control of acid phosphatase secretion into the rhizosphere of proteoid roots from phosphorus-stressed white lupin. *Plant Physiology* 127, 594–606.

Mitsukawa, N., Okumura, S., Shirano, Y., Sato, S., Kato, T., Harashima, S. and Shibata, D. (1997) Overexpression of an *Arabidopsis thaliana* high-affinity phosphate transporter gene in tobacco cultured cells enhances cell growth under phosphate-limited conditions. *Proceedings of the National Academy of Sciences USA* 94, 7098–7102.

Mudge, S.R., Rae, A.L., Diatloff, E. and Smith, F.W. (2002) Expression analysis suggests novel roles for members of the Pht1 family of phosphate transporters in *Arabidopsis*. *Plant Journal* 31, 341–353.

Mudge, S.R., Smith, F.W. and Richardson, A.E. (2003) Root-specific and phosphate-regulated expression of phytase under the control of a phosphate transporter promoter enables *Arabidopsis* to grow on phytate as a sole P source. *Plant Science* 165, 871–878.

Newman, R.H. and Tate, K.R. (1980) Soil phosphorus characterization by ^{31}P-nuclear magnetic resonance. *Communications in Soil Science and Plant Analysis* 11, 835–842.

Oberson, A., Friesen, D.K., Rao, I.M., Bühler, S. and Frossard, E. (2001) Phosphorus transformations in an oxisol under contrasting land-use systems: the role of the microbial biomass. *Plant and Soil* 237, 197–210.

Oehl, F., Oberson, A., Sinaj, S. and Frossard, E. (2001) Organic phosphorus mineralization studies using isotopic dilution techniques. *Soil Science Society of America Journal* 65, 780–787.

Otani, T. and Ae, N. (1999) Extraction of organic phosphorus in Andosols by various methods. *Soil Science and Plant Nutrition* 45, 151–161

Pant, H.K., Edwards, A.C. and Vaughan, D. (1994) Extraction, molecular fractionation and enzyme degradation of organically associated phosphorus in soil solutions. *Biology and Fertility of Soils* 17, 196–200.

Pate, J.S. and Watt, M. (2002) Roots of *Banksia* spp. (Proteaceae) with special reference to functioning of their specialized proteoid root clusters. In: Waisel, Y., Eshel, A. and Kafkafi, U. (eds) *Plant Roots: the Hidden Half.* Marcel Dekker, New York, pp. 989–1006.

Pellet, D.M., Papernik, L.A., Jones, D.L., Darrah, P.R., Grunes, D.L. and Kochian, L.V. (1997) Involvement of multiple aluminium exclusion mechanisms in aluminium tolerance in wheat. *Plant and Soil* 192, 63–68.

Polglase, P.J., Attiwill, P.M. and Adams, M.A. (1992) Nitrogen and phosphorus cycling in relation to stand age of *Eucalyptus regnans* F. Muell. III. Labile inorganic and organic P, phosphatase activity and P availability. *Plant and Soil* 142, 177–185.

Raghothama, K.G. (1999) Phosphate acquisition. *Annual Review of Plant Physiology and Plant Molecular Biology* 50, 665–693.

Randall, P.J., Hayes, J.E., Hocking, P.J. and Richardson, A.E. (2001) Root exudates in phosphorus acquisition by plants. In: Ae, N., Arihara, J., Okada, K. and Srinivasan, A. (eds) *Plant Nutrient Acquisition: New Perspectives.* Springer, Tokyo, pp. 71–100.

Rausch, C. and Bucher, M. (2002) Molecular mechanism of phosphate transport in plants. *Planta* 216, 23–37.

Read, D.B., Bengough, A.G., Gregory, P.J., Crawford, J.W., Robinson, D., Scrimgeour, C.M., Young, I.M., Zhang, K. and Zhang, X. (2003) Plant roots release phospholipid surfactants that modify the physical and chemical properties of soil. *New Phytologist* 157, 315–326.

Richardson, A.E. (1994) Soil microorganisms and phosphorus availability. In: Pankhurst, C.E., Doube, B.M., Gupta, V.V.S.R. and Grace, P.R. (eds) *Soil Biota: Management in Sustainable Farming Systems.* CSIRO, Australia, pp. 50–62.

Richardson, A.E. (2001) Prospects for using soil microorganisms to improve the acquisition of phosphorus by plants. *Australian Journal of Plant Physiology* 28, 897–906.

Richardson, A.E., Hadobas, P.A. and Hayes, J.E. (2000) Phosphomonoesterase and phytase activities of wheat (*Triticum aestivum* L.) roots and utilisation of organic phosphorus substrates by seedlings grown in sterile culture. *Plant, Cell and Environment* 23, 397–405.

Richardson, A.E., Hadobas, P.A. and Hayes, J.E. (2001a) Extracellular secretion of *Aspergillus* phytase from *Arabidopsis* roots enables plants to obtain phosphorus from phytate. *Plant Journal* 25, 641–649.

Richardson, A.E., Hadobas, P.A., Hayes, J.E., O'Hara, C.P. and Simpson, R.J. (2001b) Utilization of phosphorus by pasture plants supplied with *myo*-inositol hexaphosphate is enhanced by the presence of soil microorganisms. *Plant and Soil* 229, 47–56.

Richardson, A.E., Hadobas, P.A. and Simpson, R.J. (2001c) Phytate as a source of phosphorus for the growth of transgenic *Trifolium subterraneum*. In: Horst, W.J., Schenk, M.K., Bürkert, A., Claassen, N., Flessa, H., Frommer, W.B., Goldbach, H.E., Merback, W., Olfs, H.-W., Römheld, V., Sattelmacher, B., Schmidhalter, U., Schenk, M.K. and von Wirén, N. (eds) *Progress in Plant Nutrition: Food Security and Sustainability of Agro-Ecosystems Through Basic and Applied Research.* Kluwer Academic, Dordrecht, pp 560–561.

Robson, A.D., Abbott, L.K. and Malajczuk, N. (1994) *Management of Mycorrhizas in Agriculture, Horticulture and Forestry.* Kluwer Academic, Dordrecht.

Roelofs, R.F.R., Rengel, Z., Cawthray, G.R., Dixon, K.W. and Lambers, H. (2001) Exudation of carboxylates in Australian Proteaceae: chemical composition. *Plant, Cell and Environment* 24, 891–903.

Rojo, M.J., Carcedo, S.G. and Mateos, M.P. (1990) Distribution and characterization of phosphatase and organic phosphorus in soil fractions. *Soil Biology and Biochemistry* 22, 169–174.

Ron Vaz, M.D., Edwards, A.C., Shand, C.A. and Cresser, M.S. (1993) Phosphorus fractions in soil solution: influence of soil acidity and fertiliser additions. *Plant and Soil* 148, 175–183.

Runge-Metzger, A. (1995) Closing the cycle: obstacles to efficient P management for improved global security. In: Tiessen, H. (ed.) *Phosphorus in the Global Environment: Transfers, Cycles and Management.* John Wiley & Sons, Chichester, UK, pp. 27–42.

Ryan, P.R., Delhaize, E. and Jones, D.L. (2001) Function and mechanism of organic anion exudation from plants. *Annual Review of Plant Physiology and Plant Molecular Biology* 52, 527–560.

Saleque, M.A. and Kirk, G.J.D. (1995) Root-induced solubilization of phosphate in the rhizosphere of lowland rice. *New Phytologist* 129, 325–336.

Sample, E.C., Soper, R.J. and Racz, G.J. (1980) Reactions of phosphate fertilizers in soils. In: Khasawneh, F.E., Sample, E.C. and Kamprath, E.J. (eds) *The Role of Phosphorus in Agriculture.* American Society of Agronomy, Madison, Wisconsin, pp. 263–310.

Sanyal, S.K. and De Datta, S.K. (1991) Chemistry of phosphorus transformations in soil. *Advances in Soil Science* 16, 1–120.

Schachtman, D.P., Reid, R.J. and Ayling, S.M. (1998) Phosphorus uptake by plants, from soil to cell. *Plant Physiology* 116, 447–453.

Seeling, B. and Jungk, A. (1996) Utilisation of organic phosphorus in calcium chloride extracts of soil by barley plants and hydrolysis by acid and alkaline phosphatases. *Plant and Soil* 178, 179–184.

Seeling, B. and Zasoski, R.J. (1993) Microbial effects in maintaining organic and inorganic solution phosphorus concentrations in a grassland topsoil. *Plant and Soil* 148, 277–284.

Shand, C.A. and Smith, S. (1997) Enzymatic release of phosphate from model substrates and P compounds in soil solution from a peaty podzol. *Biology and Fertility of Soils* 24, 183–187.

Shand, C.A., Macklon, A.E.S., Edwards, A.C. and Smith, S. (1994) Inorganic and organic P in soil solutions from three upland soils. I. Effects of soil solution extraction conditions, soil type and season. *Plant and Soil* 159, 255–264.

Shand, C.A., Smith, S., Edwards, A.C. and Fraser, A.R. (2000) Distribution of phosphorus in particulate, colloidal and molecular-sized fractions of soil solution. *Water Research* 34, 1278–1284.

Shang, C., Stewart, J.W.B. and Huang, P.M. (1992) pH effect on kinetics of adsorption of organic and inorganic phosphates by short-range ordered aluminum and iron precipitates. *Geoderma* 53, 1–14.

Smith, S.E. and Read, D.J. (1997) *Mycorrhizal Symbiosis.* Academic Press, San Diego, California, 605 pp.

Stewart, J.W.B. and Tiessen, H. (1987) Dynamics of soil organic phosphorus. *Biogeochemistry* 4, 41–60.

Tadano, T., Ozawa, K., Sakai, H., Osaki, M. and Matsui, H. (1993) Secretion of acid phosphatase by the roots of crop plants under phosphorus-deficient conditions and some properties of the enzyme secreted by lupin roots. *Plant and Soil* 155/156, 95–98.

Tarafdar, J.C. and Claassen, N. (1988) Organic phosphorus compounds as a phosphorus source for higher plants through the activity of phosphatases produced by plant roots and microorganisms. *Biology and Fertility of Soils* 5, 308–312.

Tarafdar, J.C. and Claassen, N. (2003) Organic phosphorus utilization by wheat plants under sterile conditions. *Biology and Fertility of Soils* 39, 25–29.

Tarafdar, J.C. and Jungk, A. (1987) Phosphatase activity in the rhizosphere and its relation to the depletion of soil organic phosphorus. *Biology and Fertility of Soils* 3, 199–204.

Tarafdar, J.C. and Marschner, H. (1994a) Efficiency of VAM hyphae in utilisation of organic phosphorus by wheat plants. *Soil Science and Plant Nutrition* 40, 593–600.

Tarafdar, J.C. and Marschner, H. (1994b) Phosphatase activity in the rhizosphere and hyphosphere of VA mycorrhizal wheat supplied with inorganic and organic phosphorus. *Soil Biology and Biochemistry* 26, 387–395.

Tarafdar, J.C., Yadav, R.S. and Meena, S.C. (2001) Comparative efficiency of acid phosphatase originated from plant and fungal sources. *Journal of Plant Nutrition and Soil Science* 164, 279–282.

Taranto, M.T., Adams, M.A. and Polglase, P.J. (2000) Sequential fractionation and characterization (^{31}P-NMR) of phosphorus-amended soils in *Banksia integrifolia* (L.f.) woodland and adjacent pasture. *Soil Biology and Biochemistry* 32, 169–177.

Tate, K.R. (1984) The biological transformation of P in soil. *Plant and Soil* 76, 245–256.

Tate, K.R. and Salcedo, I. (1988) Phosphorus control of soil organic matter accumulation and cycling. *Biogeochemistry* 5, 99–107.

Thien, S.J. and Myers, R. (1992) Determination of bioavailable phosphorus in soil. *Soil Science Society of America Journal* 56, 814–818.

Toor, G.S., Condron, L.M., Di, H.J., Cameron, K.C. and Cade-Menun, B.J. (2003) Characterization of organic phosphorus in leachate from a grassland soil. *Soil Biology and Biochemistry* 35, 1317–1323.

Trasar-Cepeda, M.C. and Carballas, T. (1991) Liming and the phosphatase activity and mineralization of phosphorus in an acidic soil. *Soil Biology and Biochemistry* 23, 209–215.

Turner, B.L. and Haygarth, P.M. (2000) Phosphorus forms and concentrations in leachate under four grassland soil types. *Soil Science Society of America Journal* 64, 1090–1097.

Turner, B.L. and Haygarth, P.M. (2001) Phosphorus solubilization in rewetted soils. *Nature* 411, 258.

Turner, B.L., McKelvie, I.D. and Haygarth, P.M. (2002a) Characterization of water-extractable soil organic phosphorus by phosphatase hydrolysis. *Soil Biology and Biochemistry* 34, 27–35.

Turner, B.L., Papházy, M.J., Haygarth, P.M. and McKelvie, I.D. (2002b) Inositol phosphates in the environment. *Philosophical Transactions of the Royal Society, London, Series B* 357, 449–469.

Vance, C.P., Uhde-Stone, C. and Allan, D.L. (2003) Phosphorus acquisition and use: critical adaptations by plants for securing a non renewable resource. *New Phytologist* 157, 423–447.

van Veen, J.A., Ladd, J.N., Martin, J.K. and Amato, M. (1987) Turnover of carbon, nitrogen and phosphorus through the microbial biomass in soils incubated with ^{14}C-, ^{15}N- and ^{32}P-labelled bacterial cells. *Soil Biology and Biochemistry* 19, 559–565.

Wasaki, J., Omura, M., Ando, M., Dateki, H., Shinano, T., Osaki, M., Ito, H., Matsui, H. and Tadano, T. (2000) Molecular cloning and root specific expression of secretory acid phosphatase from phosphate-deficient lupin (*Lupinus albus* L.). *Soil Science and Plant Nutrition* 46, 427–437.

Wild, A. and Oke, O.L. (1966) Organic phosphate compounds in calcium chloride extracts of soils: identification and availability to plants. *Journal of Soil Science* 17, 356–371.

Williams, C.H. and Anderson, G. (1968) Inositol phosphates in some Australian soils. *Australian Journal of Soil Research* 6, 121–130.

Wyss, M., Brugger, R., Kronenberger, A., Rémy, R., Fimbel, R., Oesterhelt, G., Lehmann, M. and van Loon, A.P.R.M. (1999) Biochemical characterization of fungal phytases (*myo*-inositol hexakisphosphate phosphohydrolases), catalytic properties. *Applied and Environmental Microbiology* 65, 367–373.

Yun, S.J. and Kaeppler, S.M. (2001) Induction of maize acid phosphatase activities under phosphorus starvation. *Plant and Soil* 237, 109–115.

Zhang, C. and McManus, M.T. (2000) Identification and characterization of two distinct acid phosphatases in cell walls of roots of white clover. *Plant Physiology and Biochemistry* 38, 259–270.

Zimmermann, P., Zardi, G., Lehmann, M., Zeder, C., Amrhein, N., Frossard, E. and Bucher M. (2003) Engineering the root–soil interface via targeted expression of a synthetic phytase gene in trichoblasts. *Plant Biotechnology Journal* 1, 353–360.

Zou, X., Binkley, D. and Doxtader, K.G. (1992) A new method for estimating gross phosphorus mineralization and immobilization rates in soils. *Plant and Soil* 147, 243–250.

Zoysa, A.K.N., Loganathan, P. and Hedley, M.J. (1997) A technique for studying rhizosphere processes in tree crops: soil phosphorus depletion around camellia (*Camellia japonica* L.) roots. *Plant and Soil* 190, 253–265.

Zoysa, A.K.N., Loganathan, P. and Hedley, M.J. (1999) Phosphorus utilisation efficiency and depletion of phosphate fractions in the rhizosphere of three tea (*Camellia sinensis* L.) clones. *Nutrient Cycling in Agroecosystems* 53, 189–201.

9 Microbial Turnover of Organic Phosphorus in Aquatic Systems

Robert T. Heath

Department of Biological Sciences and Water Resources Research Institute, Kent State University, Kent, OH 44242, USA

Introduction

During the past two decades the significance of bacteria in aquatic ecosystems has undergone a reassessment. Once viewed solely as 're-mineralizers', essential only for mineralization of dissolved and particulate detritus, aquatic bacteria are increasingly viewed as essential components in aquatic food webs. Heterotrophic bacteria are now seen as potentially important links between dissolved resources and higher trophic levels (Vadstein *et al.*, 1993). This altered view of the role of bacteria has especially important consequences for conceptualizing the significance of bacteria in phosphorus dynamics at the base of the food web, because heterotrophic bacteria differ from phytoplankton in being relatively phosphorus-rich organisms (Vadstein, 2000). Heterotrophic bacterial metabolism of phosphorus is dependent not only on phosphorus availability but also on the availability of labile dissolved organic carbon in the immediate environment. Unlike carbon resources, which are largely respired by heterotrophic bacteria, phosphorus is largely retained under most conditions occurring in natural freshwater environments. Major fractions of phosphorus are taken up by heterotrophic bacteria and transported through the base of the food web by bacterivorous micrograzers of the microbial food web. The dependence on the microbial food web for phosphorus transport may become greater in oligotrophic environments of both marine and freshwater ecosystems.

But what of dissolved organic phosphorus compounds; where do they fit into this revised scheme? Phytoplankton appear to depend on them to augment their supply of phosphorus available for growth and metabolic activities, especially when phosphate concentrations become critically low. When internal phosphorus stores become dangerously low, hydrolytic enzymes are adaptively synthesized as ectoenzymes attached to the outer surface of algal cells. When extended to heterotrophic bacterioplankton, this 'phytoplanktonic' view of dissolved organic phosphorus use provides an imperfect and misleading view of the role of bacteria in the phosphorus dynamics of freshwater ecosystems. As with many misleading concepts, in being only partly true it has led to the benign neglect of important questions regarding the unique role and capabilities of bacterioplankton in using dissolved organic phosphorus in the environment. Heterotrophic bacteria in culture are capable of taking up certain phosphorylated compounds from their surroundings intact and without prior hydrolysis, but does this happen in natural bacterioplankton assemblages? Since heterotrophic bacteria in natural environments can be limited by carbon or phosphorus, which moiety is taken up when dissolved organic phosphorus compounds are hydrolysed: the phosphoryl moiety, the organic moiety, both, or does it depend upon the nutritional status of the cell? What about dissolved organic phosphorus compounds that are resistant to hydrolysis, but can release phosphate under photolytic processes; are heterotrophic bacteria uniquely able to assimilate such released phosphorus due to their high affinity for phosphate? These are the issues addressed in this chapter.

Determining the 'strategy' of bacterial use of environmental dissolved organic phosphorus requires a broader understanding of the metabolic status of bacteria. Because much of the focus on dissolved organic phosphorus compounds has been as a source of phosphorus, their potential use as a carbon source has been incompletely considered and examined. Also, the 'phosphorus-limited phytoplanktonic' view of dissolved organic phosphorus has led us to focus on it as a nutritional source only; here I suggest that some organic phosphorus compounds (e.g. adenosine 3′,5′-cyclic monophosphate) may serve an 'informational' role, such as quorum sensing, and bacteria may function to receive or attenuate that signal. I conclude by identifying 'emerging issues' in bacterial use of dissolved organic phosphorus and ecosystem regulation of nutrient availability in phosphorus-limited communities.

Phosphate Uptake by Heterotrophic Bacteria

Examination of the role of bacterioplankton in processing naturally occurring dissolved organic phosphorus depends first of all on considerations of the uptake and metabolism of phosphate from the environment and those factors that influence the rate and extent of uptake. In particular, phosphate uptake is controlled genetically and biochemically by external phosphate and by the availability of carbon sources. Much of what we know about phosphate uptake has been learned from cultures of enteric bacteria, primarily *Escherichia coli*. Studies on cultures of *Bacillus subtilis*, *Acinetobacter johnsonii* and various species of *Pseudomonas* indicate some generality to the findings from *E. coli*. Relatively few studies have been conducted on natural assemblages of freshwater bacteria, and even fewer have conducted parallel studies in both cultures and natural environments. Bacterial cultures are generally conducted under conditions that favour the growth of the organisms examined, rather than challenging them with low nutrients and suboptimal environments. Also, it is generally agreed that by far the majority of prokaryotes that occur in natural aquatic environments have yet to be cultured, indicating that conditions necessary for their successful isolation and characterization may differ widely from those of the bacteria commonly examined. Rather than being accepted at face value, studies of organisms in culture (especially those organisms that normally inhabit environments dissimilar from typical aquatic environments) demonstrate possible processes and provide testable hypotheses for field microbiologists to examine.

There are two transport systems in *E. coli* through which phosphate can be assimilated (Rosenberg *et al.*, 1977; Willsky and Malamy, 1980). The phosphate transport system (Pit) transports phosphate through a proton-motive-force-driven proton–phosphate symport; phosphate uptake through Pit exhibits Michaelis–Menten kinetics with a K_t of 25 μM for the divalent anion (Rosenberg, 1987). The Pit symport is synthesized constitutively in *E. coli* (Rosenberg *et al.*, 1977) and provides sufficient phosphate to sustain growth until phosphate decreases below 160 nM. Cytoplasmic phosphate remains high (10 mM) despite phosphorus limitation (Rao *et al.*, 1993). Instead, when cells become phosphorus-limited, an external phosphate signal is received by the high-affinity phosphate transport complex, which ultimately causes the phosphorylation of a cytoplasmic protein, PhoB, in an ATP (adenosine 5′-triphosphate)-dependent process. The phosphorylated form of PhoB

(PhoB-P) is a transcriptional activator that specifically recognizes an 18-nucleotide consensus sequence, the 'pho-box', found as part of the promotor of at least seven operons containing at least 31 genes. Collectively, this set of operons is called the *pho* regulon (Makino *et al.*, 1989). When external phosphate becomes limiting, the entire suite of genes in the *pho* regulon is activated; when external phosphate becomes sufficient, this signal sequence is reversed, dephosphorylating PhoB-P and ceasing the transcription of the *pho* regulon (Muda *et al.*, 1992; Shinagawa *et al.*, 1994).

The *pho* regulon contains genes for the second phosphate transport system, the phosphate specific transport system (Pst), which has a 100-fold higher affinity for phosphate than Pit, with a K_t of 200 nM (Webb and Cox, 1994). Phosphate is taken up by the Pst system in a process that requires one mole of ATP per mole of phosphate assimilated; Pst is one of a class of enzymes known as 'traffic ATPase transporters'. Besides the *pst* genes, the *pho* regulon includes a gene (*phoE*) for phosphoporin in the outer membrane that facilitates diffusion of phosphate into the periplasm, genes for expression of alkaline phosphatase (*phoA*), uptake and metabolism of glycerol 3-phosphate (the *ugp* operon), and the *phn* operon for phosphonate transport and degradation (Table 9.1). See Torriani-Gorini (1994) for a complete, readable and relatively recent review of the *pho* regulon of *E. coli*. The broad generality of these findings and the systematics of phosphate regulation have yet to be determined, but it is clear that similar regulatory processes occur in divergent organisms. Gram-positive *Bacillus subtilis* (Phylum: Firmicutes) is not closely related to Gram-negative *E. coli* (Phylum: Proteobacteria) (Boone *et al.*, 2001), yet a similar *pho* regulon has been noted in *B. subtilis* (Hulett *et al.*, 1994).

Phosphate uptake by natural bacterial assemblages

Whether taxa in natural bacterioplankton assemblages exhibit similar characteristics to cultured populations largely remains to be demonstrated. Examination of natural assemblages of bacterioplankton indicated very high affinities and low K_t values, considerably lower than values observed for the Pst system of *E. coli*. Examination of natural bacterioplankton assemblages grown in chemostatic culture using 0.2 μm filtered lake water amended with phosphate (final concentration 700 nM), showed half-saturation constants between 96 and 132 nM (Vadstein and Olsen, 1989). In a later study of two heterotrophic bacteria (identified only as '2g' and '3h', but otherwise uncharacterized), isolated from the epilimnion of Lake Nesjøvatn, a eutrophic lake in central Norway, and studied in culture, phosphate uptake exhibited similar K_t (~100 nM) but different V_{max}, indicating different strategies of phosphate uptake (Vadstein, 1998). One strain was seen as having a competitive advantage when phosphate concentrations were transiently elevated, while the other was viewed as the better competitor when phosphate was available at uniformly low concentrations.

By amending whole-water samples with ^{32}P-labelled phosphate and following uptake by collecting particles on 1.0 and 0.2 μm filters, bacterioplankton from a 'moderately eutrophic' temperate lake exhibited K_t values between 0.1 and 211 nM (Cotner and Wetzel, 1992), with the lowest values occurring in late summer. Using similar procedures to investigate phosphate uptake by bacterioplankton in a series of lakes ranging from 'oligotrophic' to 'highly eutrophic,' K_t of phosphate uptake was reported to range between 6.7 and 54.3 nM in mid-summer (Gao, 2002). Although this is not evidence for an *E. coli* Pst-like system, these investigations of natural bacterioplankton assemblages indicate that transport mechanisms with a high affinity for phosphate must be broadly distributed in nature. This is especially evident when it is noted that concentrations of bioavailable phosphate may frequently fall below 10 nM in freshwater ecosystems (Hudson *et al.*, 2000).

Other studies indicated that phosphate uptake was dependent on the availability of carbon sources. For example, when

Table 9.1. Summary characteristics of various processes for uptake of phosphate and selected dissolved organic phosphorus compounds, including genetic control and catabolic repression. Genetic terms refer to structures and processes in *Escherichia coli*. (ATPase = adenosine 5′-triphosphatase.)

Substance	Agent	Action	Uptake process	Gene control	Catabolic repression
Phosphate	Pit	Uptake of phosphate	H^+-symport	Constitutive	–
Phosphate	Pst	Uptake of phosphate	Traffic ATPase	*pho* regulon	No
Phosphate monoesters, pyrophosphate, tripolyphosphate	PhoA, APA	Hydrolyse phosphate monoesters	Uptake of phosphate by Pst	*pho* regulon	No
5′-mononucleotides	Ush	Hydrolyse nucleotides	Uptake of nucleosides by specific transporters	–	No
3′-mononucleotides, 2′,3′-cyclic nucleotides	CpdB	Hydrolyse nucleotides	Uptake of nucleosides by specific transporters	–	No
Phosphonates	PhnCDE	Uptake of intact dissolved organic phosphorus	Traffic ATPase ?	*pho* regulon	No
Glycerol 3-phosphate	UgpT	Uptake of intact dissolved organic phosphorus	Traffic ATPase	*pho* regulon; *intracellular* phosphate slows uptake	No
Glycerol, glycerol 3-phosphate	GlpT	Uptake of intact dissolved organic phosphorus	H^+-symport or phosphate-antiport	*glp* regulon: external glycerol 3-phosphate induces, internal glycerol 3-phosphate represses	Yes
Glucose 6-phosphate, hexose phosphates, glycerol 3-phosphate	UhpT	Uptake of intact dissolved organic phosphorus	Phosphate-antiport	External glucose 6-phosphate induces	Yes

Table 9.2. Two physiological states of natural bacterioplankton assemblages. Cellular phosphorus content and phosphate uptake in environments with low or high concentrations of labile dissolved organic carbon. Values are means (standard error in parentheses). Reproduced from Gao (2002) with permission. ND, not determined.

	Low nutrient-use efficiency	High nutrient-use efficiency
Bacterial available phosphate (nM)	0.4 (ND)	21 (±2)
Labile dissolved organic carbon (μM)	15 (ND)	109 (±3)
Average biovolume per cell (μm^3)	54 (±7) $\times 10^{-3}$	82 (±10) $\times 10^{-3}$
Growth rate (per min)	1.5×10^{-4}	4.9×10^{-4}
Phosphorus content (nmol/cell)	2.5×10^{-7}	0.7×10^{-7}
Phosphate uptake rate (nmol/cell/min)	9.3×10^{-11}	3.5×10^{-11}

Pseudomonas K7, isolated from brackish water in the pelagic zone of the Bothnian Sea in northern Sweden, was grown in batch cultures under phosphorus limitation, phosphate uptake was faster in the presence of glucose (Jansson, 1993). When phosphorus-starved cells of this pseudomonad were exposed to a 16-min pulse of ^{32}P-labelled phosphate, followed by a chase (using 10 mM phosphate), the uptake of phosphate showed no rapidly exchangeable release of phosphate recently taken up, but cells slowly released phosphate taken up during the hour-long chase, under those conditions.

Coupled culture and field studies indicate that labile dissolved organic carbon can have a significant effect on bacterioplankton cellular phosphorus composition and rate of phosphate uptake, leading to very different nutrient-use efficiencies (Gao, 2002). A recent study on *Pseudomonas fluorescens* (ATCC 13525) in batch culture, using M9 minimal medium adjusted to have very low concentrations of phosphate (25–400 nM) and glutamate (50–200 μM) to simulate conditions found in natural freshwater habitats, showed that the rate of phosphate uptake was controlled by phosphate concentration only when glutamate concentration equalled or exceeded 100 μM (Gao, 2002). In batch cultures with 100 μM glutamate as the sole carbon source, the apparent K_t for phosphate was 60 nM (Gao, 2002). Natural bacterioplankton assemblages were examined in communities having a trophic status ranging from oligotrophic to hypereutrophic. As labile dissolved organic carbon (Sondergaard *et al.*, 1995) increased, bacterioplankton growth rate and biovolume increased, but cell phosphorus content and velocity of phosphate uptake decreased (Table 9.2). Labile dissolved organic carbon was the major factor controlling phosphate uptake by bacterioplankton, explaining more than 75% of the variation in bacterial phosphate uptake velocity and cell phosphorus composition. Labile dissolved organic carbon appeared to influence the nutrient-use efficiency of phosphate (i.e. more carbon was fixed per unit of phosphate assimilated); phosphate was used most efficiently for growth in environments containing at least 50 μM labile dissolved organic carbon (Gao, 2002). Experimental additions of natural labile dissolved organic carbon 'switched' an assemblage from low to high nutrient-use efficiency. Regarding the use of dissolved organic phosphorus in natural communities, these findings imply that the relationship between carbon- and phosphorus-limitation of bacteria in natural assemblages needs to be considered, as it may well be that when cells are severely carbon-limited, dissolved organic phosphorus will be used as a source of carbon (Chróst and Overbeck, 1987; Heath and Edinger, 1990).

Variety of Dissolved Organic Phosphorus Compounds in Aquatic Environments

This topic is treated in greater detail elsewhere in this volume (see McKelvie, Chapter 1; Cade-Menun, Chapter 2; Cooper *et al.*,

Chapter 3); here I briefly describe the classes of compounds that aquatic microbial organisms can conceivably utilize. Classes of naturally occurring dissolved organic phosphorus can be distinguished according to the processes that release phosphorus in an available form, allowing its utilization by osmotrophic organisms (Francko and Heath, 1979). One class released phosphate through the photolytic action of sunlight on apparently high-molecular-weight dissolved organic matter, although it was resistant to hydrolytic enzymes. Phosphate was released slowly from these compounds (i.e. chemical detection as filterable reactive phosphorus increased) in acid bog lake water when exposed to low intensities of ultraviolet-A and ultraviolet-B at a rate similar to the rate of photoreduction of iron, from iron (III) to iron (II); when ultraviolet irradiation ceased, phosphate became undetectable as filterable reactive phosphorus, presumably by a reverse process (Francko and Heath, 1982). A subsequent study on dissolved humic matter from Finnish forest lakes showed that ^{55}Fe rapidly associated with dissolved humic matter and co-eluted with it using Sephadex G-100 size-exclusion chromatography; by contrast, ^{32}P very slowly became associated with these humic–iron materials (de Haan *et al.*, 1990). Double-labelled experiments with free humic–^{55}Fe–^{32}P materials (control: ^{55}Fe and ^{32}P salts added without dissolved humic matter) in the dark showed that the presence of dissolved humic matter maintained the material in solution (Shaw, 1994). Presumably, when phosphate is released, it is released to the general pool of phosphate and is most probably taken up by those organisms having the greatest affinity for it. The lower the phosphate concentration, the more likely that those organisms are bacteria (Currie and Kalff, 1984). The mechanism of phosphorus release from these 'colloidal' complexes remains unresolved, and will not be treated further here.

The other functional class of compounds is a group that releases phosphate through enzymatic hydrolysis; these compounds generally have a low apparent molecular weight (Francko and Heath, 1979) and are composed of materials such as sugar phosphates and mononucleotides (e.g. adenosine monophosphate), as well as adenosine di- and triphosphate (McGrath and Sullivan, 1981), and cyclic mononucleotides (Francko, 1982). High-molecular-weight DNA and RNA have been identified in natural systems (Karl and Bailiff, 1989); release of constituent mononucleotides by phosphodiesterase activity or by endo- and exonucleolytic activity is necessary before such compounds can be assimilated. Concentrations of natural substrates have been determined to range from 1–2 nM for naturally occurring ATP in the oceans (Ammerman and Azam, 1991b) to about 500 nM for total detectable phosphate monoesters in a eutrophic lake (Heath and Cooke, 1975).

Hydrolytic 'Harvest' of Phosphate from Dissolved Organic Phosphorus

Bacterial hydrolytic enzymes: culture studies

Bacterial alkaline phosphatase is the gene product of *phoA*, a member of the *pho* regulon (Table 9.1). When the *pho* regulon is induced by low external quantities of phosphate, synthesis of this alkaline phosphatase can represent as much as 6 mole% of total protein synthesis, and enzyme activity per cell can increase 1000-fold (Coleman and Gettins, 1983). The enzyme is synthesized as 43,000 Da monomers, which are transported to the periplasmic space and become active only after dimerization. As with many alkaline phosphatases, this enzyme accepts a broad range of substrates, which it hydrolyses at similar rates (Fernley and Walker, 1967; Reid and Wilson, 1971). Substrates are compounds with the general formula

$$X\text{–}PO_3$$

where X can be RO- or RS-. Direct phosphorus bonds to nitrogen or carbon are not hydrolysed by this enzyme. The enzyme can hydrolyse common phosphate monoesters such as glucose 6-phosphate, glycerol 3-

phosphate, 3′ and 5′ mononucleotides, and adenosine di- and triphosphate, hydrolysing the terminal phosphoryl group. It can also hydrolyse pyrophosphate and tripolyphosphate, but has no phosphodiesterase activity. The turnover time is 5×10^4/s. Phosphate and arsenate are potent inhibitors of alkaline phosphataase, with a K_i of 0.6 μM at pH 8.0 in 0.1 M tris-hydroxymethyl-amino methane buffer without additional salt ions (Reid and Wilson, 1971).

Other periplasmic enzymes in *E. coli* that hydrolyse 5′-mononucleotides and 3′-mononucleotides ('nucleotidases') have been identified and characterized (Heppel, 1971). The enzyme most commonly associated with hydrolysis of 5′-mononucleotides is the *udh* gene product (Beacham *et al.*, 1973). Mononucleotides are first transported through porins in the outer membrane of Gram-negative bacteria, then hydrolysed by periplasmic nucleotidases (Beacham *et al.*, 1977). The resultant mononucleosides are then actively taken up by relatively specific transporters located as integral proteins in the inner membrane (see Neuhard and Nygaard, 1987). The 3′-nucleotideases accept 2′,3′-cyclic mononucleotides as substrates, as well as several related compounds; the *E. coli* enzyme most associated with this activity is a *cpdB* gene product, 2′,3′-cyclic phosphodiesterase (Beacham and Garrett, 1980). The primary function of nucleotidases is probably the turnover of cyclic nucleotides in the environment (Beacham *et al.*, 1977; Francko, 1984). K_m for the isolated nucleotidases are 30–50 μM, but the apparent K_t of the uptake process in intact cells is 5–10 times greater, indicating that facilitated diffusion through the porin in the outer membrane may be rate-limiting. Mutants shown to be 'cryptic' for nucleotidase activity (i.e. activity is found in cell lysates but not in intact cells), have been shown to be defective in the porins in the outer membrane despite having wild-type nucleotidase activity in the periplasm (Beacham *et al.*, 1977). In contrast to PhoA alkaline phosphatase, the activity of nucleotidases is not greatly affected by phosphate. Nucleotidases appear to serve as a scavenger pathway for the recovery of nucleosides from exogenous polynucleotides and mononucleotides in the dissolved organic phosphorus pool surrounding the cell (Neuhard and Nygaard, 1987); recovery and uptake of phosphorus from the environment around the cell is incidental.

Phosphonates (compounds with direct carbon to phosphorus bonds) can be utilized by *E. coli*; the genes for utilization of phosphonates are members of the *pho* regulon, indicating that this class of compounds can be used as a source of phosphorus when phosphate becomes limiting (Table 9.1). Genes have been identified for the transport and metabolism of organophosphonates, such as α-aminoethylphosphonate (Metcalf and Wanner, 1993). Organophosphonates are degraded by several separate pathways. One is a hydrolytic pathway that depends on phosphonatase (phosphonoacetaldehyde hydrolase), which yields phosphate (Wanner, 1994). The other probably involves a redox mechanism for the oxidation of organophosphonates not readily hydrolysed to phosphate (Avila and Frost, 1989). A phosphonate-degrading pathway independent of phosphate starvation has been identified in *Pseudomonas fluorescens* (McMillan and Quinn, 1994), possibly indicating that phosphonates may be used by bacteria other than as sources of phosphorus.

Occurrence of bacterial hydrolytic exoenzymes in plankton assemblages

Several independent techniques have been used to identify bacterial alkaline phosphatase activity in natural bacterial assemblages. Phosphatase activity *per se* has often been detected using *para*-nitrophenyl phosphate as a substrate; enzyme activity is expressed as rate of formation of the hydrolytic product, *para*-nitrophenol, which is easily detected spectrophotometrically because it absorbs light strongly at $\lambda =$ 395–410 nm at pH > 9 (see Huber and Kidby, 1984, for a review of variations on this theme). Other substrates that have been used to detect alkaline phosphatase activity visually or spectrophotometrically by intense colour change upon hydrolysis include phenolphthalein phosphate (PPP; Parker, 1963), chemically by release of detectable quantities of phosphate using glucose 6-phosphate or adenosine 5′-monophosphate as substrates (Taft *et al.*, 1977), fluorometrically using 3-*O*-methylfluorescein phosphate (3-MFP; Perry, 1972), or 4-methylumbellyferyl phosphate (4-MUP; Hoppe, 1983) as substrates, and radiometrically by uptake of ^{32}P-labelled phosphate from ^{32}P-labelled glucose 6-phosphate (Taft *et al.*, 1977; Heath and Edinger, 1990) or from ^{32}P-labelled ATP (Bentzen *et al.*, 1992). Similar substrates can be used to detect phosphodiesterase activity (bis-*para*-nitrophenyl phosphate), 3′-nucleotidase (*para*-nitrophenyl-thymidine 3′-monophosphate), and 5′-nucleotidase (*para*-nitrophenyl-thymidine 5′-monophosphate). Hydrolytic end-products phenolphthalein, *para*-nitrophenol and 4-methylumbelliferol require adjustment to pH>9 for optimum detection (Chróst and Krambeck, 1986). It is not clear whether this has always been performed, but depending on the pH of standards, studies that do not adjust pH to detect the hydrolytic product may be reporting erroneous estimates of phosphatase activity.

Detection of bacterial phosphatase activity in aquatic ecosystems has often been an incidental and secondary concern to the research conducted, especially in studies conducted in the 1970s and early 1980s, when attention was focused on eutrophication and the factors that influenced phytoplankton growth. The appearance of alkaline phosphatase activity in coastal seawater was correlated with bacterial biomass; about 40% of the culturable colony-forming heterotrophic bacteria produced alkaline phosphatase detected fluorometrically using 3-*O*-methylfluorescein phosphate (Kobori and Taga, 1977). A subsequent study noted that the fraction of culturable heterotrophic bacteria capable of producing alkaline phosphatase increased along a nearshore–offshore axis (Kobori *et al.*, 1979). They also noted that much of the alkaline phosphatase activity was apparently produced constitutively (i.e. its appearance was unaffected by growth in a medium with high concentrations of phosphate) and an increasing fraction of *Pseudomonas* produced alkaline phosphatase activity along this nearshore–offshore axis. These studies showed that culturable marine heterotrophic bacteria were capable of producing alkaline phosphatase activity, but it was unclear whether all of the alkaline phosphatase activity detected in these Japanese bays was of bacterial origin. Some alkaline phosphatase activity may well have been produced by phytoplankton. Such studies were also limited to the characterization of easily culturable heterotrophic bacteria, which are currently regarded as representing perhaps less than 1% of viable bacteria *in situ* in freshwater and marine communities that are detectable only by recently developed molecular techniques (Glockner *et al.*, 2000).

The technique most commonly used to distinguish bacterial alkaline phosphatase activity from that produced by phytoplankton has been size-selective filtration, either before or after the addition of phosphatase substrates. This approach has been used by many investigators; a particularly thorough approach to size fractionation of alkaline phosphatase activity was conducted on river waters in southeastern Australia, sampled throughout an annual cycle (Boon, 1993). River water was passed through a 25 μm nylon net, sequentially through tangential-flow filters of 1.0 and 0.2 μm pore sizes, and then through filters having nominal apparent molecular weight cut-offs of 100,000 and 10,000 Da. Alkaline phosphatase activity was detected by both *para*-nitrophenyl phosphate and 4-methylumbelliferyl phosphate. The majority of the activity was usually detected in the 0.2–1.0 μm fraction, indicating the significance of bacterial alkaline phosphatase activity in these rivers. Lesser yet significant activities were detected in the 1–25 μm size range and in the 'colloidal' fraction (having an apparent molecular size ranging from 100,000 Da to 0.2 μm). Only negligible activity was detected in fractions >25 μm and in the 10,000–100,000 Da, and <10,000 Da ranges. Finding greater activity in the

100,000 Da–0.2 μm fraction than in the 10,000–100,000 Da fraction suggested that dissolved alkaline phosphatase activity was associated with 'molecular detritus' such as humic or fulvic acids. However, this interpretation should be accepted with caution, because *E. coli* alkaline phosphatase has a molecular weight of 86,000 (Coleman and Gettins, 1983), and the larger hydrated form may have an apparent hydrodynamic molecular weight that would prevent it from ready passage through hydrated pores of a tangential flow filter, especially in solutions having low ionic strength (Tanford, 1961). The extent to which such exoenzymes occur as 'free' or 'dissolved' enzymes is unclear, but it is known that phosphatases can associate with humic materials and that their activity can be affected by that association (Nannipieri *et al.*, 1988; Boavida and Wetzel, 1998).

Significant alkaline phosphatase activity in the 1–25 μm size range is generally interpreted as being 'algal' activity. A caveat regarding this interpretation is worth noting. Heterotrophic colony-forming bacteria were found in fractions >25 μm and 1–25 μm, as well as in the 0.2–1 μm fraction, with a constant proportion of 10–15% of the colonies being capable of hydrolysing phenolphthalein phosphate (Boon, 1993). Finding culturable heterotrophic bacteria in each of these fractions indicates that those bacteria were associated with living phytoplankton or detritus and that alkaline phosphatase activity detected in those size fractions may have been bacterial in origin. Positive correlations of alkaline phosphatase activity with both free and attached bacteria were noted also in English lakes (Jones, 1972), where bacteria associated 'most strongly' with filamentous blue-green algae. Bacteria attached to algae represented as much as 44% of the total bacteria detected by direct count and plate count methods. Viable bacteria attached to algae that exhibited alkaline phosphatase activity ranged from 0 to 65%. Such findings confound determination of the fraction of alkaline phosphatase activity of bacterial origin and imply that the enzyme activity detected in the 0.2–1.0 μm fraction should be taken as a conservative estimate of bacterial alkaline phosphatase activity.

Alkaline phosphatase is not the only hydrolytic exoenzyme activity that has been detected in natural plankton assemblages. Acid phosphatase activity (i.e. phosphatase activity with optimum activity at pH 6 or less) was noted in an acidic bog lake that was heavily stained with humic and fulvic acids (Cotner and Heath, 1988). The majority of acid phosphatase activity appeared to be associated with particles in the 0.2–0.45 μm size range, or that passed through a 0.2 μm filter. Only 21% of the activity was associated with the major phytoplankter, *Gonyostomum semen*. Phosphodiesterase (Boon, 1993) and phosphotriesterase activities (Rowland *et al.*, 1991) have been reported in natural aquatic communities and were determined to be largely if not exclusively of bacterial origin. Phosphodiesterase activity would be necessary for rapid degradation of the DNA and RNA that may occur in natural waters either by 'sloppy feeding' release from grazers (Johannes, 1965; Jürgens and Güde, 1990) or by lysis of algal or bacterial cells upon death. Phosphate triesters are uncommon in biochemical systems, but some organophosphorus pesticides, such as parathion and coumaphos, are phosphate triesters. Phosphotriesterases have been isolated from several bacterial species, including *Flavobacterium* spp. (Brown, 1980) and *Streptomyces lividans* (Rowland *et al.*, 1991).

Bacterial 5′-nucleotidase activity was first noted in marine systems as a means of turning over 5′-mononucleotides, conceivably released by phosphodiesterase or exonucleolytic activity (Ammerman and Azam, 1985). Although alkaline phosphatase can hydrolyse these substrates, it can be distinguished from 5′-nucleotidase activity by its sensitivity to micromolar concentrations of phosphate; 5′-nucleotidase was virtually unaffected by 100 μM phosphate, while 80% of alkaline phosphatase activity was lost under the same conditions (Ammerman and Azam, 1991a). Further, 5′-nucleotidase was competitively inhibited by 5′-guanidyl monophosphate, but almost completely unaffected by 1 μM glucose 6-phosphate; 5′-mononucleotides and glucose 6-phosphate are substrates of alkaline phosphatase and each would competitively inhibit hydrolysis of any coincident substrates (Fernley and Walker, 1967). Activity of 5′-nucleotidase was closely related to high algal and bacterial biomass and to total primary productivity, but was virtually all (>90%) associated with particles <1.0 μm. Along nearshore–offshore transects in the Southern California Bight and the Long Island Sound, 5′-nucleotidase was greatest in relatively eutrophic nearshore surface waters and declined in strata deeper than 70 m, especially in relatively oligotrophic offshore stations (Ammerman and Azam, 1991b). It was concluded that bacterial 5′-nucleotidase activity was determined by algal and bacterial biomass and growth rather than as a response to phosphate depletion. Comparable rates of ATP hydrolysis were noted in a eutrophic lake in Michigan and along a nearshore–offshore axis in Lake Michigan. About 20% of the activity was inhibited in the eutrophic lake in the presence of 1 μM phosphate, while ATP hydrolysis was virtually uninhibited by phosphate amendments to water from the nearshore Lake Michigan station, and only about 10% of the hydrolytic activity was inhibited in offshore Lake Michigan (Cotner and Wetzel, 1991). These findings imply that 5′-nucleotidase activity was more important in hydrolysis of ATP at these sites than alkaline phosphatase activity in oligotrophic environments.

Direct Uptake of Dissolved Organic Phosphorus

Although aquatic biochemical ecologists have focused nearly exclusively on lytic processes for dissolved organic phosphorus utilization, heterotrophic bacteria can directly take up certain organic phosphorus compounds from their surrounding environment without prior hydrolysis (Table 9.2). *E. coli* is known to have two different processes by which glycerol 3-phosphate can be taken up without prior hydrolysis (Table 9.1). The first process, the *ugp*-transport system, is part of the *pho* regulon and is induced when external phosphate is low (Wanner, 1994). Glycerol 3-phosphate and glycerol phosphoryl phosphate diesters specifically associate with UgpB, a periplasmic binding protein, and are transported through the membrane in an ATP-dependent process, similar to phosphate uptake through the Pst-complex (Schweizer *et al.*, 1982). Using this process alone, cells grown in minimal medium can use glycerol 3-phosphate as a sole source of phosphorus but not of carbon: additional carbon sources must be added for growth and survival (Schweizer *et al.*, 1982). Internal glycerol 3-phosphate does not inhibit uptake by the *ugp*-transport system, but *internal* phosphate concentrations >0.5 mM can greatly slow uptake by this system; apparently affecting V_{max} but not K_t of uptake (Brzoska *et al.*, 1994a, b). The second process for the uptake of glycerol 3-phosphate is the *glp*-transport system (Larson *et al.*, 1982), which is also unaffected by external phosphate. This system specifically transports glycerol, glycerol 3-phosphate, and glycerol phosphoryl phosphate diesters (Lin, 1987). It is part of the *glp* operon that is induced by *external* glycerol 3-phosphate (Lin and Iuchi, 1991) and repressed by *intracellular* glycerol 3-phosphate (Gutnick *et al.*, 1969; de Boor *et al.*, 1986). Glycerol 3-phosphate and other substrates are transported through the membrane-bound GlpT, which transports its substrates either as a phosphate-antiport (Elvin *et al.*, 1985) or as a proton-symport (Ambudkar *et al.*, 1986). When fully expressed, the V_{max} of both systems is similar, but the K_t of the Ugp system is about an order of magnitude lower (1–2 μM) than the K_t of GlpT (Schweizer *et al.*, 1982). Glycerol 3-phosphate transported exclusively through GlpT can serve as a sole source of both carbon and phosphorus (Brzoska *et al.*, 1994b).

Mechanisms for the uptake of glucose 6-phosphate without prior hydrolysis are broadly, but not uniformly, distributed among bacterial taxa (Winkler, 1973). *External* glucose 6-phosphate and 2-deoxy-D-glucose 6-phosphate induce the expression of four genes in the *uhp* operon, including the gene for transporter *uhpT*. Medium containing at least 33 μM glucose 6-phosphate is sufficient for induction (Weston and Kadner, 1988; Kadner *et al.*, 1994). In *E. coli*, glucose 6-phosphate is taken up by UhpT, an integral membrane protein (Table 9.1). UhpT is relatively non-specific; it transports hexose 6-phosphates in general but not glucose 1-phosphate, although it can take up glycerol 3-phosphate (Winkler, 1973). Glucose 6-phosphate is taken up in exchange for intracellular phosphate (Kadner *et al.*, 1994).

Can heterotrophic bacterial cells encounter 'too much of a good thing' in direct uptake of phosphorylated hexoses or glycerol from their environment? The answer seems to be yes. Catabolite repression of adaptive expression of genes involved in uptake of carbon sources from the environment is a common theme in bacterial biochemistry that has long been noted (Table 9.1). The *uhpT* promotor is activated by the catabolite-activated-protein, the cyclic adenosine monophosphate-dependent transcriptional activator (Kadner *et al.*, 1994). When glucose is present in sufficiently high quantities to support growth, cyclic adenosine monophosphate is low due to inhibition of adenylate cyclase activity through Enzyme III_{Glc} of the glucose-phosphotransferase system (see Postma, 1987); expression of *uhpT* is repressed. As glucose availability becomes limiting in the environment, adenyl cyclase activity increases and catabolite-activated protein is activated, removing catabolite repression. Only under such conditions are wild-type cells capable of the rapid synthesis of UhpT, even when glucose 6-phosphate is present in sufficiently large quantities to induce expression of *uhpT*. Mutants deficient in catabolite repression are very sensitive to glucose 6-phosphate in the environment and readily die due to the synthesis of methylglyoxal, a toxic metabolic derivative of dihydroxyacetone phosphate (Ackerman *et al.*, 1974). Similarly, the *glp* regulon is controlled by catabolite repression by cytoplasmic concentration of both glucose and glycerol (Zwaig *et al.*, 1970).

The Dual Use of Dissolved Organic Phosphorus: Bacterial Strategies

Aquatic bacteria are well adapted to assimilate phosphate from their environment, but growth and efficient use of phosphorus require sufficient carbon sources as well. Dissolved organic phosphorus offers them a menu choice depending on their metabolic needs. When bacteria have sufficient levels of carbon and are able to maintain optimum levels of glucose 6-phosphate, catabolite repression will prevent the expression of many genes that would mobilize the cell to take up carbon sources from their surroundings. Activation of genes under catabolite repression (e.g. *glpT* and *uhpT*) infers that the cells are severely carbon-limited and dissolved organic phosphorus is taken as a carbon source. The genes in the *pho* regulon are induced only when external phosphate levels become low and growth-limiting, and the activation of genes in the *pho* regulon strongly infers that dissolved organic phosphorus is taken as a source of phosphorus. The strategy involved is a function of the expression of the pathways, transporters and enzymes involved; in turn the pathways are activated as a result of the internal needs and the external environment. To understand bacterial strategies for the use of dissolved organic phosphorus, it is essential to determine whether bacterial cells are taking up the phosphoryl moiety, the organic moiety, or both. If both are taken up, investigators need to determine whether both moieties are taken up at the same rate, indicating that dissolved organic phosphorus is assimilated directly and without prior hydrolysis.

A landmark study in addressing this question was that of Taft *et al.* (1977). Chesapeake Bay seawater was pre-filtered through a 35-μm Nitex net to remove most planktonic grazers, then passed sequentially through 5-μm Nitex, 0.8-μm and 0.1-μm membrane filters to size-fractionate the plankton assemblage. Seawater samples were also incubated with substrates prior to filtration. When ^{32}P-labelled glucose 6-phosphate (10 nM) and ^{14}C-glucose 6-phosphate (45 nM) were added to the water samples, the two labels were taken up by seston at different rates, inferring that hydrolysis was necessary to assimilate either the phosphoryl or the glucosyl moiety from glucose 6-phosphate. The majority of the phosphoryl and the glucosyl moieties taken up from glucose 6-phosphate were assimilated by the 0.8–5 μm fraction, which contained the largest concentrations of particulate phosphorus and chlorophyll *a*. The ^{32}P label was assimilated into the 0.8–5 μm size fraction approximately ten times faster than the ^{14}C label. These investigators concluded appropriately that the majority of phosphorus was taken up by phytoplankton and that bacterial uptake (i.e. the 0.1–0.8 μm size fraction) was at most a 'small contribution'. It is worth noting that *when uptake was scaled for particulate phosphorus*, the bacterial-sized fraction (0.1–0.8 μm) showed the greatest specific uptake of phosphate, and of glucosyl and phosphoryl from glucose 6-phosphate (Table 9.3). That is, if particulate phosphorus can be taken as a surrogate of biomass, bacterial-sized particles were most actively involved in phosphate uptake and glucose 6-phosphate hydrolysis.

A later study in a hard-water mesotrophic lake also investigated the separate fate of ^{14}C- and ^{32}P-labelled glucose 6-phosphate into 'bacterial-sized' particles and phytoplankton (Hernandez *et al.*, 1996). Aquatic bacteria and phytoplankton assimilated the phosphoryl moiety 100 times faster than they assimilated the glucosyl moiety. Bacterial uptake conformed to Michaelis–Menten kinetics with an apparent K_t of 86 nM and a V_{max} of 1.4 nM/min. Phytoplankton took up phosphoryl with an apparent K_m of 380 nM and a V_{max} of 7.6 nM/min. These investigators determined the concentration of naturally occurring phosphate monoesters to be between 25 and 40 nM. Comparison of the rate of phosphoryl uptake from glucose 6-phosphate with the phosphate uptake rate indicated that phytoplankton could satisfy a significant portion (42–99%) of their phosphorus demand by hydrolysis of phosphate monoesters. A similar comparison indicated that bacterial assimilation of phosphorus from glucose 6-phosphate represented only 2–5% of the phosphorus taken up as phosphate. The findings by Hernandez *et al.* (1996) helped to explain apparently conflicting earlier findings that the release of phosphate from phosphate monoesters did not satisfy the phosphorus requirements for growth of the plankton community as a whole (Heath, 1986), but did appear to satisfy a significant portion of phytoplankton needs (Bentzen *et al.*, 1992).

If the use of dissolved organic phosphorus as a nutritional source of phosphorus provides such a small portion of the apparent phosphorus demand to natural assemblages of bacteria, what is the 'strategy' of bacterial use of dissolved organic phosphorus? I see this as an open question. A recent study in Lake Constance infers that under conditions of phosphorus sufficiency, bacterial alkaline phosphatase and 5′-nucleotidase activities lead to stimulation of bacterial growth by providing sources of organic carbon (Siuda and Chróst, 2001). This study suggested that the usefulness of phosphatases may depend on the sufficiency of environmental phosphorus and carbon, and may change seasonally. When carbon is sufficient, phosphatases serve to make phosphorus available; when phosphorus is sufficient, they act on the same substrates to make organic carbon available. Alternatively, dissolved organic phosphorus may not be a nutritional source of phosphorus or carbon, but may act in an 'informational' capacity, informing bacterial populations of the status of their environment (e.g. 'quorum sensing'), or may modulate community functions. For example, cyclic adenosine monophosphate is present in many aquatic environments and is known to modulate planktonic metabolic variables (Reimann, 1979; Francko and Wetzel, 1982).

Could bacteria act to attenuate or modulate informational signals through their capability to interact with dissolved organic phosphorus? Recent studies suggest the likelihood of such a possibility. When added to seawater, cyclic adenosine monophosphate can enhance the cultivation success of bacterioplankton in marine (Bruns *et al.*, 2002) and freshwater (Bruns *et al.*, 2003)

Table 9.3. Biomass and uptake rates for four size fractions of a natural phytoplankton assemblage from 1 m depth at a station in Chesapeake Bay, 19 August 1974. Whole-water samples were incubated with labelled compounds, then fractionated into size classes. Uptake rates of ^{32}P-phosphate, the ^{32}P-phosphoryl moiety of glucose 6-phosphate, and the ^{14}C-glucosyl moiety of glucose 6-phosphate. The three right-hand columns are uptake rates scaled for particulate phosphorus (nM/h/nM). Reproduced from Taft *et al*. (1977) with permission from the American Society of Limnology and Oceanography.

Size fraction (μm)	Chlorophyll *a* (μg/l)	Particulate phosphorus (ng P/l)	Phosphate uptake (nM/h)	Glucosyl uptake (nM/h)	Phosphoryl uptake (nM/h)	Phosphate uptake / particulate phosphorus (per h)	Glucosyl uptake / particulate phosphorus (per h)	Phosphoryl uptake / particulate phosphorus (per h)
>35	0	0.07	0	0	0	0	0	0
5–35	2.6	0.13	1.6	0.06	7.0	12.3	0.46	53.8
0.8–5	10.3	0.57	10.6	0.66	8.4	18.6	1.16	14.7
0.1–0.8	0.3	0.10	2.8	0.50	5.4	28.0	5.00	54.0

environments. Not all taxa respond to this stimulus. Gram-positive bacteria do not seem to use cyclic adenosine monophosphate as a signal compound (Lengeler *et al.*, 1999), and autoradiography using ^{3}H-cyclic adenosine monophosphate indicated that only 18% of the freshwater bacterioplankton in samples from a temperate eutrophic lake appeared to take up at least the labelled portion of the compound (Bruns *et al.*, 2003). Using denaturing gradient gel electrophoresis of PCR-amplified bacterioplankton DNA, these investigators also showed a differential response to cyclic adenosine monophosphate by the taxa present. The mechanism for cyclic adenosine monophosphate enhancement of culturability of bacterioplankton remains unclear, yet these recent reports indicate the use of at least this dissolved organic phosphorus compound in an informational role in natural communities.

Summary and Future Prospects

Most of what we know about the ability of bacteria to use dissolved organic phosphorus in their environment has been learned from investigation of the genetic architecture and metabolic capabilities of enteric bacteria in culture, especially *E. coli*. Several general lessons are learned from an overview of studies of bacteria in culture. The first is that bacteria, at least those most studied, have evolved mechanisms that primarily regulate a careful balance of intracellular carbon metabolites and reserves. A continued availability of phosphorus is essential to bacteria, active bacterial cells have a greater phosphorus content than phytoplankton, and they have also evolved the *pho* regulon to orchestrate an environmental scavenger hunt for phosphorus when environmental phosphate concentrations become very low and the Pit system provides insufficient resources.

The second lesson learned from a biochemical overview is that carbon and phosphorus metabolic strategies are inextricably intertwined. The rate of phosphate uptake, the cellular carbon-to-phosphorus ratio, and the growth efficiency per mole of phosphorus taken up appear to depend more on labile dissolved organic carbon availability than on phosphorus availability. Judging from the sheer numbers of genes, fraction of the genome devoted, or the moles of ATP metabolically spent, recovering carbon and maintaining its proper internal balance is the more important concern in the most well-characterized bacteria. The genetic strategy of the *pho* regulon is to gain phosphorus from the environment; whether phosphorus enters the cellular interior as phosphate, is actively transported by Pst-adenosine triphosphatase, or comes attached with a carbon handle (e.g. glycerol 3-phosphate or phosphonate), seems to make little difference, as the *pho* regulon genes are not catabolite-repressed. On the other hand, acquiring glycerol 3-phosphate or hexose phosphates by the action of genes under the *glp* regulon or the *uhp* operon is a strategy to gain carbon; if compounds enter with a phosphorus appendage, so much the better. Even synthesis of the group of gene products collectively known as 'nucleotidases' is a strategy to gain carbon, as released nucleosides are rapidly taken up and metabolized. Uptake of phosphate from hydrolysis of environmental nucleotides in the periplasm is likely to be only a metabolic bonus.

To focus on bacterial use of dissolved organic phosphorus solely as a source of phosphorus is potentially to miss a good part of the story, perhaps the better part. But that is the part which is largely still unexplored and unreported. As Kuhn (1968) reminded us, we often see only what we seek and expect to find. Since the 1970s the major focus in both marine and freshwater environments has been on nutrient availability to phytoplankton. Finding phosphorus as a limiting resource to phytoplankton in freshwater systems, and recent recognition of phosphorus limitation of primary production in coastal marine systems, has focused attention on sources of phosphorus and the mechanisms by which phosphorus becomes available to phytoplankton. I argue that this focus is too narrow when considering heterotrophic planktonic bacteria. Aquatic bacteria are capable of utilizing naturally occurring dissolved organic phosphorus, either by direct uptake or by prior hydrolysis through agents such as alkaline phosphatase activity or 5′-nucleotidase. Therefore, if aquatic bacteria are generally competitively advantaged to take up phosphate from their environment, they are probably the major beneficiaries of humic–phosphorus compounds sensitive to photolytic release of phosphate.

Understanding whether bacteria use dissolved organic phosphorus largely as a phosphorus source or as a carbon source will require continued investigations such as those of Taft *et al.* (1977) and Hernandez *et al.* (1996), which follow the fate of double-labelled organic phosphorus compounds in various environments. To date, much of that work has been performed in relatively eutrophic environments, which are more likely to be limited by phosphorus than carbon. Not surprisingly, bacteria have been found to favour uptake of phosphorus over carbon from synthetic and naturally occurring dissolved organic phosphorus. Future investigations must focus on environments where bacteria are likely to be carbon-limited. I believe we also need to think beyond dissolved organic phosphorus as simply a nutritional source and consider the possibility that it may serve an informational function in the environment, such as quorum sensing or metabolic modulation. Aquatic bacterial alterations of dissolved organic phosphorus or consumption of dissolved organic phosphorus fragments may play their part in a strategy to attenuate or modify those environmental signals.

Acknowledgements

This work was supported in part by grants from the US National Oceanic and Atmospheric Administration (NOAA), Ohio Sea Grant, and the Lake Erie Protection Fund.

References

Ackerman, R.S., Cozzarelli, N.R. and Epstein, E. (1974) Accumulation of toxic concentrations of methylglyoxal by wild-type *Escherichia coli* K-12. *Journal of Bacteriology* 119, 357–362.

Ambudkar, S.V., Larson, T.H. and Maloney, P.C. (1986) Reconstitution of sugar phosphate transport systems in transport systems of *Escherichia coli*. *Journal of Biological Chemistry* 261, 9083–9086.

Ammerman, J.W. and Azam, F. (1985) Bacterial 5′-nucleotidase in aquatic ecosystems: a novel mechanism of phosphorus regeneration. *Science* 227, 1338–1340.

Ammerman, J.W. and Azam, F. (1991a) Bacterial 5′-nucleotidase activity in estuarine and coastal marine waters: characterization of enzyme activity. *Limnology and Oceanography* 36, 1427–1436.

Ammerman, J.W. and Azam, F. (1991b) Bacterial 5′-nucleotidase activity in estuarine and coastal marine waters: role in phosphorus regeneration. *Limnology and Oceanography* 36, 1437–1447.

Avila, L.Z. and Frost, J.W. (1989) Phosphonium ion fragmentations relevant to organophosphonate biodegradation. *Journal of the American Chemical Society* 111, 8969–8970.

Beacham, I.R. and Garrett, S. (1980) Isolation of *Escherichia coli* mutants (*cpdB*) deficient in periplasmic 2′,3′-cyclic phosphodiesterase and genetic mapping of the *cpdB* locus. *Journal of General Microbiology* 119, 31–34.

Beacham, I.R., Kahana, R., Levy, L. and Yagil, E. (1973) Mutants of *Escherichia coli* K-12 'cryptic,' or deficient in 5′-nucleotidase (uridine diphosphate-sugar hydrolase) and 3′-nucleotidase (cyclic phosphodiesterase) activity. *Journal of Bacteriology* 116, 957–964.

Beacham, I.R., Haas, D. and Yagil, E. (1977) Mutants of *Escherichia coli* 'cryptic' for certain periplasmic enzymes: evidence for an alteration of the outer membrane. *Journal of Bacteriology* 129, 1034–1044.

Bentzen, E., Taylor, W.D. and Millard, E.S. (1992) The importance of dissolved organic phosphorus to phosphorus uptake by limnetic plankton. *Limnology and Oceanography* 37, 217–231.

Boavida, M.J. and Wetzel, R.G. (1998) Inhibition of phosphatase activity by dissolved humic substances and hydrolytic reactivation by natural ultraviolet light. *Freshwater Biology* 40, 285–293.

Boon, P.I. (1993) Organic matter degradation and nutrient regeneration in Australian fresh waters. III. Size fractionation of phosphatase activity. *Archiv für Hydrobiologie* 126, 339–360.

Boone, D.R., Castenholz, R.W. and Garrity, G.M. (2001) *Bergey's Manual of Systematic Bacterioogy*, 2nd edn, Vol. 1: *The Archaea,*

Photosynthetic Bacteria, and Deeply Branched Bacteria. Springer, Berlin.

Brown, K.A. (1980) Phosphotriesterases of *Flavobacterium* sp. *Soil Biology and Biochemistry* 12, 105–112.

Bruns, A., Cypionka, H. and Overmann, J. (2002) Cyclic AMP and acyl homoserine lactones increase the cultivation efficiency of heterotrophic bacteria from the central Baltic Sea. *Applied and Environmental Microbiology* 68, 3978–3987.

Bruns, A., Nübel, U., Cypionka, H. and Overmann, J. (2003) Effect of signal compounds and incubation conditions on the culturability of freshwater bacterioplankton. *Applied and Environmental Microbiology* 69, 1980–1989.

Brzoska, P., Rimmele, M., Brzostek, K. and Boos, W. (1994a) The *pho* regulon-dependent Ugp uptake system for glycerol-3-phosphate in *Escherichia coli* is *trans*-inhibited by P_i. *Journal of Bacteriology* 176, 15–20.

Brzoska, P., Rimmele, M., Brzostek, K. and Boos, W. (1994b) The Ugp paradox: the phenomenon that glycerol-3-phosphate, exclusively transported by the *Escherichia coli* Ugp system, can serve as a sole source of phosphate but not as a sole source of carbon is due to *trans* inhibition of Ugp-mediated transport by phosphate. In: Torriani-Gorini, A., Yagil, E. and Silver, S. (eds) *Phosphate in Microorganisms: Cellular and Molecular Biology*. American Society of Microbiology Press, Washington, DC, pp. 30–36.

Chróst, R.J. and Krambeck, H.J. (1986) Fluorescence correction for measurements of enzymatic activity in natural waters using methylumbelliferyl-substrates. *Archiv für Hydrobiologie* 106, 70–90.

Chróst, R.J. and Overbeck, J. (1987) Kinetics of alkaline phosphatase activity and phosphorus availability for phytoplankton and bacterioplankton in Lake Plussee (North German eutrophic lake). *Microbial Ecology* 13, 229–240.

Coleman, J.E. and Gettins, P. (1983) Alkaline phosphatase, solution structure and mechanism. *Advances in Enzymology* 55, 381–452.

Cotner, J.B., Jr and Heath, R.T. (1988) Potential phosphate release from phosphomonoesters by acid phosphatase in a bog lake. *Archiv für Hydrobiologie* 111, 329–338.

Cotner, J.B., Jr and Wetzel, R.G. (1991) 5′-Nucleotidase activity in a eutrophic lake and an oligotrophic lake. *Applied and Environmental Microbiology* 57, 1306–1312.

Cotner, J.B., Jr and Wetzel, R.G. (1992) Uptake of dissolved inorganic and organic phosphorus compounds by phytoplankton and bacterioplankton. *Limnology and Oceanography* 37, 232–243.

Currie, D.J. and Kalff, J. (1984) The relative importance of bacterioplankton and phytoplankton in phosphorus uptake in freshwater. *Limnology and Oceanography* 29, 311–321.

de Boor, M., Broekhuizen, C.P. and Postma, P.W. (1986) Regulation of glycerol kinase by enzyme III^{Glc} of the phosphoenolpyruvate:carbohydrate phosphotransferase system. *Journal of Bacteriology* 167, 393–395.

de Haan, H., Jones, R.I. and Salonen, K. (1990) Abiotic transformation of iron and phosphate in humic lake water revealed by double-isotope labeling and gel filtration. *Limnology and Oceanography* 35, 491–497.

Elvin, C.M., Hardy, C.M. and Rosenberg, H. (1985) P_i exchange mediated by the GlpT-dependent *sn*-glycerol-3-phosphate transport system in *Escherichia coli*. *Journal of Bacteriology* 161, 1054–1058.

Fernley, H.N. and Walker, P.G. (1967) Studies on alkaline phosphatase. *Biochemistry* 104, 1011–1018.

Francko, D.A. (1982) The isolation of cyclic 3′,5′-monophosphate (cAMP) from lakes of differing trophic status: correlation with planktonic metabolic variables. *Limnology and Oceanography* 27, 27–38.

Francko, D.A. (1984) Phytoplankton metabolism and cyclic nucleotides. II. Nucleotide-induced perturbations of alkaline phosphatase activity. *Archiv für Hydrobiologie* 100, 409–421.

Francko, D.A. and Heath, R.T. (1979) Functionally distinct classes of complex phosphorus compounds in lake water. *Limnology and Oceanography* 24, 463–473.

Francko, D.A. and Heath, R.T. (1982) UV-sensitive complex phosphorus: association with dissolved humic material and iron in a bog lake. *Limnology and Oceanography* 27, 564–569.

Francko, D.A. and Wetzel, R.G. (1982) The isolation of cyclic adenosine 3′:5′-monophosphate (cAMP) from lakes of differing trophic status: correlation with planktonic metabolic variables. *Limnology and Oceanography* 27, 27–38.

Gao, X. (2002) Phosphorus dynamics in freshwater lakes: examining the factors that control phosphate uptake by bacterioplankton communities. PhD dissertation, Kent State University, Ohio, USA.

Glockner, F.O., Zaichikov, E., Belkova, N., Denissova, L., Pernthaler, J., Pernthaler, A. and Amann, R. (2000) Comparative 16S rRNA analysis of lake bacterioplankton reveals globally distributed phylogenetic clusters including an abundant group of actinobacteria. *Applied and Environmental Microbiology* 66, 5053–5065.

Gutnick, D., Calvo, J.M., Klopotowski, T. and Ames, B.N. (1969) Compounds which serve as the sole source of carbon or nitrogen for *Salmonella typhimurium* LT-2. *Journal of Bacteriology* 100, 215–219.

Heath, R.T. (1986) Dissolved organic phosphorus compounds: do they satisfy planktonic phosphate demand in summer? *Canadian Journal of Fisheries and Aquatic Sciences* 43, 343–350.

Heath, R.T. and Cooke, G.D. (1975) The significance of alkaline phosphatase in a eutrophic lake. *Proceedings of the International Association of Theoretical and Applied Limnology* 19, 959–965.

Heath, R.T. and Edinger, A.C. (1990) Uptake of ^{32}P-phosphoryl from glucose-6-phosphate by plankton in an acid bog lake. *Proceedings of the International Association of Theoretical and Applied Limnology* 24, 210–213.

Heppel, L.A. (1971) The concept of periplasmic enzymes. In: Rothfield, L.I. (ed.) *Structure and Function of Biological Membranes*. Academic Press, New York, pp. 223–241.

Hernandez, I., Hwang, S.-J. and Heath, R.T. (1996) Measurement of phosphomonoesterase activity with a radiolabelled glucose-6-phosphate: role in the phosphorus requirement of phytoplankton and bacterioplankton in a temperate mesotrophic lake. *Archiv für Hydrobiologie* 137, 265–280.

Hoppe, H.-G. (1983) Significance of exoenzymatic activities in the ecology of brackish water: measurements by means of methylumbelliferyl-substrates. *Marine Ecology Progress Series* 11, 299–308.

Huber, A.L. and Kidby, D.J. (1984) An examination of the factors involved in determining phosphatase activities in estuarine water. 1. Analytical procedures. *Hydrobiologia* 111, 3–11.

Hudson, J., Taylor, W.D. and Schindler, D.W. (2000) Phosphate concentrations in lakes. *Nature* 406, 54–56.

Hulett, F.M., Sun, G. and Liu, W. (1994) The Pho regulon of *Bacillus subtilis* is regulated by sequential action of two genetic switches. In: Torriani-Gorini, A., Yagil, E. and Silver, S. (eds) *Phosphate in Microorganisms: Cellular and Molecular Biology*. American Society of Microbiology Press, Washington, DC, pp. 50–54.

Jansson, M. (1993) Uptake, exchange and excretion of orthophosphate in phosphate-starved *Scenedesmus quadricauda* and *Pseudomonas* K7. *Limnology and Oceanography* 38, 1162–1178.

Johannes, R.E. (1965) Influence of marine protozoa on nutrient regeneration. *Limnology and Oceanography* 10, 434–442.

Jones, J.G. (1972) Studies of freshwater bacteria: association with algae and alkaline phosphatase activity. *Journal of Ecology* 60, 59–75.

Jürgens, K. and Güde, H. (1990) Incorporation and release of phosphorus by planktonic bacteria and phagotrophic flagellates. *Marine Ecology Progress Series* 59, 271–284.

Kadner, R.J., Island, M.D., Merkel, T.J. and Webber, C.A. (1994) Transmembrane control of the Uhp sugar-phosphate transport system: the sensation of Glu6P. In: Torriani-Gorini, A., Yagil, E. and Silver, S. (eds) *Phosphate in Microorganisms: Cellular and Molecular Biology*. American Society of Microbiology Press, Washington, DC, pp. 78–84.

Karl, D.M. and Bailiff, M.D. (1989) The measurement and distribution of dissolved nucleic acids in aquatic environments. *Limnology and Oceanography* 34, 543–558.

Kobori, H. and Taga, N. (1977) Phosphatase activity and its role in the mineralization of organic phosphorus in coastal sea water. *Journal of Experimental Marine Biology and Ecology* 36, 23–39.

Kobori, H., Taga, N. and Simidu, U. (1979) Properties and generic composition of phosphatase-producing bacteria in coastal and oceanic seawater. *Bulletin of the Japanese Society of the Science of Fisheries* 45, 1429–1433.

Kuhn, T.S. (1968) *The Structure of Scientific Revolutions*. University of Chicago Press, Chicago, 212 pp.

Larson, T.J., Schumacher, G. and Boos, W. (1982) Identification of the *glpT*-encoded *sn*-glycerol-3-phosphate permease of *Escherichia coli*, an oligomeric integral membrane protein. *Journal of Bacteriology* 152, 1008–1021.

Lengeler, J.W., Drews, G. and Schlegel, H.G. (1999) *Biology of Prokaryotes*. Georg Thieme, Stuttgart, 955 pp.

Lin, E.C.C. (1987) Dissimilatory pathways for sugars, polyols, and carboxylates. In: Neidhardt, F.C., Ingraham, J.L., Low, K.B., Magasanik, B., Schaechter, M. and Umbarger, H.E. (eds) Escherichia coli *and* Salmonella typhimurium*: Cellular and Molecular Biology*. American Society of Microbiology, Washington, DC, pp. 244–284.

Lin, E.C.C. and Iuchi, S. (1991) Regulation of gene expression in fermentative and respiratory systems in *Escherichia coli* and related bacteria. *Annual Review of Genetics* 25, 361–387.

Makino, K., Shinagawa, H., Amemura, M., Kawamoto, T., Yamada, M. and Nakata, A. (1989) Signal transduction in the phosphate regulon of *Escherichia coli* involves phosphotransfer between PhoR and PhoB proteins. *Journal of Molecular Biology* 210, 551–559.

McGrath, S.M. and Sullivan, C.W. (1981) Community metabolism of adenylates by microheterotrophs from the Los Angeles Harbor and Southern California coastal waters. *Marine Biology* 62, 217–226.

McMillan, G. and Quinn, J.P. (1994) In vitro characterization of a phosphate-starvation-independent carbon-phosphorus bond cleavage activity in *Pseudomonas fluorescens* 23F. *Journal of Bacteriology* 176, 320–324.

Metcalf, W.W. and Wanner, B.L. (1993) Mutational analysis of an *Escherichia coli* fourteen-gene operon for phosphonate degradation using Tn*phoA'* elements. *Journal of Bacteriology* 175, 3430–3442.

Muda, M., Rao, N. and Torriani, A. (1992) Role of PhoU in phosphate transport regulation and alkaline phosphatase regulation. *Journal of Bacteriology* 174, 8054–8064.

Nannipieri, P., Ceccanti, B. and Bianchi, D. (1988) Characterization of humus-phosphatase complexes extracted from soil. *Soil Biology and Biochemistry* 20, 683–691.

Neuhard, J. and Nygaard, P. (1987) Purines and pyrimidines. In: Neidhardt, F.C., Ingraham, J.L., Low, K.B., Magasanik, B., Schaechter, M. and Umbarger, H.E. (eds) Escherichia coli *and* Salmonella typhimurium*: Cellular and Molecular Biology*. American Society of Microbiology, Washington, DC, pp. 445–473.

Parker, A.B.C. (1963) Classification of micrococci and staphylococci based on physiological and biochemical tests. *Journal of General Microbiology* 30, 409–427.

Perry, M.J. (1972) Alkaline phosphatase activity in subtropical central North Pacific waters using a sensitive fluorometric method. *Marine Biology* 15, 113–119.

Postma, P.W. (1987) Phosphotransferase system for glucose and other sugars. In: Neidhardt, F.C., Ingraham, J.L., Low, K.B., Magasanik, B., Schaechter, M. and Umbarger, H.E. (eds) Escherichia coli *and* Salmonella typhimurium*: Cellular and Molecular Biology*. American Society of Microbiology, Washington, DC, pp. 127–141.

Rao, N., Roberts, M.F., Torriani, A. and Yashphe, J. (1993) Effect of *glpT* and *glpD* mutations on expression of the *phoA* gene in *Escherichia coli*. *Journal of Bacteriology* 175, 74–79.

Reid, T.W. and Wilson, I.B. (1971) *E. coli* alkaline phosphatase. In: Boyer, P.D. (ed.) *The Enzymes*, Vol. 4. Academic Press, New York, pp. 373–415.

Reimann, B. (1979) The occurrence and ecological importance of dissolved ATP in fresh water. *Freshwater Biology* 9, 481–490.

Rosenberg, H. (1987) Phosphate transport in prokaryotes. In: Rosen, B. and Silver, S. (eds) *Ion Transport in Prokaryotes*. Academic Press, New York, pp. 205–248.

Rosenberg, H., Geerdes, R.G. and Chegwidden, K. (1977) Two systems for the uptake of phosphate in *Escherichia coli*. *Journal of Bacteriology* 131, 505–511.

Rowland, S.S., Speedie, M.K. and Pogell, B.M. (1991) Purification and characterization of a secreted recombinant phosphotriesterase (Parathion hydrolase) from *Streptomyces lividans*. *Applied and Environmental Microbiology* 57, 440–444.

Schweizer, H., Argast, M. and Boos, W. (1982) Characteristics of a binding protein-dependent transport system for *sn*-glycerol-3-phosphate in *Escherichia coli* that is part of the *pho* regulon. *Journal of Bacteriology* 150, 1154–1163.

Shinagawa, H., Makino, K., Yamada, M., Amemura, M., Sato, T. and Nakata, A. (1994) Signal transduction in the phosphate regulon of *Escherichia coli*: dual functions of PhoR as a protein kinase and a protein phosphatase. In: Torriani-Gorini, A., Yagil, E. and Silver, S. (eds) *Phosphate in Microorganisms: Cellular and Molecular Biology*. American Society of Microbiology Press, Washington, DC, pp. 285–289.

Shaw, P.J. (1994) The effect of pH, dissolved humic substances, and ionic composition on the transfer of iron and phosphate to particulate size fractions in epilimnetic lake water. *Limnology and Oceanography* 39, 1734–1743.

Siuda, W. and Chróst, R.J. (2001) Utilization of selected dissolved organic phosphorus compounds by bacteria in lake water under non-limiting orthophosphate conditions. *Polish Journal of Environmental Studies* 10, 475–483.

Sondergaard, M., Hansen, B. and Markager, S. (1995) Dynamics of dissolved organic carbon and its lability in a eutrophic lake. *Limnology and Oceanography* 40, 46–54.

Tanford, C. (1961) *Physical Chemistry of Macromolecules*. John Wiley & Sons, New York.

Taft, J.L., Loftus, M.E. and Taylor, W.R. (1977) Phosphate uptake from phosphomonoesters by phytoplankton in the Chesapeake Bay. *Limnology and Oceanography* 22, 1012–1021.

Torriani-Gorini, A. (1994) Introduction: the Pho regulon of *Escherichia coli*. In: Torriani-Gorini, A., Yagil, E. and Silver, S. (eds) *Phosphate in Microorganisms: Cellular and Molecular Biology*. American Society of Microbiology Press, Washington, DC, pp. 1–4.

Vadstein, O. (1998) Evaluation of competitive ability of two heterotrophic planktonic bacteria under phosphorus limitation. *Aquatic Microbial Ecology* 14, 119–127.

Vadstein, O. (2000) Heterotrophic, planktonic bacteria and cycling of phosphorus: phosphorus requirements, competitive ability, and food web interactions. *Advances in Microbial Ecology* 16, 115–167.

Vadstein, O. and Olsen, Y. (1989) Chemical composition and phosphate uptake kinetics of limnetic bacterial communities cultured in chemostats under phosphorus limitation. *Limnology and Oceanography* 34, 939–946.

Vadstein, O., Olsen, Y. and Reinertsen, H. (1993) The role of planktonic bacteria in phosphorus cycling in lakes: sink and link. *Limnology and Oceanography* 38, 1539–1544.

Wanner, B.L. (1994) Phosphate-regulated genes for the utilization of phosphonates in members of the Family *Enterobacteriaceae*. In: Torriani-Gorini, A., Yagil, E. and Silver, S. (eds) *Phosphate in Microorganisms: Cellular and Molecular Biology*. American Society of Microbiology Press, Washington, DC, pp. 215–221.

Webb, D.C. and Cox, G.B. (1994) Proposed mechanism for phosphate translocation by the phosphate-specific transport (Pst) system and role of the Pst system in phosphate regulation. In: Torriani-Gorini, A., Yagil, E. and Silver, S. (eds) *Phosphate in Microorganisms: Cellular and Molecular Biology*. American Society of Microbiology Press, Washington, DC, pp. 37–42.

Weston, L.A. and Kadner, R.J. (1988) Role of *uhp* genes in expression of the *Escherichia coli* sugar-phosphate transport system. *Journal of Bacteriology* 170, 3375–3383.

Willsky, G.R. and Malamy, M.H. (1980) Characterization of two genetically separable inorganic phosphate transport systems in *Escherichia coli*. *Journal of Bacteriology* 144, 356–365.

Winkler, H.H. (1973) Distribution of an inducible hexose-phosphate transport system among various bacteria. *Journal of Bacteriology* 116, 1079–1081.

Zwaig, N., Kistler, W.S. and Lin, E.C.C. (1970) Glycerol kinase, the pacemaker for the dissimilation of glycerol in *Escherichia coli*. *Journal of Bacteriology* 102, 753–759.

10 Ecological Aspects of Phosphatase Activity in Cyanobacteria, Eukaryotic Algae and Bryophytes

Brian A. Whitton,[1] Abdulrahman M. Al-Shehri,[2] Neil T.W. Ellwood[3] and Benjamin L. Turner[4]

[1]School of Biological and Biomedical Sciences, University of Durham, Durham DH1 3LE, UK; [2]Biological Sciences Department, College of Science, King Khalid University, Abha, Saudi Arabia; [3]Dipartimenti di Scienze Geologiche, Universita Roma Tre, Largo San Leonardo Murialdo, 00146 Roma, Italy; [4]Smithsonian Tropical Research Institute, Box 2072, Balboa, Ancon, Republic of Panama

Introduction

As most, if not all, phototrophs can utilize inorganic phosphate in their environment, it is widely assumed that the ability to utilize organic phosphate is more restricted. If the supply of phosphate is sufficiently restricted that an organism starts to become phosphorus-limited, one possibility to overcome this is to use inorganic phosphate more efficiently, such as by adopting different uptake pathways (Wagner and Falkner, 2001). Although overlooked by these and many other authors, another possibility is to use the organic phosphate present in the environment. This chapter sets out to assess how widespread this is in cyanobacteria, algae and bryophytes, and to review the methods involved. The focus is on soluble organic phosphate, because this has been the subject of most studies, though some cyanobacteria may be able to utilize insoluble organic phosphate (Whitton *et al.*, 1991). A further approach adopted by some chrysophytes, haptophytes, cryptomonads and dinoflagellates is to feed on other living organisms (phagotrophy; see Graham and Wilcox, 2000) and thus obtain phosphorus via this route. As 'soluble' organic phosphorus in the environment has almost always been determined following filtration and molybdate colorimetry, the values obtained can include high-molecular-weight complexes and inorganic polyphosphate (see McKelvie, Chapter 1; Mitchell and Baldwin, Chapter 14, this volume).

Although it has been shown directly that organisms can utilize organic phosphates as their sole phosphorus source (Cembella *et al.*, 1984), much of the evidence has come indirectly from studies on phosphatases and phosphatase activity. Phosphatases are represented by a 'whole bunch of enzymes' (Boavida, 1990), characterized by different half-saturation constants, temperature and pH optima, and substrate specificity (Hoppe, 2003). Activity can often be demonstrated simply by staining, although most accounts provide quantitative measurements. However, though the literature on these phototrophs is considerable, it is often difficult to interpret. Factors overlooked at one time have subsequently proved to be important, so the older literature must be consulted with caution. The following review should be considered together with Chapter 9 of this book, written by R.T. Heath, which stresses the importance of bacteria utilizing organic phosphate as a source of carbon as well as phosphorus. This has seldom been considered for phototrophs, even those known to be photoheterotrophs. There is, of course, an extensive literature on the ecological aspects of phosphatases of other organisms. This is mentioned only briefly here, and would require critical review in order to make meaningful comparisons.

Growth Studies

That organic phosphate can supply all the phosphorus requirements for some cyanobacteria, algae and mosses has been demonstrated by prolonged growth of axenic strains in media with organic, but no inorganic, phosphate, while at the same time demonstrating that no detectable hydrolysis of the organic phosphate occurs in the absence of the organism. In a study of 50 cyanobacterial strains grown in batch culture without shaking, all grew using the phosphate monoesters β-glycerophosphate or *para*-nitrophenyl phosphate as the sole phosphorus source (Whitton *et al.*, 1991). The phosphate diester bis-*para*-nitrophenyl phosphate was used by 47 strains, herring sperm DNA by 49 strains, ATP (adenosine 5′-triphosphate) by 40 strains and phytic acid (*myo*-inositol hexakisphosphate) by 35 strains. The relative growth rates in response to bis-*para*-nitrophenyl phosphate and DNA differed considerably, which the authors suggested might result from more than one enzyme being involved in DNA utilization and reflect differences in the relative amounts of these enzymes. The failure of some strains to grow with ATP was probably due to its toxicity at the concentration tested; six of the seven which failed to grow were *Calothrix* strains.

In studies of marine cyanobacteria, *Synechococcus* WH 7803 grew with inorganic phosphate, 2′-deoxycytidine 5′-triphosphate, *para*-nitrophenyl phosphate, glucose 6-phosphate, or glycerol phosphate, but not cAMP (cyclic adenosine 5′-monophosphate) (Donald *et al.*, 1997), while *Prochlorococcus marinus* PCC 9511 grew with β-glycerophosphate, pyrophosphate, glucose 6-phosphate or ATP (Rippka *et al.*, 2000). Although this and another strain of *Synechococcus* could not use cAMP, strain WH8103 can do so (L. Moore, cited in Scanlan, 2003), and Scanlan suggested that this ability may be a clade-specific trait in marine *Synechococcus*. Strains of the marine green alga *Nannochloris* and the marine eustigmatophyte *Nannochloropsis* were able to grow using β-glycerophosphate (Lubián, 1981, cited in Lubián *et al.*, 1992), but not *Nannochloris oculata* (Lubián *et al.*, 1992). For mosses, *Hydrogonium fontanum* from a Saudi Arabian tufa-forming stream (Whitton *et al.*, 1986) can grow in axenic culture with *para*-nitrophenyl phosphate, bis-*para*-nitrophenyl phosphate, β-glycerophosphate or ATP, but not phytic acid, as the sole phosphorus source (Al-Shehri, 1992).

All these studies used relatively high ambient phosphorus concentrations, for instance 1 mg P/l by Whitton *et al.* (1991) and 100 μM (~3 mg P/l) by Rippka *et al.* (2000). Many of the 50 strains tested by Whitton *et al.* (1991) grew as fast or almost as fast in the presence of organic phosphate as with inorganic phosphate, while accounts by other authors suggest that this may also have been true in their studies. Repeat experiments at much lower substrate concentrations are needed to see if differences occur when substrate concentrations are closer to those likely to occur in nature.

Freshwater *Synechococcus* strains made up seven of the 15 cyanobacteria unable to grow in the presence of phytic acid (Whitton *et al.*, 1991). Since these are planktonic organisms, the authors suggested that this might reflect the very low solubility and hence lack of availability of this compound in the natural environment. However, Reichardt (1971) reported that a bacterized strain of the cyanobacterium *Aphanizomenon*, which is also a colonial planktonic organism, preferred phytin (the calcium salt of *myo*-inositol hexakisphosphate) to inorganic phosphate, whereas the converse was true for seven other strains. Chróst and Siuda (2002) stated that phytase has an optimum pH around 5.0, but this was based on a study of organic phosphorus fractions in lake water by Herbes *et al.* (1975) using commercial soil fungal phytase with pH optima at 2.5 and 5. As the 35 strains of cyanobacteria mentioned above were cultured at pH 7.6, the pH optimum of their phytase is probably in the alkaline range. Nevertheless the study of Herbes *et al.* (1975) demonstrated that up to 50% of the organic phosphate in lake water could be hydrolysed by added phytase.

Phosphatases

Location

Most studies indicate that the use of organic phosphates depends on the presence of phosphatase enzymes in the cell wall or its immediate surroundings, leading to the release of inorganic phosphate external to the cytoplasmic membrane (Heath and Edinger, 1990; Bjoerkman and Karl, 1994; Hernández *et al.*, 1996a; Štrojsová *et al.*, 2003) and subsequent uptake of much of this into the cell. Comparison of the separate fates of ^{14}C-labelled and ^{32}P-labelled glucose 6-phosphate into bacterial-sized particles and phytoplankton showed that phosphate was assimilated 100 times faster than the glucosyl moiety (Hernández *et al.*, 1996a), suggesting that all or almost all hydrolysis occurred external to the cytoplasmic membrane. Nevertheless, the literature on bacteria indicates that the ability to take up glycerol 3-phosphate or glucose 6-phosphate is widespread (see Heath, Chapter 9, this volume). There is some circumstantial evidence that hydrolysis may sometimes also occur inside the cytoplasmic membrane of phototrophs, such as strong staining for orthophosphate immediately inside the membrane of the cyanobacterium *Calothrix parietina* exposed to β-glycerophosphate (Wood *et al.*, 1986), the slow release of *para*-nitrophenol from seaweed tissue long after the tissue has been washed free of *para*-nitrophenyl phosphate substrate (Hernández and Whitton, 1996) and the contrast in some organisms between the ability to use a phosphate diester and the lack of surface phosphodiesterase activity (see below). All these studies suggesting possible uptake of the unhydrolysed molecule were made using high substrate concentrations, so experiments with labelled substrates at much lower concentrations are needed to establish whether such uptake is likely to be important in nature.

Authors differ in the terms used to describe activity in different locations (Fig. 10.1). Activity external to the cytoplasmic membrane, but retained on a filter (typically 0.2 μm), has been termed cellular (Grainger *et al.*, 1989), whole-cell (Lubián *et al.*, 1992), ectoenzyme (Chróst, 1991; Chróst and Siuda, 2002), cell-bound (Whitton *et a1.*, 1991), surface (Whitton *et al.*, 1998) or surface-bound extracellular (Nedoma *et al.*, 2003). Extracellular has been used in several other ways, including everything external to the cytoplasmic membrane (Priest, 1984; Wetzel, 1991; Weich and Granéli, 1989; Chappell and Goulder, 1994; Lee, 2000; Nedoma *et al.*, 2003). Ammerman (1991) introduced the term ecto-phosphohydrolase activity to describe essentially the same component: 'hydrolysis of organic or other complex phosphorus compounds, soluble or particulate, in which the hydrolysed phosphate is released outside the cell'. However, at least for studies on the physiological ecology of algae, much of the literature regards extracellular as material released by cells, but subsequently in solution or perhaps re-bound to other surfaces (e.g. Wynne and Rhee, 1988; Chróst, 1991; Whitton *et al.*, 1991; Chróst and Siuda,

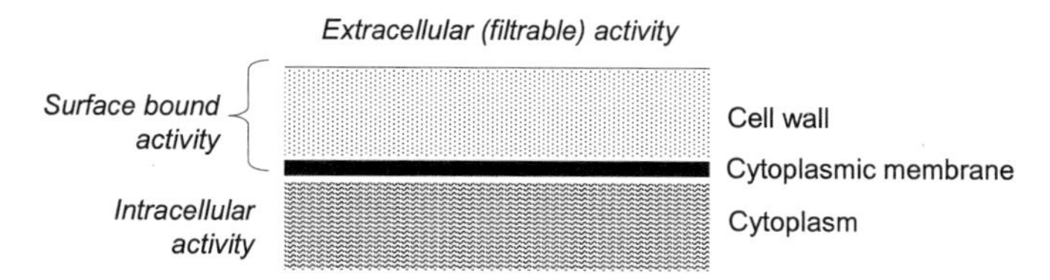

Fig. 10.1. Terms used in this chapter to describe possible enzyme location. In the case of mucilage external to the wall, it is recommended to state in each case whether it is treated as 'surface' or 'extracellular'.

2002). The term 'intracellular' has also been used in different ways. Most authors use it to indicate activity inside the cytoplasmic membrane, but a few also include in this fraction the activity associated with the cell wall (e.g. Wynne and Rhee, 1988).

In this chapter, activity inside the plasma membrane is termed intracellular, while that external to the membrane is surface-bound if retained on a filter, or extracellular if released by the organism and passing through a filter (Fig. 10.1). This definition makes it easy to distinguish practically between surface and extracellular activity when quantifying different components of an ecosystem. However, most ecophysiological studies treat mucilaginous surrounds as 'extracellular', though it is difficult to separate their contribution to phosphatase activity from that of the wall proper or, in the case of some cyanobacteria, a firm sheath. In addition, it is often difficult to distinguish rigorously between a mucilage layer and material dissolved in the surrounding liquid, because their relative distribution is influenced by environmental conditions and growth stage (Grainger *et al.*, 1989; Whitton *et al.*, 1990). It is therefore important to explain exactly how the term 'extracellular' is used in any particular study.

The presence of phosphomonoesterase in sheaths and mucilaginous material of cyanobacteria has been demonstrated chemically (Weckesser *et al.*, 1988) and by staining (Grainger *et al.*, 1989). Phosphatases may also be released by cell lysis, but information is sparse about the extent to which such enzymes can become bound to the sheaths or mucilaginous surrounds of living cells of the same or other species. The possible interactions between enzymes external to the cytoplasmic membrane, including phosphatases, were reviewed by Wetzel (1991). Extracellular enzymes in solution may bind to suspended particles and humic materials, or be exposed to a variety of inhibitors present in the water or to proteases, leading to degradation.

Staining for light microscopy has been done with a range of azo-dyes on cyanobacteria (Grainger *et al.*, 1989; Xie *et al.*, 1989), green algae (Livingstone *et al.*, 1983; Gibson and Whitton, 1987b) and mosses (Turner *et al.*, 2001). Staining by deposition of lead phosphate in response to inorganic phosphate release due to phosphomonoesterase activity is particularly useful for mucilage layers, such as the mucilage surrounding hairs of the green alga *Draparnaldia* (Gibson and Whitton, 1987b), because the effect is more obvious than with azo-dye staining. Lead staining was used to reveal phosphomonoesterase activity around the hairs of *Calothrix parietina* and also the formation of polyphosphate granules in the basal part of the filament, both of which can be present for a period subsequent to the addition of β-glycerophosphate to a phosphorus-limited culture (Livingstone and Whitton, 1983). When combined with electron microscopy (Wood *et al.*, 1986), lead phosphate precipitation can be semi-quantitative.

In the case of the mosses investigated by Turner *et al.* (2001), the azo-dye bromo-4-chloro-3-indolyl phosphate (BCIP), which forms a blue/purple colour following hydrolysis (Coston and Holt, 1958), proved to be the most useful stain. Staining occurred in different locations in different species. In some species (e.g. *Hylocomium splendens* and *Racomitrium lanuginosum*) staining occurred inside the cell wall, although, as with the study on *Calothrix parietina* by Wood *et al.* (1986), it seems likely that this may be an artefact due to the high substrate concentration. In *R. lanuginosum*, strong staining was evident in the cells near to the shoot tips, with comparatively little staining lower down. Some species showed little or no staining, so the absence of staining should therefore not be taken as evidence of lack of phosphomonoesterase activity in a species unless other studies are included to demonstrate the environmental conditions required to optimize staining. Clear evidence of staining using light microscopy requires a high substrate concentration, with the concomitant uncertainty about interpretation of the results. Furthermore, particular care is needed with BCIP; for example, staining of *Calothrix parietina* with BCIP was less successful when done on a slide under a coverslip than in shaken flasks (Grainger *et al.*, 1989). The authors suggested that this was probably due to lack of oxygen in the former, since staining with BCIP is a two-stage process, the second of which requires oxygen (Coston and Holt, 1958).

Highly sensitive molecular and immunological approaches are increasingly being adopted to detect the presence and location of phosphatase enzymes, as opposed to the products of phosphatase activity (Scanlan and Wilson, 1999), and this seems likely to prove one of the most important ways of establishing the role of surface phosphatases in the plankton, especially that of oligotrophic regions of the oceans. As pointed out by Hoppe (2003), isolation and purification of a surface phosphatase, such as achieved with the marine dinoflagellate *Prorocentrum minimum* (Dyhrman and Palenik, 1997), provides a target for developing an antibody probe for use in environmental research.

Measurement of activity

A variety of methods have been used to quantify phosphatase activity, but only a few researchers have adopted several methods at the same time, making it hard to compare different studies. The most frequent quantitative method has involved the use of *para*-nitrophenyl phosphate (or bis-*para*-nitrophenyl phosphate as a phosphate diester), the organic hydrolysis product of which, *para*-nitrophenol, can be measured spectrophotometrically. The fluorigenic methylumbelliferyl phosphate provides a more sensitive method (Pettersson, 1980; Hoppe, 1983; Chróst and Krambeck, 1986), although studies on seasonal changes in natural populations do not always give the same results as measurements made with *para*-nitrophenyl phosphate at the same time. Methylumbelliferyl phosphate indicated higher activity than *para*-nitrophenyl phosphate for the acid stream moss *Warnstorfia fluitans* (Ellwood *et al.*, 2002), and also for the marine cyanobacterium *Rivularia atra* (Yelloly and Whitton, 1996) for part of the year; though in the latter case the converse was true for another part of the year. Other possible substrates for such studies are listed by Hoppe (2003) and Heath (Chapter 9, this volume).

Enzyme-labelled fluorescence (ELF) involving the soluble substrate ELF97 phosphate (2-(5′-chloro-2′-phosphoryloxyphenyl)-6-chloro-4-(3H)-quinazolinone), also known as ELFP, was introduced by Huang *et al.* (1992). Hydrolysis leads to the fluorescent precipitate ELF97 alcohol (ELFA) at the site of the enzyme, which can be quantified by image analysis. Studies using this method include those of González-Gil *et al.* (1998), Dyhrman and Palenik (1999, 2001), Rengefors *et al.* (2001, 2003), Nedoma *et al.* (2003) and Štrojsová *et al.* (2003), although the last authors pointed out the need for caution in interpreting previous studies where the substrate had been supplied in the presence of additional alcohol.

Precipitation of ELFA seems to be a complex process, which may not necessarily be linear with time (Huang *et al.*, 1992). The lag is almost certainly not a biological effect, but due to the formation of an intermediate complex before fluorescence is detected. Nedoma *et al.* (2003) used the method to quantify fluorescence obtained with assays of individual phytoplankton cells during a study of two waterbodies. They also compared spectrofluorimetrically the use of ELFP and methylumbelliferyl phosphate for detecting phosphatase activity of whole phytoplankton samples from the same sites, but activity at one site proved to be too low to detect with ELFP in spite of the fact that some activity was indicated by image analysis. Lags occurred with ELFP, but not methylumbelliferyl phosphate. Application of saturation kinetics to the lake with high phosphomonoesterase activity gave results conforming to the Michaelis–Menten model, values of V_{max} and K_m being two to four times higher with ELFP than with methylumbelliferyl phosphate. Comparisons have not yet been made between the use of ELFP and methylumbelliferyl phosphate spectrophotometrically for populations of individual species, and it is open to question whether the enzyme-labelled fluorescence technique is sensitive enough to detect low rates of phosphatase activity (P. Gualtieri, personal communication). The method apparently detects only phosphomonoesterase, not phosphodiesterase, activity. Practical matters relating to the study of field samples are discussed in more detail below.

Surface phosphatases of cyanobacteria

All 50 cyanobacteria able to grow with phosphate monoesters showed significant surface phosphomonoesterase activity (Whitton *et al.*, 1991). However, four strains only showed activity at pH 10.3, not pH 7.6, the condition used for testing growth. The contrast between ability to grow with bis-*para*-nitrophenyl phosphate and the presence of surface phosphodiesterase activity was even more marked, since 13 of the 47 strains able to use this compound as a phosphorus source showed no phosphodiesterase activity at either pH. In these cases it remains uncertain where hydrolysis occurred or whether some other enzyme was involved. Ten strains of the cyanobacterium *Arthrospira* (Mühling, 2000) showed even more contrast, because all grew well using β-glycerophosphate or *para*-nitrophenyl phosphate, but phosphomonoesterase activity was always very low, despite being tested under a range of conditions. None of these *Arthrospira* strains was able to use bis-*para*-nitrophenyl phosphate as a phosphorus source. The low surface phosphomonoesterase may reflect the fact that this organism occurs naturally in waters with high ambient phosphate concentrations, such as African soda lakes. In culture *Arthrospira* is also routinely grown in a medium with very high phosphate concentration (Zarrouk, 1966), and thus never encounters stress due to phosphorus limitation. However, this still leaves open the question of how some *Arthrospira* strains are able to grow in culture using phosphate monoesters.

In cyanobacteria the primary location for the occurrence of cell surface phosphomonoesterase is the periplasm (Ihlenfeldt and Gibson, 1975), though phosphomonoesterase activity can apparently always be found in any mucilaginous layer external to the wall if it also present in the periplasm. Molecular studies on phosphomonoesterases in cyanobacteria likely to be involved in utilizing organic phosphate in their environment have been reviewed by Bhaya *et al.* (2000). The best-known example is that found in *Synechococcus* PCC 7942 (Block and Grossman, 1988), which shows little similarity to any of the bacterial phosphatases studied previously. The PhoA enzyme is transported to the periplasm from its site of synthesis across the cytoplasmic membrane (Ray *et al.*, 1991). The addition of high levels of inorganic phosphate to a culture of *Synechococcus* PCC 7942 led to transcription from *phoA* being inhibited almost immediately. Part of this enzyme has sequences similar to the kinases involved in binding nucleotide triphosphates. Bhaya *et al.* (2000) suggested that these might allow for the binding of phosphate groups attached to a variety of different compounds, thus expanding the substrate specificity of the enzyme, a desirable feature for an enzyme utilizing diverse phosphorylated compounds in the environment.

A second periplasmic phosphatase, PhoV, in the same organism (Wagner *et al.*, 1995) has substrate specificity for phosphate monoesters and requires zinc ions for activity. The *phoV* gene is apparently not regulated by phosphate and may be a constitutive periplasmic phosphatase (Bhaya *et al.*, 2000). A third cyanobacterial phosphatase was isolated from *Nostoc commune* UTEX 584 (Xie *et al.*, 1989; Potts *et al.*, 1993), which may possibly also be in the periplasm. This shows broad phosphomonoesterase activity towards a range of molecules, including proteins, peptides and low-molecular-weight organic phosphates. It also has measurable pyrophosphatase activity, and may have a further role in sensing the environment (Kennelly and Potts, 1996; Potts, 2000). In view of the range of organic phosphates in the environment and the need to maximize uptake where phosphorus is low or limiting, it seems probable that the ability to detect diverse molecules in the environment will prove to be an important feature of periplasmic proteins in cyanobacteria (Mann, 2000).

The ability of cyanobacteria to mobilize phosphate from insoluble inorganic materials is apparently widespread (Whitton, 2000), including calcium triphosphate (Bose *et al.*, 1971), Mussoorie rock phosphate (Roychaudhury and Kaushik, 1989) and hydroxyapatite (Cameron and Julian, 1988). Several of these authors suggested that extracellular phosphatases might be involved, the most plausible example being that of a study by Natesan and Shanmugasundaram (1989) on an *Anabaena* strain grown with phosphate immobilized in soil.

Surface phosphatases of eukaryotic algae

Among the eukaryotic algae, surface phosphatase activity has been reported widely in red, brown and green algae (Whitton, 1991; Hernández *et al.*, 2003), diatoms (Myklestad and Sakshaug, 1983; Štrojsová *et al.*, 2003), dinoflagellates (Rivkin and Swift, 1980; Dyhrman and Palenik, 2001; Štrojsová *et al.*, 2003) and yellow-green algae (B.A. Whitton, unpublished data). Some caution is needed when interpreting data from brown algae, because of the presence of a fungus in the thallus of many species. In a study of lake plankton, Štrojsová *et al.* (2003) reported that, though activity could sometimes be found in chrysophytes and cryptophytes, it was less common here. Rengefors *et al.* (2001) suggested that the lack of phosphatase activity in the cryptophytes studied by them might have been due to a switch to mixotrophy (phagotrophy). Štrojsová *et al.* (2003) found no activity in any euglenophyte, even under conditions when many other species showed activity. They did not speculate on the possible reasons, but in view of the frequent occurrence of euglenophytes in waters with rotting vegetation, it would be surprising if there is a widespread lack of ability to use organic phosphate. If phosphatase activity does prove to be widespread in euglenophytes, it is important to establish why Štrojsová *et al.* (2003) obtained negative results. Possible reasons might be that the method used was insufficiently sensitive, that organic phosphate can be taken up as an entire molecule and hydrolysed intracellularly, or that euglenophytes use phosphate diesters and other organic molecules that were not assayed. *Euglena mutabilis* was a component of algal mats showing high phosphomonoesterase activity in a highly acidic river (Sabater *et al.*, 2003). Intracellular acid phosphatase activity of *Euglena* has been reported by Blum (1965) and Palisano and Walne (1972).

Nedoma *et al.* (2003) compared values of cell-specific activity (surface phosphomonoesterase activity per cell) obtained in their own study of two waterbodies in the Czech Republic (10–2260 fmol/cell/h, methylumbelliferyl phosphate) with those in the literature. These include (expressed as fmol/cell/h): *Phaeocystis*, 2.5 (methyl fluorescin phosphate; van Boekel and Veldhuis, 1990); *Nannochloris*, 0.5–5.5 (*para*-nitrophenyl phosphate; Lubián *et al.*, 1992); *Staurastrum chaetoceras*, 10 (*para*-nitrophenyl phosphate; Spijkerman and Coesel, 1998); *Chlamydomonas reinhardtii*, 3000–22,000 (methyl fluorescin phosphate; Joseph *et al.*, 1994).

Apart from the comment by Rengefors *et al.* (2001) mentioned above, there is apparently no information on the response of phagotrophic algae to phosphorus limitation. However, numerous questions arise.

Does this lead to enhanced phagotrophy? To what extent does phagotrophy lead to engulfment of particulate organic phosphate as opposed to live organisms? Does the extent of phagotrophy influence synthesis of surface phosphatases? Are the same surface phosphatases used in the walls of vacuoles surrounding ingested particles as in the wall directly in contact with the external environment? There are several reports indicating that dinoflagellates can possess high surface phosphomonoesterase activity (e.g. Dyhrman and Palenik, 1997; Štrojsová *et al.*, 2003), but the possible importance of phagotrophy was not investigated in the populations studied.

Multicellular organisms sometimes show marked differences in the extent to which surface phosphatases are developed on different parts of the thallus. This is most striking in taxa forming multicellular hairs (see below), which enhance the surface-to-volume ratio. However, in general, the larger the organism, the lower is the value for this ratio; Hernández *et al.* (1999) assessed the implication of this for seaweeds.

Surface phosphatases of bryophytes

The extent to which species of moss differ in their phosphatase activities was investigated in 19 populations, representing terrestrial, semi-aquatic and aquatic species (Turner *et al.*, 2001). All showed surface phosphomonoesterase activity, but not all showed phosphodiesterase activity. Species expressing markedly high rates of both phosphomonoesterase and phosphodiesterase activity included *Hylocomium splendens* growing on calcareous soil, *Sphagnum cuspidatum* growing in an acidic bog, and the semi-aquatic *Palustriella commutata* var. *falcata* growing on the bank of a small stream at slightly alkaline pH. *Polytrichum commune* had the lowest rates of activity. The authors suggested that this may due to its rudimentary cuticle and primitive vascular system; the latter may allow translocation of nutrients from the substratum. Unfortunately there are no data on surface phosphatase activities of liverworts.

Extracellular phosphatases

Most, but not all, of the 50 cyanobacteria screened by Whitton *et al.* (1991) showed extracellular phosphomonoesterase. Some green algae with marked surface phosphomonoesterase also release extracellular phosphomonoesterase (Gibson and Whitton, 1987b), but not the moss *Hydrogonium fontanum* (Al-Shehri, 1992). Several studies have shown slight differences between the surface and extracellular phosphomonoesterase activity of a particular organism. It is uncertain whether this reflects a true difference between the two, or merely differences in kinetic properties between bound enzyme and enzyme in true solution (Engasser and Horvath, 1975; Thiébart-Fassey and Hervagault, 1993). However, none of these studies showed the presence of extracellular phosphodiesterase, even when there was marked surface activity, nor have any of a wide range of other cyanobacteria and filamentous green algae tested (B.A. Whitton, unpublished data). The development of suitable staining techniques is needed in order to establish whether or not phosphodiesterase can be immobilized in mucilage external to the main part of the cell wall, even if it is not present in solution.

Substrate Concentration

Phosphatase assays reported in the literature have been conducted using markedly different substrate concentrations, making it hard to compare results. However, experiments fall into broadly three types. Some have used a concentration sufficient to saturate enzyme activity, which is likely to be several orders of magnitude higher than that typically occurring in nature. This approach has been used in the majority of studies on seaweeds (Hernández *et al.*, 2003) and also for some moss studies (e.g. Press and Lee, 1983; Turner *et al.*, 2001). Other studies have used concentrations more like those in nature, although conditions such as those found in some streams (around 1 μg P/l; Livingstone and Whitton, 1984) are hard to simulate in the laboratory. The majority of

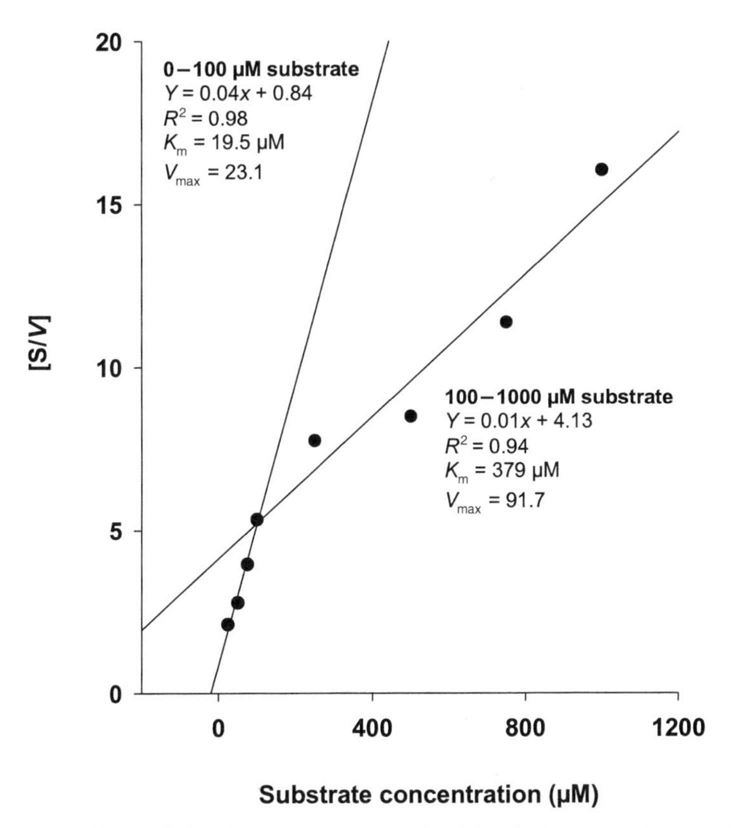

Fig. 10.2. Hanes–Woolf plot of phosphomonoesterase activity of *Fontinalis antipyretica* from Cranecleugh Burn, UK; showing two rates of reaction at 'low' (0–100 μM) and 'high' (100–1000 μM) substrate concentrations (where the *Y* axis [*S*/*V*] is substrate concentration (μM) divided by the velocity of the reaction (μmol *para*-nitrophenol (*p*NP) per g dry weight/h)). Units of V_{max} are μmol *p*NP/g dry weight/h. Reproduced from Turner *et al.* (2001).

studies on freshwater organisms have used a concentration somewhere between the two, sufficient to make it easy to detect change spectrophotometrically or fluorimetrically, but not enough to risk toxicity, at least during short-term assays.

The majority of kinetic studies on the effect of substrate concentration on phosphatase activities have shown a typical Michaelis–Menten response. However, several examples have been reported where the organisms clearly show the presence of low- and high-affinity systems. These include the red algae *Corallina elongata* (Hernández *et al.*, 1996b) and two species of *Gelidium* (Hernández *et al.*, 1995), and the aquatic mosses *Fontinalis antipyretica* and *Rhynchostegium riparioides* (Turner *et al.*, 2001; Fig. 10.2). The high affinity for the substrate found at low substrate concentrations may represent the more realistic kinetics of the enzyme, because 'low' substrate concentrations are typically found in the environment (Turner *et al.*, 2001). The authors suggested possible explanations for the severe reduction in affinity found in these two mosses at high substrate concentrations. There may be two enzyme systems, with the low-affinity system activated in the presence of high substrate concentration to prevent physiological damage or inorganic phosphate saturation. Other possibilities are that this phenomenon may result from restricted access in the cell wall, competitive product inhibition by inorganic phosphate, or some form of negative cooperativity. Two-phase kinetics were not apparent in any of the terrestrial mosses tested by Turner *et al.* (2001). However, it is possible that a low-affinity system exists in these mosses, but is activated at much greater phosphorus concentrations than were used in that study. An alternative hypothesis, possibly more likely, is that a much higher affinity system exists in these mosses, detectable only at very low substrate concentrations.

It is only safe to compare the results of studies on different organisms done at different assay concentrations if it is assumed that enzyme activity responds to environmental factors in the same way over the whole concentration range. However, it has been shown for a mammalian tissue (Fedde and Whyte, 1990) that this is not so, with the pH optimum for phosphomonoesterase (assayed with *para*-nitrophenyl phosphate) being two pH units lower using a low than a high substrate concentration. A similar effect was reported for the freshwater diatom *Synedra acus* (Hantke and Melzer, 1993). Of 16 phototrophs assayed by B.A. Whitton and A. Donaldson (unpublished data) using 1 and 250 μM methylumbelliferyl phosphate, four (three *Calothrix* strains and *Stigeoclonium* D565) showed a pH optimum between 1.5 and 2 units lower at the lower substrate concentration, this value being much closer to that of the environment from which the organism had been isolated originally. The reverse effect occurred with the aquatic mosses *Fontinalis antipyretica* and *Rhynchostegium riparioides* (Ellwood, 2002) and the semi-aquatic mosses *Palustriella commutata* var. *falcata* and *P. commutata* var. *commutata* (B.L. Turner, unpublished data), where the pH optimum for phosphomonoesterase was in the range 5.5–5.0 when assayed with 100 μM substrate, but nearer the typical field pH values of 7.0 or greater when assayed with 1 μM substrate. In contrast to phosphomonoesterase activity, there is no effect of substrate concentration on the pH optimum for phosphodiesterase activity of these mosses.

It is unclear why substrate concentration sometimes has a marked effect on the pH optimum of phosphomonoesterase activity. Among the possible reasons, not necessarily the same for each organism, is the modification of a single enzyme by changes in pH, or there being more than one enzyme involved, each with a different pH optimum. It would be interesting to know whether those organisms where the pH optimum of phosphomonoesterase shows a marked response to substrate concentration are also those showing a marked response to light (see below). Although the effects of pH were not included in a study by Rivkin and Swift (1980), characterization of phosphomonoesterase of the marine dinoflagellate *Pyrocystis noctiluca* showed three different K_m values at substrate concentrations over the range 0.1–22 μM. Another less likely possibility is that it is not a direct effect on the enzyme, but rather on activity associated with removal of phosphate subsequent to hydrolysis. It is also unclear what effect the relatively high concentration of buffer required may have on the active site of the enzyme (see below).

Kinetic studies have been made on a number of strains or field populations of a single species, treating the whole organism in a similar way to that used in measurements of Michaelis–Menten constants for pure enzymes. Such measurements of K_m and V_{max} provide a useful means of comparing strains and considering what role their phosphatase plays in the field, although in at least one case, that of a *Chroococcidiopsis* (cyanobacterium) community in an Antarctic desert rock, the kinetic constants were not the same in crushed rock and a laboratory isolate of the dominant phototroph (Banerjee *et al.*, 2000a). Ideally, such kinetic studies would be repeated over the full range of pH and other environmental conditions to which the species is subjected in nature.

Environmental Factors

These have been considered in detail with respect to seaweeds (Hernández *et al.*, 2003), so only key aspects are discussed here.

Temperature

The effects of temperature include the effect on stability of the enzyme, the effect on the velocity of breakdown of the organic phosphate and the effect on the enzyme–substrate affinity (Hernández *et al.*, 2003). The various processes do not necessarily respond to temperature in the same way, as shown in a study of the marine red alga *Porphyra umbilicalis*, where the relationship between surface phosphomonoesterase activity and total cellular phosphorus differed according to the assay temperature (Hernández, 1996). The optimum temperature for phosphomonoesterase activity of benthic marine seaweeds is usually higher than the temperatures to which the organism is typically exposed in nature, often with values ranging from 25 to >30°C (see, e.g. Hernández *et al.*, 1996b). The optima for phosphomonoesterase and phosphodiesterase activities of *Nostoc commune* UTEX 584 assayed at pH 7.6 differed, being 32 and 42°C, respectively (Whitton *et al.*, 1990).

Light

It might be expected that the activity of an enzyme external to the cytoplasmic membrane would not be closely linked to photon irradiance, at least in the short term, which probably explains why light conditions have not been controlled or even considered in many studies. Where the effect of light has been considered, some studies, such as for the cyanobacterium *Rivularia atra* (Yelloly and Whitton, 1996) and the red alga *Porphyra umbilicalis* (Hernández *et al.*, 1992), showed no effect of light on phosphomonoesterase activity. However, comparisons of light and dark conditions showed both positive (e.g. *Gelidium sesquipedale*) and negative (e.g. *Corallina elongata*) effects on phosphomonoesterase activity of red algae (Hernández *et al.*, 1996b). Samples of Antarctic dry desert rock containing *Chroococcidiopsis* showed lower phosphomonoesterase activity in the light than the dark, even when irradiance for the former was as low as 5 μmol photon/m^2/s (Banerjee *et al.*, 2000b). The mosses *Fontinalis squamosa*, *Rhynchostegium riparioides* and *Warnstorfia fluitans* showed less activity in the presence of quite low irradiance (around 15 μmol photon/m^2/s; Ellwood, 2002) than in the dark; further increases in irradiance apparently caused no further reduction in activity for these mosses. However, the difference in response between light and dark was influenced by whether the moss was kept overnight in the light or the dark prior to the experiment.

Klotz (1985) reported a particularly marked effect of light intensity on phosphomonoesterase activity of the green alga *Selenastrum capricornutum* in dialysis chambers incubated at stream sites in New York State during the 1982 growing season. The populations grown at low light showed much higher activity that those at high light. In two cases the comparison was made only a short distance apart in the same stream, so the chambers were presumably subject to the same ambient nutrient regime. In the stream where the value for irradiance in the shade was only about 1% of that in full light, phosphomonoesterase activity was 15 times higher in the former. The author suggested that the effect might be due to the more efficient uptake of phosphate at high irradiance. However, it is hard to see how the lower rate of phosphate uptake at low light would lead to such high phosphomonoesterase activity, unless light intensity had much less effect on uptake of nitrogen sources, which would lead to the low light populations having a higher intracellular nitrogen-to-phosphorus ratio. In view of the importance of *Selenastrum capricornutum* as the standard assay organism in many environmental studies, it is surprising that this problem has apparently still not been resolved.

Light quality was also shown to influence extracellular and total cellular activity (intracellular plus surface activity) in four marine phytoplankton species (Wynne and Rhee, 1988), with activity being higher under blue light than other parts of the spectrum, including white light of the same intensity. In the case of the diatom *Phaeodactylum tricornutum*, extracellular activity was higher under limiting light intensity, whereas total cellular activity was higher at saturation intensities. It is worth considering possible explanations for these results. Espeland and Wetzel (2001) found that the switch between light and dark in a biofilm caused changes in pH, which had direct effects on the surface enzyme activity. There are several reports of photochemical degradation of phosphatases in bacterial populations or communities (Herndl *et al.*, 1993; Müller-Niklas *et al.*, 1995), while Garde and Gustavson (1999) reported that ultraviolet-B inhibition of phosphomonoesterase activity enhanced phosphorus limitation in Norwegian coastal waters. In addition, ultraviolet irradiation can degrade organic phosphate complexes in nature, perhaps enhancing the availability of phosphate (Francko and Heath, 1982). It is especially difficult to assess the effects of ultraviolet irradiation, because vessels used for field assays are likely to provide a partial screen, while most laboratory light sources have a very low ultraviolet component.

Even if there is no direct effect of light on the enzyme, light is likely to enhance uptake of phosphate into the cell, enhancing the removal of phosphate from the vicinity of the enzyme and thus reducing possible substrate inhibition. This suggests that where light had a positive effect on phosphatase activity, the effect may only have occurred at higher substrate concentrations. Studies testing the effects of a range of substrate concentrations and the use of phosphorylation uncouplers should be able to resolve this. Possible reasons for the converse effect of light are discussed by Hernández *et al.* (2003), who stress the likely influence of length of incubation period. In contrast to phosphomonoesterase, light had no effect on phosphodiesterase activity of *Fontinalis antipyretica*, *F. squamosa* or *Rhynchostegium riparioides* (Ellwood, 2002).

pH and buffers

Most studies have distinguished between acid phosphatases (EC 3.1.3.2) and alkaline phosphatases (EC 3.1.3.1). Although the terms can be useful, they have encouraged some authors to treat all activity in one part of the pH range as due to one enzyme. In general, authors seem to have assumed that a pH value of 7 separates environments where acid phosphatases prevail from ones where alkaline phosphatases do so. However, Nedoma *et al.* (2003) take pH 6 as the boundary, whereas Chróst and Siuda (2002) state that alkaline phosphatases react optimally in the range 7.6–9.6. Hoppe (2003) gives the optimum pH for acid and alkaline phosphatases as 4–6 and 8.3–9.5, respectively. Reports on alkaline phosphatase activity apparently always treat this as synonymous with alkaline phosphomonoesterase activity (e.g. Feuillade *et al.*, 1990), ignoring any role of other phosphohydrolases. However, as described above, the ability to hydrolyse forms of organic phosphate other than phosphate monoesters is widespread and different activities do not necessarily respond to pH in the same way. For instance, phosphomonoesterase and phosphodiesterase of *Nostoc commune* UTEX 584 assayed using 250 μM substrate were shown (Whitton *et al.*, 1990) to differ slightly in their pH optimum (7.0 vs. 7.8). Even where only phosphomonoesterase is being considered, it is seldom clear whether or not more than one enzyme is involved. For all these reasons we discourage use of the term alkaline phosphatase activity, although we have sometimes retained it here when used by other authors.

Further difficulty in use of the terms acid and alkaline phosphatase activity arises from the uncertainty about the pH conditions under which the enzyme(s) functions in nature. The surface phosphomonoesterase activity of both freshwater and marine algae has often been reported to show an optimum pH well above the values at which the organisms typically grow in nature. However, as mentioned above, this may sometimes be an artefact resulting from the use of high substrate concentrations, which may explain some of the extreme values reported, such as pH 12.2 for *Calothrix vigueiri* isolated from a mangrove root (Grainger *et al.*, 1989). Most values for pH optima reported by Hernández *et al.* (1996b) were in the range 8.7–9.0; values normally only found in shallow pools of the upper littoral zone (Hernández *et al.*, 2003) where high pH values are often reached (Larsson *et al.*, 1997). Assays of phosphomonoesterase activity of intertidal seaweeds, such as *Fucus spiralis* (Hernández *et al.*, 1997) at typical seawater pH (around 8.3) mostly show values 70–80% of those at the optimum pH (Hernández *et al.*, 2003). The choice of substrate can sometimes also influence the results, as shown by Štrojsová *et al.* (2003), who found that the maximum phosphomonoesterase activity of phytoplankton samples was between pH 10 and 11 when assayed with methylumbelliferyl phosphate, but pH 8 when assessed using enzyme-labelled fluorescence substrate.

Some mosses show the converse effect to algae, with the pH optimum of phosphomonoesterase being about 1.5 pH units less than the typical pH of the environment, with a pH optimum of 5.5 for *Fontinalis antipyretica* (Christmas and Whitton, 1998b) and *Rhynchostegium riparioides* (Ellwood, 2002) and 6.0 for *Hydrogonium fontanum* (B.A. Whitton, A.M. Al-Shehri and A. Donaldson, unpublished data). In the last case there was no difference between assays conducted using 1 or 250 μM methylumbelliferyl phosphate substrate. As the field pH was typically above pH 7.0 in all three cases, this shows how confusing it would be to report the enzymes involved as acid phosphatases.

Hoppe (2003) stressed the importance of considering possible effects of buffers on phosphatases when designing experiments in the field and in the laboratory, especially for freshwaters. The higher buffer concentration used in assays with high substrate concentration may be a factor contributing to the differences found in some cases (see above). It is known that the pH of the charged surface of maize root cells can be lower than the ambient bulk flow (Sentenac and Grignon, 1985), but the use of a high buffer concentration during phosphatase assays would make such a difference less likely. Several authors have indicated that the choice of buffer for a particular pH value may also influence the results, though Al-Shehri (1992) found close agreement between different pairs of buffers (all at 50 mM) over the pH range 3–10 on phosphomonoesterase and phosphodiesterase activities of *Hydrogonium fontanum*. Following comparisons of several buffers, Whitton *et al.* (1999) recommended 3,3-dimethylglutaric acid for routine use in the pH range 5.0–6.5 and glycine for the range 8.5–9.5.

Metals

There are several reports of enhanced phosphomonoesterase activity in response to increased calcium. Extracellular phosphomonoesterase activity of *Calothrix parietina* rose when calcium in the medium was increased from 0.1 to 1 mM, whereas surface phosphomonoesterase activity did not (Grainger *et al.*, 1989). However, surface phosphomonoesterase and phosphodiesterase activities of *Nostoc commune* UTEX 584 both increased when calcium in the assay medium was increased from 1 to 10 mM, while 10 mM magnesium was slightly inhibitory (Whitton *et al.*, 1990); sodium and potassium had little effect over the range 0.001–10 mM. An increase in zinc from 0.001 to 0.1 mM halved surface phosphomonoesterase and phosphodiesterase activities of *N. commune*.

Durrieu *et al.* (2003) investigated the effects of chromium, nickel, copper, zinc, cadmium, mercury and lead on surface phosphomonoesterase activity of *Chlorella vulgaris*. In all cases V_{max} decreased markedly with increasing metal concentration, with values only 5% of the control at 1.26 mg/l cadmium or mercury. The sequence of toxicity towards the inhibition of V_{max} was approximately the same for each concentration studied. The situation for K_m was less clear-cut. The value was almost unchanged in response to changes in cadmium or mercury concentration, indicating that these metals can be considered as non-competitive inhibitors. However, in the case of chromium, nickel, copper, zinc and lead the results are more complicated, with K_m

apparently increasing at only the highest concentrations of the metals, suggesting the possibility of complex interactions between the metals and phosphomonoesterase. The effects of the metals on phosphomonoesterase activity were apparent only on live cells; they had no effect on purified phosphatase.

The effects of zinc on phosphomonoesterase activity of two lichens belonging to the genus *Peltigera*, one with a cyanobacterial symbiont and the other with a green alga, were investigated by Stevenson (1994). Populations of both lichens were taken from a zinc-rich mine site and a site uncontaminated by zinc. Laboratory assays required 10 M zinc to reduce phosphomonoesterase activity of the high-zinc populations by about one-quarter, whereas only 0.1 M zinc was required to have a similar effect on the low-zinc populations. This suggests that the phosphomonoesterases of these lichens had evolved resistance to zinc, though the relative contributions of the phototroph and fungus to this effect are not known.

An ability to use organic phosphate effectively is likely to be important for phototrophs in many heavy-metal-rich streams, where the phosphate salts of the relevant metals are highly insoluble. In a survey of phosphomonoesterase activity of filamentous algal populations at 14 stream sites in an old mining region in northern England, Bellos (1990) found a highly significant positive relationship between phosphomonoesterase activity and zinc in the water and a significant negative relationship between phosphomonoesterase activity and total filtrable phosphate in the water. The high aluminium concentration in acidic mine drainages, which is known to lead to phosphorus precipitation (Gross, 2000), was suggested to be an important factor influencing the high surface phosphomonoesterase activity of algal mats in the Rio Tinto, Spain (Sabater *et al.*, 2003). Elevated aluminium was also shown to lead to high phosphomonoesterase activity in *Chlamydomonas reinhardtii* (Joseph *et al.*, 1994).

An increase in the chelator EDTA (ethylenediaminetetraacetate) in the assay medium of *Nostoc commune* from 1 to 10 mM reduced phosphomonoesterase and phosphodiesterase activities to 35% and 13% of maximum activity, respectively (Whitton *et al.*, 1990). About one-third of the inhibitory effects of EDTA or zinc were reversed if the filaments were subsequently incubated in standard medium. The presence of EDTA in the assay medium had a much more pronounced effect on phosphomonoesterase activity of *Calothrix parietina*, with 1 mM EDTA being sufficient to reduce the activity below the detection rate (Grainger *et al.*, 1989). However, washing cells with EDTA prior to an assay under standard conditions had much less effect, with a 20 mM EDTA wash causing only 23% reduction in activity.

Polyphenolics

The relatively recalcitrant dissolved polyphenolic compounds resulting from breakdown of higher plants (fulvic and humic acids) complex with many bacterial and algal enzymes, but particularly phosphatases (Wetzel, 1992). The formation of such complexes inactivates phosphomonoesterase (Boavida and Wetzel, 1998), which is inhibited both competitively and non-competitively (Wetzel, 1992). Phosphorus-limited cells therefore need to expend more energy on phosphatase synthesis, and enhanced phosphomonoesterase activity has been reported for waters with increased humic materials (Stewart and Wetzel, 1992b). Wetzel (1992) pointed out these results with frequent observations (e.g. Jones, 1990) that the primary production in humic-rich waters is consistently lower than in clear waters with comparable loadings and light availability.

Competitive inhibition by inorganic and organic phosphate

Phosphomonoesterase activity is inhibited competitively by phosphate, one of the products of hydrolysis. Such inhibition has been investigated with respect to phosphate concentrations in the external environment, both for water samples and for individual species. Phosphomonoesterase activity in the Plussee, Germany, was inhibited by 15 μg P/l (Chróst, 1991). Over 80% of phosphomonoesterase activity in Polish lake water samples was inhibited when the inhibitor-to-substrate ratio was higher than 2.5 (Siuda and Güde, 1994). In the lakes chosen for study, this ratio was mostly in the range 3.3–29.1, only occasionally dropping to >1 during periods of maximum phosphorus depletion. However, the interpretation of results is complicated by the finding of Štrojsová *et al.* (2003) that the extent of inhibition of enzyme activity can be influenced markedly by the substrate. The phosphate concentration causing 50% inhibition was about 1 μM with ELFP as substrate, but 30 μM with *para*-nitrophenyl phosphate.

Competitive inhibition has also been investigated in the laboratory. In the case of *Calothrix parietina* (Grainger *et al.*, 1989), the addition of phosphate had a marked inhibitory effect on surface and extracellular phosphomonoesterase activity, with 10% inhibition at a phosphate concentration of 10 μM (~0.3 mg P/l) and 80% inhibition at 100 μM. Competitive inhibition of phosphomonoesterase has also been reported for cultures of the marine dinoflagellate *Prorocentrum micans* (Uchida, 1992) and extracts of *Ulva lactuca* (Lee, 2000). Alkaline phosphatase of *Ulva* was more sensitive than acid phosphatase, the phosphate concentrations required to reduce activity by 50% being 30 and 50 μM inorganic phosphate, respectively. Little is known about the inorganic phosphate concentrations likely to accumulate in the immediate vicinity of phosphomonoesterases should hydrolysis proceed faster than removal of phosphate by active uptake into the cell.

Factors Influencing Phosphatase Synthesis and Loss

Ambient versus internal phosphate

Several studies on lake phytoplankton have reported that high surface phosphomonoesterase activity occurs only when concentrations of ambient phosphate are very low (e.g. Pettersson, 1980; Chróst and Overbeck, 1987; Jansson *et al.*, 1988), but it is difficult to distinguish between the short-term effects of inhibition by phosphate or longer-term effects due to repression of phosphomonoesterase synthesis or the degradation of phosphomonoesterase already present. However, in the case of planktonic bacteria, the evidence (see Heath, Chapter 9, this volume) does indicate that ambient phosphate concentration has an important influence on phosphomonoesterase synthesis. Unfortunately no studies have been made which would provide unambiguous results for phototrophs of similar size to the heterotrophic bacteria. The phototrophs described below are all organisms several to many orders of magnitude larger than typical heterotrophic bacteria. In addition, there are apparently no reports for phototrophs of particular surface phosphatases being synthesized in response to the presence of the relevant substrate, so the situation remains unclear as to whether or not the enzymes needed to hydrolyse the various organic phosphates used for growth are merely a response to marked phosphorus limitation. However, Whitton *et al.* (1991) suggested that in a few cases their data would fit with the enzyme being synthesized in direct response to the presence of a substrate.

Influence of internal phosphate

An inverse relationship between phosphomonoesterase activity and internal phosphorus concentration has been shown experimentally in axenic culture for cyanobacteria such as *Calothrix parietina* (Grainger *et al.*, 1989) and *Nostoc commune* (Whitton *et al.*, 1990), green algae such as *Draparnaldia* and *Stigeoclonium* (Gibson and Whitton, 1987b), and the moss *Hydrogonium fontanum* (Al-Shehri, 1992). As the organisms were grown under standard conditions in batch culture, and assays for activity were also conducted under standard conditions, it is assumed that differences in activity largely reflect differences in enzyme

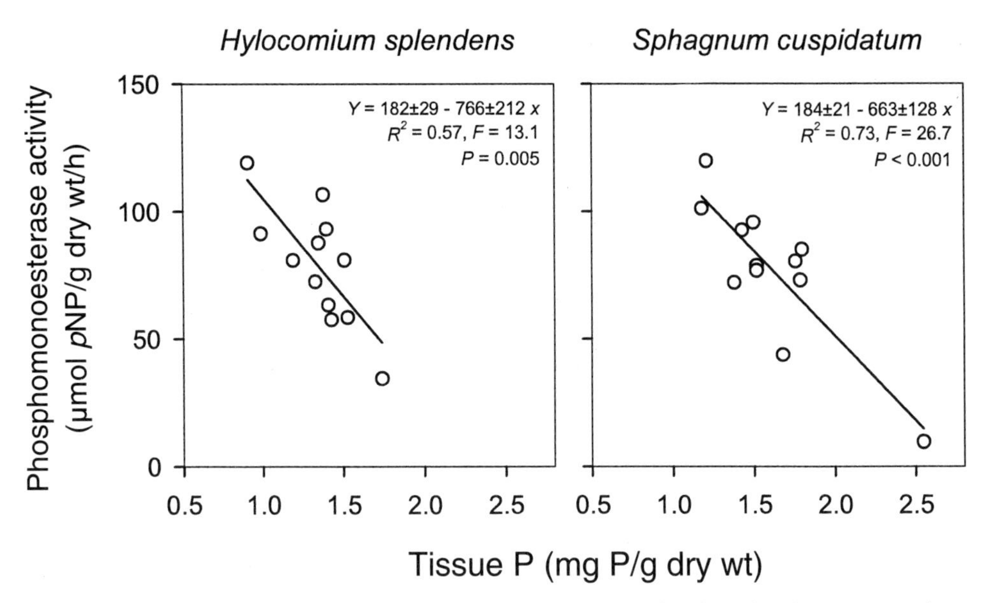

Fig. 10.3. The relationship between tissue phosphorus concentration and surface phosphatase activity for two mosses growing in Upper Teesdale, northern England. Data are from mosses sampled during an annual cycle and activity is expressed as µmol *para*-nitrophenol (*p*NP) per g dry weight/h. Data taken from Turner *et al.* (2003a).

content. In the case of the green alga *Ulva lactuca*, Lee (2000) found an inverse relation between acid phosphatase activity, but not alkaline phosphatase activity, and internal phosphorus. Both activities were reported as intracellular, but the methods indicate that the values included surface activity as well, so the relationship between phosphomonoesterase activity and internal phosphorus is uncertain here. An inverse relationship between phosphomonoesterase activity and internal phosphorus concentration has also been shown by comparing different field populations of a particular species (e.g. *Stigeoclonium tenue*; Gibson and Whitton, 1987a) and 2-cm apices of moss shoots in aquatic (*Fontinalis antipyretica* and *Rhynchostegium riparioides*; Christmas and Whitton, 1998a) and terrestrial environments (*Hylocomium splendens, Palustriella commutata* var. *commutata, Polytrichum commune* and *Sphagnum cuspidatum*; Turner *et al.*, 2003a; Fig. 10.3). Press and Lee (1983) obtained similar results using whole plants of several *Sphagnum* species.

There have been a number of investigations to establish whether there is a threshold value for phosphorus concentration below which phosphatase activities develop. *Trichodesmium* strain WH9601 showed slight phosphomonoesterase activity even under phosphorus-replete conditions (Stihl *et al.*, 2001; see below). However, surface phosphomonoesterase and phosphodiesterase activities of *Nostoc commune* UTEX 584 were not detectable when the internal phosphorus concentration was high, and only became detectable when the phosphorus concentration of the culture fell below 0.76% dry weight (Whitton *et al.*, 1990). A deep-water rice-field *Calothrix* strain showed a similar response, with phosphomonoesterase and phosphodiesterase becoming detectable when the phosphorus concentration of the culture fell below 0.95% dry weight (Islam and Whitton, 1992). However, a few hormogonia, motile filaments that only develop in the absence of phosphorus limitation, continued to form until the phosphorus concentration fell to 0.55% dry weight. This reflects the heterogeneity of the developmental stage of individual filaments characteristic of *Calothrix* and is presumably part of its strategy to deal with highly variable concentrations of inorganic and organic phosphate in its natural environment. As measurements were of the phosphorus concentration in samples of the whole culture, the phosphorus concentration of individual *Calothrix* filaments developing phosphomonoesterase activity would have been less than the mean for the population. Kumar *et al.* (1991) found that a deepwater rice-field *Anabaena* strain developed surface phosphomonoesterase and phosphodiesterase activities when the phosphorus concentration fell to 0.46% cell protein, a value probably quite similar to those above when expressed as dry weight. As the phosphorus concentration above which intracellular granules of polyphosphate start to form in *Nostoc* and *Calothrix* is typically also about 0.7–0.8% dry weight (B.A. Whitton, unpublished data), the possibility should be considered that polyphosphate formation and repression of phosphatase synthesis are linked in some way. However, Huber and Hamel (1985), when assessing the results of a study on the cyanobacterium *Nodularia spumigena*, suggested that the relationship to stored polyphosphate is indirect and that phosphatase activities probably relate directly to small changes in cellular phosphate concentration.

Some eukaryotes are similar to *Nostoc* and *Calothrix* in that surface phosphomonoesterase falls to negligible levels under phosphorus-rich conditions, as found with four marine dinoflagellates, *Amphidinium*, *Ceratium*, *Prorocentrum* and *Scrippsiella* (Sakshaug *et al.*, 1984) and the moss *Warnstorfia fluitans* (Ellwood *et al.*, 2002). In contrast, field populations of *Fontinalis antipyretica* showed slight phosphomonoesterase activity, but not phosphodiesterase activity, even when the phosphorus concentration in both environment and moss were high (Christmas and Whitton, 1998b). The occurrence of constitutive phosphatases (mostly intracellular) in microalgae was reviewed by Cembella *et al.* (1984).

Phosphodiesterase activity developed in response to increasing activity phosphorus limitation in *Nostoc commune* UTEX 584 (Whitton *et al.*, 1990) and *Warnstorfia fluitans* (Ellwood *et al.*, 2002) at about the same level of phosphorus limitation as for phosphomonoesterase activity. However, phosphodiesterase activity of an axenic culture of *Hydrogonium fontanum* developed in response to greater phosphorus limitation than phosphomonoesterase activity (phosphorus = 1.15% and 0.45% dry weight, respectively; Al-Shehri, 1992). A field population of *Fontinalis antipyretica* also required greater phosphorus limitation for phosphodiesterase activity than for phosphomonoesterase activity (Christmas and Whitton, 1998b). This applied even if the contribution of phosphomonoesterase persistent under high phosphorus conditions was removed from consideration. This led to marked differences in the phosphodiesterase-to-phosphomonoesterase ratio, with phosphodiesterase activity becoming relatively more important the more phosphorus-limited the organism was. As increasing phosphorus limitation was correlated with increasing humic concentration in the waters studied (and also increasing altitude), the authors pointed out the importance of the moss being able to utilize phosphate diesters effectively under these conditions.

Nitrogen-to-phosphorus ratio

The nitrogen-to-phosphorus ratio in tissue sometimes gives a better indication of phosphorus limitation than the absolute concentration, as suggested for the diatom *Skeletonema costatum* in Trondheimsfjord (Myklestad and Sakshaug, 1983). Phosphomonoesterase activity of a phytoplankton community dominated by cyanobacteria in Lake Nantua, France, increased when the nitrogen-to-phosphorus concentration in the community (particulate fraction) was close to or above the Redfield (molar) ratio of 16:1 (Feuillade *et al.*, 1990). Phosphomonoesterase and phosphodiesterase activities increased when the nitrogen-to-phosphorus ratio of 2-cm apices exceeded 9 (by mass) in *Fontinalis antipyretica* and *Rhynchostegium riparioides* (Christmas and Whitton, 1998a), or 10 in *Warnstorfia fluitans*

(Ellwood *et al.*, 2002), although in these cases there were no data to suggest that the ratio was a better indicator than absolute phosphorus concentration.

It seems likely that the quantitative relationship between phosphomonoesterase synthesis and internal composition may also be influenced by the phosphorus demand of the organism at the time of study. This may explain why phosphomonoesterase and phosphodiesterase activities of a deepwater rice-field *Calothrix* were higher when grown in the presence of combined nitrogen than under the more typical N_2-fixing conditions (Islam and Whitton, 1992). The influence of phosphorus demand also needs to be considered when only part of an organism is being studied, if the organism is one capable of transporting phosphate between different regions, as occurs in at least some larger seaweeds and mosses. For mosses this can occur during periods of high physiological activity, when the phosphorus demand cannot be met even if ambient phosphate is relatively high. Such periods might occur during rapid shoot growth (Bates, 1994) or at sporogenesis, with translocation from the gametophyte taking place via the conducting tissues of the seta (Bates, 2000). There is evidence from stable isotope studies that nitrogen is effectively translocated about the moss *Hylocomium splendens* growing in the Scandinavian sub-Arctic (Eckstein and Karlsson, 1999).

Severe phosphorus limitation

Several studies show that further changes may take place when organisms are highly phosphorus-limited, so the situation in nature is likely to be complex. In batch culture of *Nostoc commune*, following the period of phosphomonoesterase and phosphodiesterase synthesis subsequent to the phosphorus concentration falling below 0.76% dry weight, the activity of both enzymes decreased again when the value fell below 0.3% dry weight (Whitton *et al.*, 1990). In the case of several, but not all, strains of *Calothrix*, cultures that are sufficiently phosphorus-limited for most of the filaments to appear unhealthy sometimes start to form a few healthy hormogonia (B.A. Whitton, unpublished data). This appears to be a different response from the heterogeneity of *Calothrix* cultures to that described above for a deepwater rice-field strain, because it is probably a consequence of some filaments being sufficiently unhealthy as to cause lysis. It is unclear whether the organic phosphates released are hydrolysed by intracellular phosphatases released at the same time or by surface phosphatases of filaments that have not lysed.

Persistence of phosphatases

There is some evidence that phosphomonoesterase can persist for long periods in the absence of biological activity. For instance, particle-free samples of water from the Red Sea still retained 50% initial phosphomonoesterase activity after a 3–6-week incubation period (Li *et al.*, 1998). Preserved algal samples sometimes retain obvious phosphomonoesterase activity many months after storage, while dried samples of cyanobacteria such as *Nostoc commune* and *Calothrix parietina* subject to alternate wetting and drying in the field retain much of their original activity when dried and rewetted 2 years later; limestone rock samples from Aldabra Atoll in the Indian Ocean with abundant cyanobacterial growths showed marked activity 25 years after sampling (A. Donaldson and B.A. Whitton, unpublished data).

There is less information about how long surface phosphatases persist on a metabolizing organism when conditions no longer favour the presence of the enzymes. This applies both when ambient phosphate is sufficient that the organisms are no longer subject to phosphorus limitation and when conditions become so severe that there is breakdown of phosphatases already synthesized. There are no data on the synthesis and replacement of surface phosphatase molecules under steady-state conditions.

Based on a number of field studies,

including some transplant experiments, Whitton *et al.* (1999) suggested likely times for the phosphomonoesterase activity of various phosphorus-limited organisms in spring–summer in the UK to show a significant decrease after they became exposed to a phosphorus-rich experiment. The values are: *Cladophora glomerata*, 1 day; upstream populations of *Stigeoclonium*, 1 day; downstream populations of *Stigeoclonium*, 0.5 day; *Fontinalis antipyretica*, 7 days; *Rhynchostegium riparioides*, 4 days. These values are likely to be modified considerably by environmental conditions.

Hair Formation

The formation of multicellular hairs in some cyanobacteria and green and brown algae provides a striking morphological and physiological parallel between at least three different groups of organisms, suggesting an important evolutionary advantage for the structure. Multicellular hairs are long tapered ends of filaments, which differentiate from typical vegetative cells, and where the cells usually get progressively longer as they narrow. Often, but by no means always, the distal parts of hairs become colourless, with loss of many cell components such as chlorophyll and, at least in the cyanobacterium *Calothrix*, DNA. They become highly vacuolate, especially in the cyanobacteria. In most taxa with hairs, they differentiate at only one end of the filament, but in a few cases they do so at both ends. The earlier literature was reviewed by Whitton (1988), so only an outline is given here.

The evidence indicates that provision of a large area of wall with high surface phosphatase activity is one, if not the most important, role for these hairs. In all cases where an organism able to develop multicellular hairs was subjected to phosphorus limitation, hairs were formed, though in some taxa they were shown experimentally to form in response to one or several other element limitations; such hairs are smaller than those formed under phosphorus limitation and often look slightly different morphologically. Where hair formation occurs in response to phosphorus limitation, the data all show that the organisms develop high surface phosphomonoesterase (and almost always also phosphodiesterase) activity. The hairs are an important site for phosphomonoesterase in these organisms, though only occasionally the sole site, as in *Calothrix vigueiri* (Mahasneh *et al.*, 1990). Greater phosphomonoesterase activity on the hairs than other parts of the thallus seems to occur especially in hairs whose cells elongate and become colourless. In contrast, a Bangladesh deepwater rice-field *Calothrix* strain, in which the cells of the tapered region did not lose their chlorophyll under phosphorus limitation, showed similar azo-dye staining over the whole surface of the filament, apart from the heterocyst (Islam and Whitton, 1992).

Many cyanobacteria and eukaryotic algae with multicellular hairs form distinct colonies, with the individual filaments embedded in mucilage. Where the surface of the colony is firm, the hairs sometimes extend far beyond this into the surrounding water, whereas in other cases the hairs reach to just below the surface of the colony proper. Presumably projection beyond the surface helps to optimize use of transient pulses of inorganic and organic phosphate. However, in Hunter's Hot Spring, Oregon, it also permits an ostracod population to graze these hairs (R.W. Castenholz, personal communication).

Some cyanobacteria which form hairs in response to phosphorus limitation form less well-developed hairs in response to iron limitation (Sinclair and Whitton, 1977; Douglas *et al.*, 1986), but such hairs do not show phosphomonoesterase activity. Similarly, some green algae which form hairs in response to phosphorus limitation (Whitton and Harding, 1978) form less well-developed hairs in response to nitrogen limitation and occasionally other limitations, but these hairs also do not develop phosphomonoesterase activity (Gibson and Whitton, 1987b; Whitton, 1988). However, no example of a field population of a cyanobacterium with hairs has been found which did not show surface phosphatase activity, and only one convincing example for a green alga, which was probably nitrogen-limited (Gibson and Whitton, 1987a). With the probable exception of larger seaweeds, the addition of the limiting nutrient to an organism with well-developed hairs leads to the formation of motile reproductive structures in the part of the organism away from the hair (Whitton, 1988, 1989).

The occurrence of cyanobacteria and eukaryotic algae with multicellular hairs has often been reported from environments with highly variable ambient phosphate and where phosphorus limitation becomes pronounced intermittently (Whitton, 1987a; Pentecost and Whitton, 2000; John *et al.*, 2002); under these conditions a number of hair-forming species often occur in the same community. In view of the fact that multicellular hairs apparently nearly always develop in nature in response to phosphorus limitation, and that high phosphorus is required for reproduction in species with these hairs, it seems probable that highly variable ambient phosphorus is an essential feature of the environment for the maintenance of populations of such phototrophs. Many of the morphological differences characterizing species of *Dichothrix* and *Rivularia* (cyanobacteria) and *Chaetophora* (green alga) are related to the periodicity of hair formation and reproduction, so perhaps particular species are characteristic of a particular type of phosphorus regime. This is supported by the fact that some organisms, which are taxonomically very different but have some morphological similarity, often occur at the same site (e.g. several *Dichothrix* spp. and *Chaetophora elegans*; B.A. Whitton, unpublished data). The diatom genus *Cymbella*, which does not possess hairs, is also especially frequent at sites where hair-forming organisms are abundant, suggesting that it is adapted to make effective use of a similar phosphorus regime.

In some cases, the presence of hairs and high phosphatase activity is the predominant condition over much of the year, such as *Rivularia* in UK streams (Whitton, 1987a; Whitton *et al.*, 1998), while in others, such as some *Stigeoclonium* populations, it is the less usual condition (Gibson and Whitton, 1987b; Whitton, 1988). The evidence from these various studies, when considered together with that of Turner *et al.* (2003b), indicates that sites dominated by *Rivularia* alternate between long periods of phosphorus limitation and shorter periods of potential nitrogen limitation assuaged by N_2 fixation (Livingstone *et al.*, 1984) at a time when there is a very high ambient concentration of filtrable organic phosphorus and inorganic phosphate, whereas sites dominated by *Stigeoclonium* do not have quite such a marked range of phosphorus concentrations and do not show such a marked period of nitrogen limitation as to permit out-competition by N_2-fixers. In the case of *Rivularia* and other members of the Rivulariaceae, the organisms may persist for long periods with only very slow increases in biomass and therefore they need to reduce the risk of grazing (Pentecost and Whitton, 2000). One possible means of doing this is to be toxic to invertebrate grazers, as shown for cyanobacterial mats including *Rivularia* in two calcareous streams in Spain (Aboal *et al.*, 2003).

The periodicity of *Rivularia biasolettiana* in upland streams in Upper Teesdale, northern England, is associated with the release of phosphorus, much of it organic, from peaty soils in early spring (Livingstone and Whitton, 1984; Turner *et al.*, 2003b; see also Turner, Chapter 12, this volume), while that of *R. atra* in the upper littoral zone at Tyne Sands, southern Scotland, is associated with storm events depositing seaweed on the supralittoral zone, and its subsequent breakdown and release of phosphorus, again much of it organic (Yelloly and Whitton, 1996). While the evidence suggests that hairs are an important site for obtaining and mobilizing organic phosphorus, little is known about their role with respect to inorganic phosphate. In locations where inorganic and organic phosphate both tend to occur in pulses (Livingstone and Whitton, 1984; Yelloly and Whitton, 1996), perhaps hairs are important in acquiring both sources of phosphorus.

Among the Rivulariaceae (i.e. cyanobacteria with highly tapered filaments), hair-forming strains were found (Whitton *et al.*, 1991) to be significantly less effective at utilizing ATP than non-hair forming strains and were in some cases killed by ATP supplied at 1 mg per litre (see above). It was suggested that the sensitivity of hair-forming strains fits with the theory that they are associated especially with environments exposed to pulses of organic phosphorus and that the filaments producing these hairs may be adapted to accumulate these compounds very rapidly. High concentrations of ATP might therefore accumulate inside cells, leading to toxic effects rather than being hydrolysed by periplasmic phosphatase. In contrast, Rivulariaceae were significantly more effective at utilizing phytic acid. Although *Aphanizomenon* is not included in the Rivulariaceae, it resembles them in that the ends of filaments become markedly tapered or almost like short hairs under phosphorus limitation (B.A. Whitton, unpublished data). Taken together with the observation of Reichardt (1971) that a strain of *Aphanizomenon* preferred phytin to phosphate as a phosphorus source, in contrast to seven other organisms tested, this suggests that, in at least some cases, the formation of highly tapered filaments or proper hairs provides a means of making effective use of phytic acid.

Among the green algae, unicellular hairs (*Bulbochaete*) and the cytoplasm-containing projections from the cells of some genera known as setae (e.g. *Coleochaete*) appear to behave in response to phosphorus limitation similarly to multicellular hairs (Whitton, 1988), although they have received much less study. Unicellular hairs are also widespread in the main group of red algae, the Florideae. Several freshwater red algae with pronounced hairs show marked phosphomonoesterase activity (e.g. *Batrachospermum*) and often occur in similar environments to algae with multicellular hairs, but these have not been studied experimentally. Phosphomonoesterase activity is known to be important on macroalgae on tropical reef ridges (Schaffelke, 2001). The hairs which are characteristic features of many reef-forming red algae probably play an important role in accessing organic phosphate in this very low-phosphorus environment and it will be important to consider phosphodiesterase and other types of phosphorylase activity as well as phosphomonoesterase activity in this environment (Hernández *et al.*, 2003).

A few branched yellow-green algae, a few species of the unbranched filamentous green alga *Oedogonium*, and the non-photosynthetic bacterial genus *Thiothrix* all form multicellular hairs, but no studies have been reported on the phosphorus status of these organisms and hair formation. It is hoped that this chapter will stimulate bacteriologists to investigate *Thiothrix*, because of its potential to be an environmental indicator in environments such as deep-ocean vents, which are difficult to study by routine means.

Measuring Activity in Field Samples

Methods

Because phosphatase activity is often a good indicator of phosphorus limitation, measurements of activity in field materials have been used for diverse purposes ranging from routine monitoring to attempts to understand the phosphorus dynamics of complex communities, such as periphyton in the Everglades (Newman *et al.*, 2003). Possible substrates for assays using spectrophotometry or fluorimetry have been mentioned above. A detailed account of the procedure for using ELFP and subsequent quantification of fluorescence associated with individual cells was given by Nedoma *et al.* (2003).

Several different approaches have been adopted with respect to choice of biological material and design of the subsequent assays, so caution is needed when considering the published results. Studies have been conducted on communities or individual species, or, in the case of large seaweeds, even small parts of an individual. Measurements have been quantified with respect to a range of metrics, including volume for the whole water column, surface area for mats and unit fresh weight, dry weight, chlorophyll or ATP for communities and individual species.

Studies on communities include contributions from bacteria and other organisms and, in the case of aquatic communities, activity present in the water. When samples from four Michigan lakes were partitioned (Stewart and Wetzel, 1982a), the maximum possible contribution of phytoplankton to alkaline phosphatase activity was <34% and in some cases was as low as 5–6%. The authors concluded that filtration techniques commonly used to assess the activity in lake waters may seriously overestimate the activity directly associated with the algal cells. Cotner and Heath (1988) found that the majority of surface acid phosphomonoesterase activity in an acidic bog lake was associated with material passing through a 0.45-μm filter and that only 21% was associated with the major phytoplankton alga, the dinoflagellate *Gonyostomum semen*. Even when cells are separated from particulate material, careful checks are still required to ascertain whether epiphytic bacteria are likely to make a significant contribution. Most studies on individual species have, however, included steps to minimize any contribution from other organisms. This is easier to do with mosses than with planktonic algae, because mosses can usually be washed vigorously without affecting their activity. However, bacteria are known to invade the cell walls of at least one liverwort and can be resistant to removal by washing (Satake and Shibata, 1986).

One approach is to conduct assays on materials in the field under ambient temperature and light conditions (Hernández *et al.*, 1993, 1997; Rott *et al.*, 2000). This has been done with samples of natural populations and also with communities developing on artificial surfaces incubated at the site. The advantages and disadvantages of the two approaches for phosphomonoesterase studies were assessed by Newman *et al.* (2003). A further possibility for some studies is to transfer rocks with attached growths between sites.

If the site is far from the laboratory, the substrate used for assay may be filter-sterilized for transport and some means of terminating the assay adopted which does not interfere with the hydrolysis product to be measured. Several marine studies on bacterial populations (e.g. Christian and Karl, 1995) have used mercuric chloride for the latter, while Nedoma *et al.* (2003) used this to preserve freshwater phytoplankton prior to quantifying the fluorescence of individual cells. Field studies have usually employed high concentrations of substrate (Hernández *et al.*, 1993, 1997) and this was recommended as standard procedure by Chróst and Siuda (2002). However, Hoppe (2003) pointed out that this should be avoided if a truly ecological experiment is wanted, such as when the estimated rates of phosphatase activity are to be integrated in calculations of the dynamics and the budget of phosphorus under natural conditions. Some early studies incubated samples for many hours, but most recent studies have used much shorter periods.

Another approach is take the material to a laboratory and conduct assays there. This has usually involved one set of experimental conditions, often at a temperature near the upper end of the range likely to occur in the field in order to speed measurements, for example 25°C by Štrojsová *et al.* (2003). Light intensity has less often been standardized; for instance, Štrojsová *et al.* (2003) merely report that daylight was used.

Monitoring

Phosphatase measurements have been employed in monitoring studies in two main ways: to assess the phosphorus status of field communities and populations and to provide a sensitive technique for evaluating the toxicity of agents such as heavy metals. The use of phosphatase measurements to assess the phosphorus status of algal field populations was first described by Fitzgerald and Nelson in 1966. Rapid adoption of what seemed at the time a very promising approach failed to occur, however, because other researchers pointed out possible complications. For instance, a laboratory study by Healey and Hendzel (1980) showed that different species responded in different ways, being a sensitive indicator of phosphorus deficiency in some, but not others.

Other methods which have been applied to natural populations are the measurement of 'surplus' cellular phosphorus (Fitzgerald and Nelson, 1966) and the use of phosphate-enrichment bioassays. Following the publication of a methods book by the US Environmental Protection Agency (1971), the latter technique became increasingly popular (see Klapwijk *et al.*, 1989), sometimes together with other methods such as measurement of cellular nitrogen-to-phosphorus ratio (Chiaudani and Vighi, 1974) and the determination of V_{max} for phosphate uptake (Gotham and Rhee, 1981; Pettersson, 1980). However, a study by Elser and Kimmel (1986) showed clearly that enrichment assays done in the laboratory do not necessarily demonstrate nutrient limitation *in situ* and that phosphatase assays can provide a more direct indication of phosphorus limitation. Overall, assessment of the phosphorus status of a community is most effective when several other methods are applied at the same time (Pettersson, 1980; Vincent, 1981; Gage and Gorham, 1985; McCormick and Stevenson, 1998). However, the phosphatase assay technique is the simplest and can provide rapid information if portable equipment is available to quantify results in the field.

The phosphomonoesterase activity of periphyton communities has been used to indicate not only the phosphorus status of a community (Burkholder and Wetzel, 1990), but also to aid understanding of the higher plant community from which the periphyton was taken. Measurements of the phosphomonoesterase activity of cyanobacterial mats in calcareous marshes in northern Belize showed that the marshes were strongly phosphorus-limited (Rejmánková and Komárková, 2000); activity was correlated with the nitrogen-to-phosphorus ratio in the water, while the addition of phosphorus suppressed phosphomonoesterase activity. Similar studies on periphyton communities in the Florida Everglades are described below (Newman *et al.*, 2003). Changes in phosphomonoesterase activity on passing down a stream or river have been followed using both periphyton mats (Mulholland and Rosemond, 1992) and individual species (Christmas and Whitton, 1998a,b).

If phosphatase assays are to be adopted widely by environmental management organizations, methods should be standardized and straightforward (Whitton, 1991). Mosses are particularly useful for this purpose, because broadly the same method can be used for terrestrial and aquatic environments, and similar sets of samples can be used for monitoring phosphorus status and heavy-metal contamination (Whitton, 2003). In the case of terrestrial environments, measurements of phosphatase activities of mosses are especially useful as an aid to long-term observations at sites where atmospheric nitrogen deposition appears to be enhancing phosphorus limitation, as is probably taking place in Upper Teesdale in northern England (Turner *et al.*, 2003b, c). Ideal mosses for this purpose should be widespread and tolerant of a range of nutrient levels, as opposed to species with a narrow ecological range. *Hylocomium splendens* is excellent for this, because it is widespread in north temperate regions and appears to grow under conditions differing considerably in their phosphorus status (Turner *et al.*, 2001). The higher phosphomonoesterase activity of this moss in Upper Teesdale than at a site in Sweden probably reflects the greater impact of atmospheric nitrogen deposition at the former site. Practical methods for routine sampling and assays of mosses in terrestrial environments were summarized by Turner *et al.* (2001), who recommended 2-cm shoot apices and assays using 20 mM buffer at pH 5.0 (apart from *Sphagnum*, which requires 50 mM buffer) and 500 mM substrate.

Practical methods for applying phosphomonoesterase assays to shallow rivers were presented in a booklet prepared for the Environment Agency of England and Wales (Whitton *et al.*, 1999). These are based mainly on two mosses (*Fontinalis antipyretica* and *Rhynchostegium riparioides*), but also on two algae (*Cladophora glomerata* and *Stigeoclonium*). The recommended procedure again used 2-cm shoot apices, but only 5 mM buffer and 100 μM substrate. Turner *et al.* (2001) also recommended a much lower substrate concentration (100–200 μM) with these aquatic mosses than with terrestrial mosses, because of the presence of low- and high-affinity systems in the former (see above).

Durrieu and Tran Minh (2002) and Durrieu *et al.* (2003) used the surface phosphomonoesterase activity of *Chlorella vulgaris* to compare the toxicity of seven heavy metals and in the latter study also phenolics and seven pesticides. The authors reported a marked inhibitory effect of heavy metals (see above) and a slight one due to phenolics, but the pesticides tested showed no effect. A dialysis system was developed for carrying out the tests, which the authors suggested could be adapted for use at field sites. The ED_{50} (metal concentration at which phosphomonoesterase activity was half of control) for cadmium (0.17 mg/l) was lower than that obtained (1.5 mg/l) by Truhaut *et al.* (1989) with *C. vulgaris* towards inhibition of chlorophyll and ATP. The authors suggested that phosphomonoesterase assays may be more suitable for detecting low concentrations of cadmium than assays based on chlorophyll or ATP.

Examples of Particular Habitats

Standing water

Most studies on natural phytoplankton populations have been done on lakes and reservoirs, where phosphomonoesterase activity has been demonstrated for all the main groups of microorganisms (Siuda, 1984). Lysis and zooplankton grazing make an important contribution (Chróst and Siuda, 2002) and high phosphomonoesterase levels can occur even in eutrophic lakes during periods of breakdown of phytoplankton blooms (Heath and Cooke, 1975; Chróst and Siuda, 2002). The types of organic phosphate present in the aquatic environment are reviewed in Chapter 14 (Mitchell and Baldwin, this volume). Although the majority of studies on phytoplankton have focused on phosphomonoesterase activity, free nucleic acids are probably the most significant reservoir of phosphorus potentially available in many lakes (Chróst and Siuda, 2002), although Heath and Cooke (1975) reported 0.5 μM total phosphate monoesters in a eutrophic lake. Siuda and Güde (1996) reported that DNase activity in Lake Constance was mostly extracellular (i.e. outside cells) or coupled to the plankton size fraction 0.1–1 μm, suggesting only a minor role for algae. There is scant information about surface phosphodiesterase activity in lake phytoplankton, though it seems probable that this will prove important in view of the many other studies reporting surface phosphodiesterase activity of cyanobacteria and green algae.

Flowing waters

Phosphatase assays have been included in a number of studies to characterize features of a particular stream or river (e.g. Bothwell, 1989). If the catchment area of a small stream is relatively uniform, then the phototrophic community of that stream may be expected to reflect the drainage from that catchment and the biogeochemical processes influencing that drainage. For instance, the algal flora of the Arctic stream sites studied by Sheath and Müller (1987) includes a number of the taxa known to be capable of showing high phosphomonoesterase activity, so it seems likely that organic phosphate is an important phosphorus source here.

Stream communities dominated by hair-forming cyanobacteria and/or green algae may be expected to occur in catchments where the drainage water shows highly variable ambient phosphate, as shown for streams in northern England with abundant *Rivularia* (Livingstone and Whitton, 1984) and *Stigeoclonium* (Gibson and Whitton, 1987a). In the case of the calcareous streams in one area, Upper Teesdale, which are overwhelmingly dominated by *Rivularia*, the seasonal cycle of morphological changes in *Rivularia* colonies reflects the seasonal changes in the nitrogen-to-phosphorus ratio and phosphorus concentration of the water (Whitton *et al.*, 1998), which in turn reflects seasonal changes in the same chemical features in the soil (Turner *et al.*, 2003b). The cyanobacteria *Schizothrix* and *Phormidium autumnale* and the chrysophyte *Hydrurus* from the calcareous alpine River Isar and *Rivularia* from a spring-fed tributary all showed marked phosphomonoesterase activity (Rott *et al.*, 2000). Phosphomonoesterase activity of *Schizothrix* measured over the growing season showed a negative relationship with phosphorus concentration of the organism, but that of *Hydrurus* showed a positive relationship. The latter merits further study, but is perhaps related to the fact that assays were conducted in vessels immersed in the river and thus at different temperatures over the range 0–10°C.

A more detailed understanding of the phosphorus requirements of a particular species dominating a stream should lead to this species providing a rapid means of characterizing the environment in that stream. For instance, the green algae *Draparnaldia* and *Microthamnion* are often frequent in northern England in small seepages and streams where anoxic water emerging from an underground source such as peat becomes oxygenated and deposition of hydrated iron oxides occurs, and the only UK record for a hair-forming *Oedogonium* was also reported from a similar environment (Harris, 1933). Under these conditions most of the available phosphorus is organic (B.A. Whitton, unpublished data), and *Draparnaldia* and *Microthamnion* typically show very high phosphomonoesterase activity. Such high activity is presumably the result of removal of phosphate due to iron (III) formation and binding of phosphatases by fulvic and humic compounds. However, *Draparnaldia* forms hairs, whereas *Microthamnion* does not. It would be of interest to compare the seasonal changes in phosphorus fractions at sites dominated by one or other of these organisms, together with the ability of the organisms to use the various phosphorus fractions.

Phosphomonoesterase activity has been studied in several metal-contaminated streams at both high and low pH values. For instance, Sabater *et al.* (2003), who investigated the highly acidic, copper-polluted Rio Tinto in southwest Spain, studied the phosphomonoesterase activity of an algal mat developing in an artificial channel. This was done in the dark, using a range of substrate (methylumbelliferyl phosphate) concentrations from 0.1 to 600 μM. The mat of *Klebsormidium flaccidum*, *Pinnularia acoricola* and *Euglena mutabilis* showed relatively high phosphomonoesterase activity and this activity increased as the mat became mature. The authors suggested that the substantial difference in activity between young and mature mat indicated that phosphorus availability was reduced in the latter. However, the values for K_m at both stages were very low, indicating the very high affinity of the mat for the substrate. The authors compared their results with those of Romani (2000) for an algal biofilm in a calcareous stream in Spain showing phosphorus limitation. Both had similar values for V_{max}, but K_m was clearly lower in the acid river.

The Florida Everglades, through which there is slow movement of calcareous water, have been shown to be highly sensitive to phosphorus enrichment (McCormick *et al.*, 1996) and measurement of periphyton phosphomonoesterase activity is one of several approaches that have been used to assess the phosphorus status of a particular area. Newman *et al.* (2003) measured the effect of adding phosphate on phosphomonoesterase activity of mats over a 5-month period at an oligotrophic site dominated by cyanobacteria and diatoms. Samples on dowels in mesocosms receiving a phosphate load of 6.4 g P/m/year showed a significant reduction in activity within 2 weeks. Samples taken over the study period from mesocosms with loading rates $<$1.6 g P/m/year also averaged less than controls, though the differences were generally not significant. The mesocosm results were less clear-cut when activity was expressed per unit biomass, while the results for floating mats, expressed per unit biomass, showed no distinct response to phosphorus enrichment. The authors stated that it is unlikely that the substrate used for assays (200 μM methylumbelliferyl phosphate) penetrates far into the mat during the incubation period. They compared their results for phosphomonoesterase activity (0.1–1 nmol/cm^2/min) with those found for epiphytes on wooden tiles in an Australian pond (0.2–0.9 nmol/cm^2/min; Scholz and Boon, 1993) and on plants in a gravel pit in northeastern England (0.01–0.7 nmol/cm^2/min; Chappell and Goulder, 1994). Newman *et al.* (2003) concluded that phosphomonoesterase measurements proved to be a suitable early warning indicator for the Everglades of changes in phosphorus availability provided that artificial substrata were used and the results expressed per unit area. However, they also stressed the need to evaluate how other environmental changes might interact with the results for phosphomonoesterase activity.

Marine

Two recent reviews have dealt with phosphatase in the oceans: Hoppe (2003) mainly on plankton and processes, and Hernández *et al.* (2003) on benthic algae. Hoppe commented that studies on phosphatase in the sea are in fact relatively rare compared with freshwaters, which may be due to nitrogen, rather than phosphorus, having been considered as the growth-limiting factor for phototrophs in the sea, although modelling studies now point towards phosphorus being the ultimate limiting nutrient in ocean systems (Tyrrell, 1999). There are many areas where phosphorus is clearly the limiting nutrient, including some offshore ones (Huang and Hong, 1999; Hoppe, 2003) and many intertidal communities. Sometimes the situation changes seasonally. For instance, Paasche and Erga (1988) reported that nitrogen limitation occurred during the spring in Oslofjord, when diatoms were dominant, but phosphorus limitation in late summer when dinoflagellates were dominant.

Some information is available about the phosphorus nutrition of the three most important cyanobacteria in the oceans. Uptake of ^{32}P-labelled phosphate by *Trichodesmium* showed a surprisingly high half-saturation constant (McCarthy and Carpenter, 1979), indicating that its ability to compete for inorganic phosphate is low; possibly this is an indication that organic phosphorus is more important for this organism, which fits with the high phosphomonoesterase activity of the population studied by Yentsch *et al.* (1972). *Trichodesmium* collected in the north Pacific gyre was found to have relatively low cellular phosphorus and ATP concentrations, but high phosphomonoesterase activity (177–300 ng P μg chl-*a*/h), which diminished after addition of inorganic phosphate (Mague *et al.*, 1977). The concentration of substrate for saturation of phosphatases of a *Trichodesmium* population in the southern Baltic Sea was about 50 μM (Nausch, 1997, 1998). Stihl *et al.* (2001) compared phosphomonoesterase activity (assayed with 400 μM *para*-nitrophenyl phosphate) in natural populations of *Trichodesmium* from the Red Sea with that of an axenic strain WH9601 from another region. Open-water colonies, which had a tuft shape, showed high activity irrespective of date or origin. Coastal colonies with a tuft shape showed low activity except for the period in autumn when the population was a maximum, but the rarer ones showing a puff or bow-tie shape had higher activity, though still less than that shown by the tuft-shaped colonies of open water. Intact filaments of the axenic strain showed baseline activity of 0.5 μmol *para*-nitrophenol μg chl/h; the rate increased 10-fold under phosphorus depletion or when β-glycerophosphate was the sole phosphorus source.

The picoplanktonic cyanobacteria *Prochlorococcus* and *Synechococcus* are the dominants of oligotrophic oceanic waters and are responsible for most of the oceanic primary production. Bertilsson *et al.* (2003) and Heldal *et al.* (2003) provide considerable insight into the role of these organisms in cycling phosphorus in the oceans and a few details are included here, because they give the background for future studies on phosphatases. All three strains studied by Bertilsson *et al.* (2003) and all but one strain of *Synechococcus* studied by Heldal *et al.* (2003) had carbon-to-phosphorus and nitrogen-to-phosphorus molar ratios well above the Redfield ratio (Redfield, 1958). The former authors reported a range of 21–35 (molar ratio) for replete cultures and 59–109 (molar ratio) under phosphorus limitation. (For comparisons with other cyanobacteria, the respective ratios by mass are 9.5–15.8 and 26.5–49.) Phosphorus-limited cultures of these organisms have only about one-third the percentage phosphorus (expressed as mass) of the large filamentous cyanobacteria discussed earlier. Bertilsson *et al.* (2003) reported that their strains differed in phosphorus content by a factor of 3–4 between phosphorus-replete and phosphorus-limited cultures, which they considered to a dramatic difference. However, it is in fact considerably less than the equivalent difference for many large cyanobacteria, where it often reaches 8 or even more (e.g. *Calothrix*; Islam and Whitton, 1992). In comparison with many freshwater, terrestrial and intertidal cyanobacteria, the marine picoplankton are therefore organisms with a very low cell phosphorus concentration, but a lesser range between maximum and minimum. Heldal *et al.* (2003) concluded that these organisms had a low phosphorus storing capacity, since they found no signs of polyphosphate granule formation in any of the six *Prochlorococcus* and one of the two *Synechococcus*. While these authors found relatively large amounts of polyphosphate in cultures of *Synechococcus* WH7803, Cuhel and Waterbury (1984) reported little polyphosphate in the same strain. This suggests the possibility that this strain may have undergone changes in phosphorus metabolism during prolonged laboratory culture in a phosphorus-rich medium.

In a study of phosphomonoesterase activity in the Gulf of Aqaba, Red Sea, Li *et al.* (1998) found that the bulk of particulate surface phosphomonoesterase activity was associated with the picoplankton fraction. As there was a strong correlation between phosphomonoesterase activity and *Synechococcus* abundance, it was suggested that the genus may be a significant contributor to the removal of organic phosphorus here.

Although several studies on phosphomonoesterase activity of planktonic eukaryotic algae were mentioned above, there are too few to generalize from. Perry (1972) reported the K_m for the phosphatase of a tropical diatom to be 0.12 mM inorganic phosphate, which would indicate a high-affinity system. Lubián *et al.* (1992) found that three of the four strains of *Nannochloris* and both strains of *Nannochloropsis*, all or almost all of which were isolates from coastal waters, showed acid surface phosphomonoesterase activity. The former genus also showed alkaline phosphomonoesterase activity. Huang and Hong (1999) found a significant inverse relationship between phosphomonoesterase activity of coastal phytoplankton and total filtrable phosphorus, and concluded that the algae were more important than bacteria in mobilizing organic phosphate. Further study is needed to establish the role of surface acid phosphatase in a marine alga.

Desert rocks

The communities of cyanobacteria, algae (sometimes lichenized) and a few heterotrophs which occur inside rocks of desert regions probably obtain most of their phosphorus by recycling inside the rock and this is likely to be especially true of the communities living within the Antarctic dry deserts. For instance, long-established communities in the McMurdo Dry Valley in Southern Victoria Land are probably almost entirely dependent on phosphorus turnover, whereas they may receive further carbon and nitrogen from the atmosphere (Banerjee *et al.*, 2000b). *Chroococcidiopsis*, *Gloeocapsa–Trebouxia* and *Trebouxia*-dominated communities all showed marked phosphomonoesterase and phosphodiesterase activities, with pH optima of fragmented rock samples (assayed at 100 μM substrate) at 9.5, 5.5 and 7.5, respectively; values quite similar to aqueous extracts of the relevant rocks. The first two communities showed higher activity at 5°C than at 1 or 10°C, while all three communities showed higher activity in the dark than at 7 μmol photon/m^2/s. As Antarctic endolithic communities are known to grow extremely slowly, with evidence for one site suggesting a doubling time of 10,000 years (Nienow and Friedmann, 1993), the cycling of organic phosphorus within the communities is likely to be a key factor influencing their overall metabolic activity and hence the growth rate during the short periods when moisture is available (Banerjee *et al.*, 2000b).

Chroococcidiopsis responded differently to both temperature and light when brought into laboratory culture, with phosphomonoesterase activity showing a much higher temperature optimum and with irradiance up to 8 μmol photon/m^2/s enhancing activity (Banerjee *et al.*, 2000a). The authors suggested that this might be an environmental response occurring within a few cell generations of the organism being removed from the rock and exposed to more favourable growth conditions, rather than genetic adaptation to these conditions.

Perspective

Surface phosphatase activity apparently occurs in most cyanobacteria, algae and mosses, though typically its expression is at least partially inducible. Although some organisms like *Arthospira* apparently lack the ability to develop activity, reports of its absence in an organism should be treated with caution until rigorous studies have been done. The majority of ecologically oriented studies on samples of the whole water column in lakes and the sea have dealt only with phosphomonoesterase activity. However, phosphodiesterase activity, which has been studied in many larger cyanobacteria, filamentous green algae and mosses, has proved to be only slightly less widespread than phosphomonoesterase activity. Although relatively little study of other surface phosphorylases has been done for phototrophs, these too may be important. Caution is therefore needed in interpreting the significance of surface phosphatase activity based on studies solely with phosphomonoesterase. Previous conclusions about the role of phototrophs in aquatic ecosystems will need to be reconsidered when the quantitative roles of a range of phosphorohydrolases have been tested.

One important question still to be clarified is whether the synthesis of phosphatases is regulated directly by one or more phosphorus fractions in the environment, or by some feature of intracellular phosphorus, and thus only indirectly in response to the environment. In spite of many accounts giving the impression that the situation is clear for the organisms under study, it is doubtful whether the situation has been resolved unequivocally in any of them. All that can be said here is that it seems likely that very small unicells respond directly to the environment and that large organisms respond indirectly. If picophytoplankton behave in the way that many heterotrophic bacteria appear to do, then phosphatase regulation would be at least partly in response to the type and concentration of different forms of phosphate in the ambient environment. Possibly the response sometimes reflects a complex interaction between the type and concentration of phosphate compounds. However, phosphatase regulation in the larger organisms investigated appears to be a response to intracellular phosphorus status. The larger the organism is, the less important it is to respond rapidly to occasional deficits in ambient phosphorus. The morphological heterogeneity of some larger organisms gives scope for spatial heterogeneity in surface phosphatase activity, permitting the organisms to utilize transient pulses of organic phosphate without the need for rapid phosphatase synthesis. It is suggested that the intracellular phosphorus status of these organisms provides a means of detecting the typical phosphorus status of the environment over a period of time.

Just as much care is needed in interpreting the results of laboratory as field studies. This is partly because of the difficulty of simulating field conditions and in the past a lack of awareness of factors likely to have significant effects on phosphatase activity. Hopefully, the evidence reviewed here will provide a useful guide to what needs to be considered when planning experiments. A further problem is harder to overcome: the fact that many strains maintained in collections have already had a long history of culture in medium with high inorganic phosphate, with the consequent risk that they have undergone morphological, physiological and molecular changes in response. Unfortunately, some of the most popular culture media used for cyanobacteria (e.g. BG-11; Rippka *et al.*, 1979), or eukaryotic algae (e.g. Bold's Basal Medium; Nichols and Bold, 1965), are highly unsuitable for studies on the effects of phosphorus limitation. There is considerable evidence for individual strains that genetic changes have occurred in response to culture under high ambient phosphorus (Whitton, 1987b, 1992), but even if no such evidence exists for a strain, the possibility that changes have taken place makes it difficult to rely on quantitative data for kinetic studies on surface phosphatases or reports of the lack of ability to use a particular substrate. Genes responsible for phosphatase activities may already have been lost from strains selected for DNA sequencing, so future studies on phosphatase may be hindered by flawed molecular data. Of all the genes present in phototrophs, those responsible for detecting and utilizing forms of phosphorus are among those at most risk of being lost in culture.

Laboratory studies can nevertheless be a help in interpreting field data, such as the recognition that species forming hairs are those growing in environments with highly variable ambient phosphate. The increasing evidence that phosphodiesterase behaves differently from phosphomonoesterase should also be followed further in the field. Whereas phosphodiesterase activity is not released to the ambient medium, phosphomonoesterase activity in solution often makes a substantial contribution to the total activity external to the cell wall. If this can be shown to apply widely in nature, then any extracellular phosphodiesterase activity presumably results from lysis or grazing and might be used as an indicator for this, whereas phosphomonoesterase activity is more likely to include a contribution from actively growing phototrophs. At least in the mosses studied, phosphodiesterase activity also contrasts with phosphomonoesterase activity in the lack of influence of substrate concentration on pH optimum and the lack of effect of light on activity. It would be of interest to establish whether these differences are related to phosphodiesterase being bound more firmly within the wall, since fungal extracellular acid phosphatases have been shown to be stabilized by adsorption to clay minerals (Leprince and Quiquampoix, 1996; Quiquampoix and Mousain, Chapter 5, this volume).

A good way to increase understanding of the role of phototroph phosphatases in mobilizing organic phosphate would be to combine studies on phosphorus fractions in the environment with detailed studies on the properties of individual cells and filaments. The literature on enzyme-labelled fluorescence shows the great potential of this approach if populations of each phototroph species could be separated by laser flow cytometry or similar technique. Individual cells or filaments could be used for enzyme-labelled fluorescence and X-ray microanalysis, while bulked material of each fraction could be used for miniaturized versions of routine phosphatase assays together with element analysis. Such an approach would be especially useful for studies on the picophytoplankton of the oceans and large oligotrophic lakes.

This chapter set out to convince the reader that there is sufficient theoretical understanding about phosphatases in non-vascular phototrophs, and enough diversity of practical techniques available for measuring activity in nature, that it is now possible to measure this activity reliably. The next challenge for researchers is to quantify in detail the role of these organisms with respect to organic phosphate in particular ecosystems, and to predict how this will change in response to changes in the environment.

Acknowledgements

The authors thank P. Gualtieri (CNR, Pisa, Italy), R. Baxter (University of Durham, UK), D.J. Scanlan (University of Warwick, UK) and S. Newman (Everglades Division, South Florida Water Management District) for helpful discussions.

References

Aboal, M., Angeles Puig, M.A., Mateo, P. and Perona, E. (2002) Implications of cyanophyte toxicity on biological monitoring of calcareous streams in north-east Spain. *Journal of Applied Phycology* 14, 49–56.

Al-Shehri, A.M. (1992) An ecophysiological study of the moss *Hydrogonium fontanum* from the Asir Mountains, Saudi Arabia. PhD thesis, University of Durham, UK.

Ammerman, J.W. (1991) Role of ecto-phosphohydrolases in phosphorus regeneration in estuarine and coastal ecosystems. In: Chróst, R.J. (ed.) *Microbial Enzymes in Aquatic Environments.* Springer, New York, pp. 165–186.

Banerjee, M., Whitton, B.A. and Wynn-Williams, D.D. (2000a) Surface phosphomonoesterase activity of a natural immobilized system: *Chroococcidiopsis* in an Antarctic desert rock. *Journal of Applied Phycology* 12, 549–552

Banerjee, M., Whitton, B.A. and Wynn-Williams, D.D. (2000b) Phosphatase activities of endolithic communities in rocks of the Antarctic Dry Valleys. *Microbial Ecology* 39, 80–91.

Bates, J.W. (1994) Responses of the mosses *Brachythecium rutabulum* and *Pseudoscleropodium purum* to a mineral nutrient pulse. *Functional Ecology* 8, 686–692.

Bates, J.W. (2000) Mineral nutrition, substratum ecology, and pollution. In: Shaw, A.J. and Goffinet, B. (eds) *Bryophyte Biology.* Cambridge University Press, Cambridge, UK, pp. 248–311.

Bellos, D. (1990) Water chemistry and algal phosphatase activity in zinc-contaminated streams. MSc Ecology dissertation, University of Durham, UK.

Bertilsson, S., Berglund, O., Karl, D.M. and Chisholm, S.W. (2003) Elemental composition of marine *Prochlorococcus* and *Synechococcus*: implications for the ecological stoichiometry of the sea. *Limnology and Oceanography* 48, 1721–1731.

Bhaya, D., Schwarz, R. and Grossman, A.R. (2000) Molecular responses to environmental stress. In: Whitton, B.A. and Potts, M. (eds) *Ecology of Cyanobacteria.* Kluwer Academic, Dordrecht, pp. 397–442.

Bjoerkman, K. and Karl, D.M. (1994) Bioavailability of inorganic and organic phosphorus compounds to natural assemblages of microorganisms in Hawaiian coastal waters. *Marine Ecology Progress Series* 111, 265–275.

Block, M.A. and Grossman, A.R. (1988) Identification and purification of a derepressible alkaline phosphatase from *Anacystis nidulans* R2. *Plant Physiology* 86, 1179–1184.

Blum, J.J. (1965) Observations on the acid phosphatases of *Euglena gracilis. Journal of Cell Biology* 24, 223–234.

Boavida, M.J. (1990) Natural plankton phosphatases and the recycling of phosphorus. *Verhandlung Internationale Vereinigung Limnologie* 24, 258–259.

Boavida, M.J. and Wetzel, R.G. (1998) Inhibition of phosphatase activity by dissolved humic substances and hydrolytic reactivation by natural ultraviolet light. *Freshwater Biology* 40, 285–293.

Bose, P., Nagal, U.S., Venkataraman, G.S. and Goyal, S.K. (1971) Solubilization of tricalcium phosphate by blue-green algae. *Current Science* 7, 165–166.

Bothwell, M.L. (1989) Phosphorus-limited growth dynamics of lotic periphyton diatom communities: areal biomass and cellular growth rate responses. *Canadian Journal of Fisheries and Aquatic Sciences* 46, 1293–1301.

Burkholder, J.A. and Wetzel, R.G. (1990) Epiphytic alkaline phosphatase on natural and artificial plants in an oligotrophic lake: re-evaluation of the role of macrophytes as a phosphorus source for epiphytes. *Limnology and Oceanography* 35, 736–747.

Cameron, H.J. and Julian, G.R. (1988) Utilization of hydroxyapatite by cyanobacteria as their sole source of phosphate and calcium. *Plant and Soil* 109, 123–124.

Cembella, A.D., Antia, N.J. and Harrison, P.J. (1984) The utilization of inorganic and organic phosphorus compounds as nutrients by eukaryotic microalgae: a multidisciplinary perspective. Part 1. *CRC Critical Reviews of Microbiology* 10, 317–391.

Chappell, K.R. and Goulder, R. (1994) Seasonal variation of epiphytic extracellular enzyme activity on two freshwater plants, *Phragmites australis* and *Elodea canadensis. Archiv für Hydrobiologie* 132, 237–253.

Chiaudani, G. and Vighi, M. (1974) The N:P ratio and tests with *Selenastrum* to predict eutrophication in lakes. *Water Research* 8, 1063–1069.

Christian, J.R. and Karl, D.M. (1995) Measuring bacterial exoenzyme activities in marine waters using mercuric chloride as a preservative and a control. *Marine Ecology Progress Series* 123, 217–224.

Christmas, M. and Whitton, B.A. (1998a) Phosphorus and aquatic bryophytes in the Swale–Ouse river system, North-East England. 1. Relationship between ambient phosphate, internal N:P ratio and surface phosphatase activity. *Science of the Total Environment* 210/211, 389–399.

Christmas, M. and Whitton, B.A. (1998b) Phosphorus and aquatic bryophytes in the Swale–Ouse river system, North-East England. 2. Phosphomonoesterase and phosphodiesterase activities of *Fontinalis antipyretica*. *Science of the Total Environment* 210/211, 401–409.

Chróst, R.J. (1991) Environmental control of the synthesis and activity of aquatic microbial ectoenzymes. In: Chróst, R.J. (ed.) *Microbial Enzymes in Aquatic Environments.* Springer, New York, pp. 29–59.

Chróst, R.J. and Krambeck, H.J. (1986) Fluorescence correction for measurements of enzyme activity in natural waters using methylumbelliferyl substrates. *Archiv für Hydrobiologie* 106, 79–90.

Chróst, R.J. and Overbeck, J. (1987) Kinetics of alkaline phosphatase activity and phosphorus availability for phytoplankton and bacterioplankton in Lake Plussee (north German eutrophic lake). *Microbial Ecology* 13, 229–248.

Chróst, R.J. and Siuda, A. (2002) Ecology of microbial enzymes in lake ecosystems. In: Burns, R.G. and Dick, R.P. (eds) *Enzymes in the Environment: Activity, Ecology and Applications.* Marcel Dekker, New York, pp. 35–72.

Coston, S. and Holt, S.J. (1958) Kinetics of aerial oxidation of indolyl and some of its halogen derivatives. *Proceedings of the Royal Society of London, Series B* 158, 506–510.

Cotner, J.B., Jr and Heath, R.T. (1988) Potential phosphate release from phosphomonoesters by acid phosphatase in a bog lake. *Archiv für Hydrobiologie* 111, 329–338.

Cuhel, R.L. and Waterbury, J.B. (1984) Biochemical composition and short-term nutrient incorporation patterns in a unicellular marine cyanobacterium, *Synechococcus* (WH 7803). *Limnology and Oceanography* 29, 370–374.

Donald, K.M., Scanlan, D.J., Carr, N.G., Mann, N.H. and Joint, I. (1997) Comparative phosphorus nutrition of the marine cyanobacterium *Synechococcus* WH7803 and the marine diatom *Thalassiosira weissflogii*. *Journal of Plankton Research* 19, 1793–1813.

Douglas, D., Peat, A., Whitton, B.A. and Wood, P. (1986) Influence of iron status on structure of the cyanobacterium (blue-green alga) *Calothrix parietina*. *Cytobios* 47, 155–165.

Durrieu, V.B. and Tran Minh, C. (2002) Optimal algal biosensor using alkaline phosphatase for determination of heavy metals. *Environmental and Ecotoxicological Safety* 51, 206–209.

Durrieu, C., Badreddine, I. and Daix, C. (2003) A dialysis system with phytoplankton for monitoring chemical pollution in freshwater ecosystems by alkaline phosphatase assay. *Journal of Applied Phycology* 15, 289–295.

Dyhrman, S.T. and Palenik, B.P. (1997) The identification and purification of a cell-surface alkaline phosphatase from the dinoflagellate *Prorocentrum minimum* (Dinophyceae). *Journal of Phycology* 33, 602–612.

Dyhrman, S.T. and Palenik, B.P. (1999) Phosphate stress in cultures and field populations of the dinoflagellate *Prorocentrum minimum* detected using a single-cell alkaline phosphatase activity assay. *Applied and Environmental Microbiology* 65, 3205–3212.

Dyhrman, S.T. and Palenik, B. (2001) A single-cell immunoassay for phosphate stress in the dinoflagellate *Prorocentrum minimum* (Dinophyceae). *Journal of Phycology* 37, 400–410.

Eckstein, R. and Karlsson, P.S. (1999) Recycling of nitrogen among segments of *Hylocomium splendens* as compared with *Polytrichum commune*: implications for clonal integration in an ectohydric bryophyte. *Oikos* 86, 87–96.

Ellwood, N.T.W. (2002) Factors influencing phosphatase activities of mosses in upland streams. PhD thesis. University of Newcastle-upon-Tyne, UK.

Ellwood, N.T.W., Haile, S.M. and Whitton, B.A. (2002) Surface phosphatase activity of the moss *Warnstorfia fluitans* as an indicator of the nutrient status of an acidic stream. *Verhandlung Internationnale Vereinigung Limnologie* 28, 62–623.

Elser, J.J. and Kimmel, B.L. (1986) Alteration of phytoplankton phosphorus status during enrichment experiments: implications for interpreting nutrient enrichment bioasssay results. *Hydrobiologia* 133, 217–222.

Engasser, J.M. and Horvath, C. (1975) Electrostatic effects on the kinetics of bound enzymes. *Journal of Biochemistry* 145, 431–435.

Espeland, E.M. and Wetzel, R.G. (2001) Complexation, stabilization, and UV photolysis of extracellular and surface-bound glucosidase and alkaline phosphatase: implications for biofilm microbiota. *Microbial Ecology* 42, 572–585.

Fedde, K.N. and Whyte, M.P. (1990) Alkaline phosphatase (tissue-nonspecific isoenzyme) is a phosphoethanolamine and pyridoxal-5′-phosphate ectophosphate: normal and hypophosphatasia fibroblast study. *American Journal of Human Genetics* 47, 767–775.

Feuillade, J., Feuillade, M. and Blanc, P. (1990) Alkaline phosphatase activity fluctuations and associated factors in a eutrophic lake dominated by *Oscillatoria rubescens*. *Hydrobiologia* 207, 233–240.

Fitzgerald, G.P. and Nelson, C. (1966) Extractive and enzymatic analyses for limiting or surplus phosphorus in algae. *Journal of Phycology* 2, 32–37.

Francko, D.A. and Heath, R.T. (1982) UV-sensitive complex phosphorus: association with dissolved humic material and iron in a bog lake. *Limnology and Oceanography* 27, 564–569.

Gage, M.A. and Gorham, E. (1985) Alkaline phosphatase activity and cellular phosphorus as an index of the phosphorus status of phytoplankton in Minnesota lakes. *Freshwater Biology* 15, 227–233.

Garde, K. and Gustavson, K. (1999) The impact of UV-B radiation on alkaline phosphatase activity in phosphorus-depleted marine ecosystems. *Journal of Experimental Marine Biology and Ecology* 238, 93–105.

Gibson, M.T. and Whitton, B.A. (1987a) Hairs, phosphatase activity and environmental chemistry in *Stigeoclonium, Chaetophora* and *Draparnaldia* (Chaetophorales). *British Phycological Journal* 22, 11–22.

Gibson, M.T. and Whitton, B.A. (1987b) Influence of phosphorus on morphology and physiology of freshwater *Chaetophora, Draparnaldia* and *Stigeoclonium* (Chaetophorales, Chlorophyta). *Phycologia* 26, 59–69.

González-Gil, S., Keafer, B.A., Jovine, R.V.M., Aguilera, A., Lu, S. and Anderson, D.M. (1998) Detection and quantification of alkaline phosphatase in single cells of phosphorus-starved marine phytoplankton. *Marine Ecology Progress Series* 164, 21–35.

Gotham, I.J. and Rhee, G.-Y. (1981) Comparative kinetic studies of phosphate limited growth and phosphate uptake in phytoplankton in continuous culture. *Journal of Phycology* 17, 257–265.

Graham, L.E. and Wilcox, L.W. (2000) *Algae.* Prentice-Hall, Upper Saddle River, New Jersey, 640 pp.

Grainger, S.L.J., Peat, A., Tiwari, D.N. and Whitton, B.A. (1989) Phosphomonoesterase activity of the cyanobacterium (blue-green alga) *Calothrix parietina. Microbios* 59, 7–17.

Gross, W. (2000) Ecophysiology of algae living in highly acid environments. *Hydrobiologia* 433, 31–37.

Hantke, B. and Melzer, A. (1993) Kinetic changes in surface phosphatase activity of *Synedra acus* (Bacillariophyceae) in relation to pH variation. *Freshwater Biology* 29, 31–36.

Harris, G.T. (1933) The Oedogoniales of Devonshire. *Transactions of Devonshire Association for Advancement of Sciences, Literature, and Art* LXV, 213–226.

Healey, F.P. and Hendzel, L.L. (1980) Physiological indicators of nutrient deficiency in lake phytoplankton. *Canadian Journal of Fisheries and Aquatic Sciences* 37, 442–453.

Heath, R.T. and Cooke, G.D. (1975) The significance of alkaline phosphatase in a eutrophic lake. *Verhandlung Internationale Vereinigung Limnologie* 19, 959–965.

Heath, R.T. and Edinger, A.C. (1990) Uptake of ^{32}P-phosphoryl from glucose-6-phosphate by plankton in an acid bog lake. *Verhandlung Internationale Vereinigung Limnologie* 24, 210–213.

Herbes, S.E., Allen, H.E. and Mancy, K.H. (1975) Enzymatic characterization of soluble organic phosphorus in lake water. *Science* 187, 432–434.

Heldal, M., Scanlan, D.J., Norland, S., Thingstad, F. and Mann, N.H. (2003) Elemental composition of single cells of various strains of marine *Prochlorococcus* and *Synechococcus* using X-ray microanalysis. *Limnology and Oceanography* 48, 1732–1743.

Hernández, I. (1996) Analysis of the expression of alkaline phosphatase activity as a measure of phosphorus status in the red alga *Porphyra umbilicalis* (L.) Kützing. *Botanica Marina* 39, 255–262.

Hernández, I. and Whitton, B.A. (1996) Retention of *p*-nitrophenol and 4-methylumbelliferone by marine macrophytes and implications for measurement of alkaline phosphatase activity. *Journal of Phycology* 32, 819–825.

Hernández, I., Niell, F.X. and Fernández, J.A. (1992) Alkaline phosphatase activity in *Porphyra umbilicalis* (L.) Kützing. *Journal of Experimental Marine Biology and Ecology* 159, 1–13.

Hernández, I., Fernández, J.A. and Niell, F.X. (1993) Influence of phosphorus status on the seasonal variation of alkaline phosphatase activity in *Porphyra umbilicalis* (L.) Kützing. *Journal of Experimental and Marine Biology and Ecology* 173, 181–196.

Hernández, I., Fernández, J.A. and Niell, F.X. (1995) A comparative study of alkaline phosphatase activity of two species of *Gelidium* (Gelidiales, Rhodophyta). *European Journal of Phycology* 30, 69–77.

Hernández, I., Hwang, S.-J. and Heath, R.T. (1996a) Measurement of phosphomonoesterase activity with a radiolabelled glucose-6-phosphate: role in the phosphorus requirement of phytoplankton and bacterioplankton in a temperate mesotrophic lake. *Archiv für Hydrobiologie* 137, 265–280.

Hernández, I., Niell, F.X. and Fernández, J.A. (1996b) Alkaline phosphatase activity of the red alga *Corallina elongata* Ellis et Solander. *Scientia Marina* 60, 297–306.

Hernández, I., Christmas, M., Yelloly, J.M. and Whitton, B.A. (1997) Factors affecting surface alkaline phosphatase activity in the brown alga *Fucus spiralis* L.: studies at a North Sea intertidal site (Tyne Sands, Scotland). *Journal of Phycology* 33, 569–575.

Hernández, I., Andría, J.R., Christmas, M. and Whitton, B.A. (1999) Testing the allometric scaling of alkaline phosphatase activity to surface/volume ratio in benthic marine macrophytes. *Journal of Experimental Marine Biology and Ecology* 241, 1–14.

Hernández, I., Pérez-Pastor, A. and Pérez-Lloréns, J.L. (2000) Ecological significance of phosphomonoesters and phosphomonoesterase activity in a small Mediterranean river and its estuary. *Aquatic Ecology* 34, 107–117.

Hernández, I., Niell, F.X. and Whitton, B.A. (2003) Phosphatase activity of benthic marine algae: an overview. *Journal of Applied Phycology* 14, 475–485.

Herndl, G.J., Müller-Niklas, G. and Frick, J. (1993) Major role of ultraviolet-B in controlling bacterioplankton growth in the surface of the ocean. *Nature* 361, 717–719.

Hoppe, H.-G. (1983) Significance of exoenzymatic activities in the ecology of brackish water measurements by means of methyl-umbelliferyl substrates. *Marine Ecology Progress Series* 11, 299–308.

Hoppe, H.-G. (2003) Phosphatase activity in the sea. *Hydrobiologia* 493, 187–200.

Huang, B. and Hong, H. (1999) Alkaline phosphatase activity and utilization of dissolved organic phosphorus by algae in subtropical coastal waters. *Marine Pollution Bulletin* 39, 205–211.

Huang, Z., Terpetschnig, E., You, W. and Haugland, R.P. (1992) 2-(2′-phosphoryloxyphenyl)-4(3H)-quinazolinone derivates as fluorogenic precipitating substrates of phosphatases. *Analytical Biochemistry* 207, 32–39.

Huber, A.L. and Hamel, K.S. (1985) Phosphatase activities in relation to phosphorus nutrition in *Nodularia spumigena* (Cyanobacteriaceae). 2. Laboratory studies. *Hydrobiologia* 123, 81–88.

Ihlenfeldt, M.J.A. and Gibson, J. (1975) Phosphate utilization and alkaline phosphatase activity in *Anacystis nidulans* (*Synechococcus*). *Archives for Microbiology* 102, 23–28.

Islam, M.R. and Whitton, B.A. (1992) Phosphorus content and phosphatase activity of the deepwater rice-field cyanobacterium (blue-green alga) *Calothrix* D764. *Microbios* 69, 7–16.

Jansson, M., Olsson, H. and Pettersson, K. (1988) Phosphatases: origin, characteristics and function in lakes. *Hydrobiologia* 170, 157–175.

John, D.M., Whitton, B.A. and Brook, A.J. (eds) (2002) *The Freshwater Algal Flora of the British Isles.* Cambridge University Press, Cambridge, UK.

Jones, R.I. (1990) Phosphorus transformation in the epilimnion of humic lakes: biological uptake of phosphate. *Freshwater Biology* 23, 323–337.

Joseph, E.M., Morel, F.M.M. and Price, N.M. (1994) Effects of aluminium and fluorides on phosphorus acquisition by *Chlamydomonas reinhardtii. Canadian Journal of Fisheries and Aquatic Sciences* 52, 353–357.

Kennelly, P.J. and Potts, M. (1996) Fancy meeting you here! A fresh look at 'prokaryotic' protein phosphorylation. *Journal of Bacteriology* 178, 4759–4764.

Klapwijk, S.P., Bolier, G. and van der Does, J. (1989) The application of algal growth potential tests (AGP) to the canals and lakes of western Netherlands. *Hydrobiologia* 188/189, 189–199.

Klotz, R.L. (1985) Influence of light on the alkaline phosphatase activity of *Selenastrum capricornutum* (Chlorophyceae) in streams. *Canadian Journal of Fisheries and Aquatic Sciences* 42, 384–388.

Kumar, A., Singh, S. and Tiwari, D.N. (1991) Alkaline phosphatase activities of an *Anabaena* sp. from deep-water rice. *World Journal of Microbiology and Biotechnology* 8, 585–588.

Larsson, C., Axelsson, L., Ryberg, H. and Beer, S. (1997) Photosynthetic carbon utilization by *Enteromorpha intestinalis* (Chlorophyta) from a Swedish rockpool. *European Journal of Phycology* 32, 49–54.

Lee, T.-M. (2000) Phosphate starvation induction of acid phosphatase in *Ulva lactuca* L. (Ulvales, Chlorophyta). *Botanical Bulletin Academia Sinica* 41, 19–23.

Leprince, F. and Quiquampoix, H. (1996) Extracellular enzyme activity in soil: effect of pH and ionic strength on the interaction with montmorillonite of two acid phosphatases secreted by ectomycorrhizal fungus *Hebeloma cylindrosporum. European Journal of Soil Science* 47, 511–522.

Li, H., Veldhuis, M.J.W. and Post, A.F. (1998) Alkaline phosphatase activities among planktonic communities in the northern Red Sea. *Marine Ecology Progress Series* 173, 107–115.

Livingstone, D. and Whitton, B.A. (1983) Influence of phosphorus on morphology of *Calothrix parietina* (Cyanophyta) in culture. *British Phycological Journal* 18, 29–38.

Livingstone, D. and Whitton, B.A. (1984) Water chemistry and phosphatase activity of the blue-green alga *Rivularia* in Upper Teesdale streams. *Journal of Ecology* 72, 405–421.

Livingstone, D., Khoja, T.M. and Whitton, B.A. (1983) Influence of phosphorus on physiology of a hair-forming blue-green alga (*Calothrix parietina*) from an upland stream. *Phycologia* 22, 345–350.

Livingstone, D., Pentecost, A. and Whitton, B.A. (1984) Diel variations in nitrogen and carbon dioxide fixation by the blue-green alga *Rivularia* in an upland stream. *Phycologia* 23, 125–133.

Lubián, L.M. (1981) Crecimiento en cultivo de cuatro

cepas de *Nannochloris*. Estudio de su citologia y composicion de pigmentos como base para el esclarecimiento de su situacion taxonomica. PhD thesis, Universidad de Sevilla, Spain.

Lubián, L.M., Blasco, J. and Establier, R. (1992) A comparative study of acid and alkaline phosphatase activities in several strains of *Nannochloris* (Chlorophyceae) and *Nannochloropsis* (Eustigmatophyceae). *British Phycological Journal* 27, 119–130.

Mague, T.H., Mague, F.C. and Holm-Hansen, O. (1977) Physiology and chemical composition of nitrogen-fixing phyoplankton in the central North Pacific Ocean. *Marine Biology* 41, 213–227.

Mahasneh, I.A., Grainger, S.L.J. and Whitton, B.A. (1990) Influence of salinity on hair formation and phosphatase activities of the blue-green alga (cyanobacterium) *Calothrix viguieri* D253. *British Phycological Journal* 25, 25–32.

Mann, N.H. (2000) Detecting the environment. In: Whitton, B.A. and Potts, M. (eds) *Ecology of Cyanobacteria*. Kluwer Academic, Dordrecht, pp. 367–395.

McCarthy, J.J. and Carpenter, E.I. (1979) *Oscillatoria* (*Trichodesmium*) *thiebautii* (Cyanophyta) in the central North Atlantic Ocean. *Journal of Phycology* 15, 75–82.

McCormick, P.V. and Stevenson, R.J. (1998) Periphyton as a tool for ecological assessment and management in the Florida Everglades. *Journal of Phycology* 34, 726–733.

McCormick, P.V., Rawlik, P.S., Lurding, K., Smith, E.P. and Sklar, F.H. (1996) Periphyton–water quality relationships along a nutrient gradient in the northern Everglades. *Journal of the North American Benthological Society* 15, 433–449.

Mühling, M. (2000) Characterization of *Arthrospira* (Spirulina) strains. PhD thesis, University of Durham, UK.

Mulholland, P.J. and Rosemond, A.D. (1992) Periphyton response to longitudinal nutrient depletion in a woodland stream: evidence of upstream–downstream linkage. *Journal of the North American Benthological Society* 11, 405–419.

Müller-Niklas, G., Heissenberger, A., Pukaric, S. and Herndl, G.J. (1995) Ultraviolet-B radiation and bacterial metabolism in coastal waters. *Aquatic Microbial Ecology* 9, 111–116.

Myklestad, S. and Sakshaug, E. (1983) Alkaline phosphatase activity of *Skeletonema costatum* populations in the Trondheimsfjord. *Journal of Plankton Research* 5, 557–564.

Natesan, R. and Shanmugasundaram, S. (1989) Extracellular phosphate solubilization by the cyanobacterium *Anabaena* ARM310. *Journal of Bioscience* 14, 203–208.

Nausch, M. (1997) Microbial activities on *Trichodesmium* colonies. *Marine Ecology Progress Series* 141, 173–181.

Nausch, M. (1998) Alkaline phosphatase activities and the relationship to inorganic phosphate in the Pomeranian Bight (southern Baltic Sea). *Aquatic Microbial Ecology* 16, 87–94.

Nedoma, J, Štrojsová, A., Vrba, J., Komárková, J. and Šimek, K. (2003) Extracellular phosphatase activity of natural plankton studied with ELF97 phosphate: fluorescence quantification and labelling kinetics. *Environmental Microbiology* 5, 462–472.

Newman, S., McCormick, P.V. and Backus, J.G. (2003) Phosphatase activity as an early warning indicator of wetland eutrophication: problems and prospects. *Journal of Applied Phycology* 15, 45–59.

Nichols, H.W. and Bold, H.C. (1965) *Trichosarcina polymorpha* gen. et sp. nov. *Journal of Phycology* 1, 34–38.

Nienow, J.A. and Friedmann, E.I. (1993) Terrestrial lithophytic (rock) communities. In: Friedmann, E.I. (ed.) *Antarctic Microbiology*. Wiley-Liss, New York, pp. 353–412.

Paasche, E. and Erga, S.R. (1988) Phosphorus and nitrogen limitation in the Oslofjord (Norway). *Sarsia* 73, 229–243.

Palisano, J.R. and Walne, P. (1972) Acid phosphatase activity and ultrastructure of aged *Euglena gracilis*. *Journal of Phycology* 8, 81–88.

Pentecost, A. and Whitton, B.A. (2000) Limestones. In: Whitton, B.A. and Potts, M. (eds) *Ecology of Cyanobacteria*. Kluwer Academic, Dordrecht, pp. 233–255.

Perry, M.J. (1972) Alkaline phosphatase activity in subtropical Central North Pacific waters. *Marine Biology* 15, 113–119.

Pettersson, K. (1980) Alkaline phosphatase activity and algal surplus phosphorus as phosphorus deficiency indicators in Lake Erken. *Archiv für Hydrobiologie* 89, 54–87.

Potts, M. (2000) *Nostoc*. In: Whitton, B.A. and Potts, M. (eds) *Ecology of Cyanobacteria*. Kluwer Academic, Dordrecht, pp. 465–504.

Potts, M., Sun, H., Mockaitis, K., Kennelly, P.J., Reed, D. and Tonks, N.K. (1993) A protein-tyrosine/serine phosphatase encoded by the genome of the cyanobacterium *Nostoc commune* UTEX 584. *Journal of Biological Chemistry* 268, 7632–7635.

Press, M.C. and Lee, J.A. (1983) Acid phosphatase activity in *Sphagnum* species in relation to phosphate nutrition. *New Phytologist* 93, 567–573.

Priest, F.G. (1984) *Extracellular Enzymes: Aspects of Microbiology 9*. Van Nostrand Reinhold, Wokingham, UK, 79 pp.

Ray, J.M., Bhaya, D., Block, M.A. and Grossman, A.R. (1991) Isolation, transcription, and inactivation of the gene for an atypical alkaline phosphatase of *Synechococcus* sp. strain PCC 7942. *Journal of Bacteriology* 173, 4297–4309.

Redfield, A.C. (1958) The biological control of chemical factors in the environment. *American Scientist* 46, 205–221.

Reichardt, W. (1971) Catalytic mobilization of phosphate in lake water and by Cyanophyta. *Hydrobiologia* 38, 3–4.

Rejmánková, E. and Komárková, J. (2000) A function of cyanobacterial mats in phosphorus-limited tropical wetlands. *Hydrobiologia* 431, 135–153.

Rengefors, K., Pettersson, K., Blenckner, T. and Anderson, D.M. (2001) Species-specific alkaline phosphatase activity in freshwater spring phytoplankton: application of a novel method. *Journal of Plankton Research* 23, 425–441.

Rengefors, K., Ruttenberg, K.C., Haupert, C.L., Taylor, C. and Howes, B.L. (2003) Experimental investigation of taxon-specific response of alkaline phosphatase activity in natural freshwater phytoplankton. *Limnology and Oceanography* 48, 1167–1171.

Rippka, R., Deruelles, J.B., Waterbury, J.B., Herdman, M. and Stanier, R.Y. (1979) Generic assignments, strain histories and properties of pure cultures of cyanobacteria. *Journal of General Microbiology* 111, 1–61.

Rippka, R., Coursin, T., Hess, W., Lichtlé, C., Scanlan, D.J., Palinska, K.A., Iteman, I., Partensky, F., Houmard, J. and Herdman, M. (2000) *Prochlorococcus marinus* Chisholm *et al.* 1992 subsp. *pastoris* subsp. nov. strain PCC 9511, the first axenic chlorophyll *a2*/*b2*-containing cyanobacterium (Oxyphotobacteria). *International Journal of Systematic and Evolutionary Microbiology* 50, 1833–1847.

Rivkin, R.B. and Swift, E. (1980) Characterization of alkaline phosphatase and organic phosphorus utilization in the oceanic dinoflagellate *Pyrocystis noctiluca*. *Marine Biology* 61, 1–8.

Romani, A.M. (2000) Characterization of extracellular enzyme kinetics in two Mediterranean streams. *Archiv für Hydrobiologie* 148, 99–117.

Rott, E., Walser, L. and Kegele, M. (2000) Ecophysiological aspects of macroalgal seasonality in a gravel stream in the Alps (River Isar, Austria). *Verhandlungen Internationalen Vereinigung Limnologie* 27, 1622–1625.

Roychaudhury, P. and Kaushik, B.D. (1989) Solubilization of Mussorie rock phosphate by cyanobacteria. *Current Science* 58, 569–570.

Sabater, S., Buchaca, T., Cambra, J., Catalan, J., Guasch, H., Ivorra, N., Muñoz, I., Navarro, E., Real, M. and Romani, A. (2003) Structure and function of benthic algal communities in an extremely acid river. *Journal of Phycology* 39, 481–489.

Sakshaug, E., Graneli, E., Elbrächter, M. and Kayser, H. (1984) Chemical composition and alkaline phosphatase activity of nutrient-saturated and P-deficient cells of four marine dinoflagellates. *Journal of Experimental Marine Biology and Ecology* 77, 241–254.

Satake, K. and Shibata, K. (1986) Bacterial invasion of the cell wall of an aquatic bryophyte *Scapania undulata* (L.) Dum. in both acidic and near-neutral conditions. *Hikobia* 9, 361–365.

Scanlan, D.J. (2003) Physiological diversity and niche adaptation in marine *Synechococcus*. *Advances in Microbial Physiology* 47, 1–64.

Scanlan, D.J. and Wilson, W.H. (1999) Application of molecular techniques to addressing the role of P as a key effector in marine ecosystems. *Hydrobiologia* 401, 149–175.

Schaffelke, B. (2001) Surface alkaline phosphatase activities of macroalgae on coral reefs of the central Great Barrier Reef, Australia. *Coral Reefs* 19, 310–317.

Scholz, O. and Boon, P.I. (1993) Alkaline phosphatase, aminopeptidase and β-D glucosidase activities associated with billabong periphyton. *Archiv für Hydrobiologie* 126, 429–443.

Sentenac, H. and Grignon, C. (1985) Effect of pH on orthophosphate uptake by corn roots. *Plant Physiology* 77, 136–141.

Sheath, R.G. and Müller, K.M. (1997) Distribution of stream macroalgae in four high Arctic drainage basins. *Arctic* 50, 355–364.

Sinclair, C. and Whitton, B.A. (1977) Influence of nutrient deficiency on hair formation in the Rivulariaceae. *British Phycological Journal* 12, 297–313.

Siuda, W. (1984) Phosphatases and their role in organic phosphorus transformation in natural waters: a review. *Polish Archives of Hydrobiology* 31, 207–233.

Siuda, W. and Güde, H. (1994) The role of phosphorus and organic carbon compounds in regulation of alkaline phosphatase activity and P regeneration processes in eutrophic lakes. *Polish Archives of Hydrobiology* 41, 171–187.

Siuda, W. and Güde, H. (1996) Evaluation of dissolved DNA and nucleotides as potential sources of phosphorus for plankton organisms in Lake Constance. In: Simon, M. (ed.) *Aquatic Microbial Ecology*. Stuttgart, pp. 155–162.

Spijkerman, E. and Coesel, P.F.M. (1998) Alkaline phosphatase activity in two planktonic desmid species and the possible role of an extracellular envelope. *Freshwater Biology* 39, 503–513.

Stevenson, P.A.R. (1994) Surface phosphatase activity of *Peltigera* and *Cladonia* lichens. MSc thesis, University of Durham, UK.
Stewart, A.J. and Wetzel, R.G. (1982a) Phytoplankton contribution to alkaline phosphatase activity. *Archiv für Hydrobiologie* 93, 265–271.
Stewart, A.J. and Wetzel, R.G. (1982b) Influence of dissolved humic materials on carbon assimilation and alkaline phosphatase activity in natural algal-bacterial assemblages. *Freshwater Biology* 12, 369–380.
Stihl, A., Sommer, U. and Post, A.F. (2001) Alkaline phosphatase activities among populations of the colony-forming diazotrophic cyanobacterium *Trichodesmium* spp. (Cyanobacteria) in the Red Sea. *Journal of Phycology* 37, 310–317.
Štrojsová, A., Vrba, J., Nedoma, J., Komárková, J. and Znachor, P. (2003) Seasonal study of extracellular phosphatase expression in the phytoplankton of a eutrophic reservoir. *European Journal of Phycology* 38, 295–306.
Thiébart-Fassey, I. and Hervagault, J.-F. (1993) Combined effects of diffusional hindrances, electrostatic repulsion and product inhibition on the kinetic properties of a bound acid phosphatase. *FEBS Letters* 334, 89–94.
Truhaut, R., Ferard, J.F. and Jouany, J.M. (1989) Cadmium IC50 determination on *Chlorella vulgaris* involving different parameters. *Ecotoxicology and Environmental Safety* 4, 215–223.
Turner, B.L., Baxter, R., Ellwood, N.T.W. and Whitton, B.A. (2001) Characterization of the phosphatase activities of mosses in relation to their environment. *Plant, Cell and Environment* 24, 1165–1176.
Turner, B.L., Baxter, R., Ellwood, N.T.W. and Whitton, B.A. (2003a) Seasonal phosphatase activities of mosses from Upper Teesdale, northern England. *Journal of Bryology* 25, 189–200.
Turner, B.L., Baxter, R. and Whitton, B.A. (2003b) Nitrogen and phosphorus in soil solutions and drainage streams in Upper Teesdale, northern England: implications of organic compounds for biological nutrient limitation. *Science of the Total Environment* 314/316C, 153–170.
Turner, B.L., Chudek, J.A., Whitton, B.A. and Baxter, R. (2003c) Phosphorus composition of upland soils polluted by long-term atmospheric nitrogen deposition. *Biogeochemistry* 65, 259–274.
Tyrrell, T. (1999) The relative influences of nitrogen and phosphorus on oceanic primary production. *Nature* 400, 525–531.
Uchida, T. (1992) Alkaline phosphatase and nitrate reductase activities in *Prorocentrum micans* Ehrenberg. *Bulletin of the Plankton Society of Japan* 38, 85–92.
US Environmental Protection Agency (1971) *Algal Assay Procedure Bottle Test.* United States Environmental Protection Agency, Corvallis, Oregon.
van Boekel, W.H.M. and Veldhuis, M.J.W. (1990) Regulation of alkaline phosphatase synthesis in *Phaeocystis* sp. *Marine Ecology Progress Series* 61, 281–189.
Vincent, W.F. (1981) Rapid physiological assays for nutrient demand by the plankton. II. Phosphorus. *Journal of Plankton Research* 3, 699–710.
Wagner, F. and Falkner, G. (2001) Phosphate limitation. In: Rai, L.C. and Gaur, J.P. (eds) *Algal Adaptation to Environmental Stresses.* Springer, Heidelberg, pp. 65–110.
Wagner, K.-U., Masepohl, B. and Pistorius, E.K. (1995) The cyanobacterium *Synechococcus* sp. strain PCC 7942 contains a second alkaline phosphatase encoded by *phoV. Microbiology* 141, 3049–3058.
Weckesser, J., Hofmann, K., Jürgens, U.J., Whitton, B.A. and Raffelsberger, B. (1988) Isolation and chemical analysis of sheath from the filamentous cyanobacteria *Calothrix parietina* and *C. scopulorum. Journal of General Microbiology* 134, 629–634.
Weich, R.G. and Granéli, E. (1989) Extracellular alkaline phosphatase activity in *Ulva lactuca* L. *Journal of Experimental Marine Biology and Ecology* 129, 33–44.
Wetzel, R.G. (1991) Extracellular enzymatic interactions: storage, redistribution, and interspecific communication. In: Chróst, R.J. (ed.) *Microbial Enzymes in Aquatic Environments.* Springer, New York, pp. 6–28.
Wetzel, R.G. (1992) Gradient-dominated ecosystems: sources and regulatory functions of dissolved organic matter in freshwater ecosystems. *Hydrobiologia* 229, 181–198.
Whitton, B.A. (1987a) The biology of Rivulariaceae. In: Fay, P. and Van Baalen, C. (eds) *The Cyanobacteria: a Comprehensive Review.* Elsevier, Amsterdam, pp. 513–534.
Whitton, B.A. (1987b) Survival and dormancy of blue-green algae. In: Henis, Y. (ed.) *Survival and Dormancy of Micro-organisms.* John Wiley & Sons, New York, pp. 209–266,
Whitton, B.A. (1988) Hairs in eukaryotic algae. In: Round, F.E. (ed.) *Algae and the Aquatic Environment: Contributions in Honour of J.W.G. Lund.* Biopress, Bristol, pp. 446–460.
Whitton, B.A. (1989) *Calothrix.* In: Section 19, Oxygenic Photosynthetic Bacteria. *Bergey's Manual of Systematic Bacteriology,* Vol. 3. Lippincott, Williams and Wilkins, Philadelphia, pp. 1791–1794.
Whitton, B.A. (1991) Use of phosphatase assays with algae to assess phosphorus status of aquatic

environments. In: Jeffrey, D.W. and Madden, B. (eds) *Bioindicators and Environmental Management: Proceedings of the 6th International Bioindicators Symposium.* Academic Press, London, pp. 295–310.

Whitton, B.A. (1992) Diversity, ecology and taxonomy of the cyanobacteria. In: Carr, N.G. and Mann, N.H. (eds) *Photosynthetic Prokaryotes.* Plenum, New York, pp. 1–51.

Whitton, B.A. (2000) Soils and rice-fields. In: Whitton, B.A. and Potts, M. (eds) *Ecology of Cyanobacteria.* Kluwer Academic, Dordrecht, pp. 257–259.

Whitton, B.A. (2003) Use of plants for monitoring heavy metals in freshwaters. In: Ambasht, R.S. and Ambasht, N.K. (eds) *Modern Trends in Applied Aquatic Ecology.* Kluwer Academic/ Plenum, New York, pp. 43–63.

Whitton, B.A. and Harding, J.P.C. (1978) Influence of nutrient deficiency on hair formation in *Stigeoclonium. British Phycological Journal* 13, 65–68.

Whitton, B.A., Khoja, T.M. and Arif, I.A. (1986) Water chemistry and algal vegetation of streams in the Asir Mountains, Saudi Arabia. *Hydrobiologia* 133, 97–106.

Whitton, B.A., Aziz, A. and Rother, J.A. (1988) Ecology of deepwater rice-fields in Bangladesh. 3. Associated algae and macrophytes. *Hydrobiologia* 169, 31–42.

Whitton, B.A., Potts, M., Simon, J.W. and Grainger, S.L.J. (1990) Phosphatase activity of the blue-green alga (cyanobacterium) *Nostoc commune* UTEX 584. *Phycologia* 29, 139–145.

Whitton, B.A., Grainger, S.L.J., Hawley, G.R.W. and Simon, J.W. (1991) Cell-bound and extracellular phosphatase activities of cyanobacterial isolates. *Microbial Ecology* 21, 85–98.

Whitton, B.A., Yelloly, J.M., Christmas, M. and Hernández, I. (1998) Surface phosphatase activity of benthic algae in a stream with highly variable ambient phosphate concentrations. *Verhandlung Internationale Vereinigung Limnologie* 26, 967–972.

Whitton, B.A., Clegg, E., Christmas, M., Gemmell, J.J. and Robinson, P.J. (1999) *Development of Phosphatase Assay for Monitoring Nutrients in Rivers: Methodology Manual for Measurement of Phosphatase Activity in Mosses and Green Algae in Rivers.* Environment Agency R & D Technical Report No. E106. Environment Agency North-East Region, Newcastle upon Tyne , UK, 46 pp.

Wood, P., Peat, A. and Whitton, B.A. (1986) Influence of phosphorus status on fine structure of the cyanobacterium (blue-green alga) *Calothrix parietina. Cytobios* 47, 89–99.

Wynne, D. and Rhee, G.Y. (1988) Changes in alkaline phosphatase activity and phosphate uptake in P-limited phytoplankton, induced by light intensity and spectral quality. *Hydrobiologia* 160, 173–178.

Xie, W.Q., Whitton, B.A., Duncan, L., Jäger, K., Reed, D. and Potts, M. (1989) *Nostoc commune* UTEX 584 gene expressing indole phosphate hydrolase activity in *Escherichia coli. Journal of Bacteriology* 171, 708–713.

Yelloly, J.M. and Whitton, B.A. (1996) Seasonal changes in ambient phosphate and phosphatase activities of the cyanobacterium *Rivularia atra* in intertidal pools at Tyne Sands, Scotland. *Hydrobiologia* 325, 201–212.

Yentsch, C.M., Yentsch, C.S. and Perras, J.P. (1972) Alkaline phosphatase activity in the tropical marine blue-green algae *Oscillatoria erythrea* ('Trichodesmium'). *Limnology and Oceanography* 17, 772–774.

Zarrouk, C. (1966) Contribution à l'étude d'une cyanophycée: influence de divers facteurs physiques et chimiques sur la croissance et la photosynthèse de *Spirulina maxima* (Setchell et Gardner) Geitler. PhD thesis, Faculté des Sciences de l'Université de Paris, France.

11 Organic Phosphorus Dynamics in Tropical Agroecosystems

Generose Nziguheba[1] and Else K. Bünemann[2]

[1]*Laboratory of Soil and Water Management, Kasteelpark Arenberg 20, 3001 Heverlee, Belgium;* [2]*Institute of Plant Sciences, Swiss Federal Institute of Technology (ETH), Eschikon 33, PO Box 185, CH-8315 Lindau, Switzerland*

Introduction

The diversity of soil types in the tropics is at least as great as in the temperate zone (Richter and Babbar, 1991), but tropical soils tend to be in later stages of pedogenesis due to the absence of recent glaciations. The duration and intensity of weathering affect the concentrations and chemical forms of phosphorus contained in the soil, with total amounts of organic phosphorus increasing until calcium-bound phosphorus has been dissolved. In the later stages of soil development, the amount of organic phosphorus declines together with total soil phosphorus, while the proportion of organic phosphorus increases, especially in relation to labile phosphate. The conceptual model developed by Walker and Syers (1976) from soil sequences in New Zealand has been confirmed for a Hawaiian chronosequence (Crews *et al.*, 1995) and by a literature review of soil phosphorus fractions in natural ecosystems (Cross and Schlesinger, 1995). The prevalence of highly weathered soils in the tropics makes organic phosphorus potentially more important for plant availability than in temperate soils. The relationships between chemically extracted phosphorus fractions in 168 soils suggested that labile phosphate is derived mainly from stable soil phosphate fractions in slightly weathered soils, whereas in highly weathered soils much of the variation in labile phosphate is explained by organic phosphorus forms (Tiessen *et al.*, 1984).

The relative importance of organic phosphorus for plant nutrition increases under conditions of phosphorus deficiency. In tropical soils, phosphorus deficiency can result from low total amounts of phosphorus and/or strong sorption of phosphate, mainly on aluminium and iron oxides. According to Fairhurst *et al.* (1999) almost half of all tropical soils are phosphorus-deficient, mainly highly weathered Oxisols and Ultisols (37% and 34%, respectively). In industrialized countries of the temperate zone, phosphorus deficiency as a constraint for crop production has generally been overcome by the widespread use of phosphorus fertilizers. However, large increases in fertilizer use since 1960 in tropical regions have occurred only in East Asia and parts of South America, where cropping systems with high external inputs are commonplace (Fairhurst *et al.*, 1999).

Tropical agroecosystems cover a wide range of intensity, including agroforests, intensive plantations of perennials such as coffee, tea and oil palm, annual crops cultivated at all intensities (from slash-and-burn systems with up to 20 years of bush fallow between cropping phases and no external

 Organic Phosphorus in the Environment (eds B.L. Turner, E. Frossard and D.S. Baldwin)

inputs, to continuous cropping with up to three crops per year in the case of lowland rice), as well as animal production systems based on extensively managed grassland or intensive pastures. The tropics also comprise diverse climatic conditions, ranging from (per)humid to semi-arid and arid zones, which greatly affect biological processes. For example, the decomposition rates found in West Africa decrease along with amounts of rainfall (Buerkert *et al.*, 2000), while in the humid tropics, the decomposition of plant residues follows a pattern similar to that in the temperate zone, but four times as fast (Jenkinson and Ayanaba, 1977). The diversity of agroecological zones and management practices in the tropics may render organic phosphorus dynamics, defined as changes in amounts and forms of organic phosphorus resulting from mineralization and immobilization processes, more complex than in the temperate zone.

The dynamics of soil organic phosphorus have been summarized conceptually (Stewart and Tiessen, 1987), methodologically (Tiessen *et al.*, 1994) and functionally (Magid *et al.*, 1996). In this chapter we review recent advances in our understanding of the role of organic phosphorus in tropical agroecosystems, with special emphasis on the effects of management on soil organic phosphorus dynamics. In an introduction to the analytical methods used to determine concentrations and forms of organic phosphorus, we summarize recent results from their application to soils under native vegetation or other control sites in the tropics. Thereafter, we describe the different approaches used to study organic phosphorus dynamics in tropical soils, and assess practices favouring its decline or accumulation in tropical agroecosystems.

Concentrations and Forms of Organic Phosphorus in Tropical Soils

Soil organic phosphorus includes phosphorus in living soil organisms and dead organic matter. Soil bacteria and fungi constitute the bulk of phosphorus in soil organisms, while their turnover is decided by higher trophic groups (Magid *et al.*, 1996). Soil microorganisms can mineralize organic phosphorus as well as immobilizing phosphate from the soil solution. Likewise, plants absorb phosphate from the soil solution to produce biomass. The *in situ* addition of organic material to soil from litterfall, root decay, crop residues, animal excreta or dead soil organisms provides the starting point for soil organic phosphorus formation. However, the phosphorus contained in all these materials may be in organic as well as inorganic forms. Stabilization of organic phosphorus in the soil occurs mainly through adsorption of the phosphorus group (Magid *et al.*, 1996), and may occur rapidly following the addition of organic matter (see Celi and Barberis, Chapter 6, this volume). In the following sections, we review the analytical methods used to quantify and characterize soil organic phosphorus and discuss their application to tropical soils.

Determination of total soil organic phosphorus by ignition and sequential extraction

Total soil organic phosphorus has been determined for decades by the increase in acid-extractable phosphate following the destruction of organic matter by high-temperature ignition (Saunders and Williams, 1955). This method relies on complete organic phosphorus destruction and extraction of the released phosphate and assumes that phosphorus solubility does not change during ignition, which is often not the case. For this reason, the ignition method may overestimate soil organic phosphorus in highly weathered soils (Condron *et al.*, 1990), as compared with other methods for total organic phosphorus measurement that involve various sequential alkali- and acid-extractions (Anderson, 1960; Bowman, 1989). In the majority of recent studies from the tropics, total soil organic phosphorus was therefore presented as the sum of sequentially extracted fractions following the procedures of Hedley *et al.* (1982) and Tiessen and Moir (1993).

Sequential phosphorus extraction partitions soil phosphorus into inorganic and organic pools with different extractability. In the Hedley *et al.* (1982) fractionation scheme, soil is sequentially extracted with anion exchange resin, 0.5 M $NaHCO_3$, 0.1 M NaOH (before and after sonication) and 1 M HCl, followed by digestion of the soil residue with concentrated sulphuric acid and hydrogen peroxide. The residual fraction often contains between 20% and 60% of the total phosphorus and a significant proportion of the organic phosphorus, so Tiessen and Moir (1993) added an additional extraction step with concentrated hydrochloric acid. While extracts with 1 M HCl contain only inorganic phosphate associated with calcium, which is mostly absent in highly weathered soils, all other extracts contain both inorganic and organic phosphorus. The indirect determination of organic phosphorus as the difference between total phosphorus and inorganic phosphate constitutes a source of variation and error. For example, phosphate in sodium bicarbonate and sodium hydroxide extracts is determined after acid precipitation of organic matter, but in highly weathered soils phosphate associated with iron or aluminium hydroxides may also precipitate with organic matter, resulting in an overestimation of organic phosphorus (Tiessen and Moir, 1993). The acid-precipitation step constitutes a limitation to the application of the sequential phosphorus extraction to highly weathered soils and soils rich in organic matter.

Studies in the tropics have reported organic phosphorus in the supernatant after resin extraction, in the sodium bicarbonate extract, in the sodium hydroxide extract before sonication, in the sodium hydroxide extract after sonication, and in the concentrated hydrochloric acid extract. Sodium bicarbonate recovers easily hydrolysable organic phosphorus (Bowman and Cole, 1978), while sodium hydroxide recovers organic phosphorus associated with humic and fulvic acids. Concentrated hydrochloric acid recovers organic phosphorus from more stable pools, although it may also extract bioavailable organic phosphorus from particulate organic matter (Tiessen and Moir, 1993). Despite this, many studies have focused on organic phosphorus in bicarbonate and sodium hydroxide extracts, because these fractions are believed to become mineralized within a time frame relevant for plant growth (Cross and Schlesinger, 1995). However, it is far from clear that this is the case, and a recent study indicated that the bioavailability of organic phosphorus fractions varies depending on soil type and plant species (Chen *et al.*, 2002).

Up to 90% of soil phosphorus can be in organic form (Harrison, 1987), although in tropical soils the proportion of organic phosphorus under native vegetation and other control sites reported in recent studies ranges between 16% and 44% (Table 11.1), with an exceptional value of 65% reported for the surface 5 cm under primary forest in the Amazon (Lehmann *et al.*, 2001c). Lower proportions have sometimes been reported from the semi-arid and arid tropics, where biomass production is limited (Tiessen *et al.*, 1992; Giardina *et al.*, 2000; Solomon and Lehmann, 2000). In all the studies summarized in Table 11.1, organic phosphorus was the sum of that extracted sequentially by sodium bicarbonate and sodium hydroxide, plus in some cases the organic phosphorus extracted by sonication and sodium hydroxide, and concentrated hydrochloric acid. It therefore seems likely that the previously reported high proportions of organic phosphorus can be attributed to differences in methods as well as methodological problems. Sequential fractionation can underestimate organic phosphorus, especially if the residue after extraction with sodium hydroxide is not further separated. In a sandy Oxisol in northeastern Brazil, this residue contained an additional 10–20% organic phosphorus, reducing the reported ratios of organic carbon to organic phosphorus from 430 to about 250 (Tiessen *et al.*, 1992).

Organic carbon to organic phosphorus ratios in most studies lie between 100 and 300, but lower and higher ratios have been reported (Table 11.1). The range of total phosphorus values (61–1780 mg P/kg soil) illustrates that while many weathered soils are low in available phosphate, they are not necessarily low in total phosphorus. Total

Table 11.1. Total phosphorus concentration, proportion of organic phosphorus, and organic carbon to organic phosphorus ratio in soils under native vegetation and other controls in studies in the tropics.[a]

Reference	Vegetation	Location	Soil type	Texture	Depth (cm)	Total phosphorus (mg/kg)	Organic phosphorus (% total P)	Organic carbon to organic phosphorus ratio
Giardina *et al*. (2000)	Dry tropical forest	C Mexico	Typic Ustorthent	sandy loam	0–5	555	33	171
Szott and Melendez (2001)	a) Forest (80 year)	Costa Rica	Typic Humitropept (Inceptisol)	34% clay	0–8	750	22	341
	b) Forest (80 year)	N Peru	Typic Paleudult (Ultisol)	22% clay	0–8	178	43	169
Beck and Sanchez (1996)	Secondary forest (17 year)	N Peru	Typic Paleudult	72% sand	0–15	109	44	nd
Lehmann *et al*. (2001c)	Primary forest (*Oenocarpus*)	N Brazil	Typic Hapludox (Ferralsol)	80% clay	0–5	61	65	1149[c]
Garcia-Montiel *et al*. (2000)	a) Forest	N Brazil	Kandiudult (Ultisol)	75% sand	0–10	136	21	435
	b) Forest	N Brazil	Paleudult (Ultisol)	73% sand	0–10	243	21	224
Tiessen *et al*. (1992)	Thorn bush savanna	NE Brazil	Oxisol (Paleustox)	79% sand	0–20	123	22	430
Neufeldt *et al*. (2000)	a) Native savanna	C Brazil	Anionic Acrustox	17% sand	0–12	371	22	283
	b) Native savanna	C Brazil	Typic Haplustox	76% sand	0–12	116	34	251
Lilienfein *et al*. (2000)	Native savanna	C Brazil	Anionic Acrustox (Oxisol)	21% sand	0–15	388	29	221
Friesen *et al*. (1997)	Native savanna	E Colombia	Tropeptic Haplustox (Oxisol)	silty clay	0–10	181	35	429[d]
Oberson *et al*. (2001)	Native savanna	E Colombia	Tropeptic Haplustox (Oxisol)	silty clay	0–10	212	38	324
Solomon *et al*. (2002)	a) Native forest	S Ethiopia	Plinthic Alisol	10% sand	0–10	1343	38[b]	160
	b) Native forest	S Ethiopia	Humic Nitisol	13% sand	0–10	1426	42[b]	169
Solomon and Lehmann (2000)	Native woodland	N Tanzania	Chromic Luvisol	49% sand	?	739	19	131
Agbenin and Goladi (1998)	Native savanna	N Nigeria	Typic Haplustalf (Alfisol)	81–91% sand	0–15	337	20	104
Reddy *et al*. (1999)	Not given (initial)	C India	Typic Haplustert (Vertisol)	clay-loam	0–15	549	16	51
Sattell and Morris (1992)	Fields	Sri Lanka	Alfisols	8–31% clay	0–15	191	44	113
Dobermann *et al*. (2002)	a) Grass fallow	Leyte, Philippines	Typic Kandiudult	35% clay	0–15	420	23	131
	b) Fallow (5 year)	Sumatra, Indonesia	Typic Paleumult	54% clay	0–15	402	30	136
	c) Abandoned field (5 year)	Luzon, Philippines	Typic Haplustox	63% clay	0–15	679	41	86
Linquist *et al*. (1997)	45 year pasture	Maui, Hawaii	Typic Palehumult (Ultisol)	60% clay	0–25	1780	18	101

nd = not determined.
[a] Results on management effects in these studies are summarized in Tables 11.4 and 11.5 (wherever applicable).
[b] Value read from figure.
[c] Carbon concentrations from Lehmann *et al*. (2001a).
[d] Carbon concentrations from Bühler *et al*. (2002).

phosphorus concentrations are mainly determined by the parent material, whereas phosphate availability decreases during pedogenesis. Low total phosphorus concentrations are usually reported in South American or sandy soils (Table 11.1). The distribution of organic phosphorus among the various fractions is not directly comparable, as different extraction schemes were applied, but in all studies, organic phosphorus extractable by sodium hydroxide was the largest fraction, containing between 55% and 82% of the total organic phosphorus (data not shown).

Microbial phosphorus

Microbial phosphorus is quantified after fumigation of soil with chloroform or hexanol to kill soil microorganisms (Brookes *et al.*, 1982). The various modifications of the method, especially in terms of extractants and use of calibration factors, are described elsewhere in this volume (see Oberson and Joner, Chapter 7). For highly weathered tropical soils, the fumigation–extraction method with anion exchange resin membranes (Kouno *et al.*, 1995) or Bray (NH_4F–HCl) extraction (Oberson *et al.*, 1997) may be most suitable.

Microorganisms affect soil phosphorus dynamics not only through organic phosphorus mineralization, but also through phosphate uptake, which may prevent sorption if release during microbial turnover is synchronized with crop demand (Magid *et al.*, 1996). Thus, the microbial biomass is considered as a labile reservoir of phosphorus, as well as other nutrients (Richardson, 1994). It has also been shown to be a sensitive indicator of management changes (Sparling, 1992), and examples from the tropics are given in Chapter 7 (Oberson and Joner, this volume).

Phosphorus in soil organic matter associated with the sand fraction

The poor sensitivity in the determination of chemically extracted organic phosphorus fractions, as well as the perception that organic phosphorus and soil organic matter dynamics are often closely linked, have resulted in various attempts to quantify phosphorus in soil particle size fractions, particularly the active organic matter fractions, usually taken as those in the sand fraction of the soil. The main difficulty in this method is to ensure the efficient dispersal of soil without disrupting organic matter structures, particularly in clay soils. While the ultrasonic procedure is widely used for dispersion (Oorts *et al.*, 2000; Smestad *et al.*, 2002), there are concerns about disruption of soil organic matter and consequent effects on its distribution in the different particle size fractions. Dispersion is often improved with sodium hexametaphosphate, but this cannot be used in phosphorus studies. Therefore, Maroko *et al.* (1999) used 2 h of shaking (150 reciprocations/min) in deionized water followed by wet sieving to isolate macro-organic matter (250–2000 μm), as well as flotation in a silica suspension (Ludox) to subdivide the 150–2000 μm fraction into three density fractions. Both procedures were followed by digestion to determine the phosphorus content. In other studies, soil was dispersed with a cation exchange resin saturated with sodium (Feller *et al.*, 1991; G. Nziguheba *et al.*, unpublished data).

A recent incubation study dispersed soil with 0.05 M NaCl and used wet sieving to isolate the 53–2000 μm fraction (particulate organic matter) at different times after the addition of plant residues to an Ultisol and an Alfisol (Salas *et al.*, 2003). This provided information on the origin of soil organic phosphorus associated with the sand fraction, because about 25% of the total phosphorus added with the residues was recovered as particulate organic matter directly after addition. This was plausible, because 63–73% of the total phosphorus in the plant residues was water-extractable phosphate. The appearance of phosphorus in particulate organic matter in the further course of the incubation suggested that within 5 days after addition, organic phosphorus in plant residues was released and was either mineralized or adsorbed to mineral surfaces. A subsequent increase in particulate organic matter was due to fungal colonization of particulate organic matter, together with immobilization of phosphorus stimulated by the presence of a carbon substrate. The processes operating during the rapid decline and subsequent build-up of particulate organic matter observed by Salas *et al.* (2003) require further investigation and verification. In any case, even if all the phosphorus contained in soil organic matter in the sand fraction originated from microbial uptake, such fractions may still be valuable indicators of labile organic phosphorus.

^{31}P NMR spectroscopy

Solution ^{31}P nuclear magnetic resonance (NMR) spectroscopy was introduced by Newman and Tate (1980) to define the structural composition of soil organic phosphorus. The technique allows the identification of several inorganic phosphorus forms (phosphate, pyrophosphate, polyphosphate) and organic phosphorus forms, such as phosphate monoesters (e.g. inositol phosphates and sugar phosphates), phosphate diesters (e.g. nucleic acids, phospholipids, teichoic acids) and phosphonates.

When combined with ^{13}C NMR spectroscopy, ^{31}P NMR can be used to evaluate linkages between carbon and phosphorus transformations (Gressel *et al.*, 1996; Möller *et al.*, 2000). The main advantage of solution ^{31}P NMR spectroscopy is that it allows direct identification of organic phosphorus forms, although a proportion of the soil organic phosphorus remains unidentified due to its incomplete recovery from soil. In the future, this could be overcome by solid-state ^{31}P NMR, but this is currently impractical due to low phosphorus concentrations and the presence of paramagnetic impurities, especially in tropical soils (Condron *et al.*, 1997).

Several recent studies have applied solution ^{31}P NMR spectroscopy to tropical soils (Guggenberger *et al.*, 1996; Mahieu *et al.*, 2000; Möller *et al.*, 2000; Solomon and Lehmann, 2000; Chapuis-Lardy *et al.*, 2001; Solomon *et al.*, 2002; Cardoso *et al.*, 2003). Various extraction protocols were followed, but extraction efficiency was generally poor, ranging between 3% and 36% of the total phosphorus (Table 11.2). The proportion of organic phosphorus in the extracts ranged between 30% and 51% of extracted phosphorus if extracts were not dialysed, and between 67% and 94% after dialysis. Phosphate monoesters were the dominant form of organic phosphorus extracted from topsoil samples, comprising 54–92% of extracted organic phosphorus, which is in accordance with the range of 60–90% cited by Turner *et al.* (2002). The only exception was a natural forest soil, in which only 39% of the extracted organic phosphorus was identified as phosphate monoesters (Solomon *et al.*, 2002).

Comparison between studies is limited by the use of different extraction protocols, and by differences in signal assignments. Chemical shifts are modified by the chemical environment around the phosphorus nuclei (e.g. sample pH), but two recent studies provide reference data for future research (Makarov *et al.*, 2002; Turner *et al.*, 2003a). They also show that concentrations of phosphate diesters can be underestimated due to the rapid degradation of some compounds, especially RNA and some phospholipids, to phosphate monoesters in alkaline extracts. Further details on solution ^{31}P NMR spectroscopy can be found elsewhere in this volume (see Cade-Menun, Chapter 2).

Combination of methods

Little is known about the relationship between organic phosphorus fractions determined by different quantification methods. From the few experiments where sequential phosphorus fractionation and ^{31}P NMR have been applied together (in bulk soil and in particle size fractions) in the tropics, it appears that phosphate diesters extracted by sodium bicarbonate are labile compounds from microbial biomass and easily decomposable organic matter associated with the sand fraction, while those in sodium hydroxide extracts are microbial

Table 11.2. Application of solution ^{31}P nuclear magnetic resonance spectroscopy to topsoil samples from agroecosystems in the tropics.

Reference	Extractant (soil-to-solution ratio)	Dialysis	Land use	Phosphorus recovery (%)		Spectral area (%)		
				Total phosphorus[a]	Organic phosphorus[a]	Organic phosphorus	Phosphate monoesters	Diester-to-monoester ratio[b]
Guggenberger *et al.* (1996)	0.1 M NaOH (1:4), then separation into humic and fulvic acids	yes	Native savanna	8	11	94	58	0.61
			Pasture (grass-only)	11	15	94	54	0.75
			Pasture (grass–legume)	10	15	93	54	0.74
Solomon and Lehmann (2000)	0.5 M NaOH (1:3)	no	Native woodland[c]	31	52	32	27	0.21
			Cultivated (3 year)	22	61	47	42	0.14
			Cultivated (15 year)	22	61	45	39	0.16
			Homestead field	26	45	30	24	0.25
Möller *et al.* (2000)	Three extractions with 0.1 M NaOH + 0.4 M NaF (1:5)	yes	Primary forest	35	67	92	58	0.45
			Secondary forest	30	76	94	59	0.54
			Reforestation	27	60	93	73	0.25
			Cabbage cultivation	17	40	82	65	0.23
Mahieu *et al.* (2000)	0.25 M NaOH (1:10), then separation into mobile gumic acid and calcium humate	yes	Dryland rice	13	31	83	60	0.36
			Lowland rice–soybean	3	10	81	57	0.43
			Lowland rice (2 crops/year)	3	16	89	56	0.59
			Lowland rice (3 crops/year)	5	20	87	53	0.64
Chapuis-Lardy *et al.* (2001)	0.5 M NaOH (1:4)	no	Native savanna	20	24	48	38	0.25
			Pasture (12 year)	22	33	47	38	0.24
Solomon *et al.* (2002)	3 extractions with 0.1 M NaOH + 0.4 M NaF (1:5)	yes	a) Natural forest[c]	22–36	not given	76	41	0.81
			a) Tea plantation			73	47	0.52
			a) Cultivation			80	61	0.28
			b) Natural forest[c]			67	27	1.43
			b) *Cupressus* plantation			82	54	0.43
			b) Cultivation			83	66	0.23
Cardoso *et al.* (2003)	0.5 M NaOH + 0.1 M EDTA (1:10), treatment with Chelex-X100 resin to remove paramagnetic ions	no	Agroforestry coffee, 15 year	31	76	43	39	0.10
			Agroforestry coffee, 19 year	17	75	51	47	0.08
			Sole coffee, 15–20 year	30	95	41	37	0.12
			Sole coffee, 20–24 year	25	65	35	32	0.10

[a] Total phosphorus determined by sequential fractionation or separate digestion, with organic phosphorus determined as the sum of sequentially extracted organic phosphorus fractions.
[b] Phosphate diesters include compounds that the authors named diester-P and additionally phosphonates in the case of Guggenberger *et al.* (1996), sugar-diesters in the case of Mahieu *et al.* (2000) and teichoic acid in the cases of Möller *et al.* (2000) and Solomon *et al.* (2002).
[c] Additional information on the sites studied by Solomon and Lehmann (2000) and Solomon *et al.* (2002) can be found in Table 11.1.

compounds which dominate the clay fraction (Solomon and Lehmann, 2000; Solomon *et al.*, 2002). In weathered soils, concomitant increases in organic phosphorus and organic carbon concentrations with decreasing particle size have suggested that organic phosphorus dynamics are closely linked with those of the soil organic matter (Agbenin and Tiessen, 1995; Solomon and Lehmann, 2000; Solomon *et al.*, 2002). Sodium hydroxide-extractable organic phosphorus represents the largest fraction in all particle size separates (Neufeldt *et al.*, 2000; Solomon and Lehmann, 2000; Solomon *et al.*, 2002). In African soils, cultivation reduced the proportion of organic phosphorus in all particle size fractions, with the largest depletion of extractable organic phosphorus (sodium bicarbonate and sodium hydroxide) in sand fractions followed by silt fractions (Solomon and Lehmann, 2000; Solomon *et al.*, 2002). In continuously cultivated Ethiopian soils, organic phosphorus in the clay fraction was dominated by phosphate diesters, while the largest proportion of phosphate monoesters was found in silt-size separates (Solomon *et al.*, 2002). The occurrence of phosphate diesters in the clay fraction, even after cultivation, reveals that not all such compounds are easily hydrolysable. Rubæk *et al.* (1999) reported that sodium hydroxide extracts of clays were enriched with microbially derived teichoic acids and other phosphate diesters, as compared to the silt and sand fractions. In addition, Solomon and Lehmann (2000) found finer size separates to be enriched in products of microbial metabolism. Microbial-derived diesters can be stabilized on mineral surfaces, whereas plant-derived diesters are more labile (Miltner *et al.*, 1998).

Dynamics of Soil Organic Phosphorus

Short- and medium-term dynamics

Methods and approaches

Short- and medium-term dynamics of soil organic phosphorus involve immobilization and mineralization processes. Immobilization is estimated either as an increase in microbial phosphorus, or as a decrease in available soil phosphate. Due to the high reactivity of phosphate with the soil solid phase, organic phosphorus mineralization cannot be measured by extraction or leaching of phosphorus from soils after various times of incubation, as in the measurement of nitrogen mineralization, except for sandy soils with zero phosphate sorption (Grierson *et al.*, 1998). Isotopic dilution methods, which are based on the assumption that phosphate released from organic phosphorus will decrease the specific activity of isotopically exchangeable phosphate in a soil labelled with ^{32}P or ^{33}P (Oehl *et al.*, 2001), are not applicable to phosphorus-deficient, strongly sorbing soils.

Phosphatase activity has been used as an indicator of potential organic phosphorus mineralization in tropical soils. Phosphatases are produced by microorganisms and plant roots, and may be stabilized on the soil surface and remain active for long periods (see Quiquampoix and Mousain, Chapter 5, this volume). A preincubation of samples in the absence of plant roots is sometimes used to favour enzymes of microbial origin (Renz *et al.*, 1999). Generally, phosphatase activity is related to levels of soil organic matter and microbial activity (Feller *et al.*, 1994; Oberson *et al.*, 1996, 2001; Renz *et al.*, 1999), is influenced by seasonal fluctuations in temperature and moisture (Grierson and Adams, 2000), and can be suppressed by the addition of mineral fertilizer (Colvan *et al.*, 2001). However, phosphatase activity does not indicate actual phosphorus mineralization rates (Renz *et al.*, 1999). Likewise, carbon mineralization provides an indicator of biological activity, but does not allow rates of organic phosphorus mineralization to be calculated.

Release of phosphorus from plant residues

In soil, the release of phosphorus contained in organic materials by mineralization depends largely on the quality of the organic material (Iyamuremye and Dick, 1996; Nziguheba *et al.*, 1998, 2000; Lehmann *et al.*, 2001b; Reddy *et al.*, 2001; Kwabiah *et al.*, 2003). There is growing evidence that both carbon and phosphorus in organic materials determine phosphorus mineralization and availability; for example, resin-extractable phosphorus in soil 1 week after the incorporation of various plant residues was negatively correlated with the ratio of water-soluble carbon to total phosphorus in the residues (Nziguheba *et al.*, 2000). The balance between mineralization and immobilization of phosphorus is partly influenced by the quality of carbon, presumably because microbial activity is closely related to soluble carbon (Chen *et al.*, 2002). Attempts to establish a general threshold for phosphorus mineralization similar to that for nitrogen mineralization have not been successful, probably due to the high reactivity of mineralized phosphorus with the soil solid phase.

The carbon-to-phosphorus ratio in organic materials has been used to predict the dominant biological process after addition to soils. Table 11.3 summarizes the influence of plant residue quality on immobilization and mineralization in studies conducted across the tropics. An organic carbon to organic phosphorus ratio in plant residues of about 200 appears to be the threshold for immobilization/mineralization processes. However, such a threshold can only be defined with respect to a given period of time, because the ratio in a given material changes during decomposition.

Solomon *et al.* (2002) calculated the organic carbon to organic phosphorus ratio in semi-arid Tanzanian soils to estimate the mineralization potential of soil organic phosphorus. They concluded that in these soils, organic phosphorus mineralization could occur when the ratio was below 200, which equals the threshold given by Dalal (1977). Buresh *et al.* (1997) proposed the combined use of extractable phosphate and the ratio of labile organic phosphorus to labile phosphate to assess the importance of organic phosphorus mineralization as a source of plant-available phosphorus. According to this approach, a low concentration of extractable phosphate, high labile organic phosphorus to phosphate ratio, and high microbial phosphorus, indicated a greater contribution of organic phosphorus mineralization to plant-available phosphorus compared with phosphate.

Attempts to quantify plant phosphorus uptake from applied plant residues are complicated by the fact that plant residues also contain nutrients other than phosphorus and can improve plant growth through indirect effects such as changes in soil pH, complexation of aluminium, and stimulation of phytohormone production by soil microorganisms. Large effects of plant residues are generally reported from extreme environments. For example, crop residues applied to the surface of acid sandy soils of the Sahel decreased the maximum soil surface temperature from 44 to 40°C and increased soil pH, termite activity, surface soil penetrability, and available phosphorus (Buerkert *et al.*, 2000). The direct contribution of phosphorus applied with plant residues to plant phosphorus uptake can therefore only be quantified via isotopic labelling; an approach limited to experiments of 2 or 3 months' duration due to the short half-lives of radioactive phosphorus isotopes. In highly weathered soils of the tropics, this has so far only been tested with a reverse isotopic dilution technique, applying *Setaria* residues with three different carbon-to-phosphorus ratios to soils that were labelled with ^{32}P and received five levels of phosphate addition just after residue addition (Umrit and Friesen, 1994). Phosphorus uptake by *Panicum* planted on these soils increased after the addition of residues with a carbon-to-phosphorus ratio of 77, whereas it was depressed in the first cutting after the addition of residues with ratios of 227 and 704. The cumulative phosphorus uptake of four cuttings from the latter two residues was generally similar to the non-amended controls. However, differences in phosphorus uptake after the addition of the three residues were apparent even at high levels of phosphate addition, raising the question of possible limitation by other nutrients. This was also indicated by the labile pool of phosphorus (*L*-value) calculated from isotopic labelling, which was higher in all residue-amended soils than in the control, even at the first cutting when phosphorus

Table 11.3. Effect of plant residue quality on phosphorus mineralization and immobilization.

Reference	Location	Soil type	Material	Phosphorus (g/kg)	Carbon-to-phosphorus ratio	Result
Lehmann *et al*. (2001b)	Amazon basin	Oxisol	*Theobroma grandiflorum*	1.01	446	Initial immobilization[a]
			Bixa orellana	2.88	156	Mineralization
Nziguheba *et al*. (2000)	Kenya	Oxisol	*Tithonia diversifolia*	2.8	142	Mineralization
			Sesbania sesban	1.9	210	Initial immobilization
Iyamuremye and Dick (1996)	Rwanda	Range of soils	Alfalfa (lucerne)	2.44	164	Mineralization
			Wheat straw	0.9	444	Immobilization[b]
Reddy *et al*. (2001)	India	Vertisol	Soybean	2.7	149	Mineralization
			Wheat	0.8	558	Immobilization
Umrit and Friesen (1994)	Malaysia	Oxisol and Ultisol	*Setaria sphacelata*	5.76	77	Mineralization
				2.12	227	Initial Immobilization
				0.72	704	Initial Immobilization
Kwabiah *et al*. (2003)	Kenya	Alfisol	*Gliricidia sepium*	3.4	132	Mineralization
			Calliandra calothyrsus	2.2	204	Mineralization in 1 day followed by no changes from control
			Azadirachta indica	2.0	225	Immobilization

[a] Immobilization occurred during a short (days to weeks) period after residue incorporation and was followed by net mineralization.
[b] Immobilization was reported throughout the study.

uptake was depressed in the case of two residues. In another study with a sandy red earth from Australia, net release from ^{32}P-labelled plant residues (calculated as the sum of ^{32}P recovered in soil phosphate and growing plants) showed no clear time trend between 15 and 39 days after residue addition, indicating that most of the phosphorus release had already occurred within the first 2 weeks (Armstrong and Helyar, 1993).

Isotopic labelling approaches to studying soil organic phosphorus dynamics

Soil organic phosphorus dynamics can be studied by isotopic labelling of soil phosphorus (Friesen and Blair, 1988; Bühler *et al.*, 2002), or by addition of a ^{32}P- or ^{33}P-labelled plant residues (Friesen and Blair, 1988; Daroub *et al.*, 2000), followed by sequential fractionation after various incubation periods. However, only Bühler *et al.* (2002) investigated highly weathered soils. After 2 weeks of incubation, recovery of the label in organic phosphorus fractions ranged from <5% in the case of phosphate-fertilized soils under annual crops, to 14% and 20% in soils under grass–legume pasture and native savanna, respectively, which were highest in soil organic matter and microbial biomass (Bühler *et al.*, 2002). Much of the labelled organic phosphorus may have been in the microbial biomass at the time of extraction, but this does not imply that it would have remained in organic form after microbial death. The inclusion of a fumigation step, suggested originally in the fractionation scheme (Hedley *et al.*, 1982), may help to separate microbial from overall organic phosphorus dynamics. On the other hand, this step would give a conservative estimate of organic phosphorus accumulation, because phosphatases released during fumigation hydrolyse organic phosphorus during extraction. This approach was tested recently when comparing microbial phosphorus uptake from ^{33}P-labelled soil or added ^{33}P-labelled plant residues in an incubation experiment with a Ferralsol from western Kenya under two different crop rotations (Bünemann, 2003). The percentage of label found in the microbial biomass was higher in the soil with more organic matter and microbial biomass, where 9% and 18% of labelled soil and residue phosphorus, respectively, were recovered in the microbial biomass after 10 days. In both cases, another 70% of the label was recovered when fumigated samples were subsequently extracted with 0.1 M NaOH. Complete separation of labelled phosphate and organic phosphorus in the sodium hydroxide extracts was not achieved, but different attempts indicated that in the case of labelling soil phosphorus, about 5% of the label was recovered in the sodium hydroxide-extractable organic phosphorus pool, compared with 15–25% in the case of labelled residues. These two studies clearly illustrate the role of the soil microbial biomass in organic phosphorus dynamics. In addition, isotopic labelling approaches demonstrate that significant redistribution can occur among different fractions, even in the absence of net changes.

Pot experiments

Organic phosphorus must be mineralized before it can be taken up by plants, and pot experiments have been conducted to relate changes in organic phosphorus fractions to plant phosphorus uptake. Such experiments have, however, yielded contradictory results. An early study with Nigerian soils during four consecutive maize crops revealed strong correlations between phosphorus uptake and organic phosphorus content at the beginning of the study, as well as with the 27% decrease in organic phosphorus observed at the end (Adepetu and Corey, 1976). Phosphorus uptake by millet in Neubauer experiments was also related to bicarbonate-extractable phosphate and sodium hydroxide-extractable organic phosphorus in several Sri Lankan Alfisols (Sattell and Morris, 1992). Lucerne (alfalfa) grown in soils from traditional Mexican agroecosystems, however, depleted only labile phosphate (resin and bicarbonate-extractable), suggesting that these soils with long cultivation histories may already have lost most of their mineralizable organic phosphorus (Crews, 1996). Guo and Yost (1998) planted 14 consecutive crops on eight Hawaiian soils in various stages of weathering; plant phosphorus uptake was related to sodium hydroxide-extractable organic phosphorus in two soils, and to bicarbonate-extractable organic phosphorus in all highly weathered soils, but organic phosphorus fractions generally showed no clear time trend, and appeared to fluctuate with microbial activity. Similarly, rhizosphere studies with tropical soils have either shown an accumulation (Kamh *et al.*, 1999; Zoysa *et al.*, 1999; George *et al.*, 2002), no change (Hedley *et al.*, 1994) or a depletion (Zoysa *et al.*, 1998; George *et al.*, 2002) of organic phosphorus in the rhizosphere. Processes governed by plant phosphorus uptake and microorganisms thriving in the carbon-rich environment of the rhizosphere may be difficult to distinguish in such studies. In addition, the amount of phosphorus contained in the seed often determines plant phosphorus uptake under conditions of phosphorus limitation (Hedley *et al.*, 1994; Bühler *et al.*, 2003).

Seasonal dynamics

Microbial biomass concentrations and activity vary seasonally, which may influence phosphorus availability. In the tropics, seasons are usually distinguished by rainfall rather than temperature. The highest concentrations of microbial biomass were found during the dry season in tropical dry forests (Singh *et al.*, 1989; Campo *et al.*, 1998), whereas studies from agroforestry systems in the Amazon and Nigeria reported minima during the dry season (Marschner *et al.*, 2002; Wick *et al.*, 2002). Although microbial activity is presumably less affected by water limitation than plant growth, higher respiration rates are usually found during the wet season (Luizão *et al.*, 1992; Fernandes *et al.*, 2002; Marschner *et al.*, 2002). Alternating cycles of soil drying and re-wetting increase plant-available phosphorus (Chepkwony *et al.*, 2001), probably by affecting the microbial biomass. During slow drying of a soil, microbial cells accumulate intracellular solutes in order to increase the internal water potential, but organic carbon is released from microbial cells upon rapid re-wetting, presumably through cell lysis (Kieft *et al.*, 1987). At the same time, phosphorus is also released (Turner *et al.*, 2003b). Thus, simulated rainfall on soil cores sampled during the rainy season from Mexican forests increased phosphorus immobilization, but resulted in phosphorus release from dry-season soil cores (Campo *et al.*, 1998). In a sandy soil with zero phosphate sorption, net phosphorus mineralization in constantly moist samples followed zero-order kinetics, while in previously dry soils, an initial flush of phosphorus release was followed by a lag phase during which microbial phosphate immobilization presumably equalled phosphorus mineralization, before the soil reverted to net phosphorus mineralization rates similar to the constantly moist soil (Grierson *et al.*, 1998).

Data on changes in inorganic and organic phosphorus pools with a high temporal resolution could improve our understanding of phosphorus mineralization and immobilization patterns. However, biweekly soil sampling in four cropping systems on an Oxisol in Colombia, followed by sequential fractionation, did not result in a clear picture. Short pulses of bicarbonate-extractable organic phosphorus were observed in two systems, while sodium hydroxide-extractable organic phosphorus concentrations changed little during the year (Friesen *et al.*, 1997). Only water-soluble organic phosphorus increased consistently during the dry season, in agreement with results from less frequent sampling in a nearby trial, which suggested an accumulation of organic phosphorus during the dry season when microbial phosphorus concentrations were lowest (Oberson *et al.*, 1999). Using anion exchange resin membranes *in situ* in an Amazonian agroforest, McGrath *et al.* (2000) found the greatest levels of available phosphorus early in the rainy season when both litterfall and re-wetting of dry soil occurred. Interpretation of such data is rather difficult, however, because although fluctuations in labile phosphorus were related mostly to precipitation, the amount of phosphorus adsorbed on the membranes is a result of changes in soil phosphorus retention capacity, microbial and plant phosphorus demand, as well as litter and root inputs. When using chemical extraction procedures, it is important to eliminate the possibility that variations in phosphorus extractability are not simply due to abiotic factors (Magid *et al.*, 1996). In conclusion, studies of seasonal dynamics are laborious and challenging due to their inherent complexity, but may ultimately improve our understanding of organic phosphorus dynamics, especially if combined with experiments under controlled conditions as well as modelling.

Long-term soil organic phosphorus dynamics

Methodological considerations

Long-term dynamics of soil organic phosphorus are usually deduced from comparison of phosphorus fractions on the same soil type under different land-use or management, in long-term field experiments (which are rare in the tropics), or in chronosequences of pastures or fallows of various ages. In all cases, the choice of control can be critical for the outcome of the study. Unless soil samples from the initial site of a field experiment (before establishment of the treatments) are available, a native site nearby is usually chosen. However, any variation in particle size distribution can significantly affect phosphorus concentrations. Agbenin and Goladi (1998) normalized the concentration of phosphorus in various fractions on the basis of soil fines (silt and clay), assuming that sand represents a dilution factor, but does not significantly influence phosphorus transformations. In their case, the native control site had the highest sand content of 91% compared with 81–89% in the field experiment, and the correction affected the apparent change (decline or increase) in total phosphorus that occurred during 45 years of continuous cultivation with fertilization. Without input–output balances and information about leaching losses in these systems, it is impossible to decide which data presentation is more accurate.

Another way to correct for variations in clay content was suggested by de Moraes *et al.* (1996) for carbon stocks under chronosequences of forest-to-pasture conversion. The measured carbon stock of a given site was corrected by adjusting its clay content to the average of all sites, assuming a linear relationship between clay and carbon content. The same chronosequences were studied later for phosphorus forms without correcting for the variation in particle size distribution (Garcia-Montiel *et al.*, 2000). In the chronosequence with six pastures ranging between 3 and 41 years old, sand content was significantly higher under forest than under the pasture sites. Without correction, a large increase in total phosphorus under young pastures was observed that could not be fully accounted for by the incorporation of phosphorus from the burnt forest biomass. Correcting the total phosphorus content with the formula:

$$P_{corrected} = \frac{(P_{measured}) * (clay+silt_{average})}{(clay+silt_{site})} \qquad (1)$$

reduced total phosphorus increases to more plausible values. While the idea is the same as in the correction used by Agbenin and Goladi (1998), the unit derived from this correction (mg P/kg soil) may be more practical than their mg/kg soil fines. In any case, these examples clearly show the need to carefully select the control in this type of study, or to correct for the variation in texture.

In long-term studies, management-induced changes in organic phosphorus must be assessed in relation to changes in total phosphorus. In the remainder of this section, we summarize results from field observations on long-term organic phosphorus dynamics in tropical agroecosystems, grouping the studies according to the direction of organic phosphorus change (increase or decrease). To simplify the comparison between different management options, we mainly examine changes in total organic phosphorus, although findings concerning specific organic phosphorus fractions are included wherever possible. In addition, we examine some of the management practices that accelerate changes in organic phosphorus (in either direction), as well as the processes behind these changes.

Decrease in soil organic phosphorus

In the absence of considerable phosphorus inputs from atmospheric deposition, continuous cropping without phosphorus fertilization must decrease total phosphorus levels due to phosphorus exports in harvested products. In this case, important questions concerning organic phosphorus dynamics are:

- How much does the decrease in organic phosphorus contribute to total phosphorus loss, and can a decrease in organic phosphorus occur even when total phosphorus increases?
- Is a decrease in organic phosphorus related to soil organic carbon loss?
- Which organic phosphorus fractions and forms are preferentially mineralized and which contribute most to decreases in total organic phosphorus?
- Are there pathways of phosphorus loss other than export with harvested products?

Some studies showed a decrease in total phosphorus due to continuous cultivation without fertilization (Agbenin and Goladi, 1998; Reddy *et al.*, 1999; Solomon and Lehmann, 2000; Solomon *et al.*, 2002). In these studies, the decreases in organic phosphorus fractions constituted 20–59% of the total phosphorus loss, but without a recognizable relationship to the duration of cultivation (Table 11.4). This is an apparent rather than a real contribution, because the mineralized organic phosphorus may have remained in the system as inorganic phosphorus, unless exported in harvest products or lost through erosion or leaching. The relative decrease in organic phosphorus ranged from 12% in the case of small total phosphorus losses (Reddy *et al.*, 1999), to 43–57% in the case of higher total phosphorus losses (Agbenin and Goladi, 1998; Solomon and Lehmann, 2000; Solomon *et al.*, 2002). In the latter three studies, the relative decrease in organic carbon was of the same order of magnitude (49–63%) as the relative decrease in organic phosphorus. In a long-term trial in Nigeria (Agbenin and Goladi, 1998), the application of mineral fertilizer diminished the decrease in total phosphorus, as well as the contribution of organic phosphorus to total phosphorus losses, while relative losses of carbon showed no clear trend (Table 11.4). Residual phosphorus in the sequential fractionation scheme, which the authors suggested may largely consist of organically bound phosphorus, decreased irrespective of phosphorus fertilization.

In some studies, a decrease in organic phosphorus due to cultivation was observed even when total phosphorus increased or did not change significantly (Table 11.4). For example, continuous cultivation of a sandy Ultisol in the Amazon reduced organic phosphorus by 42% (Beck and Sanchez, 1996). In a previously cropped sandy Oxisol in northeastern Brazil, organic phosphorus remained 19–32% lower than under native vegetation, even after 4 years of bush fallow, and soil carbon concentrations also remained lower than under native vegetation until after 10 years of bush fallow (Tiessen *et al.*, 1992). Likewise, losses from the bicarbonate-extractable organic phosphorus fraction during 4 years of a continuous soybean–maize rotation on a Hawaiian high-phosphorus Ultisol were related to losses of carbon, while changes in total organic phosphorus were not significant (Linquist *et al.*, 1997). Such results confirm the general observation that soil organic matter is easily lost from tropical soils upon continuous cultivation, and indicate an often close correlation between organic phosphorus and organic carbon. On the other hand, agroforestry techniques such as short planted fallows aim to increase soil organic matter levels, and several studies showed an associated increase in organic phosphorus while total phosphorus remained unchanged (Maroko *et al.*, 1999; Hoang Fagerström *et al.*, 2002; Smestad *et al.*, 2002; Bünemann *et al.*, 2004).

Table 11.4. Literature reports of a decrease in organic phosphorus induced by cultivation of tropical soils.

Reference[a]	Crops	Phosphorus input (kg/ha/year)	Years	Change in total phosphorus (mg/kg)	Phosphorus loss as organic phosphorus (%)	Relative change (%)	
						Organic phosphorus	Organic carbon
Beck and Sanchez (1996)	Annual (rice, maize, soybean)	1 × 80 kg P	13	+2 ns	na	−42	nd
Tiessen *et al*. (1992)	Annual (sorghum, millet)	Unknown small fertilizer inputs	12	+29	na	−14	−25[a]
	1 year bush fallow after cassava		4–5	+14 ns	na	−32	−33[a]
	4 year bush fallow after cassava, beans		4–5	+14 ns	na	−19	−22[a]
Solomon *et al*. (2002)	a) Annual (maize)	none	25	−416	59[b]	−48[b]	−55
	b) Annual (maize–sorghum)	none	30	−552	57[b]	−52[b]	−63
Solomon and Lehmann (2000)	Annual (maize, beans)	none	3	−250	24	−43	−56
			15	−294	24	−50	−56
Agbenin and Goladi (1998)	Annual (cotton, maize, groundnut)	none	45	−188	20	−57	−49
		0P, +N	45	−151	12	−27	−41
		18–54 P, +N	45	−90	10	−14	−32
		18–54 P, +NK	45	−72	10	−10	−55
Reddy *et al*. (1999)	Annual (soybean, wheat)	none	4	−21	50	−12	nd

nd, not determined; na, not applicable; ns, not significantly different from control.
[a] Location, soil type, texture, sampling depth and total phosphorus content of control sites in these studies are given in Table 11.1.
[b] Value read from figure.

Relative changes in a given organic phosphorus fraction can indicate its lability. In all the studies listed in Table 11.4, the relative decrease in sodium bicarbonate-extractable organic phosphorus compared with the control was similar to, or greater than, the relative decrease in sodium hydroxide-extractable organic phosphorus, suggesting that the former organic phosphorus pool is preferentially mineralized. The greatest absolute decrease, however, occurred in most cases in sodium hydroxide-extractable organic phosphorus, as it is a larger pool than the bicarbonate-extractable pool. In studies using solution ^{31}P NMR spectroscopy, a decrease in the ratio of phosphate diesters to monoesters was generally observed under continuous cultivation and perennial plantations, while the ratio remained unchanged or increased under pasture and lowland rice (Table 11.2). Thus, under conditions of a decrease in organic phosphorus, phosphate diesters appear to be preferentially mineralized, and organic phosphorus forms remaining in a continuously cultivated soil will be relatively resistant to further mineralization. In a study by Maroko *et al.* (1999), the phosphorus that accumulated in macro-organic matter a month after the incorporation of fallow biomass had decreased significantly after three seasons of maize cropping, suggesting that phosphorus mineralization had occurred. Likewise, maize yield was related to the increase in soil organic matter fractions and associated phosphorus after fallow biomass inputs, although these fractions contained <1% of total phosphorus and <2% of sodium hydroxide-extractable organic phosphorus (non-sequential extraction). This confirmed that phosphorus contained in soil organic matter in the sand fraction may be a good indicator of phosphorus availability in soils that are low in available phosphate. For example, phosphorus in the sand fraction was significantly correlated with bicarbonate-extractable phosphate and organic phosphorus in a Colombian volcanic soil (Phiri *et al.*, 2001).

Phosphorus export data necessary to reveal phosphorus losses other than export in harvested products was reported only by Reddy *et al.* (1999). In this study, the amount of total phosphorus lost during 4 years was similar to the amount of phosphorus exported with harvest products. Loss of total phosphorus in the Tanzanian trial after 3 years of cultivation translated into an annual loss of 83 kg P/ha, whereas after 15 years an additional phosphorus loss of only 44 kg P/ha was observed (Solomon and Lehmann, 2000). Clearly, other pathways of phosphorus loss, such as erosion of phosphorus-enriched topsoil after woodland clearing, must have been relevant. Likewise, annual phosphorus losses of 17–19 kg P/ha in a study from Ethiopia (Solomon *et al.*, 2002) appear to be at the high end for phosphorus exports from unfertilized maize. In addition, significant reductions in total phosphorus under plantations of tea and *Cupressus*, despite low mineral fertilizer inputs, suggest that phosphorus loss through erosion and/or leaching occurred.

Practices favouring a decrease in soil organic phosphorus: slash-and-burn and tillage

Most agricultural areas in the tropics have been probably prepared at least once for cultivation by burning the native vegetation, and slash-and-burn practices continue to be important for resource-poor farmers throughout the tropics. Burning of vegetation increases available soil phosphate through ash inputs, but part of the phosphorus can be lost in smoke or erosion of the ash by water and wind. More recent studies confirmed previous suggestions that soil heating, besides effects on various other soil properties, converts organic phosphorus into inorganic phosphate. For example, heating the top 2.5 cm of various Ultisols from Kalimantan at 250°C for 1 h increased bicarbonate- and sodium hydroxide-extractable phosphate by an equivalent of 25 kg P/ha, while organic phosphorus fractions (sodium hydroxide and both hydrochloric acid-extractable fractions) and residual phosphorus each decreased by 4–8 kg/ha (Lawrence and Schlesinger, 2001). During slash-and-burn clearing of a dry tropical forest in Mexico, organic phosphorus and residual phosphorus in the top 5 cm decreased by 35 kg P/ha, while the various phosphate pools together increased by 45 kg P/ha. As the ash immediately after burning contained only 11 kg P/ha, part of which was lost by wind erosion before the first rain occurred, it was concluded that soil heating had a larger effect on soil phosphorus availability than ash inputs (Giardina *et al.*, 2000). From experiments involving field- and oven-burning a Ferralsol from Indonesia at various temperatures, Ketterings *et al.* (2002) concluded that changes in phosphorus availability after low intensity fires were mainly due to ash addition, whereas at maximum surface temperatures >300°C, phosphorus sorption was observed to increase. Nevertheless, microsites with high fire intensity best supported *Eucalyptus* seedling growth on a podsolic soil in Australia (Romanya *et al.*, 1994), a finding that may be relevant for large parts of the tropics.

Tillage is common in tropical agriculture, particularly in small-scale farming systems. Due to land scarcity, the frequency of tillage has increased and can occur on a plot at the start of each rainy season. Tillage decreases soil organic phosphorus inside aggregates due to exposure and mineralization of organic phosphorus that was previously inaccessible to microbial attack. Loss of organic phosphorus through water erosion can also be important in tillage systems due to decreased stability of aggregates and the removal of a vegetated soil cover, particularly in steep areas. Therefore no-till systems are of increasing importance, especially in erosion-prone areas of South America. For example, 32% of the cultivated grain area in Brazil is now under no-tillage agriculture (Six *et al.*, 2002). While no-till systems generally increase soil organic matter concentrations in topsoil, the effect on soil phosphorus fractions has been rarely studied. In phosphorus-fertilized systems, interpretation is made difficult if sampling depth under different tillage regimes is less than the deepest ploughing depth, resulting in higher phosphorus contents under no-till simply due to the absence of fertilizer incorporation. In any case, the proportion of organic phosphorus in the surface 10 cm of an Oxisol in southern Brazil increased from 19% under conventional tillage to 22% after 5 years of zero tillage (Selles *et al.*, 1997). On the other hand, maize–soybean rotations in central Brazil under no-till and conventional tillage did not significantly differ in total amounts of organic phosphorus (Lilienfein *et al.*, 2000), even though the no-till system had received higher phosphorus inputs. However, the short time since establishment of the different tillage systems (1–3 years) may have precluded the emergence of significant differences. In principle, the accumulation and dynamics of organic phosphorus in no-till systems could increase together with soil aggregation, microbial biomass and soil organic matter, but may also decrease if microbial activity is reduced by the high rates of herbicide application that are often necessary under no-till conditions.

Increase in soil organic phosphorus

In most studies with continuous cultivation, pasture, or perennial crops, moderate phosphorus fertilization occurs, resulting in higher levels of inorganic as well as organic phosphorus than under the native control. If phosphorus inputs and outputs are carefully accounted for, the comparison of phosphorus fractions in different systems can give valuable indications of soil phosphorus availability and cycling. For example, an improved grass–legume pasture in Colombia had lower phosphate sorption and maintained higher levels of organic phosphorus and labile phosphate than a fertilized grass-only pasture, even though both systems received similar phosphorus inputs, and outputs were greater under the grass–legume pasture (Oberson *et al.*, 1999). This provided strong evidence for the role of biological activity in improving phosphorus availability to plants. However, inputs and outputs of phosphorus are often not taken into account and simple comparisons of phosphorus fractions in unfertilized native and phosphorus-fertilized cultivated soils do little to improve our understanding of phosphorus dynamics.

Conclusions about the effect of land-use and management on organic phosphorus dynamics become possible when the apparent movement of added phosphorus into organic phosphorus is evaluated according to the following formula (Sattell and Morris, 1992; Oberson *et al.*, 2001):

$$\text{Organic P gain (\% of the total P gain)} = 100^* \frac{(\text{Organic P}_{\text{fertilized}} - \text{Organic P}_{\text{unfertilized}})}{(\text{Total P}_{\text{fertilized}} - \text{Total P}_{\text{unfertilized}})} \quad (2)$$

Using this formula we evaluated organic phosphorus dynamics in a range of studies, selecting all cases where total phosphorus was significantly increased over the native or unfertilized control (Table 11.5).

In most studies with annual crops, only 3–10% of applied phosphate appears to have been transformed into organic phosphorus (Beck and Sanchez, 1996; Friesen *et al.*, 1997; Lilienfein *et al.*, 2000; Neufeldt *et al.*, 2000; Solomon and Lehmann, 2000; Oberson *et al.*, 2001; Dobermann *et al.*, 2002), although the movement of added phosphate into organic phosphorus fractions amounted to 58% in a maize–bean rotation in Costa Rica (Szott and Melendez, 2001). In this case, however, it is unclear whether the control forest site was representative of pre-trial conditions, because annual phosphorus inputs were too low to explain the observed increase in total phosphorus. Likewise, the soybean–wheat rotation on a Vertisol in India (Reddy *et al.*, 1999) presents a case of unusually high transformation of applied phosphate into organic phosphorus, which may be partly explained by the low initial proportion of organic phosphorus (Table 11.1). Finally, the maximum of 18% of total phosphorus gain occurring in organic phosphorus under rice monocropping in Colombia (Friesen *et al.*, 1997) was not confirmed by subsequent samples taken from the same experiment 2 years later (Oberson *et al.*, 2001).

High proportional accumulation of organic phosphorus (17–44%) was observed under perennial crops in the Amazon (Lehmann *et al.*, 2001b), an alley-cropping system in Peru (Szott and Melendez, 2001) and under reforestation with *Pinus* and *Eucalyptus* species in the Brazilian Cerrado (Neufeldt *et al.*, 2000), while values between 5% and 30% were reported under grassland (Neufeldt *et al.*, 2000; Lehmann *et al.*, 2001c; Oberson *et al.*, 2001). In contrast to the studies summarized in Table 11.4, in which a decrease in organic phosphorus was always accompanied by a decrease in soil organic carbon, several of the studies in fertilized systems (Table 11.5) report an increase in organic phosphorus together with a decrease in soil carbon. This may indicate that in fertilized systems, organic phosphorus dynamics are less strongly coupled to carbon dynamics than in natural or unfertilized systems, but the exact processes are poorly understood. Mineral phosphate fertilizer was reported to suppress acid phosphatase activity in a temperate grassland soil, whereas the activities of alkaline phosphatase and phosphodiesterase were positively related to phosphate concentrations (Colvan *et al.*, 2001). While a suppression of phosphatase activity could theoretically favour organic phosphorus accumulation in phosphorus-fertilized systems, results on organic phosphorus accumulation in fertilized agroecosystems in the tropics (Table 11.5) point much more to the role of carbon sources in increasing organic phosphorus concentrations. Different stabilization mechanisms for organic phosphorus and soil organic carbon may then cause a continued accumulation of organic phosphorus, while for soil organic carbon, equilibrium concentrations may be rapidly reached, as biological activity and thus mineralization increase together with substrate availability. It should be remembered, however, that phosphorus availability will not necessarily be improved by large accumulation rates of organic phosphorus, but rather by increased fluxes between active pools.

Practices favouring an increase in soil organic phosphorus: the role of carbon inputs, roots, soil fauna, and the case of lowland rice

Stewart and Tiessen (1987) proposed that organic phosphorus can accumulate in

Table 11.5. Increase[a] in total and organic phosphorus concentrations with land use involving phosphorus fertilization.

Reference[b]	Crops	Input[c] (kg/ha/year)	Years	Change in total phosphorus (mg/kg)	Phosphorus gain as organic phosphorus (% total phosphorus gain)		Relative change (%) Organic P	Relative change (%) Organic C
Szott and Melendez (2001)	a) Annual (maize, beans)	39 P, +N	10	+1214	58		+421	−40
	a) Alley (*Gliricidia*)+annual	39 P, +N	10	+1290	61		+470	−37
	a) Perennial (cocoa, laurel)	34 P, +N	15	+968	65		+378	−33
	b) Alley (*Gliricidia*)+rice	25 P, −N	6.5	+135	35		+62	+15
Beck and Sanchez (1996)	Annual (rice, maize, soybean)	80 P	13	+98	9		+18	nd
Lehmann *et al*. (2001b)	Perennial (cupuassu)	7.2 P	5	+145	19		+69	−35[d]
	Perennial (peach palm)	3.4 P	5	+366	19		+173	−47[d]
	Perennial (annatto)	6.2 P	5	+422	17		+183	nd
	Perennial (brazil nut)	2.1 P	5	+391	20		+192	nd
	Cover (*Pueraria*)	Unknown	5	+60	1	ns	+1	−41[d]
	Grasses	Unknown	5	+82	25		+52	nd
	Secondary forest	Unknown	5	+44	3	ns	+4	−37[d]
Neufeldt *et al*. (2000)	a) Annual (maize, soybean)	70–80 P, +NK	10	+206	6		+16	−3
	a) Pasture (*Brachiaria decumbens*)	1x 34 kg P	9	+104	5	ns	+6	+5
	a) Reforestation (*Pinus caribaea*)	Unknown	20	+48	40		+23	−9
	b) Annual (maize, soybean)	80 P, +NK	9	+71	3	ns	+5	−28
	b) Reforestation (*Eucalyptus*)	Unknown	13	+25	44		+28	+4
Lilienfein *et al*. (2000)	Annual (maize, soybean, NT)	150 P	1–3	+179	10	ns	+15	−12
Friesen *et al*.(1997)	Annual (rice)	60 P	3	+108	18		+30	−4[e]
	Annual (rice, green manure)	100 P?	3	+119	8	?	+14	−4[e]
Oberson *et al*. (2001)	Pasture (grass–legume)	1×60, 1×20 P	5	+51	30	ns	+19	+9
	Annual (rice)	60 P	5	+151	10	ns	+19	−5
Solomon and Lehmann (2000)	Homestead field	Unknown manure	10	+216	9	?	+14	+3
Reddy *et al*. (1999)	Annual (soybean, wheat)	88 P	4	+97	32		+36	nd
	Annual (soybean, wheat)	69 P (partly manure)	4	+94	35		+39	nd
	Annual (soybean, wheat)	113 P	4	+136	39		+62	nd
Dobermann *et al*. (2002)	a) annual (rice, soybean)	466 kg P	2	+174	6	ns	+11	nd
	b) annual (rice, soybean)	340 kg P	1.5	+145	9	ns	+11	nd
	b) annual (rice, soybean)	413 kg P	1.5	+213	4	ns	+7	nd

[a] Significant unless stated otherwise.
[b] Location, soil type, texture, sampling depth and total phosphorus concentration of control sites in these studies are given in Table 11.1.
[c] Mineral phosphate fertilizer (with single superphosphate, triple super phosphate, calcium apatite, or rock phosphate) unless stated otherwise.
[d] Carbon concentrations from Lehmann *et al*. (2001a).
[e] Carbon concentrations from Bühler *et al*. (2002).

fertilized systems if adequate carbon and nitrogen sources are available. The greater apparent transformation of applied phosphate into organic phosphorus observed under perennials and pastures compared to annual crops (Table 11.5) supports this hypothesis. The addition of sole carbon sources as well as plant residues in incubation or field experiments has frequently increased microbial phosphorus as well as soil organic phosphorus (Chauhan *et al.*, 1981; Nziguheba *et al.*, 1998; Maroko *et al.*, 1999; Reddy *et al.*, 2001; Bünemann *et al.*, 2004). The overriding importance of carbon inputs for microbial and organic phosphorus was also demonstrated in agroforestry trials in Kenya, in which fallow biomass production did not respond to phosphorus fertilization, and increases in organic and microbial phosphorus were similar between unfertilized and fertilized systems (Smestad *et al.*, 2002; Bünemann *et al.*, 2004). Thus, increased inputs of organic materials have the potential to enhance levels of soil organic phosphorus.

An increase in organic phosphorus under tree crops can be related to regular litterfall, as well as nutrient pumping from lower soil layers. The latter was suggested as the main reason for an increase in organic and total phosphorus in the top 30 cm under secondary forest regrowth during up to four cultivation–fallow cycles after slash-burning the primary forest in Kalimantan, Indonesia (Lawrence and Schlesinger, 2001), supported by the fact that the maximum depth of fine roots was greater in secondary than in primary forest (50 *vs.* 20 cm). Uptake of phosphorus from below the tilled 15 cm also appeared to contribute to the observed increase in organic phosphorus in the topsoil under fallows in Kenya (Bünemann *et al.*, 2004). In two chronosequences of forest-to-pasture conversion in the Amazon (Garcia-Montiel *et al.*, 2000), the relative proportion of organic phosphorus increased under older pastures, together with a significant increase in soil carbon. Under pasture, soil carbon may be derived mainly from root inputs, which could ultimately increase soil organic phosphorus either directly through organic phosphorus contained in the roots, or through organic phosphorus in microbial metabolites.

In many systems, root inputs and their consequences for organic phosphorus dynamics remain to be investigated. In tropical forests, intensive nutrient recycling in a shallow topsoil layer has often been reported. It may be advantageous to mimic this situation in cropping systems by creating a mulch layer. For example, the phosphorus uptake of sugarcane increased with mulching of residues instead of burning, and the presence of roots within the decomposing litter was shown (Ball-Coelho *et al.*, 1993). Likewise, levels of resin-extractable phosphorus were higher in the superficial root mat of peach palm where litter was trapped, compared with the mineral soil, possibly due to the absence of phosphorus sorbing surfaces (McGrath *et al.*, 2000). However, the proportion of phosphorus taken up by roots within organic layers remains to be shown.

Evidence for the role of the soil fauna in organic phosphorus accumulation comes from the examination of phosphorus fractions in earthworm casts on a Colombian savanna, where one-third of the increased total phosphorus concentrations in casts, compared with the bulk soil, was found in organic fractions (Jimenez *et al.*, 2003). No clear trend for phosphatase activity in casts compared with bulk soil was detected, and organic phosphorus fractions in casts did not change significantly during 64 days of *in situ* aging in the field, suggesting that the organic phosphorus was rather stable. However, a decrease in organic phosphorus concentrations in a Ferralsol from the Peruvian Amazon was observed after transit through the gut of another earthworm species, suggesting that mineralization had occurred (Chapuis-Lardy *et al.*, 1998). In termite mounds from the Orinoco savannahs of Venezuela, 17% of the total phosphorus increase was in organic fractions (Lopez-Hernandez, 2001). Interestingly, the 251% relative increase in organic phosphorus in mounds compared to bulk soil was more similar to that in total phosphorus (225%) than in carbon and microbial phosphorus (379% and 367%, respectively). However, wide organic carbon to organic phosphorus ratios of about 800 and 1200 in bulk soil and mounds, respectively, suggested that the large residual phosphorus pool (60–78% of total phosphorus) may have contained additional organic phosphorus.

Lowland rice represents a special case. With longer duration of submergence due to intensified lowland rice cultivation, organic matter accumulation and an increase in the phosphate diester-to-monoester ratio has been observed (Mahieu *et al.*, 2000). Under controlled conditions, an increase in sodium hydroxide-extractable organic phosphorus was detected after 6 weeks of incubating a lowland rice soil from the Philippines at water saturation as compared to field capacity, suggesting the involvement of chelation of dissolved phosphorus by organic ligands under reducing conditions (Huguenin-Elie *et al.*, 2003). In the field, however, greater inputs of plant residues under more intensive cultivation, combined with reduced decomposition rates under anaerobic conditions, may be the main reasons for the higher proportions of organic phosphorus under intensive lowland rice cultivation.

Conclusions

Due to difficulties in the determination of phosphorus mineralization, organic phosphorus dynamics in tropical agroecosystems must be studied indirectly by assessing changes in the amounts of phosphorus in specific fractions and forms. During the last decade, sequential phosphorus fractionation has been used most often to determine the concentrations and dynamics of soil organic phosphorus, resulting in lower reported proportions of soil organic phosphorus than those presented previously. Organic phosphorus declines rapidly when soils are initially taken into cultivation, or after a fallow phase, especially in the absence of phosphorus fertilization. Bicarbonate-extractable organic phosphorus is preferentially mineralized, although the greatest absolute decrease usually occurs in the sodium hydroxide-extractable fraction. Results from studies using solution ^{31}P NMR spectroscopy suggest that phosphate diesters are more labile than monoesters. Phosphorus contained in organic matter associated with the sand fraction also appears to be readily mineralized. The greater apparent transformation of mineral phosphate fertilizer into organic phosphorus under grassland and perennials than under annual crops suggests that the availability of carbon substrates is decisive for an accumulation of organic phosphorus. In particular, roots appear to provide important carbon inputs, while significant phosphorus uptake from lower soil layers and recycling to the topsoil has also been shown. Approaches using isotopic labelling point to the role of the microbial biomass in soil organic phosphorus synthesis. A serious limitation for a better understanding of organic phosphorus dynamics is the scarcity of long-term field experiments in the tropics. Thus, the available data are insufficient to draw firm conclusions on soil organic phosphorus dynamics considering the diversity in soil types, climatic regimes and management systems, as well as the relation to soil organic carbon dynamics.

Analytical improvements in non-invasive techniques to characterize soil organic phosphorus, such as solid-state ^{31}P NMR spectroscopy, would be desirable, because less than 50% of total soil phosphorus is typically characterized using current procedures. In system-oriented research, input–output phosphorus balances provide an important but often neglected framework to interpret changes in organic phosphorus. Further promising approaches include studies with a high temporal resolution to better understand seasonal dynamics, possibly coupled with modelling. Interactions between the availability of phosphate and phosphatase, as well as microbial activity, should be rigorously tested in highly weathered soils of the tropics, where large phosphate additions may be needed in order to observe significant effects on the microbial biomass. Under conditions of low external phosphorus inputs, management options that increase organic phosphorus and stimulate a build-up of soil organic matter remain as important measures to improve phosphorus availability.

Acknowledgements

We thank Astrid Oberson for critical comments and fruitful discussions on a previous version of the manuscript.

References

Adepetu, J.A. and Corey, R.B. (1976) Organic phosphorus as a predictor of plant-available phosphorus in soils of southern Nigeria. *Soil Science* 122, 159–164.

Agbenin, J.O. and Goladi, J.T. (1998) Dynamics of phosphorus fractions in a savanna Alfisol under continuous cultivation. *Soil Use and Management* 14, 59–64.

Agbenin, J.O. and Tiessen, H. (1995) Phosphorus forms in particle-size fractions of a toposequence from Northeast Brazil. *Soil Science Society of America Journal* 59, 1687–1693.

Anderson, G. (1960) Factors affecting the estimation of phosphate esters in soil. *Journal of the Science of Food and Agriculture* 9, 497–503.

Armstrong, R.D. and Helyar, K.R. (1993) Utilization of labelled mineral and organic phosphorus sources by grasses common to semi-arid mulga shrublands. *Australian Journal of Soil Research* 31, 271–283.

Ball-Coelho, B., Salcedo, I.H., Tiessen, H. and Stewart, J.W.B. (1993) Short- and long-term phosphorus dynamics in a fertilized Ultisol under sugarcane. *Soil Science Society of America Journal* 57, 1027–1034.

Beck, M.A. and Sanchez, P.A. (1996) Soil phosphorus movement and budget after 13 years of fertilized cultivation in the Amazon basin. *Plant and Soil* 184, 23–31.

Bowman, R.A. (1989) A sequential extraction procedure with concentrated sulfuric acid and dilute base for soil organic phosphorus. *Soil Science Society of America Journal* 53, 362–366.

Bowman, R.A. and Cole, C.V. (1978) Transformations of organic phosphorus substrates in soils as evaluated by $NaHCO_3$ extraction. *Soil Science* 125, 49–54.

Brookes, P.C., Powlson, D.S. and Jenkinson, D.S. (1982) Measurement of microbial biomass phosphorus in soil. *Soil Biology and Biochemistry* 14, 319–329.

Buerkert, A., Bationo, A. and Dossa, K. (2000) Mechanisms of residue mulch-induced cereal growth increases in West Africa. *Soil Science Society of America Journal* 64, 346–358.

Bühler, S., Oberson, A., Rao, I.M., Friesen, D.K. and Frossard, E. (2002) Sequential phosphorus extraction of a ^{33}P-labeled Oxisol under contrasting agricultural systems. *Soil Science Society of America Journal* 66, 868–877.

Bühler, S., Oberson, A., Sinaj, S., Friesen, D.K. and Frossard, E. (2003) Isotope methods for assessing plant-available phosphorus in acid tropical soils. *European Journal of Soil Science* 54, 605–616.

Bünemann, E.K. (2003) Phosphorus dynamics in a Ferralsol under maize–fallow rotations: the role of the soil microbial biomass. PhD dissertation, Swiss Federal Institute of Technology, Zurich (see http://e-collection.ethbib.ethz.ch/show?type=diss&nr=15207).

Bünemann, E.K., Smithson, P.C., Jama, B., Frossard, E. and Oberson, A. (2004) Maize productivity and nutrient dynamics in maize–fallow rotations in western Kenya. *Plant and Soil* (in press).

Buresh, R.J., Smithson, P.C. and Hellums, D.T. (1997) Building soil phosphorus capital in Africa. In: Buresh, R.J., Sanchez, P.A. and Calhoun, F. (ed.) *Replenishing Soil Fertility in Africa.* Soil Science Society of America and American Society of Agronomy, Madison, Wisconsin, pp. 111–149.

Campo, J., Jaramillo, V.J. and Maass, J.M. (1998) Pulses of soil phosphorus availability in a Mexican tropical dry forest: effects of seasonality and level of wetting. *Oecologia* 115, 167–172.

Cardoso, I.M., van der Meer, P., Oenema, O., Janssen, B.H. and Kuyper, T.W. (2003) Analysis of phosphorus by ^{31}P NMR in Oxisols under agroforestry and conventional coffee systems in Brazil. *Geoderma* 112, 51–70.

Chapuis-Lardy, L., Brossard, M., Lavelle, P. and Schouller, E. (1998) Phosphorus transformations in a ferralsol through ingestion by *Pontoscolex corethrurus*, a geophagous earthworm. *European Journal of Soil Biology* 34, 61–67.

Chapuis-Lardy, L., Brossard, M. and Quiquampoix, H. (2001) Assessing organic phosphorus status of Cerrado oxisols (Brazil) using ^{31}P-NMR spectroscopy and phosphomonoesterase activity measurement. *Canadian Journal of Soil Science* 81, 591–601.

Chauhan, B.S., Stewart, J.W.B. and Paul, E.A. (1981) Effect of labile inorganic phosphate status and organic carbon additions on the microbial uptake of phosphorus in soils. *Canadian Journal of Soil Science* 61, 373–385.

Chen, C.R., Condron, L.M., Davis, M.R. and Sherlock, R.R. (2002) Phosphorus dynamics in the rhizosphere of perennial ryegrass (*Lolium perenne* L.) and radiata pine (*Pinus radiata* D. Don.). *Soil Biology and Biochemistry* 34, 487–499.

Chepkwony, C.K., Haynes, R.J., Swift, R.S. and Harrison, R. (2001) Mineralization of soil organic P induced by drying and rewetting as a source of plant-available P in limed and unlimed samples of an acid soil. *Plant and Soil* 234, 83–90.

Colvan, S.R., Syers, J.K. and O'Donnell, A.G. (2001) Effect of long-term fertiliser use on acid and alkaline phosphomonoesterase and phosphodiesterase activities in managed grassland. *Biology and Fertility of Soils* 34, 258–263.

Condron, L.M., Moir, J.O., Tiessen, H. and Stewart, J.W.B. (1990) Critical evaluation of methods for determining total organic phosphorus in tropical soils. *Soil Science Society of America Journal* 54, 1261–1266.

Condron, L.M., Frossard, E., Newman, R.H., Tekely, P. and Morel, J.-L. (1997) Use of ^{31}P NMR in the study of soils and the environment. In: Nanny, M.A., Minear, R.A. and Leenheer, J.A. (eds) *Nuclear Magnetic Resonance Spectroscopy in Environmental Chemistry*. Oxford University Press, New York, pp. 247–271.

Crews, T.E. (1996) The supply of phosphorus from native, inorganic phosphorus pools in continuously cultivated Mexican agroecosystems. *Agriculture, Ecosystems and Environment* 57, 197–208.

Crews, T.E., Kitayama, K., Fownes, J.H., Riley, R.H., Herbert, D.A., Muellerdombois, D. and Vitousek, P.M. (1995) Changes in soil phosphorus fractions and ecosystem dynamics across a long chronosequence in Hawaii. *Ecology* 76, 1407–1424.

Cross, A.F. and Schlesinger, W.H. (1995) A literature review and evaluation of the Hedley fractionation: applications to the biogeochemical cycle of soil-phosphorus in natural ecosystems. *Geoderma* 64, 197–214.

Dalal, R.C. (1977) Soil organic phosphorus. *Advances in Agronomy* 29, 83–117.

Daroub, S.H., Pierce, F.J. and Ellis, B.G. (2000) Phosphorus fractions and fate of phosphorus-33 in soils under plowing and no-tillage. *Soil Science Society of America Journal* 64, 170–176.

de Moraes, J.F.L., Volkoff, B., Cerri, C.C. and Bernoux, M. (1996) Soil properties under Amazon forest and changes due to pasture installation in Rondonia, Brazil. *Geoderma* 70, 63–81.

Dobermann, A., George, T. and Thevs, N. (2002) Phosphorus fertilizer effects on soil phosphorus pools in acid upland soils. *Soil Science Society of America Journal* 66, 652–660.

Fairhurst, T., Lefroy, R.D.B., Mutert, E. and Batjes, N. (1999) The importance, distribution and causes of phosphorus deficiency as a constraint to crop production in the tropics. *Agroforestry Forum* 9, 2–8.

Feller, C., Burtin, G., Gérard, B. and Balesdent, J. (1991) Utilisation des résines sodiques et des ultrasons dans le fractionement granulométrique de la matière organique des sols : intérêt et limites. *Science du Sol* 29, 77–93.

Feller, C., Frossard, E. and Brossard, M. (1994) Phosphatase activity in low activity tropical clay soils: distribution in the various particle-size fractions. *Canadian Journal of Soil Science* 74, 121–129.

Fernandes, S.A.P., Bernoux, M., Cerri, C.C., Feigl, B.J. and Piccolo, M.C. (2002) Seasonal variation of soil chemical properties and CO_2 and CH_4 fluxes in unfertilized and P-fertilized pastures in an Ultisol of the Brazilian Amazon. *Geoderma* 107, 227–241.

Friesen, D.K. and Blair, G.J. (1988) A dual radiotracer study of transformations of organic, inorganic and plant residue phosphorus in soil in the presence and absence of plants. *Australian Journal of Soil Research* 26, 355–366.

Friesen, D.K., Rao, I.M., Thomas, R.J., Oberson, A. and Sanz, J.I. (1997) Phosphorus acquisition and cycling in crop and pasture systems in low fertility tropical soils. *Plant and Soil* 196, 289–295.

Garcia-Montiel, D.C., Neill, C., Melillo, J., Thomas, S., Steudler, P.A. and Cerri, C.C. (2000) Soil phosphorus transformations following forest clearing for pasture in the Brazilian Amazon. *Soil Science Society of America Journal* 64, 1792–1804.

George, T.S., Gregory, P.J., Robinson, J.S. and Buresh, R.J. (2002) Changes in phosphorus concentrations and pH in the rhizosphere of some agroforestry and crop species. *Plant and Soil* 246, 65–73.

Giardina, C.P., Sanford, R.L. and Dockersmith, I.C. (2000) Changes in soil phosphorus and nitrogen during slash-and-burn clearing of a dry tropical forest. *Soil Science Society of America Journal* 64, 399–405.

Gressel, N., McColl, J.G., Preston, C.M., Newman, R.H. and Powers, R.F. (1996) Linkages between phosphorus transformations and carbon decomposition in a forest soil. *Biogeochemistry* 33, 97–123.

Grierson, P.F. and Adams, M.A. (2000) Plant species affect acid phosphatase, ergosterol and microbial P in a Jarrah (*Eucalyptus marginata* Donn ex Sm.) forest in south-western Australia. *Soil Biology and Biochemistry* 32, 1817–1827.

Grierson, P.F., Comerford, N.B. and Jokela, E.J. (1998) Phosphorus mineralization kinetics and response of microbial phosphorus to drying and rewetting in a Florida Spodosol. *Soil Biology and Biochemistry* 30, 1323–1331.

Guggenberger, G., Haumaier, L., Thomas, R.J. and Zech, W. (1996) Assessing the organic phosphorus status of an Oxisol under tropical pastures following native savanna using ^{31}P-NMR

spectroscopy. *Biology and Fertility of Soils* 23, 332–339.
Guo, F. and Yost, R.S. (1998) Partitioning soil phosphorus into three discrete pools of differing availability. *Soil Science* 163, 822–833.
Harrison, A.F. (1987) *Soil Organic Phosphorus: a Review of World Literature.* CAB International, Wallingford, UK, 257 pp.
Hedley, M.J., Stewart, J.W.B. and Chauhan, B.S. (1982) Changes in inorganic and organic soil phosphorus fractions induced by cultivation practices and by laboratory incubations. *Soil Science Society of America Journal* 46, 970–976.
Hedley, M.J., Kirk, G.J.D. and Santos, M.B. (1994) Phosphorus efficiency and the forms of soil phosphorus utilized by upland rice cultivars. *Plant and Soil* 158, 53–62.
Hoang Fagerström, M.H., Nilsson, S.I., van Noordwijk, M., Phien, T., Olsson, M., Hansson, A. and Svensson, C. (2002) Does *Tephrosia candida* as fallow species, hedgerow or mulch improve nutrient cycling and prevent nutrient losses by erosion on slopes in northern Vietnam? *Agriculture, Ecosystems and Environment* 90, 291–304.
Huguenin-Elie, O., Kirk, G.J.D. and Frossard, E. (2003) Phosphorus uptake by rice from soil that is flooded, drained or flooded then drained. *European Journal of Soil Science* 54, 77–90.
Iyamuremye, F. and Dick, R.P. (1996) Organic amendments and phosphorus sorption by soils. *Advances in Agronomy* 56, 139–185.
Jenkinson, D.S. and Ayanaba, A. (1977) Decomposition of carbon-14 labeled plant material under tropical conditions. *Soil Science Society of America Journal* 41, 912–915.
Jimenez, J.J., Cepeda, A., Decaens, T., Oberson, A. and Friesen, D.K. (2003) Phosphorus fractions and dynamics in surface earthworm casts under native and improved grasslands in a Colombian savanna Oxisol. *Soil Biology and Biochemistry* 35, 715–727.
Kamh, M., Horst, W.J., Amer, F., Mostafa, H. and Maier, P. (1999) Mobilization of soil and fertilizer phosphate by cover crops. *Plant and Soil* 211, 19–27.
Ketterings, Q.M., van Noordwijk, M. and Bigham, J.M. (2002) Soil phosphorus availability after slash-and-burn fires of different intensities in rubber agroforests in Sumatra, Indonesia. *Agriculture, Ecosystems and Environment* 92, 37–48.
Kieft, T.L., Soroker, E. and Firestone, M.K. (1987) Microbial biomass response to a rapid increase in water potential when dry soil is rewetted. *Soil Biology and Biochemistry* 19, 119–126.
Kouno, K., Tuchiya, Y. and Ando, T. (1995) Measurement of soil microbial biomass phosphorus by an anion exchange membrane method. *Soil Biology and Biochemistry* 27, 1353–1357.
Kwabiah, A.B., Palm, C.A., Stoskopf, N.C. and Voroney, R.P. (2003) Response of soil microbial biomass dynamics to quality of plant materials with emphasis on P availability. *Soil Biology and Biochemistry* 35, 207–216.
Lawrence, D. and Schlesinger, W.H. (2001) Changes in soil phosphorus during 200 years of shifting cultivation in Indonesia. *Ecology* 82, 2769–2780.
Lehmann, J., Cravo, M.D. and Zech, W. (2001a) Organic matter stabilization in a Xanthic Ferralsol of the central Amazon as affected by single trees: chemical characterization of density, aggregate, and particle size fractions. *Geoderma* 99, 147–168.
Lehmann, J., Cravo, M.S., Macedo, J.L.V., Moreira, A. and Schroth, G. (2001b) Phosphorus management for perennial crops in central Amazonian upland soils. *Plant and Soil* 237, 309–319.
Lehmann, J., Günther, D., da Mota, M.S., de Almeida, M.P., Zech, W. and Kaiser, K. (2001c) Inorganic and organic soil phosphorus and sulfur pools in an Amazonian multistrata agroforestry system. *Agroforestry Systems* 53, 113–124.
Lilienfein, J., Wilcke, W., Ayarza, M.A., Vilela, L., do Carmo Lima, S. and Zech, W. (2000) Chemical fractionation of phosphorus, sulphur, and molybdenum in Brazilian savannah Oxisols under different land use. *Geoderma* 96, 31–46.
Linquist, B.A., Singleton, P.W. and Cassman, K.G. (1997) Inorganic and organic phosphorus dynamics during a build-up and decline of available phosphorus in an Ultisol. *Soil Science* 162, 254–264.
Lopez-Hernandez, D. (2001) Nutrient dynamics (C, N and P) in termite mounds of *Nasutitermes ephratae* from savannas of the Orinoco Llanos (Venezuela). *Soil Biology and Biochemistry* 33, 747–753.
Luizão, R.C.C., Bonde, T.A. and Rosswall, T. (1992) Seasonal variation of soil microbial biomass: the effects of clearfelling a tropical rainforest and establishment of pasture in the central Amazon. *Soil Biology and Biochemistry* 24, 805–813.
Magid, J., Tiessen, H. and Condron, L.M. (1996) Dynamics of organic phosphorus in soils under natural and agricultural ecosystems. In: Piccolo, A. (ed.) *Humic Substances in Terrestrial Ecosystems.* Elsevier, Amsterdam, pp. 429–466.
Mahieu, N., Olk, D.C. and Randall, E.W. (2000) Analysis of phosphorus in two humic acid fractions of intensively cropped lowland rice soils by ^{31}P-NMR. *European Journal of Soil Science* 51, 391–402.
Makarov, M.I., Haumaier, L. and Zech, W. (2002)

Nature of soil organic phosphorus: an assessment of peak assignments in the diester region of P-31 NMR spectra. *Soil Biology and Biochemistry* 34, 1467–1477.

Maroko, J.B., Buresh, R.J. and Smithson, P.C. (1999) Soil phosphorus fractions in unfertilized fallow–maize systems on two tropical soils. *Soil Science Society of America Journal* 63, 320–326.

Marschner, P., Marino, W. and Lieberei, R. (2002) Seasonal effects on microorganisms in the rhizosphere of two tropical plants in a polyculture agroforestry system in Central Amazonia, Brazil. *Biology and Fertility of Soils* 35, 68–71.

McGrath, D.A., Comerford, N.B. and Duryea, M.L. (2000) Litter dynamics and monthly fluctuations in soil phosphorus availability in an Amazonian agroforest. *Forest Ecology and Management* 131, 167–181.

Miltner, A., Haumaier, L. and Zech, W. (1998) Transformations of phosphorus during incubation of beech leaf litter in the presence of oxides. *European Journal of Soil Science* 49, 471–475.

Möller, A., Kaiser, K., Amelung, W., Niamskul, C., Udomsri, S., Puthawong, M., Haumaier, L. and Zech, W. (2000) Forms of organic C and P extracted from tropical soils as assessed by liquid-state ^{13}C- and ^{31}P-NMR spectroscopy. *Australian Journal of Soil Research* 38, 1017–1035.

Neufeldt, H., da Silva, J.E., Ayarza, M.A. and Zech, W. (2000) Land-use effects on phosphorus fractions in Cerrado oxisols. *Biology and Fertility of Soils* 31, 30–37.

Newman, R.H. and Tate, K.R. (1980) Soil phosphorus characterization by ^{31}P nuclear magnetic resonance. *Communications in Soil Science and Plant Analysis* 11, 835–842.

Nziguheba, G., Palm, C.A., Buresh, R.J. and Smithson, P.C. (1998) Soil phosphorus fractions and adsorption as affected by organic and inorganic sources. *Plant and Soil* 198, 159–168.

Nziguheba, G., Merckx, R., Palm, C.A. and Rao, M.R. (2000) Organic residues affect phosphorus availability and maize yields in a Nitisol of western Kenya. *Biology and Fertility of Soils* 32, 328–339.

Oberson, A., Besson, J.M., Maire, N. and Sticher, H. (1996) Microbiological processes in soil organic phosphorus transformations in conventional and biological cropping systems. *Biology and Fertility of Soils* 21, 138–148.

Oberson, A., Friesen, D.K., Morel, C. and Tiessen, H. (1997) Determination of phosphorus released by chloroform fumigation from microbial biomass in high P sorbing tropical soils. *Soil Biology and Biochemistry* 29, 1579–1583.

Oberson, A., Friesen, D.K., Tiessen, H., Morel, C. and Stahel, W. (1999) Phosphorus status and cycling in native savanna and improved pastures on an acid low-P Colombian Oxisol. *Nutrient Cycling in Agroecosystems* 55, 77–88.

Oberson, A., Friesen, D.K., Rao, I.M., Bühler, S. and Frossard, E. (2001) Phosphorus transformations in an Oxisol under contrasting land-use systems: the role of the soil microbial biomass. *Plant and Soil* 237, 197–210.

Oehl, F., Oberson, A., Sinaj, S. and Frossard, E. (2001) Organic phosphorus mineralization studies using isotopic dilution techniques. *Soil Science Society of America Journal* 65, 780–787.

Oorts, K., Vanlauwe, B., Cofie, O.O., Sanginga, N. and Merckx, R. (2000) Charge characteristics of soil organic matter fractions in a Ferric Lixisol under some multipurpose trees. *Agroforestry Systems* 48, 169–188.

Phiri, S., Barrios, E., Rao, I.M. and Singh, B.R. (2001) Changes in soil organic matter and phosphorus fractions under planted fallows and a crop rotation system on a Colombian volcanic-ash soil. *Plant and Soil* 231, 211–223.

Reddy, D.D., Rao, A.S. and Takkar, P.N. (1999) Effects of repeated manure and fertilizer phosphorus additions on soil phosphorus dynamics under a soybean–wheat rotation. *Biology and Fertility of Soils* 28, 150–155.

Reddy, D.D., Subba Rao, A. and Singh, M. (2001) Crop residue addition effects on myriad forms and sorption of phosphorus in a Vertisol. *Bioresource Technology* 80, 93–99.

Renz, T.E., Neufeldt, H., Ayarza, M., da Silva, J.E. and Zech, W. (1999) Acid monophosphatase: an indicator of phosphorus mineralization or of microbial activity? A case study from the Brazilian Cerrados. In: Thomas, R.J. and Ayarza, M. (eds) *Sustainable Land Management for the Oxisols of the Latin American Savannas.* CIAT, Cali, Columbia, pp. 173–186.

Richardson, A.E. (1994) Soil microorganisms and phosphorus availability. In: Pankhurst, C.E., Doube, B.M., Gupta, V.V.S.R. and Grace, P.R. (eds) *Soil Biota: Management in Sustainable Farming Systems.* CSIRO, Melbourne, Australia, pp. 50–62.

Richter, D.D. and Babbar, L.I. (1991) Soil diversity in the tropics. *Advances in Ecological Research* 21, 315–389.

Romanya, J., Khanna, P.K. and Raison, R.J. (1994) Effects of slash burning on soil phosphorus fractions and sorption and desorption of phosphorus. *Forest Ecology and Management* 65, 89–103.

Rubæk, G.H., Guggenberger, G., Zech, W. and Christensen, B.T. (1999) Organic phosphorus in soil size separates characterized by phosphorus-31 nuclear magnetic resonance and resin extraction. *Soil Science Society of America Journal* 63, 1123–1132.

Salas, A.M., Elliott, E.T., Westfall, D.G., Cole, C.V. and Six, J. (2003) The role of particulate organic matter in phosphorus cycling. *Soil Science Society of America Journal* 67, 181–189.

Sattell, R.R. and Morris, R.A. (1992) Phosphorus fractions and availability in Sri Lankan Alfisols. *Soil Science Society of America Journal* 56, 1510–1515.

Saunders, W.M.H. and Williams, E.G. (1955) Observations on the determination of total organic phosphorus in soils. *Journal of Soil Science* 6, 254–267.

Selles, F., Kochhann, R.A., Denardin, J.E., Zentner, R.P. and Faganello, A. (1997) Distribution of phosphorus fractions in a Brazilian Oxisol under different tillage systems. *Soil and Tillage Research* 44, 23–34.

Singh, J.S., Raghubanshi, A.S., Singh, R.S. and Srivastava, S.C. (1989) Microbial biomass acts as a source of plant nutrients in dry tropical forest and savanna. *Nature* 338, 499–500.

Six, J., Feller, C., Denef, K., Ogle, S.M., Sa, J.C.D. and Albrecht, A. (2002) Soil organic matter, biota and aggregation in temperate and tropical soils: effects of no-tillage. *Agronomie* 22, 755–775.

Smestad, B.T., Tiessen, H. and Buresh, R.J. (2002) Short fallows of *Tithonia diversifolia* and *Crotalaria grahamiana* for soil fertility improvement in western Kenya. *Agroforestry Systems* 55, 181–194.

Solomon, D. and Lehmann, J. (2000) Loss of phosphorus from soil in semi-arid northern Tanzania as a result of cropping: evidence from sequential extraction and ^{31}P-NMR spectroscopy. *European Journal of Soil Science* 51, 699–708.

Solomon, D., Lehmann, J., Mamo, T., Fritsche, F. and Zech, W. (2002) Phosphorus forms and dynamics as influenced by land-use changes in the subhumid Ethiopian highlands. *Geoderma* 105, 21–48.

Sparling, G.P. (1992) Ratio of microbial biomass carbon to soil organic carbon as a sensitive indicator of changes in soil organic matter. *Australian Journal of Soil Research* 30, 195–207.

Stewart, J.W.B. and Tiessen, H. (1987) Dynamics of soil organic phosphorus. *Biogeochemistry* 4, 41–60.

Szott, L.T. and Melendez, G. (2001) Phosphorus availability under annual cropping, alley cropping, and multistrata agroforestry systems. *Agroforestry Systems* 53, 125–132.

Tiessen, H. and Moir, J.O. (1993) Characterization of available P by sequential extraction. In: Carter, M.R. (ed.) *Soil Sampling and Methods of Analysis.* CRC Press, Boca Raton, Florida, pp. 75–86.

Tiessen, H., Stewart, J.W.B. and Cole, C.V. (1984) Pathways of phosphorus transformations in soils of differing pedogenesis. *Soil Science Society of America Journal* 48, 853–858.

Tiessen, H., Salcedo, I.H. and Sampaio, E.V.S.B. (1992) Nutrient and soil organic matter dynamics under shifting cultivation in semi-arid northeastern Brazil. *Agriculture, Ecosystems and Environment* 38, 139–151.

Tiessen, H., Stewart, J.W.B. and Oberson, A. (1994) Innovative soil phosphorus availability indices: assessing organic phosphorus. In: Havlin, J.L. and Jacobsen, J.S. (eds) *Soil Testing: Prospects for Improving Nutrient Recommendations.* Soil Science Society of America and American Society of Agronomy, Madison, Wisconsin, pp. 143–162.

Turner, B.L., Papházy, M.J., Haygarth, P.M. and McKelvie, I.D. (2002) Inositol phosphates in the environment. *Philosophical Transactions of the Royal Society, London, Series B* 357, 449–469.

Turner, B.L., Mahieu, N. and Condron, L.M. (2003a) Phosphorus-31 nuclear magnetic resonance spectral assignments of phosphorus compounds in soil NaOH–EDTA extracts. *Soil Science Society of America Journal* 67, 497–510.

Turner, B.L., Driessen, J.P., Haygarth, P.M. and McKelvie, I.D. (2003b) Potential contribution of lysed bacterial cells to phosphorus solubilisation in two rewetted Australian pasture soils. *Soil Biology and Biochemistry* 35, 187–189.

Umrit, G. and Friesen, D.K. (1994) The effect of C:P ratio of plant residues added to soils of contrasting phosphate sorption capacities on P uptake by *Panicum maximum* (Jacq.). *Plant and Soil* 158, 275–285.

Walker, T.W. and Syers, J.K. (1976) The fate of phosphorus during pedogenesis. *Geoderma* 15, 1–19.

Wick, B., Kühne, R.F., Vielhauer, K. and Vlek, P.L.G. (2002) Temporal variability of selected soil microbiological and biochemical indicators under different soil quality conditions in southwestern Nigeria. *Biology and Fertility of Soils* 35, 155–167.

Zoysa, A.K.N., Loganathan, P. and Hedley, M.J. (1998) Phosphate rock dissolution and transformation in the rhizosphere of tea (*Camellia sinensis* L.) compared with other plant species. *European Journal of Soil Science* 49, 477–486.

Zoysa, A.K.N., Loganathan, P. and Hedley, M.J. (1999) Phosphorus utilisation efficiency and depletion of phosphate fractions in the rhizosphere of three tea (*Camellia sinensis* L.) clones. *Nutrient Cycling in Agroecosystems* 53, 189–201.

12 Organic Phosphorus Transfer from Terrestrial to Aquatic Environments

Benjamin L. Turner

Smithsonian Tropical Research Institute, Box 2072, Balboa, Ancon, Republic of Panama

Introduction

In recent decades there has been a considerable interest in phosphorus transfer from soils to waterbodies, because the enrichment of receiving waters with phosphorus can contribute to eutrophication and associated blooms of cyanobacteria (blue-green algae). Notable examples of waterbodies suffering severe water quality problems linked to phosphorus pollution have been recorded in the USA (e.g. Burkholder *et al.*, 1992), Europe (Lawton and Codd, 1991), and Australia (EPA, 1995). Most spectacularly, a 1000-km cyanobacterial bloom, reported to be the largest ever recorded in the world, occurred during the summer of 1991/92 in the Barwon–Darling river system, New South Wales, Australia (Bowling and Baker, 1996). This not only resulted in oxygen depletion and fish kills, but also had a detrimental effect on human health, causing influenza-type symptoms including sore throats, blistering in the mouth, abdominal and pleuritic pain, diarrhoea, fever and vomiting. Cyanobacterial blooms can also result in the production of various neuro- and hepato toxins (liver toxins) that pose severe risks to the health of humans and animals (Kotak *et al.*, 1993). Aside from hazards to health and the environment, financial costs are incurred for additional water treatment for potable supply and, in extreme cases, the requirement for alternative water supplies (Bowling and Baker, 1996). More seriously, financial impacts on the tourist industry due to the eutrophication of recreational waters can be immense. For example, in the highly eutrophic Gippsland Lakes region of Victoria, Australia, losses from tourism are estimated to be in the region of millions of Australian dollars each year (EPA, 1995).

Problems from phosphorus pollution of watercourses arise because phosphorus is one of the most important nutrients for biological growth, but is also the nutrient that most frequently limits biological productivity in terrestrial and aquatic environments (Schindler, 1977; Wild, 1988). Unlike carbon and nitrogen, the gaseous phosphorus phase is ecologically negligible, so aquatic organisms must acquire phosphorus from the surrounding water, or in some cases from sediments. This means that many organisms rely on soil-derived phosphorus for their phosphorus nutrition (Livingstone and Whitton, 1984; Whitton *et al.*, 1998; Turner *et al.*, 2003a), but also that waterbodies are susceptible to even small amounts of pollutant phosphorus transfer (Moss, 1988).

The focus of research on phosphorus transfer has been filterable phosphate, because this form is perceived as being the most immediately available for uptake by aquatic organisms (e.g. Sharpley *et al.*, 1996). However, organic phosphorus is a key component of the phosphorus transfer process in both natural and managed environments. It can account for a large proportion of the total phosphorus in drainage waters, including soil solution (Shand *et al.*, 1994), leachate (Turner and Haygarth, 2000), overland flow (Haygarth *et al.*, 1998) and stream water (Turner *et al.*, 2003a). Some organic phosphorus compounds are mobile in the soil profile, notably under heavy manure application (Rolston *et al.*, 1975; Chardon *et al.*, 1997), and can therefore be more susceptible to leaching than phosphate. Indeed, the quantitative importance and potential mobility of organic phosphorus in soil has been known for some time (Pierre and Parker, 1927; Spencer and Stewart, 1934). Importantly, organic phosphorus compounds are biologically available in surface waters, because many organisms can access them through the synthesis of phosphatase enzymes, either excreted from the cell, or associated with cell surfaces (Jansson *et al.*, 1988; Quiquampoix and Mousain, Chapter 5, this volume). Indeed, many species of cyanobacteria synthesize 'surface' phosphatase enzymes that allow them to grow well on a range of organic phosphorus compounds (Whitton *et al.*, 1991). Aquatic bacteria can also directly take up organic phosphates without the need for extracellular hydrolysis (see Heath, Chapter 9, this volume), while various abiotic hydrolysis mechanisms can release phosphate from complex organic compounds (Baldwin *et al.*, Chapter 4, this volume).

Despite the importance of organic phosphorus in phosphorus transfer, its precise role remains poorly understood (Haygarth and Jarvis, 1999), although it is clear that diffuse phosphorus pollution cannot be effectively managed without a comprehensive appreciation of the role of organic phosphorus. Similarly, understanding the ecology of natural ecosystems requires detailed information on organic phosphorus movement between soils and waterbodies (Turner *et al.*, 2003a). General principles of phosphorus transfer from agricultural land have been reviewed extensively elsewhere (Sharpley and Smith, 1990; Sharpley *et al.*, 1996; Haygarth and Jarvis, 1999) and are not discussed in detail here. Rather, this chapter provides a critical overview of specific information on organic phosphorus transfer, including the magnitude, forms, mechanisms of release from soil, and transport of organic phosphorus from soils to watercourses.

Concentrations of Organic Phosphorus in Drainage Waters

Literature information on organic phosphorus in runoff and its quantitative contribution to soluble phosphorus transfer remains limited, making it difficult to generalize the importance of this process. However, the available data show that organic phosphorus can contribute almost nothing to almost all the filterable phosphorus transfer from soils, with patterns varying markedly among soils and land-use systems (Table 12.1).

Organic phosphorus concentrations have been measured in various types of drainage water, including soil solution, leachate, tile drainage and surface runoff. Concentrations rarely reach those of phosphate leaching from heavily fertilized or manured soils, or of particulate phosphorus in surface runoff from tilled land (Haygarth and Jarvis, 1999); indeed, erosive loss of soil and associated particulate phosphorus can be catastrophic (e.g. Haith and Shoemaker, 1987). In assessing the available data, it is important to note that patterns of organic phosphorus transfer tend to exhibit marked temporal variability, and can be profoundly influenced by the experimental methodology used to obtain runoff samples for analysis (see below). The experimental conditions used to determine organic phosphorus concentrations must therefore be considered carefully when evaluating literature values.

Of the many factors that can influence organic phosphorus in runoff, the concentration of organic phosphorus in the soil is clearly important. For example, drainage from organic upland soils contains the greatest proportion of organic phosphorus. Concentrations in filtered soil solution from grazed upland pastures exceeded those of inorganic phosphate by a factor of between 5 and 20 (Shand *et al.*, 1994), while drainage from peat and calcareous soils under high rates of atmospheric nitrogen deposition in the northern Pennines contained almost exclusively organic phosphorus (Livingstone and Whitton, 1984; Turner *et al.*, 2003a). However, the importance of soil organic phosphorus concentrations must be considered in relation to a range of interacting climatic, hydrological, soil and management factors (Haygarth and Jarvis, 1999). Thus, filtered surface runoff from grazed pasture in southwest England contained approximately 50% organic phosphorus (Haygarth and Jarvis, 1997), whereas filtered surface runoff from grazed pasture with a similar concentration of soil organic phosphorus in the drier Gippsland region of southeastern Australia contained <10% organic phosphorus (Nash and Murdoch, 1997).

Land use and management have important impacts on organic phosphorus transfer. This is illustrated by the fact that pasture soils typically contain far greater concentrations of organic phosphorus than arable soils, which in turn is reflected in the proportions of organic phosphorus in associated drainage waters. Thus, filterable organic phosphorus constituted <5% of the total phosphorus in tile drainage from arable soils in southern England (Heckrath *et al.*, 1995), whereas leachate from cut grassland soils in southwest England contained between 8% and 17% organic phosphorus (Turner and Haygarth, 2000). Drain outflow from a small pasture catchment in Northern Ireland contained an even greater proportion (approximately 30%) of organic phosphorus (Jordan and Smith, 1985).

Differences in the intensity of management clearly influence the relative concentrations and importance of organic phosphorus, which is reflected in stream and river water. As an example, Christmas and Whitton (1998) reported that filterable organic phosphorus concentrations in the Swale–Ouse river system in northeastern England increased with distance downstream, from a mean of <10 μg P/l in the moorland-dominated headwaters, to >500 μg P/l at 145 km downstream where land-use was dominated by arable and dairy farming.

In some cases atmospheric phosphorus inputs can be important contributions to the overall phosphorus balance in both terrestrial and aquatic ecosystems, notably where dust deposition provides phosphorus to highly oligotrophic lakes (Gibson *et al.*, 1995; Newman, 1995). Further, modelling studies indicate that the deposition of phosphorus in dust can have a marked influence on long-term ecosystem development, at least in tropical forests (see Parton *et al.*, Chapter 15, this volume). However, organic phosphorus is rarely quantified in precipitation, and little published data exists.

Conceptual Model of Organic Phosphorus Transfer

The transfer of phosphorus from soils to waterbodies involves complex interactions between rainfall, landscape form and management, and the biological, chemical, hydrological and physical properties of soil (Haygarth and Jarvis, 1999). The conceptual separation of processes controlling organic phosphorus transfer is illustrated in Fig. 12.1. The model is based on that proposed for phosphorus transfer by Haygarth and Jarvis (1999), and subdivides the process by soil phosphorus forms, release mechanisms, forms in solution, and transport through the soil. Rainfall is the driving force behind phosphorus transfer, providing the energy and the carrier for phosphorus movement in the landscape. The model provides a framework around which the remainder of the chapter is based.

Forms and Behaviour of Organic Phosphorus in Soil

Detailed information on the chemical forms and behaviour of soil organic phosphorus

Table 12.1. Literature information on the forms and concentrations of organic phosphorus measured at a range of experimental scales from soil solution to river water. Concentrations are filterable molybdate-unreactive phosphorus unless specified.

Reference	Type of sample	Soil type/environment	Location	Land-use and management	Depth of sampling (cm)	Organic phosphorus (μg/l)	Proportion of the total filterable phosphorus (%)
Shand *et al.* (1994)	Soil solution (centrifugation)	Brown forest soils	Scotland	Upland grazed grass/clover	0–18	55–110	85–96
Chapman *et al.* (1997)	Soil solution (centrifugation)	Podsolic sandy loam	Scotland	Improved grassland (grazed) receiving mineral fertilizer and manure	0–6	88 ± 11	34
Ron Vaz *et al.* (1993)	Soil solution (centrifugation)	Iron–humus podsol	Scotland	Historically mixed land use, various fertilizer rates	0–70	80–464	42–62[b]
Hens *et al.* (2001)	Soil solution (centrifugation)	Sandy podsolic soil	Belgium	Coniferous forest	7–17	226	31
				Permanent pasture	5–15	1087	30
				Arable cropping	5–15	300	22
Magid *et al.* (1992)	Soil solution (suction cells)	Layered sandy soil	Denmark	–	90	10–250[a]	–
Chapman *et al.* (1997)	Leachate (laboratory columns)	Podsolic sandy loam	Scotland	Improved (grazed) grassland receiving mineral fertilizer	12	2–80[a]	–
Chardon *et al.* (1997)	Leachate	Sandy soil	Netherlands	Disturbed soil columns under heavy slurry application	40	20–300[a]	40–100
Jensen *et al.* (2000)	Leachate (laboratory column)	Structured clay	Denmark	Bare soil with single amendment of dairy cattle faeces; saturated and unsaturated flow conditions	50	600 (saturated) 500 (unsaturated)	8 (saturated) 25 (unsaturated)
Hannapel *et al.* (1964a,b)	Leachate (laboratory columns)	Calcareous sandy loam	Arizona, USA	Not specified. Columns amended with plant residues or sucrose	20	–	71–89 (plant residues); 90 (sucrose)
Magid *et al.* (1992)	Leachate (field lysimeters)	Layered sandy soil	Denmark	Arable cropping	90	10–500[a]	–
Turner and Haygarth (2000)	Leachate (field lysimeters)	Silty clay	Southwest England	Intact field monoliths under cut grassland receiving mineral fertilizer	130	0–47	8
		Clay loam				1–71	16
		Sandy loam				1–32	17

		Sand				0–35	6
Culley *et al*. (1983)	Tile drainage	Clay	USA	Mixed arable cropping	–	30–50	11–16
Haygarth *et al*. (1998)	Tile and mole drainage	Silty clay	Southwest England	Grazed pasture receiving mineral fertilizer	85	10–165	27[c]
Heckrath *et al*. (1995)	Tile drainage	Silty clay loam	Rothamsted, England	Arable cropping, various mineral fertilizer rates	65	–	<5
Jordan and Smith (1985)	Tile drainage	–	Northern Ireland	Intensively grazed pasture	–	23–57 (flow weighted means)	30[c]
Nash and Murdoch (1997)	Overland flow	Podsolic sandy loam	Victoria, Australia	Grazed pasture (dairy) receiving mineral fertilizer	Surface	90–320	2–9
Haygarth *et al*. (1998)	Overland flow and interflow	Silty clay	Southwest England	Undrained grazed pasture	Surface to 30 cm	10–447	29
				Drained grazed pasture		10–179	19
Preedy *et al*. (2001)	Overland flow and interflow	Silty clay	Southwest England	Unfertilized pasture	Surface to 30 cm	42 ± 8	29
				Pasture with mineral fertilizer added 24 h prior to rainfall		521 ± 195	6
				Pasture with slurry applied 24 h prior to rainfall		987 ± 68	14
Ulén and Mattson (2003)	Tile drainage	Clay	Western Sweden	Grass and cereal production	100	12–35	13–24
Beauchemin *et al*. (1998)	Tile drainage	Textural gradient of nine soil series of neutral pH	Quebec, Canada	Intensive arable cropping	>90	<20[a]	0–79
Foy *et al*. (1982)	River water	Six rivers entering Lough Neagh	Northern Ireland	Predominantly grazed pasture	–	38[c]	16[c]
Christmas and Whitton (1998)	Stream and river water	Swale–Ouse river system	Northeast England	Peat moorland in upstream reaches, arable and dairy farming in lower reaches	–	6–559 (annual means)	8–52

[a] Values were estimated from figures.
[b] Organic phosphorus determined using ultraviolet oxidation (i.e. does not include inorganic polyphosphate or colloids containing mineral phosphorus).
[c] Average value.

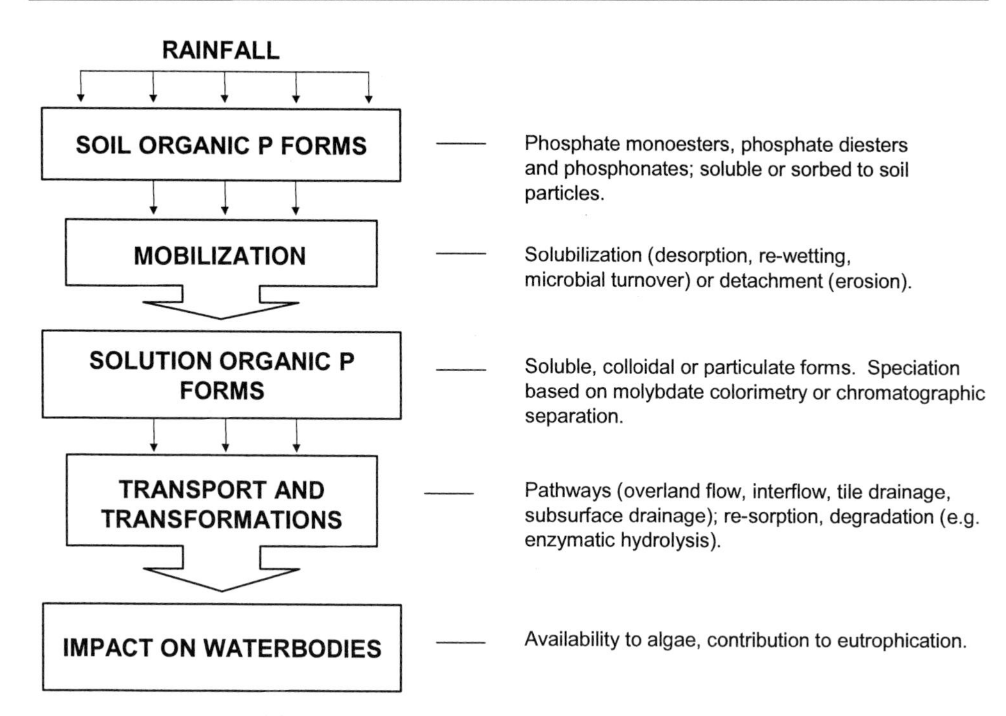

Fig. 12.1. Conceptual model of process units comprising the transfer of organic phosphorus from soils to watercourses (adapted from Haygarth and Jarvis, 1999).

can be found in several reviews (Dalal, 1977; Anderson, 1980; Harrison, 1987; Magid *et al.*, 1996; Condron *et al.*, 2005) and elsewhere in this volume (Cade-Menun, Chapter 2; Celi and Barberis, Chapter 6; Oberson and Joner, Chapter 7; Nziguheba and Bünemann, Chapter 11; Condron and Tiessen, Chapter 13). A brief overview is included here, because this is fundamental to understanding the mobilization and subsequent potential for transfer of organic phosphorus from soil.

In most soils, phosphate monoesters are the most abundant functional class of organic phosphorus compounds, often constituting >90% of the total organic phosphorus (e.g. Turner *et al.*, 2003d). They occur mainly as inositol phosphates, predominantly *myo*-inositol hexakisphosphate (phytic acid), but also in a range of lower esters and stereoisomers (Harrison, 1987; Turner *et al.*, 2002b). Phosphate diesters, such as DNA and phospholipids, occur in smaller quantities, of which a significant proportion is probably contained within live microbes. Proportions of phosphate diesters are typically <10%, although larger proportions occur in high-organic-matter soils (e.g. Tate and Newman, 1982; Bedrock *et al.*, 1994; Cade-Menun *et al.*, 2000; Turner *et al.*, 2003b). Phosphonates are also detected in some soils, but only those where conditions are unfavourable for decomposition (Tate and Newman, 1982).

Soil organic phosphorus composition does not reflect organic phosphorus inputs to soil, which are dominated by phosphate diesters derived from plant and microbial remains. For example, nucleic acids constitute around 60% of the intracellular phosphorus in fungi and bacteria (Webley and Jones, 1971), while phospholipids are the major form of organic phosphorus in fresh plant tissue (Bieleski, 1973). In contrast, inositol phosphates constitute only a small proportion of total organic phosphorus inputs, yet are the major class of soil organic phosphorus (Turner *et al.*, 2002b).

The disparity between the composition of organic phosphorus forms measured in soil and those constituting the majority of the organic phosphorus in fresh inputs arises through a differential stabilization in soil, which in turn has important implications for the potential involvement of the organic phosphorus compounds in phosphorus transfer. Sorption of organic phosphorus usually occurs through interactions with the phosphorus moiety, which represents the dominant charge on most compounds (see Celi and Barberis, Chapter 6, this volume). As a result, phosphate monoesters with a single phosphate moiety (e.g. sugar phosphates) are only weakly sorbed; similarly, the low charge density of phosphate diesters affords only weak adsorption, although they can be strongly sorbed in soils of pH <5 (Greaves and Wilson, 1969; Anderson *et al.*, 1974). However, with six phosphate groups, *myo*-inositol hexakisphosphate undergoes extensive interaction, being adsorbed to clays, or precipitated with soil minerals such as sesquioxides of iron and aluminium (Anderson *et al.*, 1974; Celi and Barberis, Chapter 6, this volume). This means that *myo*-inositol hexakisphosphate undergoes preferential stabilization and can even displace phosphate into solution by competing successfully for the same binding sites (Bowman *et al.*, 1967; Anderson *et al.*, 1974). These differences in behaviour among the various organic phosphorus compounds almost certainly influence their susceptibility to transfer in drainage water (see below).

Mobilization of Soil Organic Phosphorus

The mobilization of soil phosphorus is the first key step in the transfer process. The conceptual separation of release mechanisms from transport factors allows these processes to be studied independently (Fig. 12.1), based on the assumption that all mobilization mechanisms can potentially contribute to phosphorus transfer if they are coincident with hydrological movement.

Both solubilization and detachment mechanisms operate to release organic phosphorus from soil to water. In addition, organic phosphorus can be transferred directly to watercourses by so-called 'incidental' transfer, for example when phosphorus contained within recently applied animal manure is washed off the land surface without having a chance to interact with the soil (Preedy *et al.*, 2001). Significant amounts of organic phosphorus can be transferred in association with eroded soil particles; however, processes involved in the detachment and transfer of soil particles and associated phosphorus have been extensively reviewed elsewhere (e.g. Sharpley and Smith, 1990; Toy *et al.*, 2002) and will not be discussed in detail here. This section deals exclusively with mechanisms of organic phosphorus solubilization.These are poorly understood, although biological processes almost certainly play a central role.

The importance of microbial turnover in organic phosphorus solubilization

Biological activity seems to be of fundamental importance in maintaining concentrations of organic phosphorus in soil solution, which is clearly demonstrated by leaching studies that have manipulated the soil microbial biomass. For example, when microbial activity was eliminated by soil sterilization, organic phosphorus became undetectable in leachate water, despite being the dominant filterable fraction in leachate from a corresponding microbially active soil (Seeling and Zasocki, 1993). Similarly, when laboratory columns of a calcareous soil were amended with organic carbon sources, including crop residues and sucrose, organic phosphorus became dominant in leachate (Hannapel *et al.*, 1964b). In both these studies, elevated biological activity in the soil appeared to increase organic phosphorus turnover through the solution, which translated into organic phosphorus transfer when coupled with downward water movement.

The impact of microbes on soil organic

phosphorus solubilization is unsurprising, because they constitute a large fraction of the total phosphorus in most soils, almost all of which occurs in organic forms (Webley and Jones, 1971; Oberson and Joner, Chapter 7, this volume). Microbial phosphorus concentrations vary markedly among different land-uses and climates, although notably large concentrations occur in pasture soils (Brookes *et al.*, 1984; Turner *et al.*, 2001b), and it is from these types of soils that the greatest concentrations of organic phosphorus are often detected in runoff. Nutrients in the microbial biomass have a short turnover time relative to those in more stable organic matter (Jenkinson and Ladd, 1981), so the continuous release of low concentrations of microbial organic phosphorus to solution can be considerable on an annual basis (Martin, 1970). For example, Brookes *et al.* (1984) used a microbial turnover time of 2.5 years to estimate an annual phosphorus flux of 42 kg P/ha through a temperate grassland soil, while Cole *et al.* (1977) estimated an annual phosphorus turnover of 30–40 kg P/ha through the decomposer biomass of semi-arid grassland soil. However, turnover may be significantly greater than this; for example, Kouno *et al.* (2002) recently reported a turnover time for microbial phosphorus of approximately 40 days in soils amended with ryegrass. Interestingly, this was much more rapid than the turnover of microbial carbon (approximately 90 days), possibly because microbial phosphorus is largely confined to the cytoplasm, whereas carbon occurs mainly in cell wall structures.

The importance of microbial turnover in the maintenance of soil solution organic phosphorus is further highlighted by attempts to measure rates of organic phosphorus mineralization in soils. Using ^{32}P-labelled RNA, daily rates of organic phosphorus mineralization up to 0.19 mg P/kg soil were recorded in English woodland soils incubated at 13°C (Harrison, 1982a). Mineralization was greatest during the springtime and in more alkaline soils (Harrison, 1982b). Slightly greater rates of between 0.2 and 1.7 mg P/kg soil were recently reported in cultivated soils from Switzerland and the USA using isotopic exchange kinetics (Lopez-Hernandez *et al.*, 1998; Oehl *et al.*, 2001), although greater incubation temperatures were employed in these studies. The turnover of microbial phosphorus is reviewed in detail elsewhere in this volume (Oberson and Joner, Chapter 7; Nziguheba and Bünemann, Chapter 11).

Most microbial organic phosphorus will be rapidly degraded within hours or days of release by the ubiquitous phosphatase enzymes present in soil, including those present in lysed microbial cells. Indeed, the efficiency of this process is demonstrated by the chloroform fumigation method for estimating microbial phosphorus in soils, during which microbial phosphorus released from lysed cells is almost completely converted to phosphate within 24 h of release (Brookes *et al.*, 1982). The rapid turnover of labile organic phosphorus compounds in soil means that compounds persisting in solution are likely to be relatively recalcitrant in terms of their availability to organisms, and may accumulate in the soil solution. Indeed, a large proportion of the organic phosphorus in soil solution is resistant to hydrolysis by commercially available phosphatase enzymes (Shand and Smith, 1997; Hens and Merckx, 2001). The implication for organic phosphorus transfer is that many of the filterable compounds in runoff are likely to be of limited bioavailability, a hypothesis supported by information on the chemical composition of carbon in soil solution (Gregorich *et al.*, 2000). However, if microbial phosphorus release is coincident with hydrological movement, such as during rainfall on to dry soil (see below), then rapid transfer of labile organic phosphorus compounds could occur.

Organic phosphorus solubilization by soil drying and wetting cycles

An important process by which microbial organic phosphorus is transferred in runoff may be the re-wetting of dry soils. It has been known for some time that soil drying can render considerable concentrations of organic carbon soluble in water (Birch,

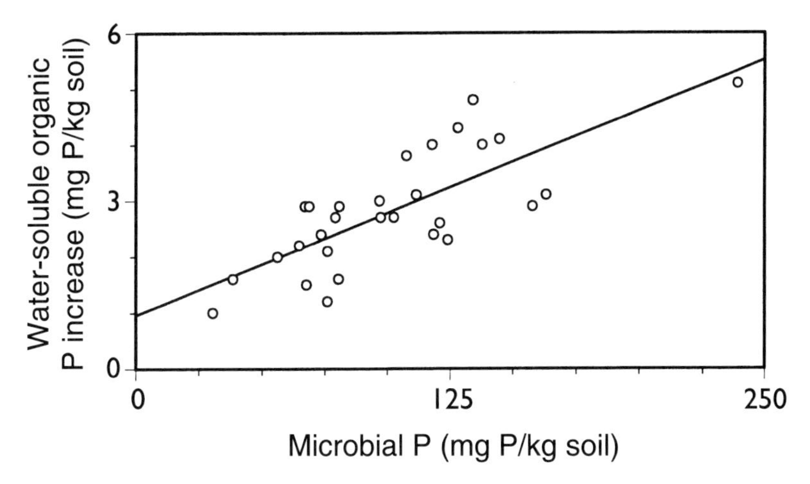

Fig. 12.2. The increase in water-soluble organic phosphorus after soil drying as a function of soil microbial phosphorus in a wide range of permanent pasture soils from England and Wales (Turner and Haygarth, 2001). Water-soluble phosphorus was determined by extracting soils at field moisture capacity with water in a 4:1 solution/soil ratio for 1 h. Subsamples were air-dried for 7 days at 30°C and extracted in an identical manner.

1959, 1960) and can also influence the solubility of inorganic nutrients (Olsen and Court, 1982; Haynes and Swift, 1985; Sparling *et al.*, 1985; Sparling and Ross, 1988; Comfort *et al.*, 1991). The effects on organic phosphorus were recently explored, but rewetting can release substantial amounts of organic phosphorus to solution and may have particular relevance for its transfer to watercourses.

In a wide range of pasture soils from England and Wales, 7 days air drying at 30°C from approximate field moisture capacity increased concentrations of water-extractable organic phosphorus by between 185% and 1900%, with organic phosphorus accounting for up to 100% of the total filterable phosphorus increase (Turner and Haygarth, 2001). The released organic phosphorus appeared to be at least partly derived from microbial cells, because a strong correlation existed between solubilized organic phosphorus and soil microbial phosphorus (Fig. 12.2). This was not unexpected, because rapid rehydration can kill between 17% and 58% of soil microbes through osmotic shock and cell rupture (Salema *et al.*, 1982; Kieft *et al.*, 1987), and was subsequently confirmed by direct bacterial cell counts in dried and re-wetted Australian pasture soils (Turner *et al.*, 2003c).

The physical stresses induced by soil drying also disrupt organic matter coatings on clay and mineral surfaces (Bartlett and James, 1980), which probably also contribute to the solubilization of organic phosphorus following re-wetting. Indeed, functional classification of water-extractable organic phosphorus in air-dried Australian pasture soils revealed similar proportions of microbially derived phosphate diesters and *myo*-inositol hexakisphosphate from the non-biomass soil organic matter (Turner *et al.*, 2002a). The release of non-biomass soil organic phosphorus was also suggested as an explanation for increases in bicarbonate-extractable organic phosphorus following re-wetting of dried pasture soils (Turner and Haygarth, 2003).

The degree of microbial death upon rewetting is primarily determined by the moisture content of the soil prior to extraction. For example, the decline in microbial carbon is linearly correlated with soil moisture, although a significant decrease occurs only after drying to $<10\%$ gravimetric moisture content (West *et al.*, 1988, 1992). The susceptibility of microorganisms to drying and wetting is also influenced by the age and type of organisms and the degree of physical protection offered by the soil matrix (van

Gestel *et al.*, 1993). For example, fast-growing (typically Gram-negative) bacteria are susceptible to drying due to their location on the outside of aggregates, whereas slow-growing (typically Gram-positive) bacteria are probably protected from anything other then severe drying events by their location within aggregates (van Gestel *et al.*, 1993, 1996). Many organisms also release compatible solutes to protect against desiccation, which may provide a further source of organic phosphorus (Halverson *et al.*, 2000).

It might be expected that microbes in temperate pasture soils, which experience intense drying on only a few occasions per year, would be less tolerant of drying than those in soils that experience more frequent and intense drying (Sparling *et al.*, 1987). However, there is evidence that the impact of drying on microbial populations is similar across a wide range of soil types and moisture regimes. For example, Sparling *et al.* (1989) found no evidence that soil microbial populations from dry areas of New Zealand were more tolerant of osmotic stress than populations from moist areas, while re-wetting can kill approximately two-thirds of the microbial biomass in xeric soils of the western USA that undergo intense summer drying (Kieft *et al.*, 1987). Furthermore, irrigation causes substantial seasonal shifts in microbial community composition in soils where tolerance to extreme fluctuations in moisture stress might be expected (Lundquist *et al.*, 1999). These changes may be linked to major shifts in population structure between moist and dry seasons, and clearly demonstrate that nutrient release from microbes following the re-wetting of dry soils occurs even in soils where microbial adaptation is expected. The release of organic phosphorus by drying and re-wetting may therefore be important for plant nutrition in tropical soils, although information remains limited from such ecosystems (see Nziguheba and Bünemann, Chapter 11, this volume).

Microbial lysis probably also occurs following freezing and thawing, because considerable changes in organic phosphorus were observed in two soils subjected to various freezing and thawing cycles (Ron Vaz *et al.*, 1994). Effects were greatest in a peaty podzol, in which acetic-acid-extractable phosphorus increased from 0.58 to 3.21 mg P/kg soil. This was partly attributed to microbial phosphorus release, although the complex effects of freezing meant that several interacting mechanisms could have been responsible.

The role of microfloral grazers in organic phosphorus solubilization

Soil fauna that graze the microflora are an additional biological factor that may contribute to organic phosphorus transfer. These organisms include bacterial grazers, such as protozoa (mainly amoebae) and nematodes, and fungal grazers, such as microarthropods (mainly Collembola). It is now recognized that microfloral grazers have a significant role in soil processes (Bradford *et al.*, 2002), including nutrient dynamics, and can have a large impact on the solubility of phosphorus (Cole *et al.*, 1978; Setälä *et al.*, 1990) and nitrogen (Woods *et al.*, 1982; Clarholm, 1985). Phosphorus is mainly excreted as phosphate from these organisms (Woods *et al.*, 1982), but their indirect effects on microbial numbers, activity and turnover (Clarholm, 1985; Ingham *et al.*, 1985) are likely to strongly influence rates of organic phosphorus solubilization. They can respond rapidly to fluctuations in biomass; for example, protozoan populations can double their size in just a few hours given sufficient bacterial numbers, and show seasonal fluctuations (Persmark *et al.*, 1996) that appear to correspond to patterns of soil microbial biomass and organic phosphorus leaching (Turner and Haygarth, 2000). The activity of microfloral grazers may therefore be an important, but overlooked, aspect of organic phosphorus solubilization in soils.

Methodological considerations for water-extractable phosphorus studies

It is important to note that results from phosphorus solubilization studies involving water-extraction can be markedly influenced by the laboratory extraction procedures employed. Clearly, drying has a considerable impact on extracted organic phosphorus, but the choice of soil to solution ratio is also critical. For example, Chapman *et al.* (1997) demonstrated that the proportion of filterable organic phosphorus in water extracts decreased from 26% to 6% as the soil-to-solution ratio increased from 1:1.5 to 1:15.4. Simultaneously, phosphate concentrations increased from 71% to 92%. Low extract ratios close to 1:1 compared most favourably with the composition of phosphorus in leachate, while wider ratios extracted a much greater proportion of filterable phosphate, diminishing the apparent importance of the organic phosphorus component. The changes in organic phosphorus were almost certainly explained by dilution of the relatively static pool of water-extractable organic phosphorus. Clearly, these factors must be considered carefully when selecting a water-extraction methodology to study organic phosphorus solubilization mechanisms.

Forms of Organic Phosphorus in Soil Solution and Runoff

Both filterable and particulate forms of organic phosphorus are transferred from soils, but the lack of suitable techniques for determining specific phosphorus compounds in aqueous samples means that there is little information available on the chemical composition of either fraction. Conceptual understanding of organic phosphorus forms and their behaviour in water is also limited by the operational definitions that arise from conventional analysis procedures (see below). For detailed reviews of speciation techniques for organic phosphorus in soil waters see McKelvie (Chapter 1, this volume) and Cooper *et al.* (Chapter 3, this volume).

Estimation of organic phosphorus in runoff by molybdate colorimetry

The analysis of organic phosphorus in drainage waters is almost exclusively based on operationally defined procedures involving crude size separation and chemical speciation (Jarvie *et al.*, 2002). Size separation is conventionally performed by membrane filtration, usually involving a pore size of 0.2 or 0.45 μm. Molybdate colorimetry is then used to estimate free phosphate (termed molybdate-reactive phosphorus), with organic phosphorus estimated as the unreactive fraction (the difference between total filterable phosphorus determined by some form of digestion, and phosphate determined by molybdate reaction). The unreactive fraction usually provides a close approximation of the filterable organic phosphorus concentration, although the true concentration may be either over- or underestimated depending on the sample in question. Underestimation occurs when acid-labile organic phosphorus compounds are hydrolysed in the acidic conditions of the colorimetric procedure, although this is negligible when reaction times are short and samples are analysed relatively soon after collection (Denison *et al.*, 1998). Overestimation is more likely, and occurs when inorganic polyphosphates and mineral colloids are included in the unreactive fraction and classified as organic phosphorus. Indeed, this can represent a significant source of error in samples that contain large concentrations of colloidal material, including phosphate associated with metal–humic complexes (Dolfing *et al.*, 1999; Hens and Merckx, 2001). Such complexes have also been implicated in the overestimation of organic phosphorus in alkaline soil extracts by molybdate colorimetry (Turner *et al.*, 2003b, d), a potentially serious problem, given the widespread use of sequential fractionation procedures that involve alkaline extraction.

Organic phosphorus analysis can be further complicated by rapid changes during storage, such as those induced by microbial activity, ultraviolet oxidation, and hydrolysis by phosphatase enzymes or mineral particles (Francko and Heath, 1979; Haygarth *et al.*, 1995; Baldwin *et al.*, Chapter 4, this volume). To avoid ambiguity associated with conventional definitions of filterable organic phosphorus, many authors prefer the term molybdate-unreactive phosphorus, accepting that this is mainly organic phosphorus, but also includes other types of compounds. A more accurate technique for organic phosphorus determination in aqueous samples involves on-line ultraviolet oxidation and molybdate reaction in a flow injection system (Ron Vaz *et al.*, 1992; Peat *et al.*, 1997). This avoids the inclusion of inorganic polyphosphates and mineral colloids, but not hydrolysis of acid-labile organic compounds.

Speciation of filterable organic phosphorus compounds in drainage water

Several analytical techniques are available for the identification of organic phosphorus species, including enzymatic, chromatographic and spectroscopic procedures. However, many of these are of limited usefulness for analysis of runoff samples, because organic phosphorus concentrations are typically several orders of magnitude less than the detection limits of the analytical techniques (Turner *et al.*, 2002b; McKelvie, Chapter 1, this volume; Cooper *et al.*, Chapter 3, this volume). This means that there is little compound-specific information on organic phosphorus in runoff water. Espinosa *et al.* (1999) reported the first information on organic phosphorus speciation in soil leachate. Using strong anion exchange resins for sample preconcentration and high-performance liquid chromatographic separation, they detected traces of *myo*-inositol hexakisphosphate, adenosine triphosphate, glucose phosphate and a phosphonate in leachate from a silty clay soil. Wild and Oke (1966) reported that filterable organic phosphorus extracted with calcium chloride from two English arable soils was dominated by inositol monophosphate, while Martin (1970) detected more than 30 different unidentified phosphate esters in water extracts of some New Zealand and US soils.

There are numerous reports of identified organic phosphorus compounds in river and lake waters. For example, DNA concentrations were determined in two Florida rivers of contrasting trophic status (Paul *et al.*, 1991), while a variety of filterable organic phosphorus compounds were detected in water from a treatment wetland by capillary-electrophoresis separation and detection by mass-spectrometry (Llewelyn *et al.*, 2002; Cooper *et al.*, Chapter 3, this volume). Similarly, enzymatic hydrolysis techniques have yielded information on functional organic phosphorus classes in natural waters (e.g. Herbes *et al.*, 1975; Shan *et al.*, 1994). However, it is difficult to infer terrestrial origins for organic phosphorus compounds measured in aquatic environments, because they are equally likely to be derived from autochthonous sources such as algal or bacterial cells.

During the last decade, information on the organic phosphorus composition of soil solutions and water extracts has been obtained by phosphatase hydrolysis. This technique not only gives structural information on filterable organic phosphorus, but also indicates its potential biological availability. In solution from Scottish upland soils, up to 64% of the filterable organic phosphorus was hydrolysed by non-specific phosphatases (Shand and Smith, 1997), while hydrolysable unreactive phosphorus in water extracts of Australian pasture soils was dominated by phosphate diesters and *myo*-inositol hexakisphosphate (Turner *et al.*, 2002a). Only small concentrations of labile monoesters were detected in the latter study, possibly due to the rapid hydrolysis of labile compounds by soil phosphatase enzymes.

In enzyme hydrolysis studies, large proportions of the filterable unreactive phosphorus are not hydrolysed by commercial phosphatase enzymes. However, some of this non-hydrolysable phosphorus may be high-molecular-weight humic–metal–phosphate complexes rather than true organic phosphorus (Gerke, 1992; Hens and Merckx, 2001). In addition, bacterial cells and cell fragments are almost certainly present (Turner *et al.*, 2003c) and were reported to dominate the phosphorus in leachate from calcareous soil columns amended with sucrose or crop residues (Hannapel *et al.*, 1964b). Stevens and Stewart (1982a) reported that filterable organic phosphorus in rivers entering Lough Neagh, Northern Ireland, consisted of both acid- and alkali-soluble components that were structurally similar to soil humic acids. Only 7% of the acid-soluble organic phosphorus and 29% of the alkali-soluble organic phosphorus were available to an axenic culture of *Oscillatoria redekei* (a cyanobacterium common in Lough Neagh), suggesting the relative recalcitrance of the compounds present. However, *O. redekei* may not be an ideal test organism for such assays, because it thrives in eutrophic waters and is unlikely to be well-adapted to access organic phosphorus compounds in its environment. A more comprehensive assay of the algal availability of filterable organic phosphorus in a particular environment would thus involve a more diverse range of the representative algal flora.

Particle-associated organic phosphorus

Particulate (particles >0.45 μm in diameter) and colloidal (particles between 1 nm and 1 μm in diameter) material contains organic phosphorus complexed with mineral and high-molecular-weight organic material, and can account for a considerable fraction of total phosphorus in runoff (Heathwaite *et al.*, 1990; Haygarth *et al.*, 1997). For example, 80% of the organic phosphorus in the soil solution of a Scottish peat was associated with particles >0.22 μm in diameter, with most being associated with particles >1.6 μm (Shand *et al.*, 2000). Particulate unreactive phosphorus constituted between 20% and 32% of the total phosphorus draining to 1.3 m depth from four soil types under cut grassland (Turner and Haygarth, 2000), and between 15% and 23% of the total phosphorus in overland flow and tile drainage from grazed pasture (Haygarth *et al.*, 1998). On a larger spatial scale, around 80% of the total phosphorus export from a small agricultural catchment in southwest England was identified as organic phosphorus bound to soil particles (Heathwaite *et al.*, 1990). However, it should be noted that, as well as organic phosphorus, the particulate unreactive fraction can also include inorganic phosphorus forms that are not reactive with molybdate, including mineral and organically complexed phosphate.

Little information exists on the chemical composition of particle-associated organic phosphorus, although much is likely to be of a similar composition to that in the soil (Suzumura and Kamatani, 1995b). However, processes removing and transporting soil particles in runoff are selective for fine clays (Sharpley, 1985), which tend to be enriched in organic phosphorus, especially labile phosphate diesters (Amelung *et al.*, 2001). In river and lake waters, a significant component of the particulate organic phosphorus may also be contained in live algal cells. For example, Gibson *et al.* (1995) calculated that as much as 30% of the particulate phosphorus export from Loughgarve, an upland lake in Antrim, Northern Ireland, occurred in this form. In contrast, Stevens and Stewart (1982b) reported that particulate organic phosphorus in river water entering Lough Neagh, Northern Ireland, resembled humic acid structures possibly involving iron. These almost certainly included inositol phosphates in high-molecular-weight complexes (Herbes *et al.*, 1975), a hypothesis supported by the detection of several isomers of inositol hexakisphosphate attached to suspended particles in Japanese riverine and estuarine waters (Suzumura and Kamatani, 1995b). This clearly indicates that inositol phosphates are involved in the phosphorus transfer process, despite their perceived stability in soils. They must also be regarded as potentially bioavailable, even when transferred in complexed forms, because there is evidence of their rapid breakdown in marine environments (Suzumura and Kamatani, 1995a) and more than 30 cyanobacterial strains can use *myo*-inositol hexakisphosphate as their sole phosphorus source (Whitton *et al.*, 1991).

Transport and Mobility of Organic Phosphorus in Soils

The driving force behind phosphorus transfer is rainfall, which provides the energy and the carrier for phosphorus movement in the landscape (Haygarth and Jarvis, 1999). The route taken by rainfall as it carries phosphorus off the land describes the hydrological pathway. This can have important implications for phosphorus transfer, in particular by determining the extent of interaction between drainage water and the soil matrix. Rainfall can take a variety of pathways as it drains from the land, including overland or shallow interflow in a lateral direction across or through the soil surface, and matrix or preferential flow in a more downward direction through the soil profile. Rapid movement of rainfall through soil can also be promoted by land drainage with tile or mole drains. Comprehensive details and definitions of the myriad pathways and their relation to the transfer of phosphorus and other pollutants can be found elsewhere (Ryden *et al.*, 1973; Haygarth *et al.*, 2000; Haygarth and Jarvis, 2002).

For many years phosphorus was considered to be immobile in the soil and was not expected to leach to depth in the profile (Sample *et al.*, 1980). This is generally true for phosphate (although preferential flow pathways can promote phosphate transfer; Grant *et al.*, 1996; Haygarth *et al.*, 1998), but is not necessarily the case for organic phosphorus. In particular, the weak sorption of some organic phosphorus compounds, including sugar phosphates and phosphate diesters, means that they are mobile in the soil and can potentially escape in runoff to surface waters (Anderson and Arlidge, 1962; Frossard *et al.*, 1989; McKercher and Anderson, 1989). This is reflected in the composition of filterable phosphorus draining through different hydrological pathways, because pathways that facilitate interaction between drainage water and the soil matrix contain greater proportions of organic phosphorus. This was highlighted by Chapman *et al.* (1997), who showed that concentrations of total filterable phosphorus in soil solution were twice those in free-draining leachate and contained a greater proportion of organic phosphorus. Similar results were obtained by Magid *et al.* (1992), who compared phosphorus in leachate from free-draining lysimeters to that in suction cups installed at 90 cm depth in a sandy arable soil. In addition, Haygarth *et al.* (1998) reported a considerably greater proportion of filterable organic phosphorus in drain flow to 85 cm (46%) compared with surface runoff and interflow to 30 cm (27%) from grazed pasture plots in southwest England, due to a fourfold reduction in filterable phosphate in drain flow.

The weak retention of organic phosphorus compared to phosphate is not true of all organic phosphorus compounds, because higher-order inositol phosphates are strongly retained in soil. This probably prevents them from leaching through soil in a soluble form, but does not preclude their movement in high-molecular-weight complexes (see above). Clearly the differences in organic phosphorus concentrations between drainage waters and soil solutions are partly linked to the pathway being sampled, because suction cups and centrifugation both measure relatively immobile pore water, while lysimeters also include mobile water moving through preferential flow pathways. It is therefore arguable that lysimeters provide a more realistic measure of phosphorus forms and concentrations in water leaving the field. However, this means that considerable difficulties arise when comparing these various experimental scales, because it only seems possible to compare data for soils sampled and analysed in a similar manner. This means that the amount of useful published data is small.

Organic phosphorus compounds expected to be mobile in soil are also the compounds most susceptible to rapid degradation by phosphatase enzymes, because most phosphate monoesters (excluding the higher-order inositol phosphates) and phosphate diesters are degraded within hours or days of release in soil (Bowman and Cole, 1978; Dick and Tabatabai, 1978; Harrison, 1982a). Phosphatase activity may therefore partly regulate organic phosphorus transfer, because a high phosphatase activity may reduce phosphorus leaching by converting mobile organic phosphorus to relatively immobile phosphate. This is likely to vary among compounds, because phosphate diesters degrade more slowly than labile monoesters (Harrison, 1982a). Given that water movement from hill slopes can be rapid during storm events, organic phosphorus compounds should have ample time to escape before they are hydrolysed. This means that hydrological flow pathways that facilitate rapid movement of rainfall from the land, either across the surface or through preferential flow pathways, will facilitate the transfer of labile soil organic phosphorus. For example, organic phosphorus solubilized by soil drying and re-wetting has high potential for transfer to watercourses, because rainfall on to dry soils moves rapidly through cracks and other preferential flow pathways (Simard *et al.*, 2000; Preedy *et al.*, 2001), with little chance for organic phosphorus to interact with the soil or enzymes. This is especially likely where a high degree of macropore connectivity exists, such as the extensive pipeflow networks within blanket peats of the northern English uplands (Holden and Burt, 2002), and may at least partly explain elevated concentrations of nutrients observed in drainage water when rainfall follows a dry period. Indeed, several studies reported a pronounced seasonal pattern in organic phosphorus transfer in a range of environments (Livingstone and Whitton, 1984; Magid *et al.*, 1992; Hooda *et al.*, 1997; Turner and Haygarth, 2000, 2003a). For example, organic phosphorus concentrations in leachate from monolith lysimeters under cut grassland were greatest during the spring period (Turner and Haygarth, 2000), while pulses of organic phosphorus occurred in the spring and early summer in soil solutions and first-order streams in the English uplands (Livingstone and Whitton, 1984; Turner *et al.*, 2003a). Elevated concentrations of other nutrients, including organic carbon (Chittleborough *et al.*, 1992) and nitrogen (Garwood and Tyson, 1973; Jordan and Smith, 1985), have also been reported in drainage water when rainfall followed a dry period.

Even if organic phosphorus escapes from the soil, it will be continually degraded by soluble phosphatase enzymes or abiotic processes, resulting in concentrations decreasing during transport. Therefore, the domination of phosphate in streams and rivers does not preclude the possibility that organic phosphorus facilitated the initial escape of phosphorus from the soil profile by protecting it from sorption or biological uptake. There is no information on this aspect of the phosphorus transfer process, which requires detailed studies of phosphatase activity and organic phosphorus speciation in drainage water, perhaps involving radiolabelled organic phosphorus tracers.

Ecological Role of Organic Phosphorus Transfer

There is increasing recognition of the importance of organic forms of nutrients in drainage from unmanaged environments (e.g. van Breemen, 2002; Turner *et al.*, 2003a). For example, speciation of nitrogen in rivers draining temperate old-growth forest in South America revealed that filterable organic nitrogen was the dominant form, which contrasts markedly with the dominance of inorganic nitrogen in drainage from northern-hemisphere temperate forests that have been influenced by high rates of atmospheric nitrogen deposition (Perakis and Hedin, 2002). In terms of phosphorus, studies of North American lakes revealed that dissolved phosphate concentrations determined by a sensitive bioassay approach were at least two orders of magnitude smaller than those determined by standard molybdate colorimetry (Hudson *et al.*, 2000), suggesting that the turnover of organic phosphate has far greater importance for phosphorus nutrition in such ecosystems than was previously realized (Karl, 2000).

Organic phosphorus appears to have particular importance in the English uplands, which have received relatively high rates of nitrogen deposition from the atmosphere. This enhances biological phosphorus limitation and, therefore, the importance of organic phosphorus compounds to plants and microorganisms (Aber *et al.*, 1989; Turner *et al.*, 2001a). A particularly well-studied region is the Upper

Teesdale National Nature Reserve, an upland area of northern England that has received considerable atmospheric nitrogen deposition during the last 150 years. Drainage from peat and calcareous soils in Upper Teesdale contains filterable phosphate concentrations that are too low to detect using conventional procedures, so the only phosphorus available to stream organisms is organic phosphorus that occurs in discrete pulses during the spring and early summer (Livingstone and Whitton, 1984; Whitton *et al.*, 1998; Turner *et al.*, 2003a). During these pulses, biological productivity in the streams probably switches from phosphorus to nitrogen limitation (Turner *et al.*, 2003a), a similar situation to that reported in the upper reaches of the Dee catchment in northeast Scotland (Edwards *et al.*, 2000). The seasonal pattern of nutrient limitation in Upper Teesdale streams exerts a marked influence on the ecology of the streams, because organisms must adapt to access phosphorus from organic compounds occurring in short pulses during the year. For example, the dominant stream phototroph, *Rivularia*, is well-suited for survival in low-phosphorus environments, because it displays high phosphatase activity for much of the year and has the ability to accumulate phosphorus rapidly from pulses and store it for long periods (Whitton *et al.*, 1998). Similarly, semi-aquatic bryophytes growing in the streams display marked phosphatase activity, suggesting adaptation to intermittent pulses of organic phosphorus during the growing season (Turner *et al.*, 2001a).

Such ecological adaptations are not restricted to the northern English uplands, suggesting that a similar pattern of phosphorus transfer occurs in other environments. For example, studies in the Canadian sub-Arctic indicate that hair-forming algae comprise an important component of the flora of drainage streams (Sheath and Müller, 1997). Such algae are notable producers of phosphatase (Whitton, 1988), suggesting that Arctic streams receive infrequent organic phosphorus in pulses from the surrounding tundra. Clearly, patterns of organic phosphorus transfer in unmanaged environments can exert a marked influence on stream ecology, although more detailed information is required from a wider range of environments to confirm this.

Animal Manures and Organic Phosphorus Transfer

The accumulation of soil phosphorus associated with manure application from intensive animal operations ranks amongst the greatest threats to water quality in agricultural regions (Sims *et al.*, 2000). Organic phosphorus can be a large proportion of the total phosphorus in animal manures (Barnett, 1994), and may play a key role in determining the impact of manure application on phosphorus transfer. In particular, there is considerable current interest in the role of manure-derived *myo*-inositol hexakisphosphate in the phosphorus transfer process (see below).

Organic phosphorus concentrations in leachate appear to increase following manure application to soils. Liquid manures often contain only a relatively small proportion of organic phosphorus (<10%; Fordham and Schwertmann, 1977), but this fraction is highly mobile in the soil and can lead to elevated concentrations of organic phosphorus in drainage water. For example, organic phosphorus in filtered leachate from a sandy soil amended with pig slurry constituted >90% of the total phosphorus, and dominated the soil-solution phosphorus at depth in field soils (Chardon *et al.*, 1997). Similarly, organic phosphorus was the dominant form in unsaturated flow leaching through intact blocks of a heavy clay soil amended with cattle manure, although this was not mirrored under saturated flow conditions, when filterable phosphate dominated (Jensen *et al.*, 2000). Elevated concentrations of filterable organic phosphorus were also reported in tile drainage following slurry application to pasture in Northern Ireland (Jordan and Smith, 1985). Transfer of manure-derived organic phosphorus can also occur by direct loss of surface-applied manure when application is immediately followed by rainfall. Few data

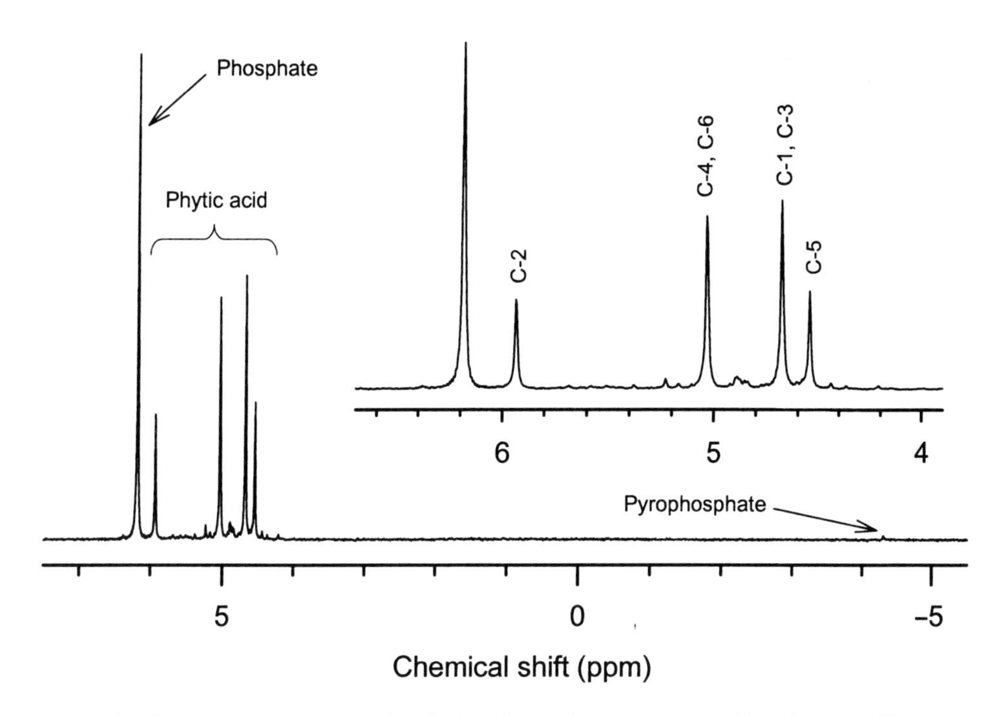

Fig. 12.3. The phosphorus composition of broiler litter from a farm in Delaware, USA, determined by extraction with a solution containing sodium hydroxide and ethylenediaminetetraacetate (EDTA) and solution ^{31}P nuclear magnetic resonance spectroscopy. The extract contained 97% of the total phosphorus in the litter, of which 40% was phosphate and 60% was *myo*-inositol hexakisphosphate (phytic acid). Traces of pyrophosphate were also detected. The inset spectrum shows the four signals from *myo*-inositol hexakisphosphate, corresponding to the positions of the phosphates on the inositol ring (C-1 to C-6). Details of analytical methodology can be found in Turner (2004).

exist on such 'incidental' transfers, but surface runoff and interflow from pasture plots in southwest England contained up to 1 mg/l filterable organic phosphorus when rainfall occurred 24 h after slurry application (Preedy *et al.*, 2001).

An important contemporary issue involves the impact of dietary manipulations to reduce *myo*-inositol hexakisphosphate in manures on the fate of manure phosphorus in the environment. Inositol phosphates form a major component of the total phosphorus in many manures (Peperzak *et al.*, 1959; Fig. 12.3), because although *myo*-inositol hexakisphosphate is the dominant phosphorus compound in most cereal grains, it cannot be digested by monogastric animals, such as poultry and swine, due to low levels of intestinal phytase (Taylor, 1965; Pallauf and Rimbach, 1997). As a result, strategies to reduce *myo*-inositol hexakisphosphate in feeds are being widely adopted to address the hyperaccumulation of soil phosphorus in areas with high livestock densities. Such strategies include the isolation of mutant grains that produce low inositol phosphate concentrations (Raboy *et al.*, 2000) and supplementation of animal feeds with phytase enzymes to hydrolyse *myo*-inositol hexakisphosphate in the gut (Simons *et al.*, 1990; Poulsen, 2000). Both these techniques increase the availability of inositol phosphates to the animal, allowing inorganic phosphate supplements to be minimized. Indeed, phytase supplements can reduce total phosphorus concentrations in pig faeces by 35–47%, and allow mineral

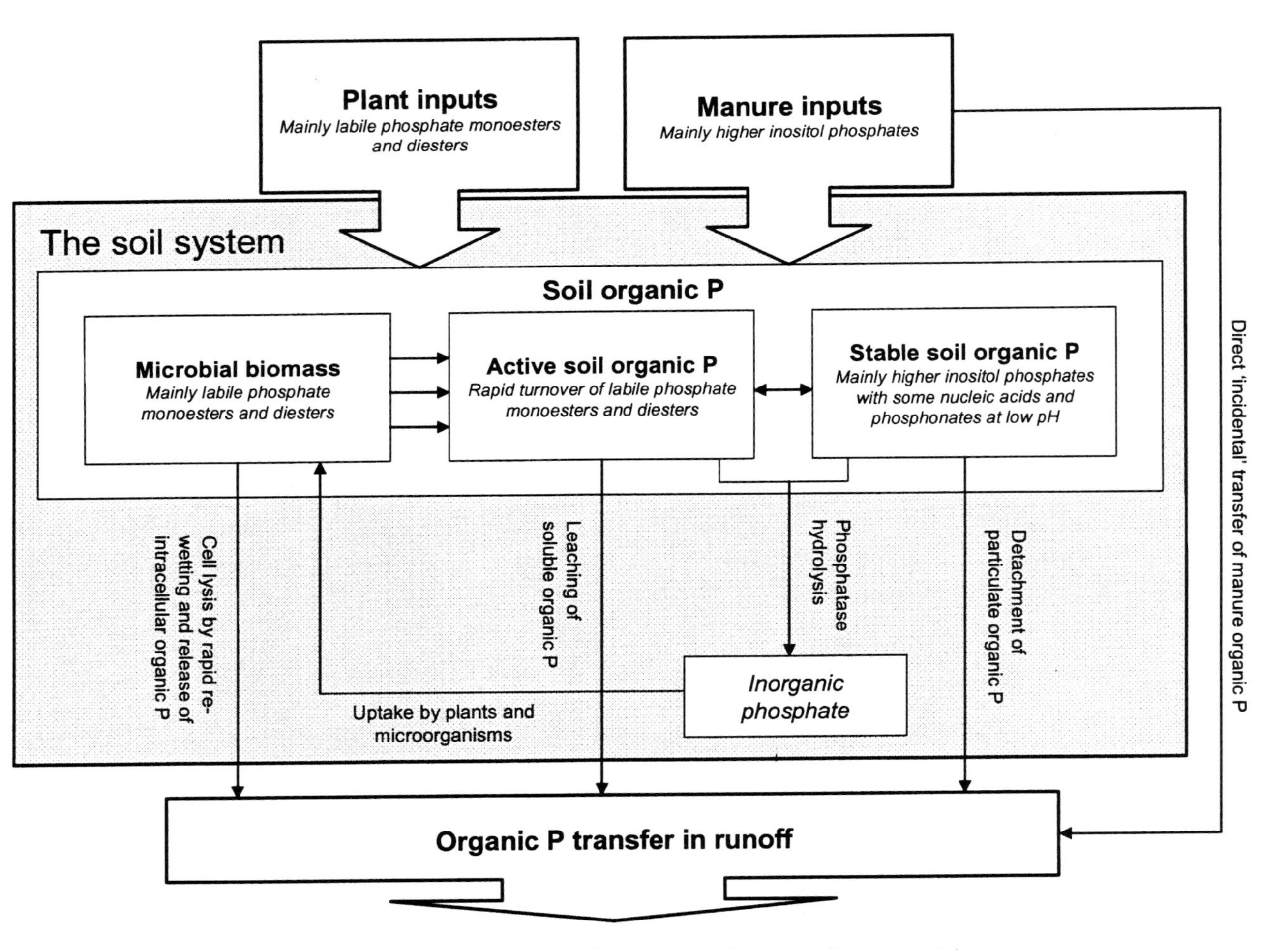

Fig. 12.4. Simplified conceptual summary of processes involved in the transfer of organic phosphorus from terrestrial to aquatic environments.

phosphate supplements in broiler chicken feed to be reduced by up to 50% with no effect on feed conversion, growth rate, bodyweight or bird performance (Simons *et al.*, 1990).

Despite the many agronomic advantages of strategies to reduce *myo*-inositol hexakisphosphate in animal feeds, the environmental benefits are less conclusive. Such strategies assume that a reduction in total manure phosphorus equates to a corresponding reduction in the risk of phosphorus transfer, but changes in the phosphorus composition of manure may induce the opposite effect. In particular, the strong fixation of *myo*-inositol hexakisphosphate in soil prevents it being leached in soluble form, so it would be expected that phosphorus in manures containing large quantities of *myo*-inositol hexakisphosphate would be relatively immobile compared with those containing mainly phosphate. This means that land application of manures from animals fed diets containing phytase or mutant grain varieties could increase phosphorus mobility in the soil and facilitate its transfer to watercourses.

In soil, *myo*-inositol hexakisphosphate can displace phosphate from binding sites and reduce the potential for further additions of phosphorus to be sorbed in the soil. This phenomenon may only become apparent following multiple applications of manure to the same soil, during which time manure-derived *myo*-inositol hexakisphosphate would progressively occupy phosphate-binding sites in the soil. Therefore, an alternative hypothesis is that reductions in *myo*-inositol hexakisphosphate in manure could reduce phosphorus solubility by minimizing the occupation of binding sites. Clearly this is a complex issue, and further detailed studies are required before it can be fully understood. Such studies should include mechanistic experiments and the development of robust techniques for the determination of inositol phosphates in animal manures and runoff waters (information becomes available, the benefits of reduced concentrations of total phosphorus in manures must be considered alongside possible increases in phosphorus mobility in the landscape.

Conclusions and Recommendations for Future Research

The processes involved in the transfer of organic phosphorus from terrestrial to aquatic environments are summarized in Fig. 12.4. Organic phosphorus clearly plays a fundamental, but under-recognized, role being mobile in the soil and potentially available to algae and other aquatic microorganisms in waterbodies. Organic phosphorus transfer has an important ecological function in unmanaged environments, where it can dominate phosphorus concentrations in drainage. In other environments it is a pollutant, although its contribution to phosphorus transfer from agricultural land varies markedly among land-uses and management systems. Despite its importance, organic phosphorus transfer remains poorly understood, and there are many aspects where further research is required. An inexhaustive list includes:

- *Speciation of the trace concentrations of organic phosphorus compounds in aqueous samples.* Robust yet sensitive techniques for organic phosphorus speciation are fundamental to a comprehensive understanding of all aspects of the phosphorus transfer process (see McKelvie, Chapter 1, this volume; Cooper *et al.*, Chapter 3, this volume).
- *Transformations of organic phosphorus during movement through the landscape.* Organic phosphorus may facilitate the escape of phosphorus from soils by protecting it from sorption or biological uptake as it moves through the profile. Understanding this requires detailed studies possibly involving radiolabelled organic phosphorus compounds. Such studies may significantly influence how soil phosphorus is managed to protect the environment.
- *The role of manure-derived organic phosphorus in phosphorus transfer.* In particular, it is unclear how strategies to reduce *myo*-inositol hexakisphosphate in manure influence the mobility of phosphorus in the environment. In addition to collecting data from field

trials, studies should address this issue at a mechanistic level, and consider the long-term consequences of multiple manure applications.

- *The ecological function and importance of organic phosphorus transfer in unmanaged and nitrogen-polluted environments.* Organic phosphorus availability has an important ecological function in such environments, and some organisms possess complex mechanisms to access organic phosphorus compounds. More information on the processes that influence organic phosphorus transfer is therefore required to effectively understand and manage these sensitive ecosystems.
- *The biological availability of terrestrially derived organic phosphorus in aquatic systems.* It can be assumed that all organic phosphorus is potentially bioavailable in waterbodies, but the common misconception that organic phosphorus is relatively unavailable to algae persists. This is reinforced by algal bioassays that involve species unlikely to be adapted to utilize organic phosphorus from their environment. Comprehensive information on the availability of naturally occurring organic phosphorus to a wide range of cyanobacteria and eukaryotic algae would allow the potential impact of organic phosphorus transfer to be considered in management decisions, especially when combined with information on organic phosphorus speciation and aquatic flora.
- *Impacts of climate change on organic phosphorus transfer.* In the coming decades, many areas will experience increased precipitation, enhanced dry periods, or more frequent wetting and drying cycles (IPCC, 1995), which may all have marked impacts on nutrient release from soils. The magnitude and geographical extent of such impacts are largely unknown, but may become increasingly important in terms of pollutant phosphorus transfer and the nutrient ecology of unmanaged environments.

References

Aber, J.D., Nadelhoffer, K.J., Steudler, P. and Melillo, J.M. (1989) Nitrogen saturation in northern forest ecosystems. *Bioscience* 39, 378–386.

Amelung, W., Rodionov, A., Urusevskaja, I.S., Haumaier, L. and Zech, W. (2001) Forms of organic phosphorus in zonal steppe soils of Russia assessed by ^{31}P NMR. *Geoderma* 103, 335–350.

Anderson, G. (1980) Assessing organic phosphorus in soils. In: Khasawneh, F.E., Sample, E.C. and Kamprath, E.J. (eds) *The Role of Phosphorus in Agriculture.* ASA-CSSA-SSSA, Madison, Wisconsin, pp. 411–431.

Anderson, G. and Arlidge, E.Z. (1962) The adsorption of inositol phosphates and glycerophosphate by soil clays, clay minerals, and hydrated sesquioxides in acid media. *Journal of Soil Science* 13, 216–224.

Anderson, G., Williams, E.G. and Moir, J.O. (1974) A comparison of the sorption of inorganic orthophosphate and inositol hexaphosphate by six acid soils. *Journal of Soil Science* 25, 51–62.

Barnett, G.M. (1994) Phosphorus forms in animal manure. *Bioresource Technology* 49, 139–147.

Bartlett, R. and James, B. (1980) Studying dried, stored soil samples: some pitfalls. *Soil Science Society of America Journal* 44, 721–724.

Beauchemin, S., Simard, R.R. and Cluis, D. (1998) Forms and concentration of phosphorus in drainage water of twenty-seven tile-drained soils. *Journal of Environmental Quality* 27, 721–729.

Bedrock, C.N., Cheshire, M.V., Chudek, J.A., Goodman, B.A. and Shand, C.A. (1994) Use of ^{31}P-NMR to study the forms of phosphorus in peat soils. *Science of the Total Environment* 152, 1–8.

Bieleski, R.L. (1973) Phosphate pools, phosphate transport and phosphate availability. *Annual Review of Plant Physiology and Plant Molecular Biology* 24, 225–252.

Birch, H.F. (1959) Further observations on humus decomposition and nitrification. *Plant and Soil* 9, 262–286.

Birch, H.F. (1960) Nitrification in soils after different periods of dryness. *Plant and Soil* 12, 81–96.

Bowling, L.C. and Baker, P.D. (1996) Major cyanobacterial bloom in the Barwon–Darling River, Australia, in 1991, and underlying limnological conditions. *Marine and Freshwater Research* 47, 643–657.

Bowman, B.T., Thomas, R.L. and Elrick, D.E. (1967) The movement of phytic acid in soil cores. *Soil Science Society of America Proceedings* 31, 477–481.

Bowman, R.A. and Cole, C.V. (1978) Transformations of organic phosphorus substrates in soils as evaluated by $NaHCO_3$ extraction. *Soil Science* 125, 49–54.

Bradford, M.A., Jones, T.H., Bardgett, R.D., Black, H.I.J., Boag, B., Bonkowski, M., Cook, R., Eggers, T., Gange, A.C., Grayston, S.J., Kandeler, E., McCaig, A.E., Newington, J.E., Prosser, J.I., Setälä, H., Staddon, P.L., Tordoff, G.M., Tscherko, D. and Lawton, J.H. (2002) Impacts of soil faunal community composition on model grassland ecosystems. *Science* 298, 615–618.

Brookes, P.C., Powlson, D.S. and Jenkinson, D.S. (1982) Measurement of microbial biomass phosphorus in soil. *Soil Biology and Biochemistry* 14, 319–329.

Brookes, P.C., Powlson, D.S. and Jenkinson, D.S. (1984) Phosphorus in the soil microbial biomass. *Soil Biology and Biochemistry* 16, 169–175.

Burkholder, J.M., Noga, E.J., Hobbs, C.W., Glasgow, H.B. Jr and Smith, S.A. (1992) New 'phantom' dinoflagellate is the causative agent of major estuarine fish kills. *Nature* 358, 407–410.

Cade-Menun, B.J., Berch, S.M., Preston, C.M. and Lavkulich, L.M. (2000) Phosphorus forms and related soil chemistry of Podzolic soils on northern Vancouver Island. I. A comparison of two forest types. *Canadian Journal of Forest Research* 30, 1714–1725.

Chapman, P.J., Edwards, A.C. and Shand, C.A. (1997) The phosphorus composition of soil solutions and soil leachates: influence of soil:solution ratio. *European Journal of Soil Science* 48, 703–710.

Chardon, W.J., Oenema, O., del Castilho, P., Vriesema, R., Japenga, J. and Blaauw, D. (1997) Organic phosphorus in solutions and leachates from soils treated with animal slurries. *Journal of Environmental Quality* 26, 372–378.

Chittleborough, D.J., Smettem, K.R.J., Cotsaris, E. and Leaney, F.W. (1992) Seasonal changes in pathways of dissolved organic carbon through a hillslope Xeralf soil with contrasting texture. *Australian Journal of Soil Research* 30, 465–476.

Christmas, M. and Whitton, B.A. (1998) Phosphorus and aquatic bryophytes in the Swale–Ouse river system, north-east England. 1. Relationship between ambient phosphate, internal N:P ratio and surface phosphatase activity. *Science of the Total Environment* 210/211, 389–399.

Clarholm, M. (1985) Interactions of bacteria, protozoa and plants leading to mineralization of soil nitrogen. *Soil Biology and Biochemistry* 17, 181–187.

Cole, C.V., Innis, G.S. and Stewart, J.W.B. (1977) Simulation of phosphorus cycling in semi-arid grassland. *Ecology* 58, 1–15.

Cole, C.V., Elliot, E.T., Hunt, H.W. and Coleman, D.C. (1978) Trophic interactions in soils as they affect energy and nutrient dynamics. V. Phosphorus transformations. *Microbial Ecology* 4, 381–387.

Comfort, S.D., Dick, R.P. and Baham, J. (1991) Air-drying and pre-treatment effects on sulfate sorption. *Soil Science Society of America Journal* 55, 968–973.

Condron, L.M., Turner, B.L. and Cade-Menun, B.J. (2005) Chemistry and dynamics of soil organic phosphorus. In: Sims, T. and Sharpley, A.N. (eds) *Phosphorus: Agriculture and the Environment.* American Society of Agronomy, Madison, Wisconsin (in press).

Culley, J.L.B., Bolton, E.F. and Bernyk, V. (1983) Suspended solids and phosphorus loads from a clay soil. 1. Plot studies. *Journal of Environmental Quality* 12, 493–498.

Dalal, R.C. (1977) Soil organic phosphorus. *Advances in Agronomy* 29, 89–117.

Denison, F.H., Haygarth, P.M., House, W.A. and Bristow, A.W. (1998) The measurement of dissolved phosphorus compounds: evidence for hydrolysis during storage and implications for analytical definitions in environmental analysis. *International Journal of Environmental Analytical Chemistry* 69, 111–123.

Dick, W.A. and Tabatabai, M.A. (1978) Hydrolysis of organic and inorganic phosphorus compounds added to soils. *Geoderma* 21, 175–182.

Dolfing, J., Chardon, W.J. and Japenga, J. (1999) Association between colloidal iron, aluminium, phosphorus, and humic acids. *Soil Science* 164, 171–179.

Edwards, A.C., Cook, Y., Smart, R. and Wade, A.J. (2000) Concentrations of nitrogen and phosphorus in streams draining the mixed land-use Dee catchment, north-east Scotland. *Journal of Applied Ecology* 37, 159–170.

EPA (1995) *Protecting Water Quality in Central Gippsland.* Report No. 144. Environment Protection Authority, Melbourne, Australia.

Espinosa, M., Turner, B.L. and Haygarth, P.M. (1999) Pre-concentration and separation of trace phosphorus compounds in soil leachate. *Journal of Environmental Quality* 28, 1497–1504.

Fordham, A.W. and Schwertmann, U. (1977) Composition and reactions of liquid manure (Gülle), with particular reference to phosphate. 1. Analytical composition and reaction with poorly crystalline iron oxide (ferrihydrite). *Journal of Environmental Quality* 6, 133–136.

Foy, R.H., Smith, R.V., Stevens, R.J. and Stewart, D.A. (1982) Identification of factors affecting

nitrogen and phosphorus loadings to Lough Neagh. *Journal of Environmental Management* 15, 109–129.
Francko, D.A. and Heath, R.T. (1979) Functionally distinct classes of complex phosphorus compounds in lake water. *Limnology and Oceanography* 24, 463–473.
Frossard, E., Stewart, J.W.B. and St Arnaud, R.J. (1989) Distribution and mobility of phosphorus in grassland and forest soils of Saskatchewan. *Canadian Journal of Soil Science* 69, 401–416.
Garwood, E.A. and Tyson, K.C. (1973) Losses of nitrogen and other plant nutrients to drainage from soil under grass. *Journal of Agricultural Science* 80, 303–312.
Gerke, J. (1992) Orthophosphate and organic phosphate in the soil solution of four sandy soils in relation to pH: evidence for humic–Fe–(Al) phosphate complexes. *Communications in Soil Science and Plant Analysis* 23, 601–612.
Gibson, C.E., Wu, Y. and Pinkerton, D. (1995) Substance budgets of an upland catchment: the significance of atmospheric inputs. *Freshwater Biology* 33, 385–392.
Grant, R., Laubel, A., Kronvang, B., Anderson, H.E., Svendsen, L.M. and Fugslang, A. (1996) Loss of dissolved and particulate phosphorus from arable catchments by subsurface drainage. *Water Research* 30, 2633–2642.
Greaves, M.P. and Wilson, M.J. (1969) The adsorption of nucleic acids by montmorillonite. *Soil Biology and Biochemistry* 1, 317–323.
Gregorich, E.G., Liang, B.C., Drury, C.F., Mackenzie, A.F. and McGill, W.B. (2000) Elucidation of the source and turnover of water soluble and microbial biomass carbon in agricultural soils. *Soil Biology and Biochemistry* 32, 581–587.
Haith, D.A. and Shoemaker, L.L. (1987) Generalized watershed loading functions for stream flow nutrients. *Water Resources Bulletin* 23, 471–478.
Halverson, L.J., Jones, T.M. and Firestone, M.K. (2000) Release of intracellular solutes by four soil bacteria exposed to dilution stress. *Soil Science Society of America Journal* 64, 1630–1637.
Hannapel, R.J., Fuller, W.H., Bosma, S. and Bullock, J.S. (1964a) Phosphorus movement in a calcareous soil. I. Predominance of organic forms of phosphorus in phosphorus movement. *Soil Science* 97, 350–357.
Hannapel, R.J., Fuller, W.H. and Fox, R.H. (1964b) Phosphorus movement in a calcareous soil. II. Soil microbial activity and organic phosphorus movement. *Soil Science* 97, 421–427.
Harrison, A.F. (1982a) ^{32}P-method to compare rates of mineralization of labile organic phosphorus in woodland soils. *Soil Biology and Biochemistry* 14, 337–341.
Harrison, A.F. (1982b) Labile organic phosphorus mineralization in relationship to soil properties. *Soil Biology and Biochemistry* 14, 343–351.
Harrison, A.F. (1987) *Soil Organic Phosphorus: a Review of World Literature.* CAB International, Wallingford, UK, 257 pp.
Haygarth, P.M. and Jarvis, S.C. (1997) Soil derived phosphorus in surface runoff from grazed grassland lysimeters. *Water Research* 31, 140–148.
Haygarth, P.M. and Jarvis, S.C. (1999) Transfer of phosphorus from agricultural soils. *Advances in Agronomy* 66, 195–249.
Haygarth, P.M. and Jarvis, S.C. (eds) (2002) *Agriculture, Hydrology and Water Quality.* CAB International, Wallingford, UK, 502 pp.
Haygarth, P.M., Ashby, C.D. and Jarvis, S.C. (1995) Short-term changes in the molybdate reactive phosphorus of stored soil waters. *Journal of Environmental Quality* 24, 1133–1140.
Haygarth, P.M., Warwick, M.S. and House, W.A. (1997) Size distribution of colloidal molybdate reactive phosphorus in river waters and soil solution. *Water Research* 31, 439–442.
Haygarth, P.M., Hepworth, L. and Jarvis, S.C. (1998) Forms of phosphorus transfer in hydrological pathways from soil under grazed grassland. *European Journal of Soil Science* 49, 65–72.
Haygarth, P.M., Heathwaite, A.L., Jarvis, S.C. and Harrod, T.R. (2000) Hydrological factors for phosphorus transfer from agricultural soils. *Advances in Agronomy* 69, 153–178.
Haynes, R.J. and Swift, R.S. (1985) Effects of air-drying on the adsorption and desorption of phosphate and levels of extractable phosphate in a group of New Zealand soils. *Geoderma* 35, 145–157.
Heathwaite, A.L., Burt, T.P. and Trudgill, S.T. (1990) The effect of land use on nitrogen, phosphorus and suspended sediment delivery to streams in a small catchment in southwest England. In: Boardman, J., Foster, L.D.L. and Dearing, J.A. (eds) *Soil Erosion on Agricultural Land.* John Wiley & Sons, Chichester, pp. 161–177.
Heckrath, G., Brookes, P.C., Poulton, P.R. and Goulding, K.W.T. (1995) Phosphorus leaching from soils containing different phosphorus concentrations in the Broadbalk Experiment. *Journal of Environmental Quality* 24, 904–910.
Hens, M. and Merckx, R. (2001) Functional characterization of colloidal phosphorus species in the soil solution of sandy soils. *Environmental Science and Technology* 35, 493–500.
Herbes, S.E., Allen, H.E. and Mancy, K.H. (1975) Enzymatic characterization of soluble organic

phosphorus in lake water. *Science* 187, 432–434.

Holden, J. and Burt, T.P. (2002) Piping and pipeflow in a deep peat catchment. *Catena* 48, 163–199.

Hooda, P.S., Moynagh, M., Svoboda, I.F., Thurlow, M., Stewart, M., Thomson, M. and Anderson, H.A. (1997) Soil and land use effects on phosphorus in six streams draining small agricultural catchments. *Soil Use and Management* 13, 196–204.

Hudson, J.J., Taylor, W.D. and Schindler, D.W. (2000) Phosphate concentrations in lakes. *Nature* 406, 54–56.

Ingham, R.E., Trofymow, J.A., Ingham, E.R. and Coleman, D.C. (1985) Interactions of bacteria, fungi, and their nematode grazers: effects on nutrient cycling and plant growth. *Ecological Monographs* 55, 19–140.

IPCC (1995) *Second Scientific Assessment Report from the Intergovernmental Panel on Climate Change, Working Group 1, World Meteorological Organization.* Cambridge University Press, Cambridge, UK, 572 pp.

Jansson, M., Olsson, H. and Pettersson, K. (1988) Phosphatase; origin, characteristics and function in lakes. *Hydrobiologia* 170, 157–175.

Jarvie, H.P., Withers, P.J.A. and Neal, C. (2002) Review of robust measurement of phosphorus in river water: sampling, storage, fractionation and sensitivity. *Hydrology and Earth System Sciences* 6, 113–131.

Jenkinson, D.S. and Ladd, J.N. (1981) Microbial biomass in soil: measurement and turnover. In: Paul, E.A. and Ladd, J.N. (eds) *Soil Biochemistry,* Vol. 5. Marcel Decker, New York, pp. 415–471.

Jensen, M.B., Olsen, T.B., Hansen, H.C.B. and Magid, J. (2000) Dissolved and particulate phosphorus in leachate from structured soil amended with fresh cattle faeces. *Nutrient Cycling in Agroecosystems* 56, 253–261.

Jordan, C. and Smith, R.V. (1985) Factors affecting leaching of nutrients from an intensively managed grassland in County Antrim, Northern Ireland. *Journal of Environmental Management* 20, 1–15.

Karl, D.M. (2000) Phosphorus, the staff of life. *Nature* 406, 31–33.

Kieft, T.L., Soroker, E. and Firestone, M.K. (1987) Microbial biomass response to a rapid increase in water potential when dry soil is wetted. *Soil Biology and Biochemistry* 19, 119–126.

Kotak, B.G., Kenefick, S.L., Fritz, D.L., Rousseaux, C.G., Prepas, E.E. and Hrudey, S.E. (1993) Occurrence and toxicological evaluation of cyanobacterial toxins in Alberta lakes and farm dugouts. *Water Research* 27, 495–506.

Kouno, K., Wu, J. and Brookes, P.C. (2002) Turnover of biomass C and P in soil following incorporation of glucose or ryegrass. *Soil Biology and Biochemistry* 34, 617–622.

Lawton, L.A. and Codd, G.A. (1991) Cyanobacterial (blue-green algal) toxins and their significance in UK and European waters. *Journal of the Institute of Water and Environmental Management* 5, 460–465.

Livingstone, D. and Whitton, B.A. (1984) Water chemistry and phosphate activity of the blue-green alga *Rivularia* in Upper Teesdale streams. *Journal of Ecology* 72, 405–421.

Llewelyn, J.M., Landing, W.M., Marshall, A.G. and Cooper, W.T. (2002) Electrospray ionization Fourier transform ion cyclotron resonance mass spectrometry of dissolved organic phosphorus species in a treatment wetland after selective isolation and concentration. *Analytical Chemistry* 74, 600–606.

Lopez-Hernandez, D., Brossard, M. and Frossard, E. (1998) P-isotopic exchange values in relation to Po mineralisation in soils with very low P-sorbing capacities. *Soil Biology and Biochemistry* 30, 1663–1670.

Lundquist, E.J., Scow, K.M., Jackson, L.E., Uesugi, S.L. and Johnson, C.R. (1999) Rapid response of soil microbial communities from conventional, low input, and organic farming systems to a wet/dry cycle. *Soil Biology and Biochemistry* 31, 1661–1675.

Magid, J., Christensen, N. and Nielsen, H. (1992) Measuring phosphorus fluxes through the root zone of a layered sandy soil: comparisons between lysimeter and suction cell solution. *Journal of Soil Science* 43, 739–747.

Magid, J., Tiessen, H. and Condron, L.M. (1996) Dynamics of organic phosphorus in soils under natural and agricultural ecosystems. In: Piccolo, A. (ed.) *Humic Substances in Terrestrial Ecosystems.* Elsevier Science, Amsterdam, pp. 429–466.

Martin, J.K. (1970) Organic phosphate compounds in water extracts of soils. *Soil Science* 109, 362–375.

McKercher, R.B. and Anderson, G. (1989) Organic phosphate sorption by neutral and basic soils. *Communications in Soil Science and Plant Analysis* 20, 723–732.

Moss, B. (1988) *Ecology of Fresh Waters: Man and Medium.* Blackwell Scientific, London, 417 pp.

Nash, D. and Murdoch, C. (1997) Phosphorus in runoff from a fertile dairy pasture. *Australian Journal of Soil Research* 35, 419–429.

Newman, E.I. (1995) Phosphorus inputs to terrestrial ecosystems. *Journal of Ecology* 83, 713–726.

Oehl, F., Oberson, A., Sinaj, S. and Frossard, E. (2001)

Organic phosphorus mineralization studies using isotopic dilution techniques. *Soil Science Society of America Journal* 65, 780–787.

Olsen, R.G. and Court, M.N. (1982) Effect of wetting and drying of soils on phosphate adsorption and resin extraction of soil phosphate. *Journal of Soil Science* 33, 709–717.

Pallauf, J. and Rimbach, G. (1997) Nutritional significance of phytic acid and phytase. *Archives of Animal Nutrition* 50, 301–319.

Paul, J.H., Cazares, L.H., David, A.W., DeFlaun, M.F. and Jeffrey, W.H. (1991) The distribution of dissolved DNA in an oligotrophic and a eutrophic river of Southwest Florida. *Hydrobiologia* 218, 53–63.

Peat, D.M.W., McKelvie, I.D., Matthews, G.P., Haygarth, P.M. and Worsfold, P.J. (1997) Rapid determination of dissolved organic phosphorus in soil leachates and runoff waters by flow injection analysis with on-line photo-oxidation. *Talanta* 45, 47–55.

Peperzak, P., Caldwell, A.G., Hunziker, R.R. and Black, C.A. (1959) Phosphorus fractions in manures. *Soil Science* 87, 293–302.

Perakis, S.S. and Hedin, L.O. (2002) Nitrogen losses from unpolluted South American forests mainly via dissolved organic compounds. *Nature* 415, 416–419.

Persmark, L., Banck, A. and Jansson, H. (1996) Population dynamics of nematophagous fungi and nematodes in an arable soil: vertical and seasonal fluctuations. *Soil Biology and Biochemistry* 28, 1005–1014.

Pierre, W.H. and Parker, F.W. (1927) Soil phosphorus studies. II. The concentration of organic and inorganic phosphorus in the soil solution and soil extracts and the availability of the organic phosphorus to plants. *Soil Science* 24, 119–128.

Poulsen, H.D. (2000) Phosphorus utilization and excretion in pig production. *Journal of Environmental Quality* 29, 24–27.

Preedy, N., McTiernan, K., Matthews, R., Heathwaite, L. and Haygarth, P.M. (2001) Rapid incidental phosphorus transfers from grassland. *Journal of Environmental Quality* 30, 2105–2112.

Raboy, V., Gerbasi, P.F., Young, K.A., Stoneberg, S.D., Pickett, S.G., Bauman, A.T., Murthy, P.P.N., Sheridan, W.F. and Ertl, D.S. (2000) Origin and seed phenotype of maize *low phytic acid 1-1* and *low phytic acid 2-1*. *Plant Physiology* 124, 355–368.

Rolston, D.E., Rauschkolb, R.S. and Hoffman, D.L. (1975) Infiltration of organic phosphate compounds in soil. *Soil Science Society of America Proceedings* 39, 1089–1094.

Ron Vaz, M.D., Edwards, A.C., Shand, C.A. and Cresser, M. (1992) Determination of dissolved organic phosphorus in soil solutions by an improved automated photo-oxidation procedure. *Talanta* 39, 1479–1487.

Ron Vaz, M.D., Edwards, A.C., Shand, C.A. and Cresser, M.S. (1993) Phosphorus fractions in soil solution: influence of soil acidity and fertilizer additions. *Plant and Soil* 148, 179–183.

Ron Vaz, M.D., Edwards, A.C., Shand, C.A. and Cresser, M.S. (1994) Changes in the chemistry of soil solution and acetic-acid extractable P following different types of freeze/thaw episodes. *European Journal of Soil Science* 45, 353–359.

Ryden, J.C., Syers, J.K. and Harris, R.F. (1973) Phosphorus in runoff and streams. *Advances in Agronomy* 25, 1–45.

Salema, M.P., Parker, C.A., Kidby, D.K., Chatel, D.L. and Armitage, T.M. (1982) Rupture of nodule bacteria on drying and rehydration. *Soil Biology and Biochemistry* 14, 15–22.

Sample, E.C., Soper, R.J. and Racz, G.J. (1980) Reactions of phosphate fertilizers in soils. In: Khasawneh, F.E., Sample, E.C. and Kamprath, E.J. (eds) *The Role of Phosphorus in Agriculture.* American Society of Agronomy, Madison, Wisconsin, pp. 263–310.

Schindler, D.W. (1977) Evolution of phosphorus limitation in lakes. *Science* 195, 260–262.

Seeling, B. and Zasocki, R.J. (1993) Microbial effects in maintaining organic and inorganic solution phosphorus concentrations in a grassland topsoil. *Plant and Soil* 148, 277–284.

Setälä, H., Matikainen, E., Tyynismaa, M. and Huhta, V. (1990) Effects of soil fauna on leaching of nitrogen and phosphorus from experimental systems simulating coniferous forest floor. *Biology and Fertility of Soils* 10, 170–177

Shan, Y., McKelvie, I.D. and Hart, B.T. (1994) Determination of alkaline phosphatase-hydrolysable phosphorus in natural water systems by enzymatic flow injection. *Limnology and Oceanography* 39, 1993–2000.

Shand, C.A. and Smith, S. (1997) Enzymatic release of phosphate from model substrates and P compounds in soil solution from a peaty podzol. *Biology and Fertility of Soils* 24, 183–187.

Shand, C.A., Macklon, A.E.S., Edwards, A.C. and Smith, S. (1994) Inorganic and organic P in soil solutions from three upland soils. I. Effect of soil solution extraction conditions, soil type and season. *Plant and Soil* 159, 255–264.

Shand, C.A., Smith, S., Edwards, A.C. and Fraser, A.R. (2000) Distribution of phosphorus in particulate, colloidal and molecular-sized fractions of soil solution. *Water Research* 34, 1278–1284.

Sharpley, A.N. (1985) The selective erosion of plant nutrients in runoff. *Soil Science Society of America Journal* 49, 1527–1534.

Sharpley, A.N. and Smith, S.J. (1990) Phosphorus transport in agricultural runoff: the role of soil erosion. In: Boardman, J., Foster, L.D.L. and Dearing, J.A. (eds) *Soil Erosion on Agricultural Land.* John Wiley & Sons, Chichester, pp. 351–366.

Sharpley, A.N., Hedley, M.J., Sivbbesen, E., Hillbricht-Ilkowska, A., House, W.A. and Ryszkowski, L. (1996) Phosphorus transfers from terrestrial to aquatic systems. In: Tiessen, H. (ed.) *Phosphorus in the Global Environment.* John Wiley & Sons, Chichester, pp. 171–199.

Sheath, R.G. and Müller, K.M. (1997) Distribution of stream macroalgae in four high Arctic drainage basins. *Arctic* 50, 355–364.

Simard, R., Beaucheim, S. and Haygarth, P.M. (2000) Potential for preferential pathways for phosphorus transport. *Journal of Environmental Quality* 29, 97–105.

Simons, P.C.M., Versteegh, H.A.J., Jongbloed, A.W., Kemme, P.A., Slump, P., Bos, K.D., Wolters, M.G.E., Beudeker, R.F. and Verschoor, G.J. (1990) Improvement of phosphorus availability by microbial phytase in broilers and pigs. *British Journal of Nutrition* 64, 525–540.

Sims, J.T., Edwards, A.C., Schoumans, O.F. and Simard, R.R. (2000) Integrating soil phosphorus testing into environmentally based agricultural management practices. *Journal of Environmental Quality* 29, 60–71.

Sparling, G.P. and Ross, D.J. (1988) Microbial contributions to the increased nitrogen mineralization after air-drying. *Plant and Soil* 105, 163–167.

Sparling, G.P., Whale, K.N. and Ramsay, A.J. (1985) Quantifying the contribution from the soil microbial biomass to the extractable P levels of fresh and air dried soils. *Australian Journal of Soil Research* 23, 613–621.

Sparling, G.P., Milne, J.D.G. and Vincent, K.W. (1987) Effect of soil moisture regime on the microbial contribution to Olsen phosphorus values. *New Zealand Journal of Agricultural Research* 30, 79–84.

Sparling, G.P., West, A.W. and Reynolds, J. (1989) Influence of soil moisture regime on the respiration response of soils subjected to osmotic stress. *Australian Journal of Soil Research* 27, 161–168.

Spencer, V.E. and Stewart, R. (1934) Phosphate studies. I. Soil penetration of some organic and inorganic phosphates. *Soil Science* 38, 65–78.

Stevens, R.J. and Stewart, B.M. (1982a) Concentration, fractionation and characterisation of soluble organic phosphorus in river water entering Lough Neagh. *Water Research* 16, 1507–1519.

Stevens, R.J. and Stewart, B.M. (1982b) Some components of particulate phosphorus in river water entering Lough Neagh. *Water Research* 16, 1591–1596.

Suzumura, M. and Kamatani, A. (1995a) Mineralization of inositol hexaphosphate in aerobic and anaerobic marine-sediments: implications for the phosphorus cycle. *Geochimica et Cosmochimica Acta* 59, 1021–1026.

Suzumura, M. and Kamatani, A. (1995b) Origin and distribution of inositol hexaphosphate in estuarine and coastal sediments. *Limnology and Oceanography* 40, 1254–1261.

Tate, K.R. and Newman, R.H. (1982) Phosphorus fractions of a climosequence of soils in New Zealand tussock grassland. *Soil Biology and Biochemistry* 14, 191–196.

Taylor, T.G. (1965) The availability of the calcium and phosphorus of plant materials for animals. *Proceedings of the Nutrition Society* 24, 105–112.

Toy, T.J., Foster, G.R. and Renard, K.G. (eds) (2002) *Soil Erosion: Processes, Prediction, Measurement, and Control.* John Wiley & Sons, New York, 352 pp.

Turner, B.L. (2004) Optimizing phosphorus characterization in animal manures by solution phosphorus-31 nuclear magnetic resonance spectroscopy. *Journal of Environmental Quality* 33, 757–766.

Turner, B.L. and Haygarth, P.M. (2000) Phosphorus forms and concentrations in leachate under four grassland soil types. *Soil Science Society of America Journal* 64, 1090–1097.

Turner, B.L. and Haygarth, P.M. (2001) Phosphorus solubilization in rewetted soils. *Nature* 411, 258.

Turner, B.L. and Haygarth, P.M. (2003) Changes in bicarbonate-extractable inorganic and organic phosphorus following soil drying. *Soil Science Society of America Journal* 67, 344–350.

Turner, B.L., Baxter, R., Ellwood, N.T.W. and Whitton, B.A. (2001a) Characterization of the phosphatase activity of mosses in relation to their environment. *Plant, Cell and Environment* 24, 1165–1176.

Turner, B.L., Bristow, A.W. and Haygarth, P.M. (2001b) Rapid estimation of microbial biomass in grassland soils by ultra-violet absorbance. *Soil Biology and Biochemistry* 33, 913–919.

Turner, B.L., McKelvie, I.D. and Haygarth, P.M. (2002a) Characterisation of water-extractable soil organic phosphorus by phosphatase hydrolysis. *Soil Biology and Biochemistry* 34, 27–35.

Turner, B.L., Papházy, M., Haygarth, P.M. and McKelvie, I.D. (2002b) Inositol phosphates in the environment. *Philosophical Transactions of the Royal Society London, Series B* 357, 449–469.

Turner, B.L., Baxter, R. and Whitton, B.A. (2003a) Nitrogen and phosphorus fractions in soil solutions and drainage streams in Upper Teesdale National Nature Reserve, northern England: implications of organic compounds for biological nutrient limitation. *Science of the Total Environment* 314/316, 153–170.

Turner, B.L., Chudek, J.A., Whitton, B.A. and Baxter, R. (2003b) Phosphorus composition of upland soils polluted by long-term atmospheric nitrogen deposition. *Biogeochemistry* 65, 259–274.

Turner, B.L., Driessen, J.P., Haygarth, P.M. and McKelvie, I.D. (2003c) Potential contribution of lysed bacterial cells to phosphorus solubilisation in rewetted Australian pasture soils. *Soil Biology and Biochemistry* 35, 187–189.

Turner, B.L., Mahieu, N. and Condron, L.M. (2003d) Phosphorus composition of temperate pasture soils determined by NaOH–EDTA extraction and solution ^{31}P NMR spectroscopy. *Organic Geochemistry* 34, 1199–1210.

Ulén, B. and Mattson, L. (2003) Transport of phosphorus forms and of nitrate through a clay soil under grass and cereal production. *Nutrient Cycling in Agroecosytems* 65, 129–140.

van Breemen, N. (2002) Nitrogen cycle: natural organic tendency. *Nature* 415, 381–382.

van Gestel, M., Merckx, R. and Vlassak, K. (1993) Microbial biomass responses to soil drying and rewetting: the fate of fast- and slow-growing microorganisms in soils from different climates. *Soil Biology and Biochemistry* 25, 109–123.

van Gestel, M., Merckx, R. and Vlassak, K. (1996) Spatial distribution of microbial biomass in microaggregates of a silty-loam soil and the relation with the resistance of microorganisms to soil drying. *Soil Biology and Biochemistry* 28, 503–510.

Webley, D.M. and Jones, D. (1971) Biological transformation of microbial residues in soil. In: McLaren, A.D. and Skujins, J. (eds) *Soil Biochemistry,* Vol. II. Marcel Dekker, New York, pp. 446–485.

West, A.W., Sparling, G.P., Spier, T.W. and Wood, J.M. (1988) Comparison of microbial C, N-flush and ATP, and certain enzyme activities of different textured soils subject to gradual drying. *Australian Journal of Soil Research* 26, 217–229.

West, A.W., Sparling, G.P., Feltham, C.W. and Reynolds, J. (1992) Microbial activity and survival in soils dried at different rates. *Australian Journal of Soil Research* 30, 209–222.

Whitton, B.A. (1988) Hairs in eukaryotic algae. In: Round, F.E. (ed.) *Algae and the Aquatic Environment: Contributions in Honour of J.W.G. Lund.* Biopress, Bristol, pp. 446–460.

Whitton, B.A., Potts, M., Simon, J.W. and Grainger, S.L.J. (1990) Phosphatase activity of the blue-green alga (cyanobacterium) *Nostoc commune* Utex 584. *Phycologia* 29, 139–145.

Whitton, B.A., Grainger, S.L.J., Hawley, G.R.W. and Simon, J.W. (1991) Cell–bound and extracellular phosphatase activities of cyanobacterial isolates. *Microbial Ecology* 21, 85–98.

Whitton, B.A., Yelloly, J.M., Christmas, M. and Hernandez, I. (1998) Surface phosphatase activity of benthic algae in a stream with highly variable ambient phosphate concentrations. *Verhandlung Internationale Vereinigung Limnologie* 26, 967–972.

Wild, A. (1988) Plant nutrients in soil: phosphate. In: Wild, A. (ed.), *Russell's Soil Conditions and Plant Growth,* 10th edn. Longman Scientific and Technical, Harlow, UK, pp. 695–742.

Wild, A. and Oke, O.L. (1966) Organic phosphate compounds in calcium chloride extracts of soils: identification and availability to plants. *Journal of Soil Science* 17, 356–371.

Woods, L.E., Cole, C.V., Elliott, E.T., Anderson, R.V. and Coleman, D.C. (1982) Nitrogen transformations in soil as affected by bacterial–microfaunal interactions. *Soil Biology and Biochemistry* 14, 93–98.

13 Interactions of Organic Phosphorus in Terrestrial Ecosystems

Leo M. Condron[1] and Holm Tiessen[2]

[1]*Agriculture and Life Sciences, PO Box 84, Lincoln University, Canterbury 8150, New Zealand;* [2]*Department of Tropical Agronomy, Göttingen University, Grisebachstrasse 6, 37077 Göttingen, Germany*

Phosphorus Dynamics in Terrestrial Ecosystems

Soil plays a critical role in the function of natural and managed terrestrial ecosystems. In addition to providing water and physical support for plants, soil properties and processes regulate the cycling, retention and bioavailability of the major elements (Daily *et al.*, 1997). Most of the phosphorus in terrestrial ecosystems is present in the soil. On a global scale it has been estimated that soil contains $96\text{–}160 \times 10^9$ Mg of phosphorus compared with only 2.6×10^9 Mg in biota (Stevenson and Cole, 1999). The proportion of total ecosystem phosphorus present in soil varies widely. For example, in a grazed permanent pasture, herbage phosphorus represented only 1% of the total phosphorus in topsoil (0–20 cm) (Williams and Haynes, 1992), while above- and below-ground plant pools accounted for 38% of total phosphorus in a temperate forest system (Hart *et al.*, 2003). In many dystrophic tropical forests, the largest reservoir of nutrient elements is in the plant biomass. In the total aboveground biomass of 300 Mg/ha of a Colombian tropical rainforest, Rodriguez-Jimenez (1988) measured 2000 kg of nitrogen and 40 kg of phosphorus. This forest is very poor in phosphorus compared with typical tropical forests, which have nitrogen-to-phosphorus ratios closer to 10:1 (Shanmughavel *et al.*, 2001). These differences in phosphorus contents and elemental ratios reflect plant community adaptation to geochemical constraints.

Dividing the biomass by the estimate of Rodriguez-Jimenez (1988) for primary productivity of 8 Mg/ha/year, gives a turnover time of the biomass and its associated phosphorus of approximately 40 years. A major portion of the phosphorus flow and availability in dystrophic and oligotrophic tropical rainforest is thus tied into the biotic cycle of plant biomass. Beyond the tie-up in long-lived boles, even the recycling through litter to the soil organic matter may be reduced by biotic sequestration. Lal *et al.* (2001) estimated that 20–91% of the phosphorus demands for leaf production of several dry forest species in India were satisfied by re-translocation prior to abscission. This means that phosphorus cycling outside the plant biomass is greatly reduced. The extent of re-translocation reflects soil nutrient availability (Tiessen *et al.*, 1994) and other factors such as fungal symbioses. Chuyong *et al.* (2000) showed twofold greater re-translocation (i.e. phosphorus conservation) in non-ectomycorrhizal trees than in mycorrhizal trees. The structure of the ecosystem thus influences linkages between phosphorus

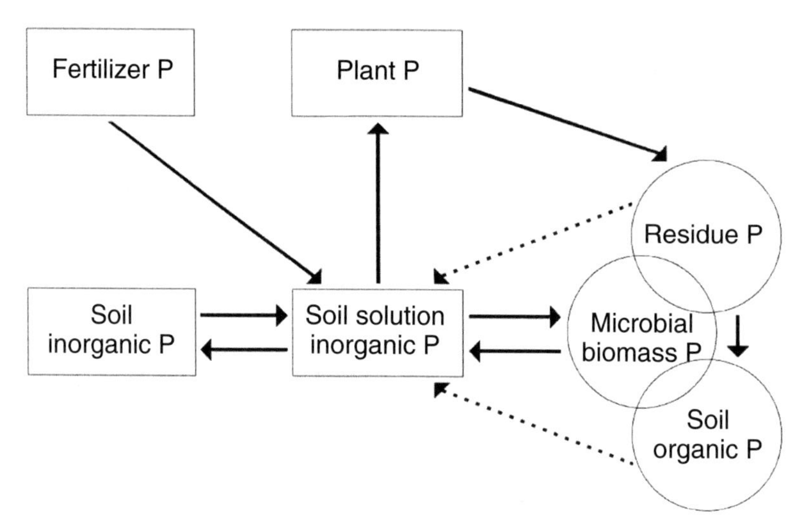

Fig. 13.1. Conceptual diagram of phosphorus dynamics in the soil–plant system.

and other elements, and phosphorus turnover in the ecosystem.

In most soils, phosphorus associated with carbon (i.e. organic phosphorus) accounts for between 30% and 65% of total phosphorus, although some soils contain up to 90% organic phosphorus (Harrison, 1987). The majority of the organic phosphorus is present as phosphate esters (C–O–P bonds) including inositol phosphates, nucleic acids and phospholipids, together with small quantities of phosphonates (C–P bonds) (Magid *et al.*, 1996; Condron *et al.*, 2005). The high proportion of organic phosphorus in soil can be directly attributed to the fact that significant quantities of organic phosphorus are added to soil in the form of plant (litter, root), animal (excreta, manure) and microbial residues (detritus). Organic phosphorus commonly makes up 10–60% of total phosphorus in plant material and animal manure (Harrison, 1987), and >90% of total phosphorus in microorganisms (Magid *et al.*, 1996). It has been shown that between 60% and 90% of the phosphorus taken up by grazed pasture annually is returned to the soil in plant residues and animal excreta (Haynes and Williams, 1991). Annual recycling of phosphorus to soil in above- and below-ground plant residues represents 18–38% and 15–80% (mean 55%) of plant phosphorus uptake in temperate crop and forest ecosystems, respectively (Hanway and Olsen, 1980; Pritchett and Fisher, 1987).

The major pools and associated transformations of phosphorus in the soil–plant system are presented in Fig. 13.1. These show that the distribution, dynamics and availability of phosphorus in soil are controlled by a combination of biological, chemical and physical processes. Soil solution is the primary source of phosphorus for plants and microorganisms, and most phosphorus is taken up as phosphate (HPO_4^{2-}, $H_2PO_4^-$). The equilibrium concentration of phosphate present in soil solution is very low (<5 μM) and phosphate removed by plant and microbial uptake must be continually replenished from the inorganic, organic and microbial phosphorus pools. This is especially important in agroecosystems where demand for phosphorus is high and annual off-farm transfer of phosphorus in produce commonly exceeds 20 kg P/ha (Stevenson and Cole, 1999).

The turnover of organic phosphorus in soil is primarily determined by rates of immobilization and mineralization. Immobilization is the collective term for the biological conversion of inorganic phosphorus to organic phosphorus in the soil–plant system. This can occur in two ways. Firstly, phosphate removed from the soil by plant uptake is returned as organic phosphorus in litter and root debris, as well as in animal excreta and manure. Secondly, removal of phosphate from the soil solution and associated solid-phase pools can occur during the microbial decomposition of organic residues with a high carbon-to-phosphorus ratio, and some of this phosphorus will subsequently be released as organic phosphorus in detritus. Mineralization is the process whereby phosphate is released from organic phosphorus in the soil. Phosphate may be released from organic phosphorus associated with soil organic matter components, or during decomposition of plant and microbial residues (Fig. 13.1). Mineralization of organic phosphorus in soil has been shown to make a significant contribution to plant phosphorus requirements (Magid *et al.*, 1996; Frossard *et al.*, 2000). Phosphorus cycling in tropical soils is governed to a greater extent by biological processes than in most temperate environments. The high biological potential of the humid tropics is seen in a large biomass production, fast organic matter turnover, and very intensive land use. Up to four crops are planted and harvested per year on a single plot. These biological processes closely link transformations of phosphorus with those of organic materials. The high phosphorus fixation capacity of sesquioxide-rich weathered soils further increases the relative importance of biological and organic phosphorus reservoirs and transformations.

Organic Phosphorus Interactions in Soil

In terms of key interactions at the ecosystem level, it is expected that plant production and its associated effects on the amounts and transformations of organic carbon in the soil–plant system will have the greatest overall impact on the dynamics of organic phosphorus. The amounts, forms, distribution and associated dynamics of organic carbon and phosphorus in soil are influenced by a combination of factors including soil type, environmental conditions and vegetation, together with land use and management (including organic matter and nutrient inputs, and nutrient availability) (e.g. Stewart and Tiessen, 1987; Gressel and McColl, 1997; Tiessen and Shang, 1998; Raich *et al.*, 2000; Conant *et al.*, 2001). Phosphorus availability often limits plant growth in natural ecosystems, although phosphorus is generally recycled and retained efficiently (Cole *et al.*, 1977; Attiwell and Adams, 1993). However, in managed ecosystems, continued inputs of phosphorus in the form of fertilizers are required in order to sustain high levels of production, which in turn can have a profound effect on the amounts and dynamics of inorganic and organic phosphorus in the soil. The nature and extent of organic carbon–phosphorus interactions that affect the dynamics of organic phosphorus in soil have been examined over different time scales in various natural and managed ecosystems.

Soil development and landscape processes

Interactions between the overall dynamics of organic carbon and phosphorus in natural ecosystems should be reflected in the respective quantities of these compounds in the soil and associated changes over time. In addition to phosphorus, nitrogen and sulphur are the major plant nutrients associated with organic carbon in soil, and their relative quantities differ from one ecosystem to another. Soil organic carbon to organic phosphorus ratios vary between 40 and 650, which is considerably wider than the corresponding ranges determined for carbon-to-nitrogen (10–20) and carbon-to-sulphur (60–120) ratios (Smeck, 1985; Stevenson, 1994). This is confirmed by data from numerous studies which show that organic carbon accounts for only 44% of the determined variations in soil organic phosphorus (Harrison, 1987).

The wider variation in organic carbon to organic phosphorus ratio compared with carbon-to-nitrogen and carbon-to-sulphur may reflect fundamental differences in the mechanisms that control organic phosphorus dynamics in the soil (Barrow, 1961; Kowalenko, 1978). McGill and Cole (1981) conducted an extensive review of the available information on changes in the stoichiometry of organic carbon, nitrogen, phosphorus and sulphur determined during soil development. They concluded that the mechanisms involved in organic phosphorus cycling in soil were separate from those for other organic matter components, including carbon. These workers proposed that organic nutrient cycling in soil is controlled by two independent mechanisms:

1. Organic nitrogen and a variable proportion of organic sulphur are directly bonded to carbon in soil organic matter (i.e. C–N, C–S), and these nutrients are released (mineralized) during the microbial oxidation of organic carbon to provide energy ('biological mineralization').
2. Organic phosphorus and the remainder of the organic sulphur are associated with soil organic matter in the form of esters (i.e. C–O–P, C–O–S), and these nutrients are mineralized by the actions of various extracellular phosphatase and sulphatase enzymes ('biochemical mineralization').

According to this paradigm, the amount and turnover of organic phosphorus in soil are controlled by phosphatase enzyme activity, which in turn is determined by the availability of phosphate. Phosphatase enzymes in the soil environment are produced by plant roots, mycorrhizae and other microorganisms (Speir and Ross, 1978; Sinsabaugh, 1994; Vance *et al.*, 2003). Long-term changes in carbon-to-phosphorus ratios were observed in soil chrono-, climo- and toposequences (Walker and Syers, 1976; McGill and Cole, 1981; Smeck, 1985; Crews *et al.*, 1995; Cross and Schlesinger, 1995), which show some independence between organic carbon and phosphorus levels. During the early stages of soil development, the availability of phosphate is relatively high due to weathering of primary mineral apatite contained in the parent material – a process that is enhanced by continued leaching and increasing acidity. The relative abundance of phosphorus in the initial stages of development leads to an enrichment of phosphorus in organic matter and possibly to a suppression of phosphatase activity. As soil development continues, phosphate availability declines due to a combination of loss by leaching and erosion and the formation of adsorbed, sparingly soluble and occluded forms of secondary inorganic phosphorus associated with iron and aluminium. This results in an impoverishment of biological materials, and therefore of phosphorus, in the residual organic matter. Based on biochemical substrate activation or product suppression on phosphatase activity, one would expect an increase in phosphatase activity under such conditions, which might explain the widening carbon-to-phosphorus ratio (i.e. increased biochemical mineralization of organic phosphorus). On the other hand, Olander and Vitousek (2000) reported that acid phosphatase activities were high across a chronosequence of soils in Hawaii and did not appear to be affected by differences in phosphorus supply.

It should be noted that at least part of the greater variability in organic carbon to organic phosphorus ratio determined in soil may be attributed to the fact that chemical processes play a more important role in determining organic phosphorus dynamics compared with organic forms of nitrogen and sulphur. In particular, inositol phosphates, which constitute >50% of the identified organic phosphorus in most soils, have been shown to be stabilized in soil by precipitation or adsorption on to soil colloids (Turner *et al.*, 2002; Celi and Barberis, Chapter 6, this volume). It is therefore likely that the dynamics of inositol phosphates in soil will be influenced by chemical as well as biological and biochemical processes. In addition, variations in organic carbon to organic phosphorus ratio may also be partly due to errors associated with the determination of organic phosphorus, especially in tropical soils (Condron *et al.*, 1990).

Links between soil organic carbon and phosphorus vary with climate and landscape. Silver *et al.* (1994) showed a close link between organic carbon and phosphorus at upper slope positions but not at lower slopes, probably because hydromorphic conditions activate the inorganic phosphorus cycle. Tiessen *et al.* (1994) similarly showed much greater phosphorus availability in hydromorphic lower slopes in the Amazon rainforest, despite low total phosphorus levels. At such sites, phosphorus availability was governed by its linkage with iron, while at higher slopes, organic links were more important. On the other hand, in semi-arid environments the sequestration of phosphorus in organic forms follows patterns of plant productivity, with greater amounts of organic phosphorus found down catenary sequences (Roberts *et al.*, 1985).

Land use and management

Information on the nature of organic carbon–phosphorus interactions in soil has been obtained from examining changes in the intensity of input, output, transfer and transformation processes in various managed agricultural and forest ecosystems. These include the effects of cultivation and afforestation of grassland.

Cultivation

Under land-use change or soil degradation, transformations of organic phosphorus are closely linked to decreases in soil organic matter. Less organic matter input and acceleration of mineralization results in a net mineralization of soil organic matter. Such biological mineralization of organic matter and associated phosphorus has been shown in numerous studies of cultivation effects. In most cases carbon loss is proportionally greater than that of organic phosphorus, indicating that phosphorus is mineralized as a result of overall organic matter mineralization.

Long-term cultivation and cropping of native grassland soils in North America without phosphorus inputs has been shown to result in a marked decrease in soil organic carbon and phosphorus. For example, Tiessen and Stewart (1983) found that similar decreases in organic carbon and organic phosphorus (67–81%) occurred in particulate organic matter and sand fractions (>50 μm) of A-horizon soils as a result of 60–90 years of cultivation. The data also showed that while the magnitude of the decreases was lower in the silt- and clay-sized fraction (<5 μm), the decreases in organic carbon were consistently greater than the corresponding decreases in organic phosphorus. This was attributed to preferential stabilization of organic phosphorus by interaction with mineral colloids, as discussed above. The latter is also consistent with other data which showed that 60 years of cultivation caused greater mineralization of organic phosphorus in a sandy loam soil (27%) than in two clay loam soils (17–18%) (Tiessen *et al.*, 1982). The importance of interactions between organic constituents and sesquioxides in determining the resilience of soil organic matter and associated microaggregate structure has also been demonstrated in tropical ecosystems (Tiessen and Shang, 1998).

One problem in quantifying the effects of soil degradation is that mineralization and erosion commonly occur at the same time. Reduced vegetation cover limits organic matter returns and shifts the balance towards soil organic matter reduction. The increased soil exposure under reduced vegetation and reduced aggregation at lower organic matter concentrations promotes erosion, which removes organic-matter-rich topsoil and thus has a similar net effect *in situ* as mineralization. In terms of organic phosphorus linkages in the ecosystem, however, the effect is very different. In degrading alpine grassland, Wu and Tiessen (2002) showed approximately equal impact of erosion and mineralization using ^{137}Cs budgets, textural gradients, organic matter analysis and phosphorus fractionations. Over 30–40 years of cultivation, some 40% of organic phosphorus, 60% of organic carbon and >70% of the ^{137}Cs marker were lost. This ^{137}Cs loss translated into a topsoil loss estimated at between 3 and 5 kg/m^2/year. Calculating the incorporation of subsoil due to tillage of the eroding soil, and the resultant dilution of topsoil carbon content, a contribution of around 50% each was attributed to mineralization and erosion effects on soil carbon loss. Mineralized carbon is lost from the soil as carbon dioxide, but mineralized phosphorus remains in the soil as inorganic phosphorus (minus any crop export). Wu and Tiessen (2002) calculated that the increase in calcium-bound phosphorus observed in these Chinese cultivated soils was attributable to fertilizer additions (24%), the incorporation of subsoil phosphorus (31%), and mineralization of organic phosphorus (45%). Observation of the close linkage of carbon and phosphorus in the mineralization process is thus made difficult by erosion, which shifts the balance of soil phosphorus from organic to inorganic forms. The combined effects of carbon mineralization and erosion fully account for the release of phosphate as a by-product of the overall soil degradation.

Land-use change modifies organic matter and phosphorus stocks. Mailly *et al.* (1997) measured the element budget through a 6-year cycle of cultivation followed by bamboo fallow in Java. Of the 790 kg of nitrogen and 130 kg of phosphorus accumulated in plants during the entire cycle, half was removed in harvested materials. Fertilization replenished only 60% of the harvested phosphorus outputs. This illustrates a typical phosphorus-loss budget for most low-input agricultural settings, without even taking erosion into consideration. Under shifting cultivation, a sufficiently long fallow phase is likely to re-establish a partial nutrient balance, but with continuing imbalance of inputs and outputs, and shortened fallow, declining yields will eventually force an increase in the use of phosphorus fertilizer. To avoid fixation of added phosphate, this should be combined with organic matter or organic phosphorus management.

Agricultural management increasingly moves phosphorus around agricultural land, even under minimum-input agriculture. In a study on phosphorus fluxes in low-input agriculture in northeast Brazil, Menezes and Sampaio (2002) indicated that the major phosphorus flow was associated with erosion from cultivated fields (6 kg/ha/year) followed by the erosive flux from (generally overstocked) pastures. Thus, phosphorus movement is linked largely to overall soil movement and less to managed flows that might be linked to organic matter or produce. Compared with the large erosive flows, the outflow of phosphorus from the farm by crop (2 kg/ha/year) and animal (0.2 kg/ha/year) exports is minor. The total phosphorus losses represented 10% of the value of farm production. Minimum-input management as part of subsistence farming typically relies on local or regional recycling of nutrient elements in plant residues or animal manure. Hoffmann *et al.* (2001) estimated recycling of nutrients under subsistence farming by the Hausa People of Nigeria to be 87 kg N/ha/year and 33 kg P/ha/year. This may be an optimistic estimate for phosphorus, since many plant residues, particularly in the tropics, may have greater capacities to provide crop nitrogen than phosphorus demand (see Nziguheba and Bünemann, Chapter 11, this volume), and recycled phosphorus even in managed fallow systems will only supply <20% of crop needs (Smestad *et al.*, 2002).

Afforestation

Afforestation involves the conversion of land from agricultural production to forestry (including agroforestry). The development of short-rotation production or plantation forestry is being encouraged as a means of protecting the remaining native forests. For example, extensive afforestation of grassland mostly with radiata pine (*Pinus radiata*) has occurred in New Zealand over the past 75 years. Currently, plantation forests cover 1.8 million ha of New Zealand (12% of the productive land area), and this is projected to increase to 2.2 million ha by 2010. The afforestation has occurred on hill and high country areas of grassland and has been mainly driven by declining returns from traditional pastoral farming and the expectation of higher future returns from forestry (Maclaren, 1996). Despite some inputs of phosphorus and sulphur fertilizer, and consequent biological fixation of nitrogen, the fertility status of most hill and high country grassland soils in New Zealand is low. Afforestation involves the direct planting of trees into pasture (no cultivation), and in most cases no phosphorus fertilizer is applied at or following establishment. In general, trees are planted at 1000–1200 stems/ha, and are subsequently thinned and pruned to a final stand density of 200–250 trees/ha. The forests are harvested by clear-felling after 20–35 years, and in most cases the land is replanted with trees.

In the absence of replicated field experiments, a significant number of retrospective paired-site comparison studies have been carried out in New Zealand to assess the effects of afforestation on soil fertility and quality (e.g. Davis and Lang, 1991; Davis, 1994; Belton *et al.*, 1995; Condron *et al.*, 1996; Giddens *et al.*, 1997; Parfitt *et al.*, 1997; Yeates *et al.*, 1997; Alfredsson *et al.*, 1998; Yeates and Saggar, 1998; Ross *et al.*, 1999; Chen *et al.*, 2000, 2001, 2003a; Davis, 2001; Groenendijk *et al.*, 2002). The findings of these studies have consistently demonstrated that, despite increases in soil acidity, levels of organic carbon and associated nutrients (nitrogen, phosphorus, sulphur) were generally significantly lower in mineral soil under recently established forest compared with grassland. At many sites, significant increases in levels of plant available nutrients (especially phosphate) were determined in the forest soils.

Davis and Condron (2002) assembled and compared data on changes in organic carbon from a total of 51 paired-site studies on afforestation on a wide range of soil types in New Zealand. The forests were a minimum of 10 years old, and data for soil carbon and bulk density were available for 28 sites. Results showed that most of the changes in organic carbon occurred in the top 10 cm of soil, and that decreases in organic carbon under forest compared with grassland ranged from 3% to 47%. The average decrease in topsoil organic carbon during the 20 years following establishment was 9.5%, which is equivalent to 4.5 Mg/ha. There is no comparable data-set for changes in soil organic phosphorus. Nonetheless, concentrations of organic phosphorus in mineral soil (0–10 cm) were found to have decreased by 13–39% under recently established forest compared with grassland (Davis and Lang, 1991; Davis, 1994; Chen *et al.*, 2000). Belton *et al.* (1995) reported that concentrations of bicarbonate-extractable phosphate (Olsen phosphorus) were 2–4 times higher in a range of soils under pine forest compared with adjacent grassland. This is consistent with reports of enhanced organic phosphorus mineralization, which should increase extractable phosphate concentrations.

The trends observed in the paired-site field studies described above have been confirmed in short-term pot experiments carried out on a wide range of soils (Davis, 1995; Condron *et al.*, 1996; Chen *et al.*, 2003b). For example, Chen *et al.* (2003b) showed that levels of organic carbon, nitrogen and phosphorus in 15 grassland soils were consistently and significantly lower after only 40 weeks growth of radiata pine compared with perennial ryegrass (*Lolium perenne* L.). The overall mean decreases in soil organic carbon, nitrogen and phosphorus were 6.7% (2.4–22.9%), 7.6% (5.8–21.7%) and 12.3% (5.8–19.8%), respectively. Levels of readily exchangeable phosphate determined by isotopic exchange kinetics also increased by an average of 15% in soil planted with radiata pine compared with ryegrass. The latter occurred despite the fact that phosphorus uptake by radiata pine seedlings was consistently two-fold greater than by ryegrass.

The dramatic short-term decreases in organic forms of carbon and phosphorus and increases in available phosphate described above indicate that afforestation resulted in the coincident mineralization of organic carbon and phosphorus in the soil. The findings of the short-term rhizosphere and pot experiments indicated that increased mineralization of organic matter under conifers may be at least partly attributed to differences in rhizosphere microflora compared with grasses, and the influence of ectomycorrhizae in particular (e.g. Chen *et al.*, 2002). However, it is more likely that the enhanced mineralization reflects differences in the amounts, forms and spatial and temporal distribution of organic carbon and phosphorus returns to soil in above- and below-ground plant residues under forests compared with grassland. Condron and Newman (1998) used solid-state ^{13}C nuclear magnetic resonance (NMR) spectroscopy to examine the chemical nature of organic carbon in mineral soil taken from adjacent areas under grassland and recently established pine forest (10–17 years old) at two sites. Results indicated a shift towards more recalcitrant forms of organic carbon in soil as a consequence of afforestation, which was attributed to lower inputs of fresh organic matter under forest compared with grassland. Yeates *et al.* (1997) also suggested that the greater release of substrates such as sugars and amino acids from roots and foliage in a grassland environment may have been responsible for the higher levels of microbial carbon determined in grassland soil compared with forest. Chen *et al.* (2000) found that conversion from grassland to forest caused a dramatic change in the distribution of organic matter through the soil profile. Results showed that the total amount of organic carbon in mineral soil (0–30 cm) decreased from 124 Mg/ha under grassland to 121 Mg/ha under pine forest after 19 years. However, the forest litter (litter and fermentation layers) contained 15.5 Mg/ha organic carbon. Accordingly, when the contribution from the forest litter was included, the total quantity of organic carbon in the soil profile had actually increased under forest (137 Mg/ha) compared with grassland (124 Mg/ha). Continued incorporation of surface litter carbon might account for the observation that levels of organic carbon in mineral soil were similar under forest and grassland after 20 years (Davis and Condron, 2002). It is also important to note that forest litter contains significant quantities of phosphorus, which directly contribute to plant phosphorus requirements (Attiwell and Adams, 1993; Gressel and McColl, 1997; Chen *et al.*, 2003a).

Confirmation of changes in the nature and dynamics of carbon and organic phosphorus come from studies that have examined the effects of afforestation on a variety of soil biological and biochemical properties and processes. These revealed that levels of microbial carbon and phosphorus were consistently and significantly lower in soils under forests compared with grassland (Yeates *et al.*, 1997; Yeates and Saggar, 1998; Ross *et al.*, 1999; Chen *et al.*, 2000). Chen *et al.* (2000) also showed that phosphatase activities were significantly lower in a forest soil compared with grassland. None the less, Chen *et al.* (2003a) found that seasonal variations in soil phosphorus were similar under forest and grassland, and that organic phosphorus mineralization increased during spring and summer in response to plant phosphorus demand, while organic phosphorus accumulated during late autumn and winter. It was concluded that plant phosphorus requirements were satisfied by phosphorus release from leaf litter inputs in the forest system and from root litter inputs under grassland. This is consistent with other data from the same site, which showed that the forest litter layer contains significant quantities of organic carbon and organic phosphorus, and that the fine root biomass (<2 mm) in topsoil (0–10 cm) was 2.5 times greater under grassland compared with forest (Chen *et al.*, 2000). Chen *et al.* (2003a) also discovered that despite the fact that levels of microbial carbon and phosphorus were lower in the forest soil, the calculated annual rate of phosphorus flux through the microbial biomass was higher in the forest soil (16.1 kg P/ha) than in the grassland soil (13.9 kg P/ha). Furthermore, the calculated annual turnover rate of microbial phosphorus was markedly higher in the forest soil (1.28) than in the grassland soil (0.80).

In addition to the investigations described above, a number of mainly process-based studies have also examined the nature of interactions between organic carbon and phosphorus in soil. The conceptual model proposed by McGill and Cole (1981) asserts that soil organic phosphorus cycling is controlled independently of organic carbon via the actions of extracellular phosphatase enzymes. This selective mineralization of phosphorus without concurrent carbon mineralization has been shown to occur in rhizosphere and microbial process studies (e.g. Tarafdar and Jungk, 1987; Asmar *et al.*,

1995; Richardson *et al.*, 2001; Chen *et al.*, 2002). On the other hand, results of other short-term experiments have failed to establish a link between phosphatase enzyme activity and depletion of organic phosphorus (e.g. Hedley *et al.*, 1983; Adams, 1992; Joner *et al.*, 1995).

Findings reported by Tarafdar and Claassen (1988) further challenge the contention that mineralization of organic phosphorus in soil is regulated by phosphatase activity. These workers added three organic phosphorus compounds (sodium glycerophosphate, phosphatidyl choline and *myo*-inositol hexakisphosphate) to soil and confirmed that their availability to plants was closely related to the activities of acid and alkaline phosphatase. However, the data from this experiment also revealed that mineralization of the added organic phosphorus compounds exceeded plant phosphorus demand by 20-fold, leading the authors to conclude that the mineralization of organic phosphorus in soil is limited by substrate availability rather than phosphatase activity. This in turn suggests that the availability (i.e. solubility) of organic phosphorus is linked to decomposition of organic carbon constituents in soil.

Information from other studies appears to confirm that the dynamics of organic phosphorus and organic carbon decomposition are more closely linked than suggested by the McGill and Cole (1981) model. In a glasshouse study, Dalal (1979) examined the fate of labelled carbon and phosphorus in plant residues following addition to soil and clearly demonstrated that there was a close relationship between phosphorus uptake and organic residue decomposition. Data from long-term field experiments in Europe indicated that organic phosphorus extracted by macroporous anion exchange resin represented 'mineralizable' organic phosphorus in soil (Rubæk and Sibbesen, 1995; Guggenberger *et al.*, 1996). The fact that the organic carbon to organic phosphorus ratio in resin extracts was consistently narrower compared with whole soil and alkali extracts also suggested that the dynamics of organic carbon and phosphorus were closely related. Furthermore, Gressel *et al.* (1996) used a combination of solid-state and solution ^{13}C and solution ^{31}P NMR spectroscopy to investigate links between organic carbon and phosphorus dynamics in a soil under a mixed conifer forest in California. They determined changes in the chemical nature of carbon and phosphorus throughout the soil profile (litter and mineral soil), and from the effect of tree harvesting. Results revealed inverse correlations between phosphate and carbohydrates (alkyl carbon, $r=-0.87$; *O*-alkyl carbon, $r=-0.76$), while positive correlations were observed between phosphate and more recalcitrant forms of carbon (carbonyl carbon, $r=0.82$; aromatic carbon, $r=0.81$). These data indicated that there was a close relationship between organic phosphorus mineralization and litter decomposition. This was confirmed by the observation that signals for phosphate diesters and *O*-alkyl carbon were closely related ($r=0.63$), while there was an inverse correlation between phosphate diesters and aromatic carbon ($r=-0.74$).

The findings described above strongly suggest that the mineralization of organic phosphorus in the soil is closely related to the availability of carbon substrate. Biological activity in the soil determines mineralization of organic phosphorus both through biochemical hydrolysis by extracellular phosphatase (Gressel and McColl, 1997; Joner *et al.*, 2000) and organic matter mineralization. In addition, exudation of low-molecular-weight organic anions (e.g. citrate, oxalate) by plant roots and mycorrhizae may enhance the solubility and consequent mineralization of soil organic phosphorus via chelation of aluminium, iron and calcium (Fox and Comerford, 1990; Braum and Helmke, 1995; Hayes *et al.*, 2000; Chen *et al.*, 2002).

Any factor(s) that alters the level of primary production in natural and managed ecosystems will affect the dynamics of organic carbon and phosphorus in soil. This includes changes in inputs of major plant nutrients (including phosphorus) in the form of mineral fertilizers and manure, as well as inputs of nitrogen and sulphur from atmospheric pollution. In addition, predicted changes in climate linked to continued increases in greenhouse gas emissions are likely to have a profound impact on ecosystem productivity and organic nutrient dynamics in the soil (Schlesinger, 1991).

Conclusions

Transformations of organic phosphorus in the soil play an important role in determining the availability of phosphorus in terrestrial ecosystems. The balance of evidence from a wide range of studies indicates that the dynamics of soil organic phosphorus are most closely linked to inputs and transformations of organic carbon as affected by the type and level of plant production. More detailed knowledge on the processes involved might be derived from an integration of appropriate techniques for examining the chemical nature of organic carbon and phosphorus, including solid-state ^{13}C NMR (e.g. Smernik and Oades, 2003) and solution ^{31}P NMR (e.g. Turner *et al.*, 2003a,b,c) with physico-chemical fractionation of organic matter (e.g. Six *et al.*, 2002). It is clear that microbial activity plays a central role in the transformations and linkages of carbon and phosphorus. At an ecosystem level this role is probably defined through overall mineralization processes rather than specific enzymatic hydrolysis reactions.

References

Adams, M.A. (1992) Phosphatase activity and phosphorus fractions in Karri (*Eucalyptus diversicolor* F. Muell.) forest soils. *Biology and Fertility of Soils* 14, 200–204.

Alfredsson, H., Condron, L.M., Clarholm, M. and Davis, M.R. (1998) Changes in soil acidity and organic matter following the establishment of conifers on former grasslands in New Zealand. *Forest Ecology and Management* 112, 245–252.

Asmar, F., Gahoonia, T.S. and Nielsen, N.E. (1995) Barley genotypes differ in activity of soluble extracellular phosphatase and depletion of organic phosphorus in rhizosphere soil. *Plant and Soil* 172, 117–122.

Attiwell, P.M. and Adams, M.A. (1993) Nutrient cycling in forests. *New Phytologist* 124, 561–582.

Barrow, N.J. (1961) Phosphorus in soil organic matter. *Soils and Fertilizers* 24, 169–173.

Belton, M.C., O'Connor, K.F. and Robson, A.B. (1995) Phosphorus levels in topsoils under conifer plantations in Canterbury high country grasslands. *New Zealand Journal of Forestry Science* 25, 265–282.

Braum, S.M. and Helmke, P.A. (1995) White lupin utilizes soil phosphorus that is unavailable to soybean. *Plant and Soil* 176, 95–100.

Chen, C.R., Condron, L.M., Davis, M.R. and Sherlock, R.R. (2000) Effects of afforestation on phosphorus dynamics and biological properties in a New Zealand grassland soil. *Plant and Soil* 220, 151–163.

Chen, C.R., Condron, L.M., Davis, M.R. and Sherlock, R.R. (2001) Effects of land-use change from grassland to forest on sulfur availability and arylsulfatase activity in New Zealand. *Australian Journal of Soil Research* 39, 749–757.

Chen, C.R., Condron, L.M., Davis, M.R. and Sherlock, R.R. (2002) Phosphorus dynamics in the rhizosphere of perennial ryegrass (*Lolium perenne* L.) and radiata pine (*Pinus radiata* D. Don). *Soil Biology and Biochemistry* 34, 487–499.

Chen, C.R., Condron, L.M., Davis, M.R. and Sherlock, R.R. (2003a) Seasonal changes in soil phosphorus and associated microbial properties under adjacent grassland and forest in New Zealand. *Forest Ecology and Management* 177, 539–557.

Chen, C.R., Condron, L.M., Sinaj, S., Davis, M.R., Sherlock, R.R. and Frossard, E. (2003b) Effects of plant species on phosphorus dynamics in a range of grassland soils. *Plant and Soil* 256, 115–130.

Chuyong, G.B., Newberry, D.M. and Songwe, N.C. (2000) Litter nutrients and retranslocation in central African rain forest dominated by ectomycorrhizal trees. *New Phytologist* 148, 493–510.

Cole, C.V., Innis, G.S. and Stewart, J.W.B. (1977) Simulation of phosphorus cycling in semi-arid grassland. *Ecology* 58, 1–15.

Conant, R.T., Paustian, K. and Elliott, E.T. (2001) Grassland management and conversion into grassland: effects on soil carbon. *Ecological Applications* 11, 343–355.

Condron, L.M. and Newman, R.H. (1998) Chemical nature of soil organic matter under grassland and forest. *European Journal of Soil Science* 49, 597–604.

Condron, L.M., Moir, J.O., Tiessen, H. and Stewart, J.W.B. (1990) Critical evaluation of methods for determining total organic phosphorus in tropical soils. *Soil Science Society of America Journal* 54, 1261–1266.

Condron, L.M., Davis, M.R., Newman, R.H. and Cornforth, I.S. (1996) Influence of conifers on the forms of phosphorus in selected New Zealand grassland soils. *Biology and Fertility of Soils* 21, 37–42.

Condron, L.M., Turner, B.L. and Cade-Menun, B.J. (2005) Chemistry and dynamics of soil organic phosphorus. In: Sims, J.T. and Sharpley, A.N. (eds) *Phosphorus: Agriculture and the Environment.* Soil Science Society of America, Madison, Wisconsin (in press).

Crews, T.E., Kitayama, K., Fownes, J.H., Riley, R.H., Herbert, D.A., Mueller-Dombois, D. and Vitousek, P.M. (1995) Changes in soil phosphorus fractions and ecosystem dynamics across a long chronosequence in Hawaii. *Ecology* 76, 1407–1424.

Cross, A.F. and Schlesinger, W.H. (1995) A literature review and evaluation of the Hedley fractionation scheme: applications to the biogeochemical cycle of soil phosphorus in natural ecosystems. *Geoderma* 64, 197–214.

Daily, G.C., Matson, P.A. and Vitousek, P.M. (1997) Ecosystem services supplied by soil. In: Daily, G.C. (ed.) *Nature's Services: Societal Dependence on Natural Ecosystems.* Island, Washington, DC, pp. 113–132.

Dalal, R.C. (1979) Mineralization of carbon and phosphorus from carbon-14 and phosphorus-32 labelled plant material added to soil. *Soil Science Society of America Journal* 43, 913–916.

Davis, M.R. (1994) Topsoil properties under tussock grassland and adjoining pine forest in Otago, New Zealand. *New Zealand Journal of Agricultural Research* 37, 465–469.

Davis, M.R. (1995) Influence of radiata pine seedlings on chemical properties of some New Zealand montane grassland soils. *Plant and Soil* 176, 255–262.

Davis, M.R. (2001) Soil properties under pine forest and pasture at two hill country sites in Canterbury. *New Zealand Journal of Forestry Science* 31, 3–17.

Davis, M.R. and Condron, L.M. (2002) Impact of grassland afforestation on soil carbon in New Zealand: a review of paired-site studies. *Australian Journal of Soil Research* 40, 675–690.

Davis, M.R. and Lang, M.H. (1991) Increased nutrient availability in topsoils under conifers in the South Island high country. *New Zealand Journal of Forestry Science* 21, 165–179.

Fox, T.R. and Comerford, N.B. (1990) Low molecular weight organic acids in selected forest soils of the southeastern USA. *Soil Science Society of America Journal* 54, 1139–1144.

Frossard, E., Condron, L.M., Oberson, A., Sinaj, S. and Fardeau, J.C. (2000) Processes governing phosphorus availability in temperate soils. *Journal of Environmental Quality* 29, 15–23.

Giddens, K.M., Parfitt, R.L. and Percival, H.J. (1997) Comparison of some soil properties under *Pinus radiata* and improved pasture. *New Zealand Journal of Agricultural Research* 40, 409–416.

Gressel, N. and McColl, J.G. (1997) Phosphorus mineralization and organic matter decomposition: a critical review. In: Cadisch, G. and Giller, K.E. (eds) *Driven by Nature: Plant Litter Quality and Decomposition.* CAB International, Wallingford, UK, pp. 297–309.

Gressel, N., McColl, J.G., Preston, C.M., Newman, R.H. and Powers, R.F. (1996) Linkages between phosphorus transformations and carbon decomposition in a forest soil. *Biogeochemistry* 33, 97–123.

Groenendijk, F.M., Condron, L.M. and Rijkse, W.C. (2002) Effects of afforestation on organic carbon, nitrogen and sulfur concentrations in New Zealand hill country soils. *Geoderma* 108, 91–100.

Guggenberger, G., Christiansen, B.T., Rubæk, G. and Zech, W. (1996) Land-use and fertilization effects on P forms in two European soils: resin extraction and ^{31}P-NMR analysis. *European Journal of Soil Science* 47, 605–614.

Hanway, J.J. and Olsen, R.A. (1980) Phosphate nutrition of corn, soya beans, and small grains. In: Khasawneh, F.E., Sample, E.C. and Kamprath, E.J. (eds) *The Role of Phosphorus in Agriculture.* ASA–CSSA–SSSA, Madison, Wisconsin, pp. 681–692.

Harrison, A.F. (1987) *Soil Organic Phosphorus: a Review of World Literature.* CAB International, Wallingford, UK, 257 pp.

Hart, P.B.S., Clinton, P.W., Allen, R.B., Nordmeyer, A.H. and Evans, G. (2003) Biomass and macronutrients (above- and below-ground) in a New Zealand beech (*Nothofagus*) forest ecosystem: implications for carbon storage and sustainable forest management. *Forest Ecology and Management* 174, 281–294.

Hayes, J.E., Richardson, A.E. and Simpson, R.J. (2000) Components of organic phosphorus in soil extracts that are hydrolysed by phytase and acid phosphatase. *Biology and Fertility of Soils* 32, 279–286.

Haynes, R.J. and Williams, P.H. (1991) Nutrient cycling and soil fertility in the grazed pasture ecosystem. *Advances in Agronomy* 49, 119–199.

Hedley, M.J., Nye, P.H. and White, R.E. (1983) Plant-induced changes in the rhizosphere of rape (*Brassica napus* var. Emerald) seedlings. IV. The effect of rhizosphere phosphorus status on the pH, phosphatase activity and deletion of phosphorus fractions in the rhizosphere and on cation-anion balance in the plants. *New Phytologist* 95, 69–82.

Hoffmann, I., Gerling, D., Kyiogwom, U.B. and Mané-Bielfeldt, A. (2001) Farmers' management strategies to maintain soil fertility in a remote area in northwest Nigeria. *Agriculture, Ecosystems and Environment* 86, 263–275.

Joner, E.J., Magid, J., Gahoonia, T.S. and Jakobsen, I. (1995) Phosphorus depletion and activity of phosphatases in the rhizosphere of mycorrhizal and non-mycorrhizal cucumber (*Cucumis sativus* L.). *Soil Biology and Biochemistry* 27, 1145–1150.

Joner, E.J., van Aarde, I.M. and Vosatka, M. (2000) Phosphatase activity of extra-radical arbuscular mycorrhizal hyphae: a review. *Plant and Soil* 226, 199–210.

Kowalenko, C.G. (1978) Organic nitrogen, phosphorus and sulfur in soil. In: Schnitzer, M. and Khan, S.U. (eds) *Soil Organic Matter.* Elsevier, Amsterdam, pp. 95–136.

Lal, C.B., Annarurna, C., Raghubanshi, A.S. and Singh, J.S. (2001) Foliar demand and resource economy of nutrients in dry tropical forest species. *Journal of Vegetation Science* 12, 5–14.

Maclaren, J.P. (1996) *Environmental Effects of Planted Forests in New Zealand.* FRI Bulletin 198. New Zealand Institute of Forest Research, Rotorua, 142 pp.

Magid, J., Tiessen, H. and Condron, L.M. (1996) Dynamics of organic phosphorus in soils under natural and agricultural ecosystems. In: Piccolo, A. (ed.) *Humic Substances in Terrestrial Ecosystems.* Elsevier, Amsterdam, pp. 429–466.

Mailly, D., Chritanty, L. and Kimmins, J.P. (1997) Without bamboo, the land dies: nutrient cycling and biogeochemistry of a Javanese bamboo talun-kebun system. *Forest Ecology and Management* 91, 155–173.

Menezes, R.S.C. and Sampaio, E.V.S.B. (2002) Simulação dos fluxos e balanços de fósforo em uma unidade de produção agrícola familiar no semi-árido paraibano. In: Silveira, L.M., Petersen, P. and Sabourin, E. (eds) *Agricultura Familiar e Agroecologia no Semi-árido: Avanços a Partir do Agreste da Paraíba.* Rio de Janeiro, pp. 249–260.

McGill, W.B. and Cole, C.V. (1981) Comparative aspects of cycling of organic C, N, S and P through soil organic matter. *Geoderma* 26, 267–286.

Olander, L.P. and Vitousek, P.M. (2000) Regulation of soil phosphatase and chitinase activity by N and P availability. *Biogeochemistry* 49, 175–190.

Parfitt, R.L., Percival, H.J., Dahlgren, R.A. and Hill, L.F. (1997) Soil and solution chemistry under pasture and radiata pine in New Zealand. *Plant and Soil* 191, 279–290.

Pritchett, W.L. and Fisher, R.F. (1987) Nutrient cycling in forest ecosystems. In: *Properties and Management of Forest Soils,* 2nd edn. John Wiley & Sons, New York, pp. 180–204.

Raich, J.W., Parton, W.J., Russell, A.E., Sanford, R.L. and Vitousek, P.M. (2000) Analysis of factors regulating ecosystem development on Mauna Loa using the Century model. *Biogeochemistry* 51, 161–191.

Richardson, A.E., Hadobas, P.A. and Hayes, J.E. (2001) Extracellular secretion of *Aspergillus* phytase from *Arabidopsis* roots enables plants to obtain phosphorus from phytate. *Plant Journal* 25, 1–10.

Roberts, T.L., Stewart, J.W.B. and Bettany, J.R. (1985) The influence of topography on the distribution of organic and inorganic soil phosphorus across a narrow environmental gradient. *Canadian Journal of Soil Science* 65, 651–665.

Rodriguez-Jimenez, L.V.A. (1988) Consideraciones sobre la biomasa, composicion quimica y dinamica del bosque pluvial tropical de colinas bajas, Bajo Calima, Buenaventura, Colombia. In: *Serie Documentacion, Corporacion Nacional de Investigacion y Fomento Forestal.* Bogota, Colombia, 36 pp.

Ross, D.J., Tate, K.R., Scott, N.A. and Feltham, C.W. (1999) Land-use change: effects on soil carbon, nitrogen and phosphorus pools and fluxes in three adjacent ecosystems. *Soil Biology and Biochemistry* 31, 803–813.

Rubæk, G. and Sibbesen, E. (1995) Soil phosphorus dynamics in a long-term experiment at Askov. *Biology and Fertility of Soils* 20, 86–92.

Schlesinger, W.H. (1991) *Biogeochemistry: Analysis of Global Change.* Academic Press, San Diego, 443 pp.

Shanmughavel, P., Sha, L.Q., Zheng, Z. and Cao, M. (2001) Nutrient cycling in a tropical seasonal rainforest of Xishuangbanna, Southwest China. Part 1. Trees species, nutrient distribution and uptake. *Bioresource Technology* 80, 163–170.

Silver, W.L., Scatena, F.N., Johnson, A.H., Siccama, T.G. and Sanchez, M.J. (1994) Nutrient availability in a montane wet tropical forest: spatial patterns and methodological considerations. *Plant and Soil* 164, 129–145.

Sinsabaugh, R.L. (1994) Enzymic analysis of microbial pattern and process. *Biology and Fertility of Soils* 17, 69–74.

Six, J., Conant, R.T., Paul, E.A. and Paustian, K. (2002)

Stabilization mechanisms of soil organic matter: implications for C-saturation of soils. *Plant and Soil* 241, 155–176.

Smeck, N.E. (1985) Phosphorus dynamics in soils and landscapes. *Geoderma* 36, 185–199.

Smernik, R.J. and Oades, J.M. (2003) Spin accounting and RESTORE: two new methods to improve quantitation in solid-state ^{13}C NMR analysis of soil organic matter. *European Journal of Soil Science* 54, 103–116.

Smestad, B.T., Tiessen, H. and Buresh, R. (2002) Short fallows of *Tithonia diversifolia* and *Crotalaria grahamiana* for soil fertility improvement in western Kenya. *Agroforestry Systems* 55, 181–194.

Speir, T.W. and Ross, D.J. (1978) Soil phosphatase and sulphatase. In: Burns, R.G. (ed.) *Soil Enzymes.* Academic Press, London, pp. 197–250.

Stevenson, F.J. (1994) Organic phosphorus and sulfur compounds. In: *Humus Chemistry: Genesis, Composition, Reactions,* 2nd edn. John Wiley & Sons, New York, pp. 113–140.

Stevenson, F.J. and Cole, M.A. (1999) Phosphorus. In: Stevenson, F.J. and Cole, M.A. (eds) *Cycles of Soil: Carbon, Nitrogen, Phosphorus, Sulfur, Micronutrients.* John Wiley & Sons, New York, pp. 279–329.

Stewart, J.W.B. and Tiessen, H. (1987) Dynamics of soil organic phosphorus. *Biogeochemistry* 4, 41–60.

Tarafdar, J.C. and Claassen, N. (1988) Organic phosphorus compounds as a phosphorus source for higher plants through the activity of phosphatase produced by plant roots and microorganisms. *Biology and Fertility of Soils* 5, 308–312.

Tarafdar, J.C. and Jungk, A. (1987) Phosphatase activity in the rhizosphere and its relation to the depletion of soil organic phosphorus. *Biology and Fertility of Soils* 3, 199–204.

Tiessen, H. and Shang, C. (1998) Organic-matter turnover in tropical land-use systems. In: Bergström, L. and Kirchmann, H. (eds) *Carbon and Nutrient Dynamics in Natural and Agricultural Tropical Ecosystems.* CAB International, Wallingford, UK, pp. 1–14.

Tiessen, H. and Stewart, J.W.B. (1983) Particle size fractions and their use in studies of soil organic matter. II. Cultivation effects on organic matter composition in size fractions. *Soil Science Society of America Journal* 47, 507–514.

Tiessen, H., Stewart, J.W.B. and Bettany, J.R. (1982) Cultivation effect on the amounts and concentrations of carbon, nitrogen and phosphorus in grassland soils. *Agronomy Journal* 74, 831–835.

Tiessen, H., Chacon, P. and Cuevas, E. (1994) Phosphorus and nitrogen status in soils and vegetation along a toposequence of dystrophic rainforests on the upper Rio Negro. *Oecologia* 99, 145–150.

Turner, B.L., Papházy, M.J., Haygarth, P.M. and McKelvie, I.D. (2002) Inositol phosphates in the environment. *Philosophical Transactions of the Royal Society London, Series B* 357, 449–469.

Turner, B.L., Mahieu, N. and Condron, L.M. (2003a) Phosphorus-31 nuclear magnetic resonance spectral assignments of phosphorus compounds in soil NaOH–EDTA extracts. *Soil Science Society of America Journal* 67, 497–510.

Turner, B.L., Mahieu, N. and Condron, L.M. (2003b) The phosphorus composition of temperate grassland soils determined by NaOH–EDTA extraction and solution ^{31}P NMR spectroscopy. *Organic Geochemistry* 34, 1199–1210.

Turner, B.L., Mahieu, N. and Condron, L.M. (2003c) Quantification of *myo*-inositol hexakisphosphate in alkaline soil extracts by solution ^{31}P NMR spectroscopy and spectral deconvolution. *Soil Science* 168, 469–478.

Vance, C.P., Uhde-Stone, C. and Allen, D.L. (2003) Phosphorus acquisition and use: critical adaptations by plants for securing a nonrenewable resource. *New Phytologist* 157, 423–447.

Walker, T.W. and Syers, J.K. (1976) The fate of phosphorus during pedogenesis. *Geoderma* 15, 1–19.

Williams, P.H. and Haynes, R.J. (1992) Balance sheet of phosphorus, sulphur, and potassium in a long-term grazed pasture supplied with superphosphate. *Fertilizer Research* 31, 51–60.

Wu, R. and Tiessen, H. (2002) Effect of land use on soil degradation in alpine grassland soil, China. *Soil Science Society of America Journal* 66, 1648–1655.

Yeates, G.W. and Saggar, S. (1998) Comparison of soil microbial properties and fauna under tussock-grassland and pine plantation. *Journal of the Royal Society of New Zealand* 28, 523–535.

Yeates, G.W., Saggar, S. and Daly, B.K. (1997) Soil microbial C, N, and P, and microfaunal populations under *Pinus radiata* and grazed pasture land-use systems. *Pedobiologia* 41, 549–565.

14 Organic Phosphorus in the Aquatic Environment: Speciation, Transformations and Interactions with Nutrient Cycles

Alison M. Mitchell and Darren S. Baldwin

Murray–Darling Freshwater Research Centre, Cooperative Research Centre for Freshwater Ecology, PO Box 921, Albury, NSW 2640, Australia

Introduction

Phosphorus dynamics in aquatic environments have been the subject of intense research for several decades (see, e.g. Boström *et al.*, 1988; Benitez-Nelson, 2000). This level of response has been driven by the recognition that high concentrations of phosphorus in aquatic environments can contribute to excessive algal and macrophyte growth and associated water-quality problems (Boström *et al.*, 1988; Cullen and Forsberg, 1988; Tyrrell, 1999). Organic phosphorus can comprise a large proportion of the total phosphorus in aquatic systems (Herbes *et al.*, 1975; McKelvie *et al.*, 1995; Suzumura *et al.*, 1998; Monaghan and Ruttenberg, 1999), but notwithstanding the attention that phosphorus dynamics has received in the aquatic-themed literature, there has been relatively little attention paid to the role and importance of organic phosphorus. This is due partly to a lack of adequate methods for the speciation and quantification of organic phosphorus in the aquatic environment. Phosphorus research has been driven by the implied role of phosphorus in primary production. This emphasis on phosphorus to predict productivity arose principally with the publication of a model by Vollenweider (1968), which revealed a relationship between total phosphorus concentration and chlorophyll *a* (an indicator of algal biomass). Because phosphate is considered the most biologically available phosphorus species, it is often used to predict productivity in aquatic systems, but a number of methodology problems throw doubt on the value of this approach.

Hudson *et al.* (2000) and Baldwin *et al.* (2003) recently questioned the importance of measuring phosphate as an indicator for production in the aquatic environment. Hudson *et al.* (2000) used a novel steady-state bioassay method for phosphate measurement and found that only picomolar concentrations were present over a wide range of total phosphorus concentrations. They concluded that a rapid turnover of organic phosphorus was supporting high rates of production. Baldwin *et al.* (2003) used weak anion-exchange chromatography and algal growth bioassays to assess temporal variability in the speciation of filterable phosphorus in a eutrophic reservoir. Two blooms occurred during their study: a dinoflagellate bloom that was phosphorus-limited or co-limited, and a green algal bloom that was not. Phosphorus speciation revealed barely detectable levels of filterable phosphorus in the water column during the dinoflagellate bloom, yet all of the phosphorus (below the detection limit using standard techniques) during the green bloom was phosphate. Baldwin *et al.* (2003) concluded that phosphate concentrations could not be equated with the probability of an algal bloom.

The relationship between productivity and phosphorus concentration in the Vollenweider (1968) model was with total phosphorus, so one may expect that total phosphorus will provide a measure of the probability of an algal bloom. However, recent observations suggest that measuring total phosphorus in the aquatic environment provides little information about potential productivity or trophic status of an aquatic system. For example, Hudson *et al.* (2000) observed a strong relationship between concentrations of total phosphorus and phosphate ($r^2=0.66$; $n=53$; $P<0.0001$), yet total phosphorus concentrations were not related to rates of production. Not only do these findings indicate an important role for organic phosphorus in the aquatic phosphorus cycle, they give particular importance to the necessity for separation and characterization of organic phosphorus in the aquatic environment before we can understand the bioavailability and biogeochemistry of phosphorus.

In this chapter, we discuss the various approaches that have been used to determine organic phosphorus in the water column and sediments of aquatic ecosystems. We then review the limited number of studies that have examined transformations of organic phosphorus in aquatic ecosystems, and finally assess how organic phosphorus interacts with the aquatic cycles of other nutrients.

Characterization of Phosphorus in the Aquatic Environment

There are a number of methodology problems associated with the speciation and characterization of phosphorus in the aquatic environment, and no method has yet been wholly successful. To date, most studies of phosphorus speciation in the aquatic environment have been limited to operationally defined phosphorus fractions that are often chemically ill-defined. Speciation methods are different for sediments and the water column, due to the inherent analytical problems associated with the sediment matrix. Organic phosphorus has received far less research effort than inorganic forms of phosphorus *per se*, due to research assumptions and methodology problems associated with phosphorus speciation and characterization.

Methods for characterizing phosphorus in the water column

Conventional approaches

Traditionally, phosphorus research in the water column has been limited to measuring concentrations of operationally defined, but not chemically well-characterized, phosphorus fractions. Aquatic phosphorus is traditionally divided into particulate and dissolved fractions by filtration, with further division into phosphate, inorganic condensed, organic condensed, organic, mineral, and mixed-phase phosphorus species (Holtan *et al.*, 1988; Maher and Woo, 1998, and references therein; see also McKelvie, Chapter 1; Cade-Menun, Chapter 2; Cooper *et al.*, Chapter 3, this volume).

The phosphorus concentration in each operationally defined fraction is typically detected using the phosphomolybdate-blue colorimetric reaction and spectrophotometric detection (American Public Health Association, 1995), with or without pretreatment and/or filtration. The term 'total phosphorus' applies to samples subjected to hydrolysis and/or oxidation pretreatment intended to convert all forms of phosphorus to phosphate (Maher and Woo, 1998; Monaghan and Ruttenberg, 1999). Phosphorus that reacts with molybdate in untreated samples is referred to as 'reactive phosphorus'. Samples may be unfiltered or filtered. When samples are unfiltered, the operationally defined fractions are total phosphorus and total reactive phosphorus. The fraction remaining after the subtraction of total reactive phosphorus from total phosphorus is conventionally termed total organic phosphorus (American Public Health Association, 1989), but we prefer the term total unreactive phosphorus, because particulate inorganic phosphorus can also be included. When samples are filtered, the operationally defined fractions are total filterable phosphorus and filterable reactive phosphorus. The fraction of phosphorus that remains when filterable reactive phosphorus is subtracted from total filterable phosphorus is filterable unreactive phosphorus. Of course, the terminology found in the literature varies. For example, unreactive phosphorus is also termed non-reactive or organic phosphorus, and filterable phosphorus is also termed dissolved or soluble phosphorus. However, the terms as defined above are used throughout this chapter.

Problems with methodology used to characterize phosphorus in the water column

The traditional methods of determining phosphorus speciation in the water column are operationally simple and therefore convenient for routine analysis, but suffer from a number of important problems, which ultimately throw doubt on the validity of this approach. One problem is that there is no current consensus on which pore size of filter to use, but different pore sizes will exclude different species of phosphorus. For example, one may expect no algal or bacterial-associated organic phosphorus to be included in a sample filtered with small pore size filters (e.g. 0.2 μm), yet they will be well represented in a 0.7-μm-filtered sample. Studies use various filtration techniques, ranging from ultrafiltration with a cut-off of 500 Da, to glass-fibre filters with a nominal pore size ranging from 0.1 to 0.7 mm. Thus, the use of different pore sizes may alter the measured proportions of phosphorus in the operationally defined fractions, which may be compounded by other analytical problems. For example, digestion techniques may not convert all complexed phosphorus to phosphate, thereby underestimating the total phosphorus concentration (McKelvie, Chapter 1, this volume). Monaghan and Ruttenberg (1999) found that phosphonates and phospholipids were under-recovered with the acid persulphate pretreatment by as much as 26% and 44%, respectively. Likewise, Ridal and Moore (1990) found that organic phosphorus compounds with polyphosphate moieties (e.g. adenosine triphosphate) were under-recovered when ultraviolet oxidation was used as a pretreatment, although such pretreatment is known not to influence polyphosphate bonds (McKelvie, Chapter 1, this volume).

In contrast to the problem of organic phosphorus underestimation, phosphomolybdate-blue analysis may overestimate free phosphate. Kuenzler and Ketchum (1962) and Rigler (1966, 1968) found a large difference between biological uptake kinetics determined with radioactive phosphate and the concentration of filterable reactive phosphorus. They attributed this difference to an overestimation of true phosphate by the molybdate-blue method. More recently, Baldwin (1998) used weak anion-exchange chromatography to show that only 20% of filterable reactive phosphorus in the water column of a billabong (oxbow lake) was phosphate, with some complexed phosphorus species hydrolysed or displaced from colloids in the presence of transition metals and in the acid environment of the phosphorus detection procedure. Furthermore, several model organic phosphorus compounds were susceptible to hydrolysis during phosphomolybdate-blue analysis, ranging from 100% for adenosine 5′-diphosphate to 10% for *myo*-inositol monophosphate. However, other studies showed no such hydrolysis. For example, Monaghan and Ruttenberg (1999) found that only two model compounds (adenosine 5′-diphosphate and phosphocreatine, a biochemical intermediate) were hydrolysed by more than 1% during the analysis, while Denison *et al.* (1998) detected hydrolysis of some organic phosphorus compounds only after several days' storage at ambient laboratory temperature. Clearly, more research is required to resolve these analytical difficulties in order to determine the character of phosphorus in aquatic systems.

Specific methods for separating and characterizing organic phosphorus in the water column

A number of tools have been used to specifically separate and quantify organic phosphorus species in the water column, including ^{31}P nuclear magnetic resonance (NMR) spectroscopy, chromatography and enzyme hydrolysis studies. Detailed reviews of phosphorus speciation techniques for environmental samples can be found elsewhere in this volume (McKelvie, Chapter 1; Cade-Menun, Chapter 2; Cooper *et al.*, Chapter 3).

^{31}P NUCLEAR MAGNETIC RESONANCE SPECTROSCOPY Although ^{31}P NMR spectroscopy has been used extensively to speciate organic phosphorus in the terrestrial environment, there have been few applications in aquatic ecosystems, particularly in the water column. In her review of the literature, Cade-Menun (Chapter 2, this volume) notes only a few studies where the technique was applied to lake water or seawater samples. One of the problems is the poor resolution of signals; for example, solid-state ^{31}P NMR spectra of freeze-dried marine phytoplankton and sediment extracts exhibited broad and overlapping signals that were difficult to assign unequivocally (Ingall *et al.*, 1990). Another problem encountered in applying NMR to natural water samples is the low concentration of phosphorus. This necessitates preconcentration, but procedures such as ultrafiltration and reverse osmosis often introduce irreversible experimental artefacts, particularly aggregation and precipitation, which result in broad NMR peaks with poor analytical sensitivity (see Nanny and Minear, 1994). Lanthanide shift reagents following preconcentration have been used to increase peak resolution (Nanny and Minear, 1994), although it has been suggested that artefacts arising from concentrating samples are only confined to the higher concentration factors (~500). For example, Suzumura *et al.* (1998) found no aggregation with a concentration factor up to 20 during stirred cell ultrafiltration of seawater.

CHROMATOGRAPHY Physical separation of phosphorus fractions with anion-exchange and high-performance liquid chromatography has not been an entirely successful method for separating and characterizing organic phosphorus species. Compounds are identified by comparing the elution time of native sample peaks with those of model phosphorus species. Minear *et al.* (1988) developed a high-performance, strong anion–ion exchange chromatography technique for inositol phosphate speciation. They used model phosphorus compounds, a sodium chloride eluent gradient, and generated intermediates of *myo*-inositol hexakisphosphate by phytase hydrolysis to determine whether these could be identified and quantified in their samples. Although inositol tetra- and pentakisphosphate were identified with this technique, base-line resolution, and hence quantification, was not achieved (Minear *et al.*, 1988).

Clarkin *et al.* (1992) further developed the technique of Minear *et al.* (1988) by adding an on-line detector and a different eluent gradient. They used a number of model phosphorus compounds and applied the analysis to preconcentrated stream water. These modifications improved peak separation at the expense of efficiency (160-min chromatogram), but DNA could not be separated from *myo*-inositol hexakisphosphate. This prompted a recommendation to separate these compounds using ultrafiltration followed by chromatographic analysis of both the filtrate and the residual (Clarkin *et al.*, 1992). Nanny *et al.* (1995) added those modifications to their analysis of forest stream and lake water samples and potentially identified and quantified a number of organic phosphorus species, including choline phosphate, cyclic adenosine monophosphate, *myo*-inositol hexakisphosphate, *myo*-inositol monophosphate and DNA. However, a large unretained peak could not be identified despite a number of attempts, including extractions for phospholipids, humic and fulvic acid-bound phosphorus, hypobromite oxidation, and hydrolysis with alkaline phosphatase. They suggested that this fraction was an artefact of the concentration method, suggesting that this technique suffers from similar methodological problems as ^{31}P NMR for analysis of water samples.

Low-pressure, weak anion-exchange chromatography has also been used to determine phosphorus speciation in aquatic samples (Baldwin, 1998, 1999; Baldwin *et al.*, 2003). The advantage of this procedure is that it allows concentration of the sample directly on the column, which eliminates the need to preconcentrate samples. Furthermore, dissolved organic matter tends not to bind irreversibly to weak anion-exchange materials. However, using this technique, none of these studies could unequivocally assign phosphorus peaks to known structures, with the exception of free phosphate.

Gel-permeation chromatography has also been used to characterize organic phosphorus species in aquatic samples. For example, Eisenreich and Armstrong (1977) used gel-permeation chromatography to characterize inositol phosphates from lake foam and concentrated lake water samples that had been pretreated by hypobromite oxidation. Their study showed that up to 26% of the total phosphorus was inositol phosphate esters. However, their study was based on the premise that hypobromite oxidation will destroy all organic phosphorus compounds except inositol phosphates, yet this has not been thoroughly tested using modern analytical techniques. Therefore, it is possible that the hypobromite oxidation is not as robust as the authors assumed (as noted for other digestion procedures; see Monaghan and Ruttenberg, 1999).

Gel-permeation chromatography has also been used to elucidate the interactions between dissolved humic substances and phosphate. Gel-permeation studies have clearly shown that phosphorus is often associated with relatively high-molecular-weight humic substances. The humic matter itself is not thought to retain much phosphate, because humics are normally negatively charged within the typical range of pH in aquatic systems (Brinkman, 1993). Rather, it has been suggested that phosphorus interacts with cations (particularly iron, aluminium and calcium) bound to the dissolved humic materials (Brinkman, 1993).

ENZYME HYDROLYSIS A number of researchers have added various combinations of enzymes to water samples to characterize the organic phosphorus, although interpretation is often limited by the fact that not all enzymes are specific for a particular species of organic phosphorus (McKelvie *et al.*, 1995; Turner *et al.*, 2002). Herbes *et al.* (1975) were one of the first groups to adopt this approach. They showed that up to 50% of organic phosphorus in lake water could be hydrolysed by added phytase, and this was found in both low- and high-molecular-weight fractions. The authors attributed the low-molecular-weight fraction to free inositol phosphates and the high-molecular-weight fraction to inositol phosphates bound to proteins, lipids or fulvic acids.

Other studies have combined enzymatic hydrolysis with other separation techniques, including gel-permeation chromatography (e.g. Hino, 1989) and ultrafiltration (e.g. Suzumura *et al.*, 1998). For example, Suzumura *et al.* (1998) used alkaline phosphatase and a combined alkaline phosphatase/phosphodiesterase treatment to determine the speciation of high-molecular-weight (10,000 Da to 0.1 μm) unreactive phosphorus isolated by ultrafiltration of water samples from a transect of Tokyo Bay. They found that their high-molecular-weight fraction was up to 36% of total unreactive phosphorus, of which 67% was not readily hydrolysed. Further chemical extraction (chloroform and Polymyxin B) suggested that phospholipids were the major component of the unreactive fraction, because phosphonates would not be detected with the standard acid-persulphate method of phosphorus analysis used. However, this conclusion may be invalid, because Monaghan and Ruttenberg (1999) found that a model phospholipid was significantly under-recovered (44%) with the acid-persulphate method and more so than a phosphonate (26%). Suzumura *et al.* (1998) did not use phytase in their study, presumably because they had previously found that *myo*-inositol hexakisphosphate was rapidly degraded in Tokyo Bay (Suzumura and Kamatani, 1995) and was as low as 0.1% of the total organic phosphorus in the coastal sediments.

OTHER TECHNIQUES A number of other techniques have been used to determine specific organic phosphorus compounds in the environment, including a firefly luciferase assay for adenosine di- and triphosphate (Kaplan and Bott, 1985), a protein assay for cyclic adenosine monophosphate (Francko and Wetzel, 1982) and cetryltrimethylammonium bromide precipitation of dissolved RNA and DNA in marine and freshwaters (Karl and Bailiff, 1989). However, none of these are regularly used in studies of organic phosphorus dynamics in aquatic ecosystems.

Methods for characterizing phosphorus in aquatic sediments

Sediments generally represent both the largest sink and the largest source of phosphorus in aquatic systems (Baldwin *et al.*, 2002). Therefore, information on sediment phosphorus speciation is important in understanding the aquatic biogeochemistry of phosphorus. However, like soils, sediments typically consist of a complex mixture of clay, silt, sand, organic matter, various minerals, micro- and macro-organisms and water and therefore represent a potentially difficult medium in which to study phosphorus speciation.

Traditionally, phosphorus speciation in sediments has been determined by sequential extraction techniques (e.g. Hieltjes and Lijklema, 1980; Ruttenberg, 1992). Sequential extraction, as the name implies, involves repeated extraction of a sediment sample with solutions of increasing chemical strength until (putatively) all the material of interest has been removed from the sediment. Phosphorus is quantified in each extract using molybdate colorimetry (American Public Health Association, 1995), with or without digestion. Sequential extraction schemes for soils are well-established, but unfortunately the chemicals and techniques for the sequential extraction of phosphorus from sediments vary between operators and there is no current consensus on the most appropriate method.

Problems with methods used to characterize phosphorus in aquatic sediments

The use of sequential extraction techniques to determine analyte speciation has been criticized (Martin *et al.*, 1987). The largest problem is that each step is operationally defined. What one reagent extracts in the sequence will depend on which reagent was previously used on that sediment and the conditions (e.g. temperature and pH) of the extraction (Golterman, 1996). Furthermore, the specificity of a reagent for a given class of compound may be questionable. In the development of sequential extraction techniques, many studies have tested the veracity of their technique using known, usually synthetic, compounds or mixtures of compounds (e.g. Ruttenberg, 1992). However, many of the extractants used are not as specific as studies on artificial mixtures would suggest. For example, in the often-used extraction scheme developed by Hieltjes and Lijklema (1980) the sediment is treated with ammonium chloride to remove soluble phosphorus, 0.1 M NaOH to remove iron-bound phosphorus, 0.5 M HCl to remove calcium-bound phosphorus, and finally with 1 M NaOH to remove any residual phosphorus. However, de Groot (1990) demonstrated that the 0.5 M HCl extraction step for calcium-bound phosphorus would also extract some organic phosphorus. This problem may considerably underestimate the size of the organic phosphorus pool, because acid-soluble organic phosphorus accounted for up to 54% of total phosphorus in river sediments (de Groot, 1990; de Groot and Golterman, 1993). Similarly, Baldwin (1996) showed that the citrate–bicarbonate–

dithionite reagent used to extract iron-bound phosphorus in the Ruttenberg (1992) SEDEX procedure could also extract a large amount of organic matter attached to the sediments. When a bicarbonate extraction step was included before the citrate–bicarbonate–dithionite extraction, however, the extracted total phosphorus and organic carbon were strongly correlated. Furthermore, ^{31}P NMR analysis of the bicarbonate extract showed the presence of both free phosphate and phosphate monoesters (Baldwin, 1996).

Specific methods for separating and characterizing organic phosphorus in aquatic sediments

Most sequential extraction schemes currently used have at least one target phase labelled as organic phosphorus (Bonzongo *et al.*, 1989; Ruttenberg, 1992; Golterman, 1996). However, as noted above, in the absence of corroborating evidence these fractions can only be considered as operationally defined. A number of studies have explored the nature of organic phosphorus in sediment extracts. In particular, extraction of sediments followed by solution ^{31}P NMR spectroscopy has proved helpful in understanding the nature of sedimentary organic phosphorus (Delgardo *et al.*, 2000; Cade-Menun, Chapter 2, this volume). For example, Hupfer *et al.* (1995) clearly showed the presence of significant quantities of polyphosphate in sodium hydroxide extracts of surficial lake sediments, which they believed to be of microbial origin. It is therefore possible that polyphosphate plays a role in phosphorus dynamics in sediments, because facultative anaerobic bacteria are known to take up and store phosphorus as polyphosphate under aerobic conditions and to release this luxury store of phosphorus as phosphate under anaerobic conditions (Wentzel *et al.*, 1986; Gächter *et al.*, 1988; Davelaar, 1993; Gächter and Meyer, 1993). However, Golterman *et al.* (1998) demonstrated that phosphorus extracted with sodium hydroxide was not related to the presence of polyphosphate in lake sediments. It is of note that solid-state ^{31}P NMR techniques have not yet been applied directly to sediments, presumably because of the high concentration of interfering metals (particularly iron; see Cade-Menun, Chapter 2, this volume). However, the technique has been applied to marine plankton to study early diagenesis of organic matter (Clark *et al.*, 1999).

One class of well-characterized organic phosphorus compounds found in sediments is the phospholipid fatty acids (Virtue *et al.*, 1996). These are found in the cell membranes of sediment microorganisms and the pattern of fatty acids is used to characterize the diversity of microorganisms present (e.g. Rajendran *et al.*, 1994; Boon *et al.*, 1996). In a typical experiment (e.g. Findlay and Dobbs, 1993) total lipids are extracted from the sediment for 24 h in a chloroform/methanol/buffer solution. Following drying, the lipids are separated on a silicic-acid column, methylated, and specific phospholipids assigned using gas chromatography and mass spectrometry.

Transformation of Organic Phosphorus in Aquatic Environments

As organic phosphorus compounds represent the largest single pool of phosphorus in many aquatic ecosystems, a critical area of understanding is represented by the rates and mechanisms of conversion of organic phosphorus to other forms, particularly phosphate. The lability of organic phosphorus compounds determines their ultimate fate in aquatic ecosystems. Although abiotic pathways exist for the breakdown of organic phosphorus, little is known about their quantitative importance (Baldwin *et al.*, Chapter 4, this volume) and there has been significantly more research on biotic, particularly enzymatic, degradation of organic phosphorus in aquatic sediments (for further details see Quiquampoix and Mousain, Chapter 5; Heath, Chapter 9; Whitton *et al.*, Chapter 10, this volume). However, such studies are often more concerned with using enzymes to determine the nature of the organic phosphorus species present than the rates and mechanisms of transformations *per se*.

Conclusions about the hydrolytic potential and/or character of organic phosphorus vary between studies. One of the earliest studies into the use of enzymes to determine the lability of dissolved organic phosphorus was by Herbes *et al.* (1975). They used three enzyme treatments: (i) alkaline phosphatase; (ii) phytase; and (iii) phosphodiesterase and alkaline phosphatase incubated sequentially, in their preconcentrated water from a mesotrophic and a eutrophic lake. They found that up to 50% of the filterable unreactive phosphorus was hydrolysed by phytase, while the other two enzyme treatments liberated no phosphate. Similarly, Cooper *et al.* (1991) found that no organic phosphorus was hydrolysed by alkaline phosphatase alone in the waters of Lough Neagh, a shallow eutrophic lake in Northern Ireland, although a combination of alkaline phosphatase and phosphodiesterase hydrolysed 14% of organic phosphorus over 30 days at 15°C. These authors concluded that most of the filterable unreactive phosphorus in the 500 to <100,000 molecular weight range was recalcitrant to bacterial mineralization in the waters of Lough Neagh (Cooper *et al.*, 1991). In contrast, Klotz (1991) found that between 9% and 100% of the organic phosphorus in central New York streams was hydrolysable by alkaline phosphatase. These variable results may be explained partly by differences in the specificity of commercially available enzyme preparations (see above).

Enzymes have also been used to study organic phosphorus in sediments. For example, from phytase hydrolysis of the residual organic phosphorus fraction in freshwater and brackish sediments, de Groot and Golterman (1993) inferred that almost half the residual organic phosphorus was labile compounds consisting of humic-bound phosphorus, nucleic acids and *myo*-inositol hexakisphosphate. It is interesting to note that phytase hydrolysed up to 34% of the residual organic phosphorus in these sediments, but that alkaline phosphatase hydrolysed <5%. Alkaline phosphatase production in algae is inhibited by high filterable reactive phosphorus concentrations (Berman, 1970), which may lead to underestimation of the hydrolytic potential of organic phosphorus in the sample if filterable reactive phosphorus concentrations are high. However, Boon (1990) found strong correlations between alkaline phosphatase activity and filterable reactive phosphorus, suggesting either that bacteria were mainly responsible for production of the alkaline phosphatase, or that production by algae in Australian freshwaters was less affected by filterable reactive phosphorus concentrations than species in the northern hemisphere. Another process that may lead to misinterpretation of the hydrolytic potential of organic phosphorus is adsorption; *myo*-inositol hexakisphosphate strongly sorbs to Fe(OOH) and also readsorbs to reduced iron rather than being released (de Groot and Golterman, 1993). None of the reports noted above mentioned the possibility that phosphate released during enzymatic hydrolysis may have adsorbed to metal or humic phases present in the samples (see below), yet this process may at least partly explain the varying results. Golterman *et al.* (1998) noted the importance of removing the inorganic sediment matrix along with phosphate to avoid the adsorption problem.

Other methods have been used to elucidate the lability of organic phosphorus in aquatic environments with relative success. Hudson *et al.* (2000) used steady-state bioassay to conclude that rapid turnover of organic phosphorus could support high production rates (see above). Bentzen *et al.* (1992) added labelled substrate and monitored the appearance of the label in freshwater plankton to determine the bioavailability of organic phosphorus. They also investigated a possible preference for uptake of inorganic or organic phosphorus in phosphorus-limited freshwater systems using ^{32}P-labelled compounds (phosphate and adenosine triphosphate), concluding that limnetic plankton had a high affinity for both substrates and that organic phosphorus contributed significantly to total phosphorus uptake.

Interactions of Organic Phosphorus with Other Biogeochemical Cycles

Clearly, organic phosphorus in aquatic ecosystems comprises a large proportion of the total phosphorus, plays an important role in regulating the phosphorus cycle, and it is putatively bioavailable. If organic phosphorus is readily bioavailable, its abundance and processing will be important in regulating not only the phosphorus cycle, but also other nutrient cycles. Certain studies that focus on aquatic productivity or N_2-fixation have inferred the interaction of organic phosphorus with other nutrient cycles. In addition, as phosphate is strongly coupled to the iron cycle, it would be reasonable to expect important interactions between iron and organic phosphorus. In this section we explore possible interactions between organic phosphorus and other nutrient cycles.

Organic phosphorus and productivity

Because phosphorus is an essential nutrient, one may expect that organic phosphorus will interact most strongly with the carbon and nitrogen cycles, particularly when production is phosphorus-limited. Although phosphate is the most readily bioavailable form (e.g. Björkman and Karl, 1994), organic phosphorus is also bioavailable after enzymatic hydrolysis (see above). A number of studies have principally aimed at elucidating the role of organic phosphorus in the potential for productivity in aquatic environments and, in so doing, infer interactions with the carbon and/or nitrogen cycles.

Enzymes that cleave phosphate from certain organic molecules may not be produced principally to satisfy a phosphorus limitation (Björkman and Karl, 1994). For example, the enzyme 5′-nucleotidase recognizes the C–N group of DNA and RNA, yet it has been shown to release significant amounts of phosphate (Benitez-Nelson and Buesseler, 1999). Benitez-Nelson and Buesseler (1999) found rapid rates of turnover for filterable reactive phosphorus and filterable unreactive phosphorus in the Gulf of Maine, using natural tracers. They also demonstrated that picoplankton in the coastal marine environment preferentially utilized nucleotides and released measurable amounts of phosphate. They postulated that because 5′-nucleotidase is not inhibited by phosphate, unlike alkaline phosphatase, it may be produced by bacteria to satisfy a carbon and/or nitrogen limitation, releasing large amounts of phosphate as an auxiliary. Thus, carbon and/or nitrogen limitation in bacteria may result in large releases of bioavailable phosphorus that could support significant primary production (Benitez-Nelson and Buesseler, 1999), constituting a strong interaction between organic phosphorus and the carbon and/or nitrogen cycles.

Björkman *et al.* (2000) also found that the organic phosphorus pool in the North Pacific subtropical gyre turned over rapidly and that the phosphate released from nucleotides was up to 50 times greater than the net rate of phosphorus uptake. That is, phosphate was not released to satisfy phosphorus limitation. In direct contrast, Siuda and Chróst (2001) found that the highest rate of phosphorus uptake amongst a number of model organic phosphorus species was in freshwater samples with added nucleotides. In addition, they found that rates of mineralization of added organic phosphorus compounds did not always correlate with assimilation of liberated phosphate in their mesotrophic lake water samples. Overall, they postulated that their data indicated that 5′-nucleotidase was mainly responsible for satisfying phosphorus demand by bacteria. While this report appears in direct contrast to those above, Siuda and Chróst (2001) also noted that observed bacterial growth rates were highest in samples enriched with nucleotides. It is possible that 5′-nucleotidase was produced in response to a carbon and/or nitrogen limitation and released large amounts of phosphorus as an auxiliary, but this was either removed in this instance to satisfy the phosphorus requirement of the bacteria, or was adsorbed to, for example, Fe(OOH).

Organic phosphorus and atmospheric nitrogen fixation

Interactions between the carbon and nitrogen cycles have received much recent attention in marine studies. This attention principally results from the paradigms that: (i) marine production has the potential to act as a long-term sink for atmospheric carbon dioxide (Karl *et al.*, 1997); and (ii) dissolved inorganic nitrogen limits production (Benitez-Nelson, 2000). However, the previous discussion suggests that the role of organic phosphorus in production warrants greater consideration. This may have particular importance during active N_2-fixing events when phosphate becomes limiting. For example, Karl *et al.* (1997) highlighted the role of biological N_2-fixation and the alterations and interactions with marine phosphorus and carbon dynamics. They synthesized information from a number of long-term studies conducted at a single site in the subtropical North Pacific Ocean (HOTS, station ALOHA) to conclude that marine fixation of N_2 can contribute significantly to production (up to 50% of the nitrogen requirement) and, therefore, may partly control carbon and phosphorus cycles. They also conjectured that N_2-fixation might periodically increase the amount of carbon sequestered above predicted values for the region, which may have important consequences for greenhouse gas emissions (Karl *et al.*, 1997).

The role of organic phosphorus in N_2-fixation and greenhouse gas emissions from the oceans was demonstrated by Sañudo-Wilhelmy *et al.* (2001). Many studies have reported that N_2-fixation is limited principally by the availability of iron (e.g. Falkowski, 1997), an essential coprotein for nitrogenase (Atlas and Bartha, 1993). However, Sañudo-Wilhelmy *et al.* (2001) recently reported that rates of N_2-fixation were highly correlated with the phosphorus content of a cyanobacterial colony rather than iron availability or intracellular iron content. The study of Sañudo-Wilhelmy *et al.* (2001) provides further evidence of the coupling of the phosphorus cycle with the carbon and nitrogen cycles. Indeed, each of these studies would indicate that, to date, the role of organic phosphorus in the potential regulation of greenhouse gas emissions has been underestimated.

Interaction between organic phosphorus and the cycling of iron and sulphur

The cycling of phosphorus in many aquatic ecosystems is closely associated with the iron and sulphur cycles (e.g. Einsele, 1936; Mortimer, 1941; Roden and Edmonds, 1997; Mitchell and Baldwin, 1998). Phosphate binds strongly to ferric oxides. Under anaerobic conditions, ferric oxides can be reduced either directly by iron-reducing bacteria, or indirectly by sulphides produced by the bacterially mediated reduction of sulphate. Phosphorus associated with the iron surface is released into solution, so the bacterially mediated reductive dissolution of phosphorus is potentially an important process in the biogeochemical cycling of phosphorus. Baldwin *et al.* (1996) showed that model organic phosphorus compounds can also bind to iron minerals, and more importantly that the mineral surface facilitates the hydrolysis of the organic phosphorus compound. Under oxic conditions, the resultant free phosphate ion irreversibly binds to the mineral surface. However, microbially mediated reductive dissolution of the mineral surface under anaerobic conditions presents a pathway for recycling of the phosphate group.

The Challenges Ahead

Organic phosphorus is ubiquitous in the aquatic environment and may constitute a significant proportion of total phosphorus, yet the role of organic phosphorus in aquatic ecosystems is still poorly understood. Methodology problems contribute significantly to this lack of understanding, yet a number of studies have clearly demonstrated the importance of organic phosphorus in the aquatic phosphorus cycle. In addition, the limited number of studies on organic phosphorus have inferred that organic phosphorus is strongly coupled to the carbon and nitrogen cycles, particularly when phosphate is limiting. Organic phosphorus may therefore play an important role in regulating processes such as the emission of greenhouse gases.

In addition to methodology problems, a number of assumptions restrict advances in our understanding of the role and importance of organic phosphorus in the aquatic environment. For example, it is becoming increasingly clear that filterable reactive phosphorus does not necessarily provide an adequate prediction of productivity in all aquatic systems (Hudson *et al.*, 2000; Baldwin *et al.*, 2003). Furthermore, it is difficult to conceptually separate the nutrient cycles and it is becoming increasingly clear that we can no longer rely on reductionist methods for their study (Harris, 1999; Mitchell, 2002). These failings (methodology problems, questionable assumptions, and a lack of understanding of interactions between nutrient cycles) impede research into phosphorus cycling and, in turn, prediction of the outcomes of management and environmental disturbance in the aquatic environment.

An improved understanding of the role and importance of organic phosphorus in the aquatic environment first requires the development of simple, robust and inexpensive methods for speciating and quantifying the myriad organic phosphorus species found in these systems. Solution ^{31}P NMR spectroscopy has shown some promise, particularly for sediments, because organic phosphorus can be extracted simply from the inorganic matrix and analysed. However the relatively long analysis time for each sample, coupled with the poor resolution within functional groups (e.g. the phosphate monoesters) and the requirement for expensive specialist equipment does present challenges (see Cade-Menun, Chapter 2, this volume). The development of multi-dimensional NMR techniques incorporating ^{31}P NMR may address some of the resolution issues. Chromatographic techniques tend to be robust and relatively inexpensive, but it remains difficult to assign a given peak to a specific compound. Developments in liquid chromatography and mass spectrometry may overcome many of the problems encountered in studying organic phosphorus compounds in the environment, but much analytical development is still required (Cooper *et al.*, Chapter 3, this volume).

Once the analytical difficulties associated with the measurement of organic phosphorus compounds have been overcome, it will then be possible to track their movement and transformations in the environment. However, a number of issues need consideration when designing such studies. For example, organic phosphorus sorbs to iron by one transformation process, yet the release of that organic phosphorus requires an alternative, anaerobic pathway (reduction of iron). Even then, the anaerobic process does not necessarily result in the release of phosphorus if it readsorbs to the reduced iron (Roden and Edmonds, 1997). Anaerobic nutrient cycling processes are strongly interrelated with hysteresis effects and non-linear responses, which complicates the study of nutrient cycling and limits our understanding of nutrient transformations, their rates and importance for aquatic ecosystem function (Mitchell, 2002). In order to overcome these difficulties, researchers must consider issues such as accurate tracking of transformation processes, transformation rates, and the effects of availability of other nutrient species for the processing and importance of organic phosphorus in aquatic ecosystems.

While mechanistic studies of transformation processes are important, they must be complemented by kinetic data. The kinetics of these processes provides key information for interpretation of the importance of these processes for aquatic ecosystem ecology. Transformation rates of organic phosphorus need to be determined accurately with labelled substrates or other appropriate methods, because net rates of processing do not allow accurate interpretation of the cycling and importance of certain nutrient species. For example, Hudson *et al.* (2000) used a novel method to observe rapid turnover of organic phosphorus supporting high rates of production in the absence of phosphate. Traditional measures of net phosphorus concentrations would have found no relationship between phosphorus and production, and therefore are inadequate for ascertaining the importance of organic phosphorus for production. Even then, transformation processes and kinetics will not complete the picture. At present, the study of nutrient cycles concentrates mainly on the arbitrarily divided processes (the *mechanisms*) but pays little attention to the connections (the *interactions*). Interactions between nutrient cycles suggest that the availability of critical nutrients may alter nutrient-cycling phenomena. For example, as yet there is no consensus on whether the supply of dissolved inorganic nitrogen, iron, or organic phosphorus regulates biological N_2-fixation. This last point, in particular, is crucial and suggests that an alternative reductionist/systems-oriented approach is now required. Such an approach, along with adequate analytical methodology, is necessary if we are to fully understand the role and importance of organic phosphorus in aquatic ecosystems.

References

American Public Health Association (1995) *Standard Methods for the Examination of Water and Wastewater,* 19th edn. American Public Health Association, American Water Works Association and Water Environment Federation, Washington, DC.

Atlas, R.M. and Bartha, R. (1993) *Microbial Ecology Fundamentals and Applications,* 3rd edn. Benjamin/Cummings, Redwood, California, 563 pp.

Baldwin, D.S. (1996) The phosphorus composition of a diverse series of Australian sediments. *Hydrobiologia* 335, 1–11.

Baldwin, D.S. (1998) Reactive 'organic' phosphorus revisited. *Water Research* 32, 2265–2270.

Baldwin, D.S. (1999) Dissolved organic matter and phosphorus leached from fresh and 'terrestrially' aged river red gum leaves: implications for assessing river–floodplain interactions. *Freshwater Biology* 41, 675–685.

Baldwin, D.S., Beattie, J.K. and Jones, D.R. (1996) Hydrolysis of an organic phosphorus compound by iron-oxide impregnated filter papers. *Water Research* 30, 1123–1126.

Baldwin, D.S., Mitchell, A.M. and Olley, J. (2002). Pollutant–sediment interactions: sorption, reactivity and transport of phosphorus. In: Haygarth, P.M. and Jarvis, S.C. (eds) *Agriculture, Hydrology and Water Quality.* CAB International, Wallingford, UK, pp. 265–280.

Baldwin, D.S., Whittington, J. and Oliver, R. (2003) Temporal variability of dissolved P speciation in a eutrophic reservoir: implications for predicting algal growth. *Water Research* 37, 4595–4598.

Benitez-Nelson, C.R. (2000) The biogeochemical cycling of phosphorus in marine systems. *Earth-Science Reviews* 51, 109–135.

Benitez-Nelson, C.R. and Buesseler, K.O. (1999) Variability of inorganic and organic phosphorus turnover rates in the coastal ocean. *Nature* 398, 502–505.

Bentzen, E., Taylor, W.D. and Millard, E.S. (1992) The importance of dissolved organic phosphorus to phosphorus uptake by limnetic plankton. *Limnology and Oceanography* 37, 217–231.

Berman, T. (1970) Alkaline phosphatase activity and phosphorus availability in Lake Kinneret. *Limnology and Oceanography* 15, 663–674.

Björkman, K. and Karl, D.M. (1994) Bioavailability of inorganic and organic phosphorus compounds to natural assemblages of microorganisms in Hawaiian coastal waters. *Marine Ecology Progress Series* 111, 265–273.

Björkman, K., Thomson-Bulldis, A.L. and Karl, D.M. (2000) Phosphorus dynamics in the North Pacific subtropical gyre. *Aquatic Microbial Ecology* 22, 185–198.

Boon, P.I. (1990) Organic matter degradation and nutrient regeneration in Australian freshwaters. II. Spatial and temporal variation, and relation with environmental conditions. *Archiv für Hydrobiologie* 117, 405–436.

Boon, P.I., Virtue, P. and Nichols, P.D. (1996) Microbial consortia in wetland sediments: a biomarker analysis of the effects of hydrological regime, vegetation and season on benthic microbes. *Marine and Freshwater Research* 47, 27–41.

Boström, B., Andersen, J.M., Fleischer, S. and Jansson, M. (1988) Exchange of phosphorus across the sediment–water interface. *Hydrobiologia* 170, 229–244.

Bonzongo, J.C., Bertru, G. and Martin, G. (1989) Speciation phosphorus techniques in the sediments: discussion and propositions to evaluate inorganic and organic phosphorus. *Archiv für Hydrobiologie* 116, 61–69.

Brinkman, A.G. (1993) A double-layer model for ion adsorption on to metal oxides, applied to experimental data and to natural sediments of Lake Veluwe, The Netherlands. *Hydrobiologia* 253, 31–45.

Clark, L.L., Ingall, E.D. and Benner, R. (1999) Marine

organic phosphorus cycling: novel insights from nuclear magnetic resonance. *American Journal of Science* 299,724–737.

Clarkin, C.M., Minear, R.A., Kim, S. and Elwood, J.W. (1992) An HPLC postcolumn reaction system for phosphorus-specific detection in the complete separation of inositol phosphate congeners in aqueous samples. *Environmental Science and Technology* 26, 199–204.

Cooper, J.E., Early, J. and Holding, A.J. (1991) Mineralization of dissolved organic phosphorus from a shallow eutrophic lake. *Hydrobiologia* 209, 89–94.

Cullen, P. and Forsberg, C. (1988) Experiences with reducing point sources of phosphorus to lakes. *Hydrobiologia* 170, 321–336.

Davelaar, D. (1993) Ecological significance of bacterial polyphosphate metabolism in sediments. *Hydrobiologia* 253, 179–192.

de Groot, C.J. (1990) Some remarks on the presence of organic phosphates in sediments. *Hydrobiologia* 207, 303–309.

de Groot, C.J. and Golterman, H.L. (1993) On the presence of organic phosphate in some Camargue sediments: evidence for the importance of phytate. *Hydrobiologia* 252, 117–126.

Delgardo, A., Ruíz, J.R., del Campillo, M.D., Kassem, S. and Andreu, L. (2000) Calcium- and iron-related phosphorus in calcareous and calcareous marsh soils: sequential chemical fractionation and ^{31}P nuclear magnetic resonance study. *Communications in Soil Science and Plant Analysis* 31, 2483–2499.

Denison, F.H., Haygarth, P.M., House, W.A. and Bristow, A.W. (1998) The measurement of dissolved phosphorus compounds: evidence for hydrolysis during storage and implications for analytical definitions in environmental analysis. *International Journal of Environmental Analytical Chemistry* 69, 111–123.

Einsele, W. (1936). Uber die Beziehungen des Eisenkreislaufs zum Phosphatkreislauf im eutrophen See. *Archiv für Hydrobiologie* 29, 664–686.

Eisenreich, S.J. and Armstrong, D.E. (1977) Chromatographic investigation of inositol phosphate esters in lake water. *Environmental Science and Technology* 11, 497–501.

Falkowski, P.G. (1997) Evolution of the nitrogen cycle and its influence on the biological sequestration of CO_2 in the ocean. *Nature* 387, 272–275.

Findlay, R.H. and Dobbs, F.C. (1993) Quantitative description of microbial communities using lipid analysis. In: Kemp, P.F., Sherr, B.F., Sherr, E.B. and Cole, J.J. (eds) *Handbook of Methods in Aquatic Microbial Ecology*. Lewis, Boca Raton, Florida, pp. 347–358.

Francko, D.A. and Wetzel, R.G. (1982) The isolation of cyclic adenosine 3′:5′ monophosphate (c-AMP) from lakes of different trophic status: correlation with planktonic variables. *Limnology and Oceanography* 27, 27–38.

Gächter, R. and Meyer, J.S. (1993) The role of microorganisms in mobilization and fixation of phosphorus in sediments. *Hydrobiologia* 253, 103–121.

Gächter, R., Meyer, J.S. and Mares, A. (1988) Contribution of bacteria to release and fixation of phosphorus in lake sediments. *Limnology and Oceanography* 33, 1542–1558.

Golterman, H.L. (1996) Fractionation of sediment phosphate with chelating compounds: a simplification, and comparison with other methods sediment. *Hydrobiologia* 335, 87–95.

Golterman, H., Paing, J., Serrano, L. and Gomez, E. (1998) Presence of and phosphate release from polyphosphates or phytate phosphate in lake sediments. *Hydrobiologia* 364, 99–104.

Harris, G.P. (1999) Comparison of the biogeochemistry of lakes and estuaries: ecosystem processes, functional groups, hysteresis effects and interactions between macro- and microbiology. *Marine and Freshwater Research* 50, 791–811.

Herbes, S.E., Allen, H.E. and Mancy, K.H. (1975) Enzymatic characterization of soluble organic phosphorus in lake water. *Science* 187, 432–434.

Hieltjes, A. and Lijklema, L. (1980) Fractionation of inorganic phosphates in calcareous sediments. *Journal of Environmental Quality* 9, 405–407.

Hino, S. (1989) Characterization of orthophosphate release from dissolved organic phosphorus by gel filtration and several hydrolytic enzymes. *Hydrobiologia* 174, 49–55.

Holtan, H., Kamp-Nielsen, L. and Stuanes, A.O. (1988) Phosphorus in soil, water and sediment: an overview. *Hydrobiologia* 170, 19–34.

Hudson, J.J., Taylor, W.D. and Schindler, D.W. (2000) Phosphate concentrations in lakes. *Nature* 406, 54–56.

Hupfer, M., Gächter, R. and Rüegger, H. (1995) Polyphosphate in lake sediments: ^{31}P NMR spectroscopy as a tool for its identification. *Limnology and Oceanography* 40, 610–617.

Ingall, E.D., Schroeder, P.A. and Berner, R.A. (1990) The nature of organic phosphorus in marine sediments: new insights from ^{31}P NMR. *Geochimica et Cosmochimica Acta* 54, 2617–2620.

Kaplan, L.A. and Bott, T.L. (1985) Adenylate energy charge in streambed sediments. *Freshwater Biology* 15, 133–138.

Karl, D.M. and Bailiff, M. (1989) The measurement

and distribution of dissolved nucleic acids in aquatic environments. *Limnology and Oceanography* 34, 543–558.

Karl, D., Letelier, R., Tupas, L., Dore, J., Christian, J. and Hebel, D. (1997) The role of nitrogen fixation in biogeochemical cycling in the subtropical North Pacific Ocean. *Nature* 388, 533–538.

Klotz, R.L. (1991) Cycling of phosphatase hydrolysable phosphorus in streams. *Canadian Journal of Fisheries and Aquatic Sciences* 48, 1460–1467.

Kuenzler, E.J. and Ketchum, B.H. (1962) Rate of phosphorus uptake by *Phaeodactylum tricornutum. Biological Bulletin (Woods Hole)* 123, 134–145.

Maher, W. and Woo, L. (1998) Procedures for the storage and digestion of natural waters for the determination of filterable reactive phosphorus, total filterable phosphorus and total phosphorus. *Analytica Chimica Acta* 375, 5–47.

Martin, M., Nierel, P. and Thomas, A.J. (1987) Sequential extraction techniques: promises and problems. *Marine Chemistry* 22, 313–341.

McKelvie, I.D., Hart, B.T., Cardwell, T.J. and Cattrall, R.W. (1995) Use of immobilized 3-phytase and flow injection for the determination of phosphorus species in natural waters. *Analytica Chimica Acta* 316, 277–289.

Minear, R.A., Segars, J.E., Elwood, J.W. and Mulholland, P.J. (1988) Separation of inositol phosphates by high performance ion-exchange chromatography. *Analyst* 113, 645–649.

Mitchell, A.M. (2002) Anaerobic nutrient cycles in freshwater sediments. PhD thesis, Charles Sturt University, Wagga Wagga, Australia.

Mitchell, A. and Baldwin, D.S. (1998) Effects of desiccation/oxidation on the potential for bacterially mediated P release from sediments. *Limnology and Oceanography* 43, 481–487.

Monaghan, E.J. and Ruttenberg, K.C. (1999) Dissolved organic phosphorus in the coastal ocean: reassessment of available methods and seasonal phosphorus profiles from the Eel River Shelf. *Limnology and Oceanography* 44, 1702–1714.

Mortimer, C. (1941) The exchange of dissolved substances between mud and water in lakes. *Journal of Ecology* 29, 280–329.

Nanny, M.A. and Minear, R.A. (1994) Use of lanthanide shift reagents with ^{31}P FT-NMR spectroscopy to analyze concentrated lake water samples. *Environmental Science and Technology* 28, 1521–1527.

Nanny, M.A., Kim, S. and Minear, R.A. (1995) Aquatic soluble unreactive phosphorus: HPLC studies on concentrated water samples. *Water Research* 29, 2138–2148.

Rajendran, N., Matsuda, O., Urushigawa, Y. and Simidu, U. (1994) Characterization of microbial community structure in the surface sediment of Osaka Bay, Japan, by phospholipid fatty acid analysis. *Applied and Environmental Microbiology* 60, 248–257.

Ridal, J.J. and Moore, R.M. (1990) A re-examination of the measurement of dissolved organic phosphorus in seawater. *Marine Chemistry* 29, 19–31.

Rigler, F.H. (1966) Radiobiological analysis of inorganic phosphorus in lake water. *Verhandlung Internationale Vereinigung Limnologie* 16, 465–470.

Rigler, F.H. (1968) Further observations inconsistent with the hypothesis that the molybdenum-blue method measures inorganic phosphorus in lake waters. *Limnology and Oceanography.* 13, 7–13.

Roden, E.E. and Edmonds, J.W. (1997) Microbial Fe(III) oxide reduction versus iron-sulfide formation. *Archiv für Hydrobiologie* 139, 347–378.

Ruttenberg, K.C. (1992) Development of a sequential extraction method for different forms of phosphorus in marine sediments. *Limnology and Oceanography* 37, 1460–1482.

Sañudo-Wilhelmy, S.A., Kustka, A.B., Gobler, C.J., Hutchins, D.A., Yang, M., Lwiza, K., Burns, J., Capone, D.G., Raven, J.A. and Carpenter, E.J. (2001) Phosphorus limitation of nitrogen fixation by *Trichodesmium* in the central Atlantic Ocean. *Nature* 411, 66–69.

Siuda, W. and Chróst, R.J. (2001) Utilization of selected dissolved organic phosphorus compounds by bacteria in lake water under nonlimiting orthophosphate conditions. *Polish Journal of Environmental Studies* 10, 475–483.

Suzumura, M. and Kamatani, A. (1995) Origin and distribution of inositol hexaphosphate in estuarine and coastal sediments. *Limnology and Oceanography* 40, 1254–1261.

Suzumura, M., Ishikawa, K. and Ogawa, H. (1998) Characterization of dissolved organic phosphorus in coastal seawater using ultrafiltration and phosphohydrolytic enzymes. *Limnology and Oceanography* 43, 1553–1564.

Turner, B.L., McKelvie, I.D. and Haygarth, P.M. (2002) Characterisation of water-extractable soil organic phosphorus by phosphatase hydrolysis. *Soil Biology and Biochemistry* 34, 27–35.

Tyrrell, T. (1999) The relative influences of nitrogen and phosphorus on oceanic primary production. *Nature* 400, 525–531.

Virtue, P., Nichols, P.D. and Boon, P.I. (1996) Simultaneous estimation of microbial phospholipid fatty acids and diether lipids by capillary gas chromatography. *Journal of Microbiological Methods* 25, 177–185.

Vollenweider, R.A. (1968) Scientific fundamentals of the eutrophication of lakes and flowing waters, with particular reference to nitrogen and phosphorus as factors in eutrophication. *OECD Technical Report* GP OE/515. OECD, Paris.

Wentzel, M.C., Lotter, L.H., Loewenthal, R.F. and Marais, G.V.R. (1986) Metabolic behaviour of *Acinetobacter* spp. in enhanced biological phosphorus removal: a biochemical model. *Water South Africa* 12, 209–224.

15 Modelling Phosphorus, Carbon and Nitrogen Dynamics in Terrestrial Ecosystems

William J. Parton,[1] Jason Neff[2] and Peter M. Vitousek[3]

[1]*Natural Resource Ecology Laboratory, Colorado State University, Fort Collins, CO 80523, USA;* [2]*US Geological Survey, MS980 Denver Federal Center, Denver, CO 80225, USA;* [3]*Department of Biological Sciences, Stanford University, Stanford, CA 94305, USA*

Introduction

A large number of integrated carbon and nutrient cycling ecosystem models have been developed during the last 20 years (VEMAP members, 1995). These models have been used to simulate plant production, nutrient cycling and soil carbon dynamics at site, regional and global scales (Running, 1994; Schimel *et al.*, 1997). The majority of these models include carbon and nitrogen cycling, with few of the models including linked cycling of carbon, nitrogen and phosphorus. Many of the models have been used to investigate the impact of climate change and management practices on temperate agriculture, grassland, and forest systems where nitrogen is the primary nutrient limiting plant growth. However, in tropical and subtropical ecosystems, phosphorus may be more limiting to plant growth than nitrogen (Vitousek and Sanford, 1986; Vitousek and Farrington, 1997). The CENTURY model is one of the few large-scale ecosystem models that include the linked cycling of carbon, nitrogen and phosphorus. However, global use of CENTURY has not included phosphorus cycling, since regional phosphorus cycling data are scarce. A number of plant–soil models with linked carbon, nitrogen and phosphorus cycles have been developed for predicting short-term (10–20 years) inorganic and organic nitrogen and phosphorus dynamics in agricultural systems. These models include the EPIC model (Jones *et al.*, 1984), the GLEAMS model (Knisel *et al.*, 1993), the ANIMO model (Kroes and Rijtema, 1998), the phosphorus model developed by Daroub *et al.* (2003) and the CREAMS model (Knisel, 1980).

Given the potential differences between nitrogen- and phosphorus-limited ecosystems, it is surprising that the phosphorus cycle is not considered in most major models of ecosystem biogeochemistry. Model-based analyses are becoming the cornerstone of our understanding of global and regional processes and are widely used to evaluate the impacts of land-use change on biogeochemical fluxes (e.g. Parton *et al.*, 1994; Running, 1994; VEMAP members, 1995; McGuire *et al.*, 1997). The lack of representation of phosphorus cycling in these models is important, because they cannot currently address some of the most important questions in nutrient ecology. Will phosphorus-limited and nitrogen-limited ecosystems respond similarly to an unusually warm and wet year? Will phosphorus-limited and nitrogen-limited ecosystems in the same environment exhibit similar rates of regrowth following disturbance? Do fires affect phosphorus-limited and nitrogen-limited ecosystems similarly? These are central questions for understanding the behaviour of natural and human-impacted systems in much of the world, yet we cannot address them with many of the models currently available. This inattention to the phosphorus cycle is due in part to several difficult issues, including reconciling modelling and experimental data on phosphorus cycling in tropical ecosystems, linking processes with broadly varying time steps, and simulating basic aspects of the biogeochemical dynamics of tropical ecosystems (Gijsman *et al.*, 1996; Huffman, 1996; Silver *et al.*, 2000).

There have been successful models of phosphorus cycling in temperate ecosystems, but there are a number of issues that make it difficult to adapt these models to tropical ecosystems (e.g. Cole *et al.*, 1977; Gijsman *et al.*, 1996; Huffman, 1996; Silver *et al.*, 2000). Much of the complexity of the phosphorus cycle revolves around the suite of biological and physical transformations that can affect the availability of the various forms of phosphorus in soils. Phosphorus is cycled through various inorganic and organic forms (Tiessen and Stewart, 1983; Tate, 1984; Wood, 1984; Yanai, 1998) and, unlike nitrogen, can be occluded by soil minerals and effectively removed from biotic cycling. As a result, the availability of phosphorus in an ecosystem over time is highly dependent on the ability of biological activity to maintain an actively cycling, non-sequestered reservoir of phosphorus (Cole *et al.*, 1978). The importance of the competition between biological and geochemical processes in the phosphorus cycle has been considered for temperate forests (Wood, 1984; Yanai, 1992), but in most temperate ecosystems it is nitrogen rather than phosphorus that tends to limit plant growth, at least on the time scale of fertilization experiments (Vitousek and Howarth, 1991). The situation is different, however, in the highly weathered soils of tropical and subtropical regions. In these areas, the presence of iron and aluminium oxides and their tendency to sequester phosphate creates a situation where geochemical and biological processes compete strongly for available phosphate and where soils can fix large amounts of phosphate added as fertilizer in short periods of time (Boswinkle, 1961; Warren, 1994a, b; Olander and Vitousek, 2000).

Gijsman *et al.* (1996) present a useful conceptual model for phosphorus cycling in terrestrial ecosystems (Fig. 15.1). The diagram shows that phosphorus cycles through both organic and inorganic pools in soil, microbes and plants. Phosphorus uptake by plants occurs primarily from solution phosphate, which is in equilibrium with labile soil phosphate and less soluble sorbed phosphate associated with calcium, iron and aluminium minerals. The solubility of soil phosphate is a function of the soil pH, with low phosphorus solubility for low-pH soils (pH <5), maximum solubility for soils with pH around 6.5, and low solubility for high-pH soils (Chapin *et al.*, 2002). The main inputs to solution phosphorus are weathering of primary phosphate minerals, fertilizer phosphate application in managed soils, and atmospheric deposition of phosphorus, mostly as dust. Another important source of solution phosphate is mineralization of labile organic phosphorus, which can occur by enzymatic hydrolysis of ester-bound organic phosphorus without the release of carbon (biochemical mineralization), or through decomposition of organic material where carbon is released (biological mineralization). These processes, originally proposed by McGill and Cole (1981), are discussed in more detail elsewhere in this volume (see Condron and Tiessen, Chapter 13). The generalized model of Walker and Syers (1976) for soil development suggests that phosphorus release due to weathering of soil minerals is relatively rapid initially and results in the formation of organic and secondary mineral phosphorus compounds. As soil development proceeds with time, phosphorus precipitates into iron and aluminium compounds (occluded phosphorus) that are insoluble and essentially unavailable for plant growth. Phosphorus availability peaks early during the soil formation time sequence and declines with time due to leaching and erosion losses of organic phos-

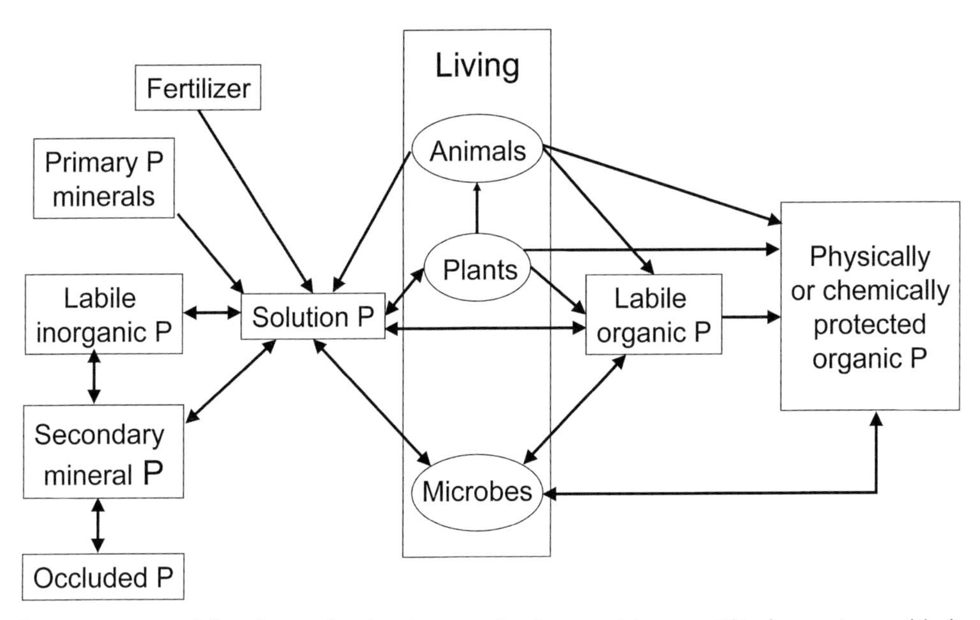

Fig. 15.1. Conceptual flow diagram for phosphorus cycling in terrestrial systems. This diagram is a modified version of a figure presented by Gijsman *et al.* (1996).

phorus compounds and stabilization of phosphorus in insoluble occluded compounds. This long-term soil weathering and soil development pattern results in plant production that is primarily limited by nitrogen in young soils and by phosphorus availability in highly weathered old soils (>1 million years). The conceptual model presented in Fig. 15.1 is the theoretical basis for most of the existing phosphorus cycling submodels (e.g. CENTURY, EPIC, GLEAMS).

The review of phosphorus models by Lewis and McGechan (2002) showed that one of the major differences in phosphorus cycling models is the number of inorganic and organic phosphorus pools and the ability to measure them in the laboratory. Models such as CENTURY have included phosphorus pools that are conceptually important for the model structure, but are difficult to measure. For example, the CENTURY model includes separate pools for parent phosphorus, strongly sorbed phosphate and occluded phosphorus, while ANIMO, GLEAMS and the phosphorus model developed by Daraub *et al.* (2003) combine these pools into a single stable phosphate pool. The CENTURY model includes these pools because they have different turnover times (30–30,000 years), which can have a large impact on phosphorus availability to the plants. Unfortunately, phosphorus fractionation techniques can only evaluate the total stable phosphate pool with any confidence (Daroub *et al.*, 2003), which complicates validation of the CENTURY model pools. The CENTURY model also identifies two stable organic phosphorus pools (turnover times of approximately 30 and 1000 years), while the other models have only a single stable organic phosphorus pool. All of the five phosphorus models are similar in terms of the other inorganic and organic phosphorus pools and include active and slow residue pools, labile and active inorganic phosphorus pools, plant uptake of phosphorus, and an active organic phosphorus pool. The controls on the flows of phosphorus among the different pools are somewhat similar in the different models and include soil water and temperature controls on decomposition, soil pH impacts on phosphorus solubility, and the impact of soil mineralogy and soil weathering state on phosphorus solubility and flows. A detailed comparison of how these variables impact the phosphorus flows for the different models is presented by Lewis and McGechan (2002). An important factor limiting the development of phosphorus cycling models is the lack of adequate data-sets to test the model performance. It would be ideal to have data-sets that show how the different phosphorus pools change in response to phosphate fertilizer, soil cultivation and crop rotations. The data-sets developed by Chadwick *et al.* (1999) and Daroub *et al.* (2003) represent the type of data needed to test phosphorus model dynamics.

This chapter will use a revised version of the CENTURY ecosystem model (Metherell *et al.*, 1993; Parton *et al.*, 1993) to simulate plant production, soil carbon, phosphorus and nitrogen dynamics, and soil nitrogen and phosphorus mineralization for the 4.1-million-year-old ecosystem development in Hawaii. This is the first application of the CENTURY model to simulate long-term ecosystem development. The modelling work will test the validity of the generalized conceptual model for long-term soil phosphorus changes (Fig. 15.1) using observed soil phosphorus data from the 4.1-million-year ecosystem development chronosequence in Hawaii (Crews *et al.*, 1995), and evaluate how phosphorus inputs from dust and loss via dissolved organic matter leaching influence phosphorus cycling and ecosystem development. The CENTURY model results will be compared with plant production, soil carbon, nitrogen and phosphorus, and foliar nutrient data-sets collected for the different soil age chronosequence sites (referred to as LSAG sites in the text) in Hawaii (Vitousek, 2004). Model runs were also set up to evaluate the ecosystem impact of adding inorganic nitrogen and phosphorus fertilizer to some of the LSAG sites in an attempt to quantify the relative importance of nitrogen and phosphorus availability on plant production for different aged soils. The observed data come from sites that developed on volcanic ash where the climate was similar. The chapter discusses how well the model simulated the observed changes in ecosystem dynamics with time, problems associated with using the model, and the necessary changes to the model structure.

CENTURY Model Description and Model Runs

Model description

The CENTURY model simulates the long-term dynamics of carbon, nitrogen and phosphorus for generalized plant–soil systems (Fig. 15.2). CENTURY is set up to simulate agricultural cropping systems, grassland systems, forest systems and savanna systems. The grass/crop and forest systems have different plant production submodels, which are linked to a common soil organic matter submodel. The Savanna version of CENTURY simulates grass/crop and forest subsystems separately and allows the systems to interact through shading effects and competition for soil nitrogen. The soil organic matter submodel simulates the flow of carbon, nitrogen and phosphorus through plant litter and the different inorganic and organic phosphorus pools in the soil. The model runs using a monthly time step (see CENTURY 4.5 version at http://www.nrel.colostate.edu/projects/century/nrel.htm) with the major input variables being monthly average daily maximum and minimum air temperature, monthly precipitation, plant lignin, nitrogen and phosphorus concentrations, soil texture, atmospheric and soil nitrogen inputs, and phosphorus inputs. The model has been tested extensively using observed plant production, soil carbon and nitrogen data, and soil nitrogen mineralization data from grasslands (Parton *et al.*, 1993), cropping systems (Paustian *et al.*, 1992, Kelly *et al.*, 1997), and forest systems (Sanford *et al.*, 1991; Kelly *et al.*, 1997; Raich *et al.*, 2000) in temperate, subtropical and tropical regions.

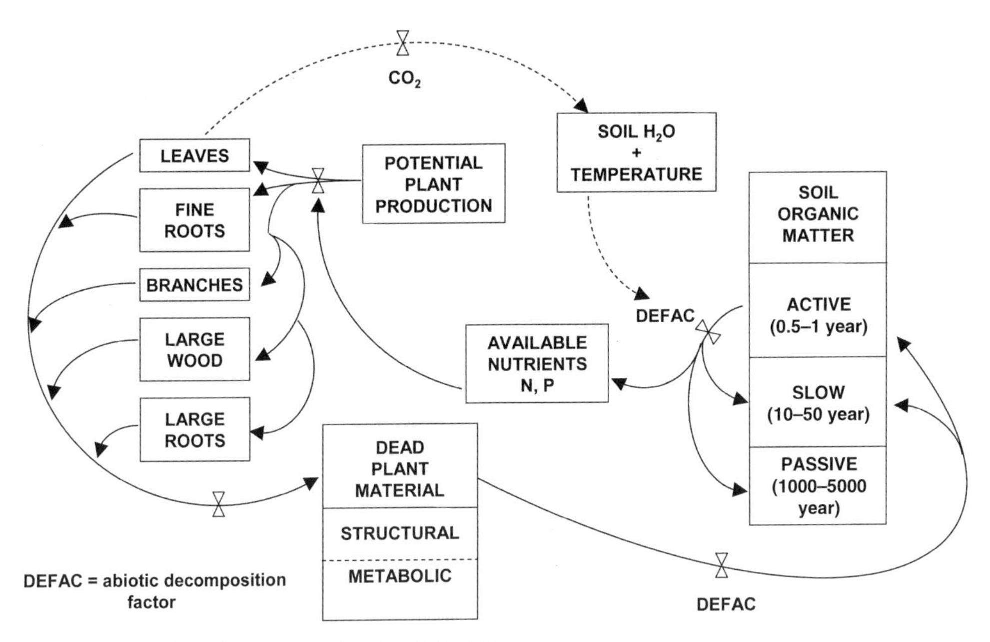

Fig. 15.2. Generalized flow diagram for the CENTURY ecosystem model.

This chapter is an extension of the Raich *et al.* (2000) use of the CENTURY model to simulate carbon, nitrogen and phosphorus dynamics during soil development in Hawaii.

The soil organic matter submodel is based on multiple compartments for soil organic matter and is similar to other soil organic matter models (Paul and van Veen, 1978). The model includes three soil organic matter pools (active, slow, passive) with different potential decomposition rates, surface microbes, and above- and below-ground litter pools. Decomposition of these pools is a function of an abiotic decomposition factor (calculated as a function of soil water and temperature) and their inherent turnover rate. The model has nitrogen and phosphorus pools analogous to all the carbon pools, with dynamic carbon-to-element ratios that change as a function of the labile mineral nitrogen and phosphorus levels (higher nutrient contents with higher mineral nitrogen and phosphorus levels). The flow of organically bound nitrogen and phosphorus between soil organic matter pools is a function of the carbon flow and the carbon-to-element ratio of the pool receiving the carbon flux. Mineralization of nitrogen and phosphorus occurs if carbon flows from soil organic matter pools with low carbon-to-element ratios, while immobilization occurs for decomposition of soil organic matter pools that have high carbon-to-element ratios.

The CENTURY model also includes a simplified water budget model that calculates monthly evaporation and transpiration water loss, water content of the soil layers, snow water content, and water flows between the soil layers. The main abiotic driver for the water budget model is the potential evapotranspiration rate, which is calculated as a function of the average monthly maximum and minimum air temperature. Near surface soil temperature is calculated as a function of maximum and minimum air temperatures, litter and standing plant biomass. The plant production model calculates potential plant production as a function of soil temperature and the ratio of actual water loss to potential water loss. Potential plant production is reduced when nutrients limit growth and the element that is most limiting for plant growth controls plant production. The nutrient content of live plant parts floats between specified minimum and maximum nutrient contents, which are a function of the ratio of plant available nitrogen or phosphorus to the maximum potential uptake of the specific nutrient. A detailed description of the CENTURY model is presented in Metherell *et al.* (1993) and Parton *et al.* (1988, 1993).

Phosphorus submodel description

The phosphorus submodel flow diagram (Fig. 15.3) shows that there are phosphorus pools for all the plant and soil organic matter compartments and five mineral phosphate pools, including labile, sorbed, strongly sorbed, occluded, and parent phosphorus. Labile phosphorus is equivalent to resin-extractable phosphorus in the sequential phosphorus fractionation scheme and is assumed to be in equilibrium with sorbed phosphorus. The equilibrium between sorbed and labile phosphorus is a function of soil texture and mineralogy, with phosphorus-fixing soil having low concentrations of labile phosphorus and higher concentrations of sorbed phosphorus. Plant uptake, immobilization into organic phosphorus, and leaching of organic phosphorus are controlled by soil labile phosphorus concentrations. Sorbed phosphorus is in dynamic equilibrium with strongly sorbed phosphorus, which can then be stabilized into insoluble occluded phosphorus compounds. Phosphorus enters the ecosystem through weathering of the parent material (typically apatite), fertilizer inputs, and atmospheric dust deposition. The equilibrium between sorbed phosphorus and strongly sorbed phosphorus is a function of soil pH. The weathering rate of parent phosphorus was specified based on observed data from Hawaii and the abiotic decomposition index (function of soil temperature and water) which also controls the flow of phosphorus between the different organic phosphorus pools. The carbon-to-phosphorus ratios for the soil organic matter fractions are variable and based on observed data from Hawaii (30–100, 100–150 and 150–300 for active, slow and passive soil organic matter pools, respectively). The main losses of phosphorus from the system are through soil erosion and leaching of dissolved organic phosphorus and labile phosphate. Two major limitations of the current phosphorus model are that the equilibrium between sorbed and labile phosphorus remains fixed during the model run and that enzymatic release of phosphate from organic phosphorus is not represented. These limitations are discussed below and illustrated by model results from the Hawaiian chronosequence sites.

CENTURY model runs

The CENTURY model was set up to simulate the long-term (4.1-million-year) soil and ecosystem development for tropical forest ecosystems in Hawaii. We simulated a site that had a mean annual temperature of 16°C and about 2500 mm of annual precipitation using the observed climate from a site near Hawaii Volcanoes National Park. The model runs were set up to simulate the observed soil and ecosystem dynamics for the six sites that had the same climate and soil parent material, but with soil ages ranging from 300 to 4,100,000 years old (Crews *et al.*, 1995; Chadwick *et al.*, 1999). The sites were all developed on a mixture of tephra, pumice and basaltic rock that had similar initial chemistry. The observed data-sets provide information about temporal changes in plant production, soil organic carbon, nitrogen and phosphorus, soil phosphate fractions, live leaf nitrogen and phosphorus concentrations, nitrogen and phosphorus losses, soil nitrogen and phosphorus mineralization, and soil respiration rates (Vitousek, 2004) along the 4.1-million-year chronosequence. The simulated model results were compared with the observed data to test how well the model represented the long-term changes in soil and ecosystem properties. In the following sections we highlight the comparison of the observed and simulated 4.1-million-year changes in organic and inorganic phosphorus, phosphorus losses, and soil phosphorus mineralization, and suggest changes to the phosphorus submodel that would enhance

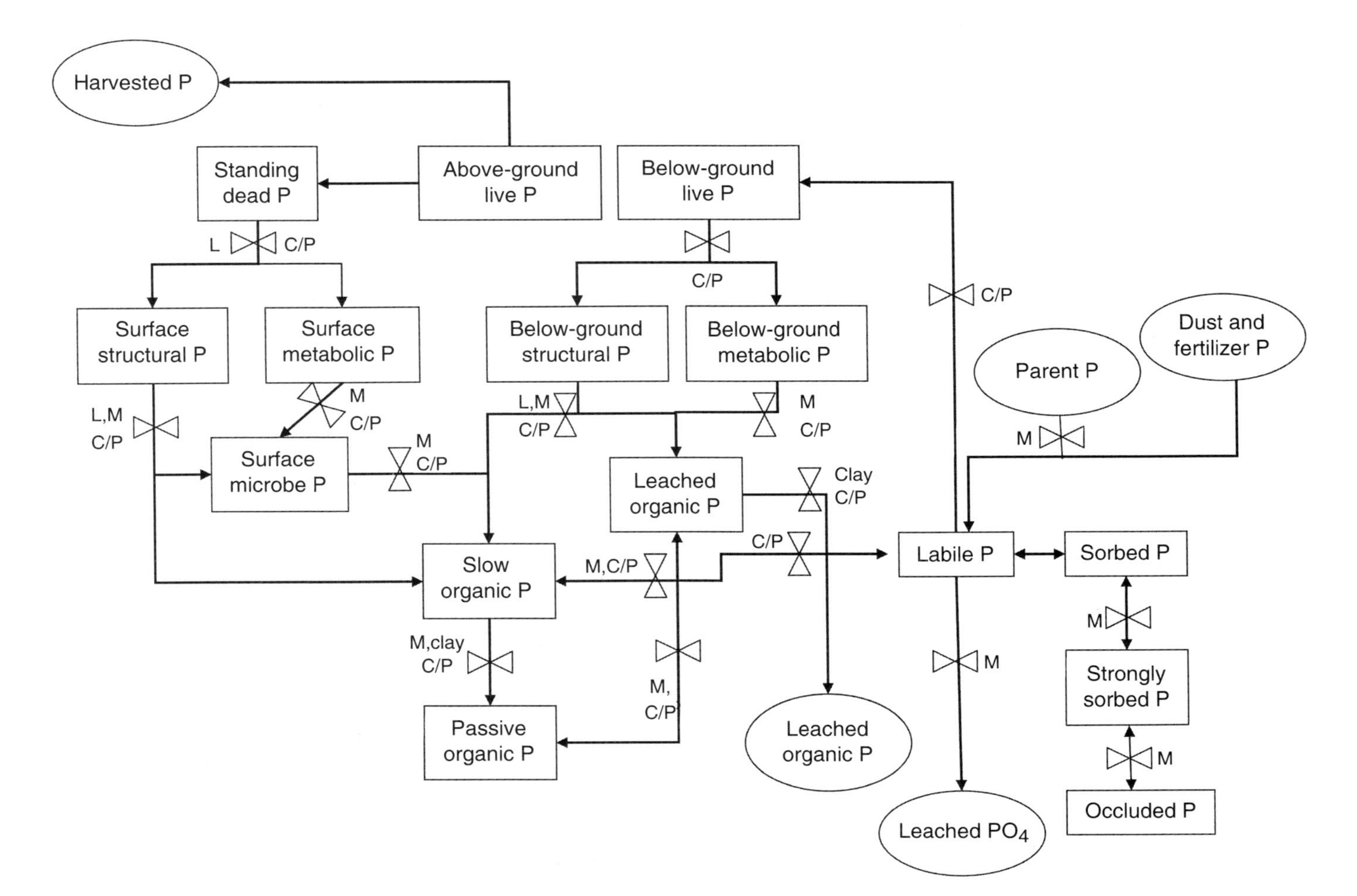

Fig. 15.3. Flow diagram for the phosphorus cycling subroutine in the CENTURY model. M = multiplier for effects of moisture and temperature; L = lignin/N.

the performance of the model and our understanding about phosphorus cycling in tropical systems.

Model runs were set up to evaluate the relative importance of nitrogen and phosphorus availability in controlling plant production for soils of different ages. We simulated the soil fertilizer experiments described by Harrington *et al.* (2001), which examined the separate effects of adding nitrogen (10 g N/m^2/year) and phosphorus (10 g P/m^2/year) fertilizer and the combined impact of both nitrogen and phosphate fertilizer (10 g/m^2/year of each). The model was run for 300, 2000, 20,000, 150,000, 250,000 and 1,400,000 years, and then fertilizer treatments (nitrogen, phosphorus, and nitrogen plus phosphorus) were added for 30 years. Simulated responses of plant production to the different fertilizer treatments for the 1–5-year treatment period were then compared with the observed plant production data (Vitousek *et al.*, 1993; Harrington *et al.*, 2001).

Model simulations were set up to simulate the impact of changing dust phosphorus inputs from Central Asia on soil and ecosystem dynamics. Constant dust phosphorus inputs were assumed for three model runs (phosphorus inputs of 0.020, 0.025 and 0.030 kg P/ha/year) with plant production, and soil organic carbon and organic phosphorus levels simulated for a 400,000-year run (most of the model responses equilibrated after 400,000 years). Model runs also evaluated the impact of changing dissolved organic matter loss rates (+ and −30% change in dissolved organic matter leaching rates, with carbon:nitrogen:phosphorus ratios dynamically determined by the model) and assuming constant dust phosphorus inputs (0.025 kg P/ha/year). These sensitivity analysis runs assumed that weathering of phosphorus from the initial volcanic ash material is the only other source of phosphorus in the system.

Model parameterization and input data

We used the mean annual climate data (mean monthly maximum and minimum air temperature and monthly precipitation) from the Thurston site, which is located in Hawaii Volcano National Park (Crews *et al.*, 1995). The simulated potential evapotranspiration rates were adjusted to match the observed data (Raich *et al.*, 2000). The soil classification was Typic Hydrandept (30% sand, 30% silt, 40% clay). The soil pH data from Crews *et al.* (1995) were assumed to change with time (pH equal to 5.4 before 2000 years, dropping linearly to 3.8 at 20,000 years and equal to 3.8 for the remainder of the run) in accordance with observed data. The annual input of nitrogen from the atmosphere was 10.2 kg N/ha for the first 2000 years and 8.0 kg N/ha for the remainder of the computer run (based on data modified by Vitousek, 2004). The two sources of phosphorus to the system were dust deposition and weathering of volcanic rock under the ash layer. The assumed total phosphorus inputs from weathering of rocks and atmospheric dust change with time (0.16, 0.10, 0.06, 0.04, 0.025, 0.02 and 0.018 kg P/ha, respectively, from 2000–40,000, 40,000–90,000, 90,000–150,000, 150,000–250,000, 250,000–350,000 and >350,000 years) based on the data presented by Vitousek (2004). Phosphorus inputs to the system are added to the labile phosphorus pool. Dust added to the system probably contains various organic and inorganic phosphorus compounds, but we assume that phosphorus release from dust is rapid enough to ensure that dust phosphorus does not accumulate in the system (this assumption is likely to be valid given the 4.1-million-year time scale). The soil rooting depth was assumed to be 60 cm and the model was set up to simulate the dynamics of soil carbon, nitrogen and phosphorus in the surface 50 cm soil layer (measurement depth for the observed soil carbon, nitrogen and phosphorus data). Another key assumption in the model runs is that soil erosion rates are minimal.

The model parameters for tree growth were primarily derived from Raich *et al.* (2000) and slightly modified based on plant nutrient content and biomass data from fertilizer experiments in Hawaii (Vitousek *et al.*, 1993; Vitousek and Farrington, 1997;

Table 15.1. Changes in observed ecosystem properties during soil development chronologically in Hawaii. The observed data are derived from results published by Crews *et al.* (1995), Vitousek (2004) and Hedin *et al.* (2003).

	Time (thousands of years)					
	0.3	2	20	150	1400	4100
Soil carbon (kg/m^2)	15.3	14.6	32.4	33.6	28.0	24.1
Soil respiration (g C/m^2/year)	798	–	908.0	1192.0	1004	896
Soil nitrogen (kg/m^2)	0.98	0.98	1.56	1.46	1.38	1.13
Resin-extractable nitrate and ammonium (mg/bag/day)	3.31	–	12.37	5.21	14.55	14.31
Organic phosphorus (kg/m^2)	0.036	0.095	0.155	0.202	0.106	0.112
Occluded phosphorus (kg/m^2)	0.025	0.04	0.10	0.180	0.20	0.21
Secondary phosphorus (kg/m^2)	0.10	0.10	0.110	0.20	0.050	0.10
Resin-extractable phosphorus (mg/bag/day)	0.20	–	1.21	2.19	0.51	0.41
Leaf carbon-to-phosphorus ratio	700	428	412	382	494	700
Leaf carbon-to-nitrogen ratio	48	38.2	30.0	37.0	40.0	89
Max tree canopy height (m)	16.5	20.0	24.7	11.6	8.2	13.7
Gaseous nitrogen loss (kg/ha/year)	0.29	0.20	2.0	0.36	1.80	3.0
Nitrate loss (kg/ha/year)	0.10	0.08	4.0	3.0	6.0	6.2
Dissolved organic nitrogen loss (kg/ha/year)	2.0	1.5	2.0	2.5	2.7	3.0
Dissolved organic phosphorus loss (kg/ha/year)	0.015	0.015	0.020	0.014	0.014	0.015
Phosphate loss (kg/ha/year)	0.065	0.090	0.010	0.010	0.014	0.024

Harrington *et al.*, 2001). Initial values for volcanic ash soil parent phosphorus concentrations, weathering rate of parent phosphorus, and soil carbon-to-nitrogen and carbon-to-phosphorus ratios were based on the observed data (Crews *et al.*, 1995). Other model parameters that were changed include the decay rate for passive soil organic matter, the effect of inorganic nitrogen on soil phosphatase activity, loss of dissolved phosphate and the carbon-to-nitrogen and carbon-to-phosphorus ratios for leached dissolved organic matter. These ratios for dissolved organic matter were assumed to be equal to 3.5 times the carbon-to-phosphorus ratio and twice the carbon-to-nitrogen ratio of the active soil organic matter (these ratios vary dynamically in the model) based on observed dissolved organic phosphorus and dissolved organic nitrogen data from soils in Hawaii (Vitousek, 2004). The decay rate of the passive soil organic matter was adjusted (reduced by 50%) to simulate the 0–50 cm soil depth (normal CENTURY runs simulate soil organic matter dynamics in the surface 20 cm of soil). This is the standard CENTURY model correction used to simulate deeper soil depths (Metherell *et al.*, 1993). The solubility of phosphate was altered to match the observed phosphate losses during the first 10,000 years of soil development. The other important model change was to add a back-flow of phosphorus from the occluded phosphorus pool to the strongly sorbed phosphorus pool (see Fig. 15.3). This flow was added because preliminary model runs showed that occluded phosphorus pools would continue to increase with time, while the observed data show that occluded phosphorus levels stabilize after 150,000 years and then increase only slightly with time. The model parameters that control the flow of strongly sorbed phosphorus to occluded phosphorus, and occluded phosphorus to strongly sorbed phosphorus, were adjusted to match the observed occluded phosphorus data (see Table 15.1; Crews *et al.*, 1995). The model assumes that mineral nitrogen levels impact the availability of phosphorus for plant growth. Adding inorganic nitrogen fertilizer was shown to increase soil phosphatase activity and reduce phosphorus sorption (Olander and Vitousek, 2000, 2004). Therefore, we assumed that adding nitrogen fertilizer increases phosphorus availability to plants, and used the observed increase in plant phosphorus uptake resulting from adding nitrogen fertilizer to the 4.1-million-year site to parameterize the model (a higher fraction of labile phosphorus is available for plant growth with high mineral nitrogen levels).

Model results

There are three major modelling activities presented in the results section. The first activity was to simulate long-term soil development in Hawaii and compare the model results with the observed soil and plant production data (see Table 15.1). The second activity was to simulate soil fertilizer experiments conducted at several of the LSAG sites (Vitousek *et al.*, 1993; Herbert and Fownes, 1995; Vitousek and Farrington 1997; Harrington *et al.*, 2001). The model runs simulate the impact of adding nitrogen and phosphorus fertilizer on plant production and evaluate the relative impact of nitrogen and phosphorus limitation on plant growth as a function of soil age. The third activity was to simulate the impact of changing phosphorus dust deposition rates (observed historical rates range from 0.003 to 0.025 kg P/ha/year) and dissolved organic matter loss rates. The runs demonstrate the sensitivity of tropical forest to dust phosphorus deposition rates and the loss of phosphorus and nitrogen in dissolved organic matter.

Figures 15.4, 15.5 and 15.6 show the simulated changes in the soil–plant system during 4.1-million-year soil development. The model results are compared with observed data from the LSAG sites with a focus on plant production, inorganic phosphorus fractions, soil organic carbon, nitrogen and phosphorus, nutrient mineralization rates, soil respiration, nitrogen and phosphorus losses, and live leaf nutrient ratios. A key assumption in the model run is that climate remained constant during the 4.1-million-year simulation; in fact the older sites were probably drier for much of their history (Hotchkiss *et al.*, 2000). Consequently an assumption of constant climate probably overstates the actual rate of soil and ecosystem development. Model results show that soil carbon, soil respiration and plant production (Fig. 15.4a,b) all follow the same pattern of increasing values during the first 10,000 years of soil development, stable values from 10,000 to 150,000 years, decreasing values from 150,000 to 700,000 years, and stable values from 700,000 to 4,100,000 years. This simulated pattern is mostly supported by the observed data, although there is a considerable amount of uncertainty in the plant production data (Vitousek, 2004), which makes it difficult to test model predictions. The rapid increase in plant production during the first 10,000 years was most probably caused by the increases in nitrogen and phosphorus soil mineralization (Fig. 15.4c). The decline in plant production from 150,000 to 700,000 years and stabilization at lower levels is a result of the decrease in soil phosphorus mineralization (see Fig. 15.4c) from 20,000 to 700,000 years and stabilization of phosphorus mineralization after 700,000 years. These results reflect the general pattern of nitrogen limitation of plant growth during early soil development and phosphorus limitation on old soils (Vitousek, 2004), which will be discussed in more detail below. The observed plant production, soil respiration and soil carbon data in Fig. 15.4 suggest that peak plant production occurs at 150,000 years, while model results suggest that the peak plant production occurs between 10,000 and 150,000 years. The maximum canopy height data (Table 15.1) suggest that maximum plant production could occur at 20,000 years.

The simulated results for soil organic phosphorus and nitrogen (Fig. 15.5a) show that these fractions accumulate rapidly during the first 20,000–30,000 years and then decrease to stable equilibrium levels after 700,000 years. The drop in organic phosphorus concentrations is much larger than the drop in organic nitrogen concentrations from 80,000 to 700,000 years. The observed data (Fig. 15.5a and Table 15.1)

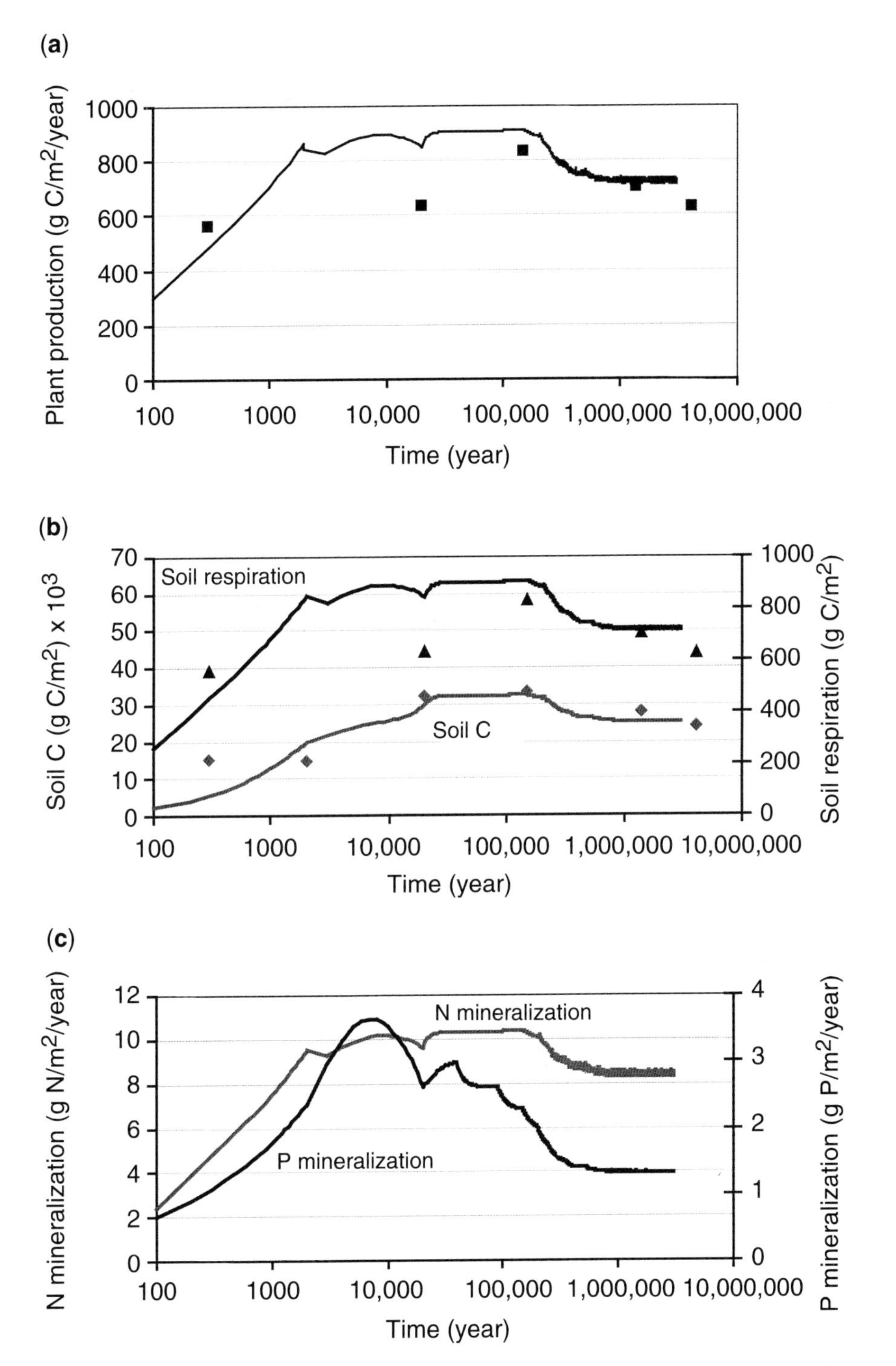

Fig. 15.4. Simulated plant production (a), soil carbon (0–50 cm depth) and soil respiration (b), and nitrogen and phosphorus soil mineralization (c) for Hawaii humid tropical forest systems during 4.1 million years of soil development. Observed data are plotted on the graphs, and observed net primary production is assumed to be equal to 0.7 times annual soil respiration.

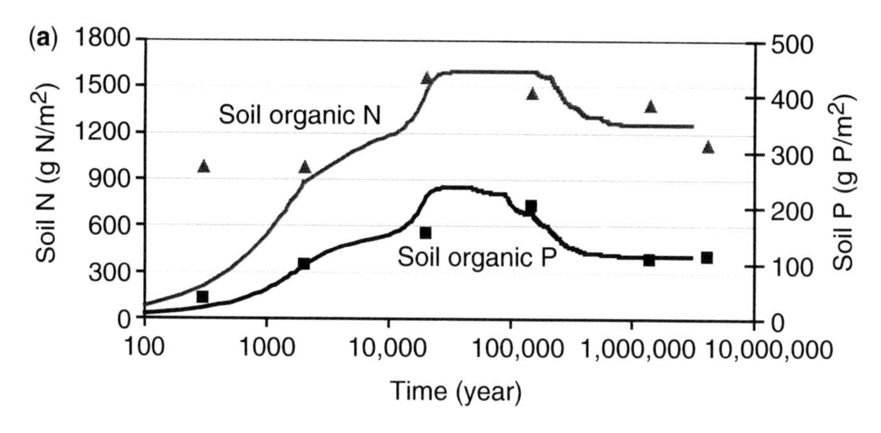

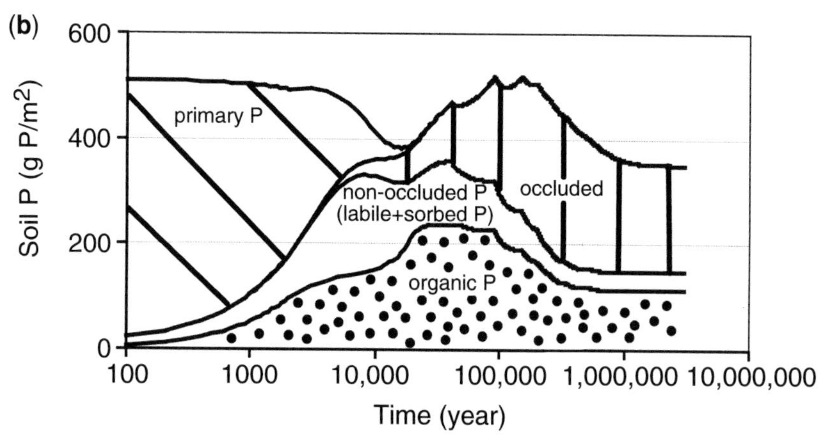

Fig. 15.5. CENTURY model-simulated patterns for soil organic nitrogen and phosphorus (a) and occluded phosphorus, parent phosphorus, non-occluded mineral phosphorus (labile plus sorbed phosphorus) and organic phosphorus (b) during 4.1 million years of soil development (0–50 cm soil depth). Observed data are plotted on the figure for organic nitrogen and phosphorus.

follow the same pattern, with the major exception that peak organic phosphorus levels occur at 20,000 years in the model, but at 150,000 years for the observed data.

The simulated change in the different soil phosphorus fractions (occluded, non-occluded mineral (labile plus sorbed), primary and organic) with time (Fig. 15.5b) show that volcanic ash parent phosphorus weathers quite rapidly during the first 20,000 years, with an accumulation of organic phosphorus, and non-occluded mineral phosphorus. Occluded phosphorus increases rapidly from 10,000 to 200,000 years and then slowly reaches a new equilibrium value. Both organic phosphorus and non-occluded mineral phosphorus decrease rapidly from 60,000 to 500,000 years and reach stable equilibrium levels after 700,000 years. Total soil phosphorus decreases during the first 20,000 years, increases from 20,000 to 150,000 years and then decreases to a new equilibrium after 700,000 years. The initial loss of total phosphorus during the first 20,000 years occurs when volcanic ash soils weather rapidly and there are large leaching losses of phosphate (see Fig. 15.6). The increase in total soil phosphorus from 20,000 to 150,000 years results from the weathering of phosphorus from underling volcanic rock material.

Model results are consistent with the general model of Walker and Syers (1976) for soil phosphorus development. However,

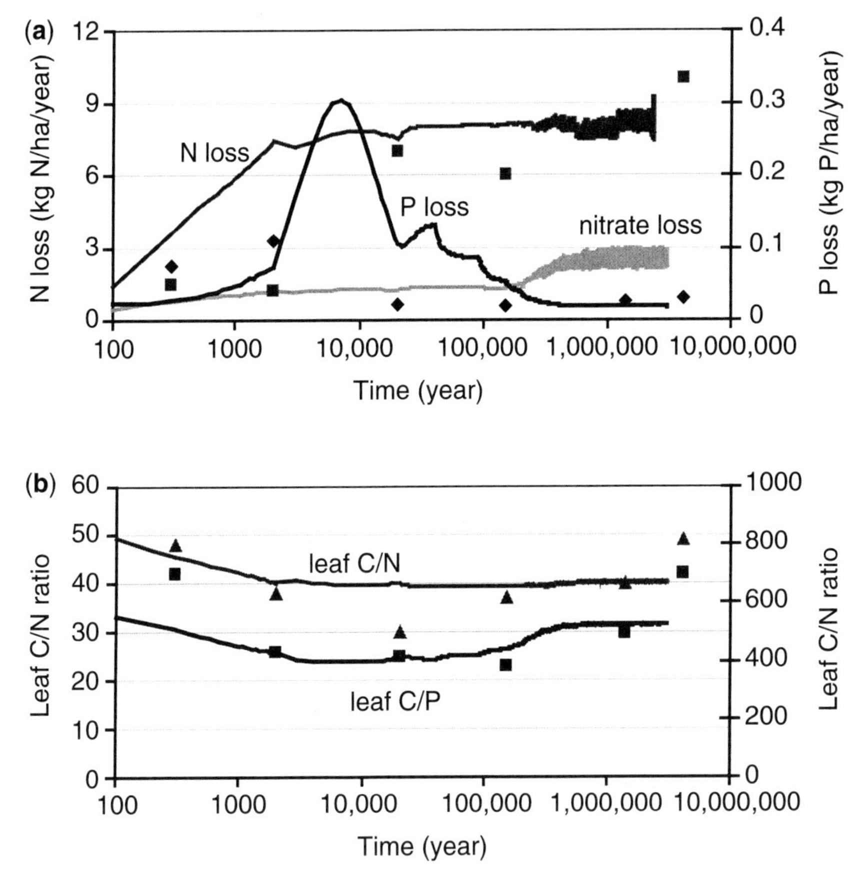

Fig. 15.6. CENTURY model-simulated results for soil phosphorus loss (organic and inorganic phosphorus) total nitrogen loss (nitrate, gaseous nitrogen and dissolved organic nitrogen) and nitrate loss (a) and change in live leaf carbon-to-nitrogen and carbon-to-phosphorus ratios (b) for the Hawaiian 4.1 million year soil chronosequence.

model results show that stable equilibrium levels of phosphorus in the different soil fractions are reached after 700,000 years, while the Walker and Syers (1976) model has total phosphorus and the different phosphorus fractions continuing to decrease during soil development. Comparison of the CENTURY model results with the observed Hawaii soil phosphorus data (Table 15.1) show that they follow the same pattern of changes in total soil phosphorus and the different fractions during soil development. Both the observed data and model results suggest that soil phosphorus pools reach new equilibrium levels after 700,000 years, which is inconsistent with the Walker and Syers (1976) general conceptual model showing total soil phosphorus decreasing during soil development. The most likely reasons for this discrepancy are that: (i) the Walker and Syers (1976) model does not account for the inputs of phosphorus from dust and other sources and (ii) erosion rates were substantial for the site represented by Walker and Syers (1976), but were assumed to be minimal for Hawaii.

The CENTURY model simulates losses of nitrogen via gas fluxes (N_2, N_2O and NO_x), dissolved organic loss, and nitrate leaching loss, while phosphorus is lost via dissolved organic and inorganic phosphorus. Model results (Fig. 15.5b) show that total phosphorus loss is quite high during the first 10,000 years of soil development and then decreases rapidly from 10,000 to 150,000 years, with a stable equilibrium reached after 150,000 years of soil development. During the first 10,000 years, simulated soluble phosphate is the dominant phosphorus loss (>80% of the total loss), while after 100,000 years 76% of the total phosphorus loss occurs as dissolved organic phosphorus loss (Table 15.1). The simulated general pattern in total phosphorus loss data and high initial inorganic phosphorus loss (<10,000 years) followed by low losses after 150,000 years are verified by the observed data (Hedin *et al.*, 2003; Fig. 15.5a; Table 15.1). It is important to note that both the model and the observed data show patterns of high phosphorus loss during the first 10,000 years of soil development, compared with the assumed inputs of phosphorus from dust (six times higher than phosphorus input rates), while the phosphorus loss rates decrease to the assumed dust phosphorus input rate (0.018 kg P/ha/year) after 700,000 years of soil development. Clearly these large losses of phosphorus during the first 20,000 years of soil development contributed to total system phosphorus decreases during this time period and the onset of phosphorus limitation for plant growth after 150,000–300,000 years of soil development.

Initially, the simulated total nitrogen loss is low and then increases to a steady state after 20,000 years. Simulated dissolved organic matter and nitrogen gas fluxes are the dominant nitrogen losses during the first 20,000 years, while nitrate loss increases after 150,000 years when phosphorus limitation of plant growth becomes important (see previous discussion). The observed data (Table 15.1 and Fig. 15.5) support the general trend for increasing total nitrogen loss with time and the increase in nitrate leaching loss after 20,000 years of soil development. The model tends to overestimate observed nitrogen gas fluxes during the whole model run, with greater overestimation of nitrogen gas loss during the first 20,000 years of soil development. The observed data also show that nitrate leaching loss increases rapidly between 2000 and 20,000 years, while the model results show nitrate leaching increasing rapidly after 100,000 years of soil development.

Leaf nutrient ratios can be used as an index of the availability of nitrogen and phosphorus for plant growth, with lower carbon-to-element ratios indicating a greater availability of nutrients for plant growth. The model results (Fig. 15.6b) show that leaf carbon-to-nitrogen and carbon-to-phosphorus ratios are high initially, decrease to minimum values around 20,000 years, and then start to increase with time. The observed data (Fig. 15.6b) generally support the model results for carbon-to-nitrogen ratios; however, the lowest leaf carbon-to-phosphorus ratio occurs at 150,000 years, while the model results have the lowest carbon-to-phosphorus ratios occurring at 20,000 years. The significance of this difference is unclear, since the observed soil phosphorus level for the 150,000-year site is quite high in phosphorus compared to the initial, 20,000 and 1,400,000 year phosphorus levels (see Table 15.1).

Model runs were set up to simulate nitrogen and phosphorus fertilization for all of the LSAG sites (Fig. 15.7) and compared with observed data for the sites where fertilizer response data was collected (3000, 20,000 and 4,100,000 year sets). Results show that adding nitrogen to sites with soil ages up to 150,000 years results in substantial increases in plant production, while adding phosphorus has little impact. The 300-year site shows that adding nitrogen plus phosphorus results in substantial increases in plant production compared with the nitrogen-alone fertilizer run and a slight response to adding both nitrogen and phosphorus fertilizer for the 2000-year site. Results at 150,000 years show a positive response to adding nitrogen fertilizer, no response to adding phosphorus fertilizer, and a slight increase associated with adding both nitrogen and phosphorus together. The simulated 250,000-year-old site (not an LSAG site) shows slight increases in plant production associated with adding either nitrogen or phosphorus and a large response to adding nitrogen and phosphorus together. The 1,400,000 and 4,100,000-year-old sites have slight increases in net primary productivity for nitrogen fertilizer addition, a substantial increase for the phosphorus fertilizer addition and a large increase in net primary productivity associated with the nitrogen plus phosphorus fertilizer addition (model results are identical for the 1,400,000 and 4,100,000-year-old sites). The model results (not shown) show that increases in net primary productivity due to phosphorus fertilizer are highest immediately after application and then decrease with time. There is a general pattern of substantial increases in net primary productivity associated with nitrogen fertilizer for soils that are ≤150,000 years old, and minimal response to phosphate fertilizer. The 250,000-year-old site appears to be equally limited by both nitrogen and phosphorus, while the 1,400,000 and 4,100,000-year-old sites are primarily limited by phosphorus.

These results are consistent with the observed data from fertilizer experiments (Vitousek *et al.*, 1993; Vitousek and Farrington, 1997; Harrington *et al.*, 2001) for young soil development sites where nitrogen is the primary limiting nutrient. However, the model tends to overestimate the positive impact of adding nitrogen fertilizer to the 20,000-year site and to underestimate the increase in plant production associated with adding both nitrogen and phosphorus fertilizer to the 20,000-year site. Analysis of model results suggests that CENTURY overestimates phosphorus availability for plant growth at the 20,000-year site. The observed fertilizer responses for the 4,100,000-year-old site show that the model underestimates the increases associated with phosphate fertilizer addition and overestimates the positive impact of adding nitrogen and phosphorus on plant production. The model also appears to underestimate nitrogen availability for plant growth when phosphate fertilizer is added to the 4,100,000-year-old site. The observed fertilizer response data (Harrington *et al.*, 2001) suggest that nitrogen uptake has to be increased by at least 1 g N/m^2/year when phosphate fertilizer is added. Model results show that nitrate leaching is reduced when phosphate fertilizer is added, but the reduction is not large enough to support the

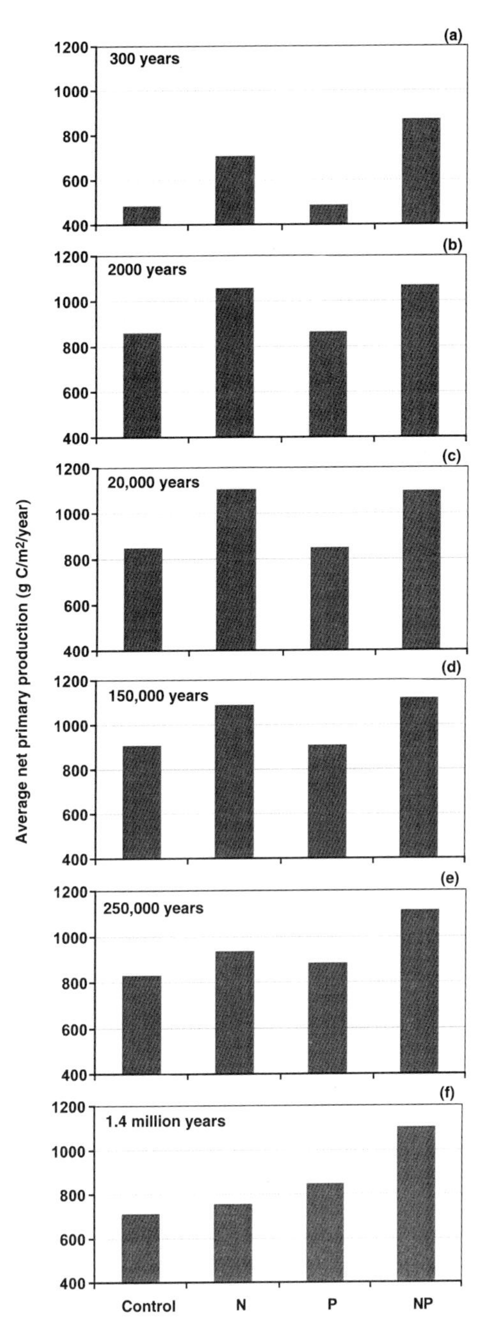

Fig. 15.7. CENTURY model results for the impact of adding nitrogen and phosphorus fertilizer on tropical soils after 300 years (a), 2000 years (b), 20,000 years (c), 150,000 years (d), 250,000 years (e), and 1.4/4.1 million years (f) of soil development.

observed increase in nitrogen uptake. Model experimental runs (not shown) suggest that the CENTURY model would correctly simulate the observed response to phosphate fertilizer addition if nitrogen inputs to the system were increased by 1–2 g N/m^2/year.

Preliminary sensitivity analysis suggests that the model is sensitive to changes in phosphorus dust deposition rates and dissolved organic matter loss. Model runs simulated the impact of increasing (+0.005 g P/m^2/year) and decreasing (−0.005 g P/m^2/year) dust phosphorus (loose dust phosphorus deposition of 0.025 g P/m^2/year) on plant production and soil organic carbon and phosphorus. The results (Fig. 15.8) show that plant production, soil organic carbon and phosphorus are not sensitive to phosphorus inputs during the first 30,000 years of soil development, because nitrogen limits plant growth during this period. However, the equilibrium levels of plant production and soil carbon and phosphorus after 150,000 years of soil development are quite sensitive to phosphorus inputs, with equilibrium levels of all of these variables being directly proportional to phosphorus input rates (higher plant production and soil carbon and phosphorus with higher phosphorus inputs). All of the model runs show the same pattern of decreasing plant production and decreasing soil carbon and phosphorus, from the peak values at 10,000–30,000 years, reaching different equilibrium levels after 150,000 years. These results suggest a very strong sensitivity of tropical forest ecosystems to the level of phosphorus inputs from dust.

Previous studies suggest that the losses of nitrogen and phosphorus associated with dissolved organic matter are important factors controlling plant production in humid tropical ecosystems (Raich *et al.*, 2000). We evaluated the sensitivity of Hawaiian tropical forest systems to dissolved organic matter losses by changing the loss rates by ±30% from the control run. These runs were designed to simulate the impact of increasing or decreasing the amount of water leaching through humid tropical forest systems. Previous modelling results from Raich *et al.* (2000) suggest that increasing dissolved organic matter loss and the associated phosphorus and nitrogen loss results in decreased plant production. The model results (Fig. 15.9) show that plant production and soil carbon and phosphorus are all reduced by increasing dissolved organic matter loss and vice versa. These results differ from the dust phosphorus sensitivity runs, because peak levels of plant production and concentrations of soil carbon and phosphorus (occurring between 10,000 and 30,000 years) are altered by changing dissolved organic matter loss, while altering phosphorus dust inputs has little impact on peak values of plant production and soil carbon and phosphorus. The likely reason for these results is that changing dissolved organic matter loss alters its loss from the system during the 10,000–150,000 year time period, when nitrogen limits plant growth and phosphorus availability is high. Simulated nitrate loss during this time period is fairly low.

Discussion

One of the major objectives of this chapter was to evaluate how well the CENTURY model could simulate the long-term (4.1-million-year) changes in plant production and soil carbon, nitrogen and phosphorus for tropical forests in Hawaii. Model results show that soil carbon, nitrogen and phosphorus, plant production, and nitrogen and phosphorus soil mineralization increase rapidly with soil development and reach maximum values after 20,000–50,000 years of soil formation. Most of the phosphorus in the volcanic ash soil parent material has been weathered out of the soil during this time period and transformed into organic and inorganic material with associated high phosphorus loss rates and phosphorus availability for plant growth. Model results from 150,000 to 600,000 years of soil formation show that plant production, soil organic carbon, nitrogen and phosphorus, nitrogen and phosphorus soil loss, and soil nitrogen and phosphorus mineralization all decrease, while the formation of occluded phosphorus increases. After 600,000 years

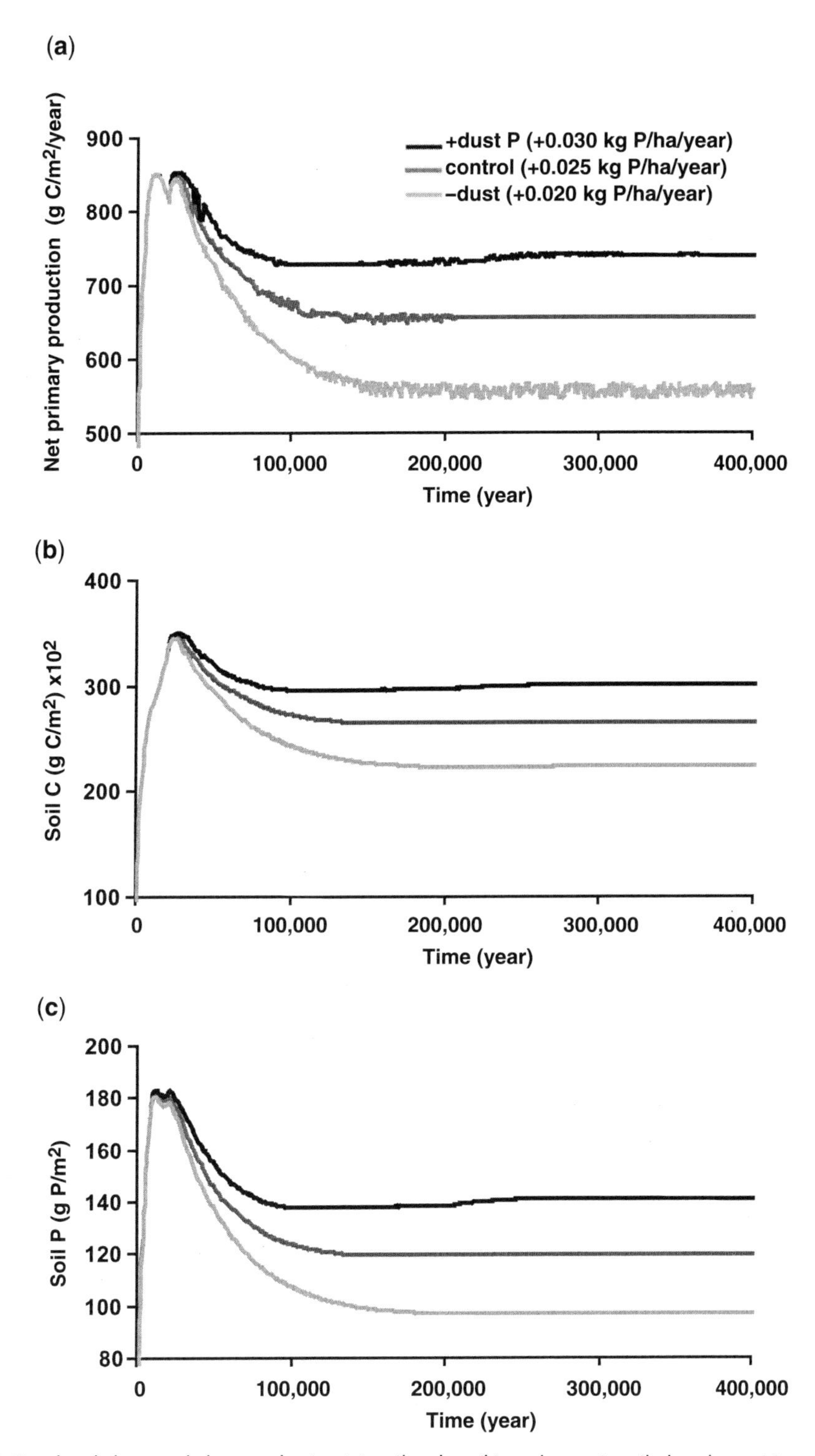

Fig. 15.8. Simulated change of plant production (a), soil carbon (b), and organic soil phosphorus (c) resulting from altered dust phosphorus deposition rates.

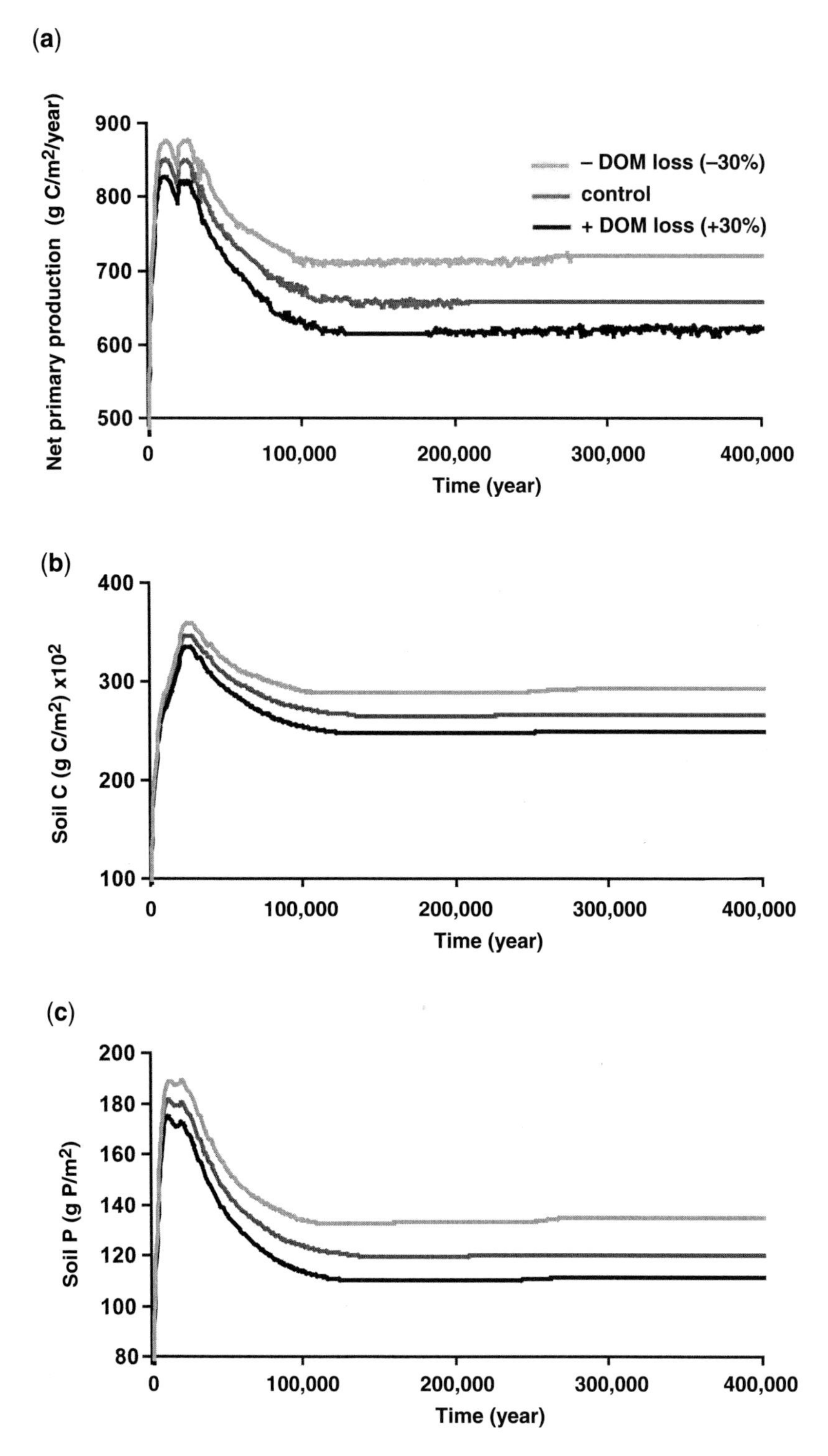

Fig. 15.9. Simulated response of plant production (a), soil carbon (b), and soil organic phosphorus (c) to an increase (+30%) and decrease (−30%) in dissolved organic matter (DOM) loss.

of soil development, near-equilibrium levels are simulated for ecosystem variables. These simulated patterns are mostly consistent with observed data (see Table 15.1), although some of the observed data suggest that peak plant production and soil carbon and phosphorus levels occur around 150,000 years, while other data show peak values at 20,000 years. The observed data that support the latter peak in plant production are soil carbon and phosphorus, soil respiration, resin-extractable phosphorus, and live leaf carbon-to-phosphorus ratio; while data that support maximum production around 20,000 years are patterns of nitrate leaching, soil nitrogen, resin-extractable nitrogen, live leaf carbon-to-nitrogen ratio, and maximum tree height.

The simulated impact of adding nitrogen and phosphorus fertilizer on plant production show that the young sites (<200,000 years) are primarily nitrogen-limited, have no response to phosphorus fertilizer and achieve maximum plant production by combining nitrogen and phosphorus fertilizer. The simulated results for the 250,000-year site (not an LSAG site) show that plant production responds equally to nitrogen and phosphorus fertilizer and exhibits a large increase in response to combined nitrogen and phosphorus addition. Results from the old sites (>1 million years) show that phosphorus is the primary limiting nutrient for plant growth. Comparison of the model results with the observed response of plant production to fertilizer additions shows that the model results are generally compatible with the observed data, although the model tends to overestimate the positive impact of adding nitrogen to the younger sites (<200,000 years) and to underestimate the positive impact of adding phosphorus to the older sites (>1 million years). The most likely reason for model overestimation of nitrogen fertilizer response for the young sites is that phosphorus availability for plant growth is overestimated (phosphorus solubility is too high), which allows for larger responses to nitrogen fertilizer. Model results for the older sites suggest that nitrogen availability for plant growth is underestimated when phosphate fertilizer is added, and that nitrogen inputs need to be increased substantially to match the observed increases (>1 g N/m^2/year) in nitrogen uptake when phosphorus is added to the old site. Model results show that nitrate leaching was reduced by adding phosphate fertilizer, but the reduction in nitrate loss is insufficient to match the observed increase in plant nitrogen uptake.

Model sensitivity analysis confirms previous modelling studies (Raich *et al.*, 2000) indicating the importance of dissolved organic phosphorus and nitrogen losses in controlling plant production in humid tropical systems (lower production with higher dissolved organic matter and phosphorus loss). This suggests that phosphorus input from dust deposition is the primary factor controlling the equilibrium levels of plant production and soil carbon, nitrogen and phosphorus levels for well-weathered humid tropical systems (higher dust phosphorus inputs leading to greater plant production, soil carbon, nitrogen and phosphorus). The results from Raich *et al.* (2000) and this study suggest that increasing precipitation in humid tropical systems would reduce plant production, while reducing precipitation could increase plant production. Similarly, changes in phosphorus inputs from dust deposition are positively correlated with changes in plant production. All of these results are consistent with those of Hedin *et al.* (2003), which showed that dissolved organic phosphorus loss is an important mechanism contributing to phosphorus limitation for humid tropical forests with low phosphorus inputs.

From a conceptual standpoint, the phosphorus cycle would be relatively simple to understand and model if phosphorus sorption isotherms could simply be added to ecosystem models, parameterized using the abundance of phosphorus sorption data from the scientific literature, and allowed to run alongside the carbon and nitrogen submodels that are so familiar to the ecological community. Unfortunately, there are two central issues that make this approach untenable. First, phosphorus sorption dynamics are not straightforward, and simple representations fail to capture the complexity of sorption/precipitation dynamics (Huffman, 1996). Secondly, if soil chemical processes operated at laboratory determined rates in tropical ecosystems, there would not be sufficient phosphorus to support the observed rates of tropical productivity (Silver *et al.*, 2000). The apparent simulated overestimate of phosphorus availability for the 20,000-year-old soil (see discussion of fertilizer experiment results) is probably due to an incorrect parameterization of the phosphorus sorption isotherm. It is also unlikely that the model assumption of a constant phosphorus sorption isotherm during the whole 4-million-year period is correct (Olander and Vitousek, 2004). Clearly we need to develop a model that captures the dynamic changes in the phosphorus sorption isotherm during the 4.1-million-year simulation.

A further limitation of the CENTURY phosphorus model is the lack of an explicit representation of the impact of soil enzymes on the mineralization of organic phosphorus compounds. The results presented in this book indicate that there is limited data with which to quantify the relative importance of enzymatic hydrolysis of organic phosphorus (*biochemical mineralization*) vs. organic phosphorus mineralization associated with the decomposition of organic compounds (*biological mineralization*). It is very difficult to develop models that include enzymatic mineralization of organic phosphorus without data to quantify the relative importance of this process.

A central and long-standing conundrum of tropical biogeochemistry is the existence of high rates of terrestrial productivity on highly weathered soils that can fix large amounts of phosphorus. The obvious answer is that biological systems in subtropical and tropical ecosystems compete effectively for available phosphorus (and thus limit occlusion of phosphorus) and/or that occluded phosphorus is not necessarily unavailable on the time scale of ecosystem development. In order to model the time scales of ecosystem development with the CENTURY model, we had to modify the representation of phosphorus occlusion so that there is a reverse flow of occluded phosphorus to strongly sorbed phosphorus. The observed soil phosphorus data from Hawaii show that the amount of occluded phosphorus is fairly stable after 150,000 years and then increases slowly with time. The only way the model could represent the observed occluded phosphorus dynamics was to add the back-flow of phosphorus into the strongly sorbed phosphorus pool and thus allow for near-equilibrium values of occluded phosphorus to be established during soil development.

An important issue in matching phosphorus models with the available data involves model and data comparison. A variety of chemical extraction techniques (Hedley and Stewart, 1982; Tiessen and Stewart, 1983; Tiessen and Moir, 1993) have been developed to extract the different inorganic and organic phosphorus pools identified in Fig. 15.1. Unfortunately, these techniques do not precisely measure the phosphorus pools identified in phosphorus cycling models and data only exist for a limited number of sites in temperate and tropical systems around the world. Gijsman *et al.* (1996) critically reviewed the structure of the CENTURY phosphorus cycling submodel and identified the major factors limiting our understanding about phosphorus cycling and phosphorus model development in terrestrial systems. These included:

1. Measurement of inorganic and organic fractions used in phosphorus cycling models.
2. Measurement of phosphorus transfer rates among the different phosphorus pools.
3. Data that would allow the quantification of phosphorus mineralization from soil organic phosphorus pools (relative importance of biological vs. biochemical mineralization).
4. Quantification of solution phosphorus concentrations in strongly phosphorus-sorbing (tropical) soils and organic phosphorus leaching from high rainfall sites.

Lewis and McGechan (2002) recently compared existing phosphorus cycling models and concluded that the approaches used to model are similar. However, they suggested that new phosphorus model development should focus on combining the best features of the existing models. Gijsman *et al.* (1996) suggested that one of the primary factors limiting phosphorus model development is the existence of adequate data-sets to develop and test phosphorus cycling models. Addressing this deficiency should be a focus of future research on phosphorus biogeochemistry in tropical forests.

References

Boswinkle, E. (1961) Residual effects of phosphorus fertilizers in Kenya. *Empire Journal of Experimental Agriculture* 29, 136–142.

Chadwick, O.A., Derry, L.A., Vitousek, P.M., Huebert, B.A. and Hedin, L.O. (1999) Changing sources of nutrients during four million years of ecosystem development. *Nature* 397, 491–497.

Chapin, F.S., III, Matson, P.A. and Mooney, H.A. (eds) (2002) *Principles of Terrestrial Ecosystem Ecology*. Springer, New York, 436 pp.

Cole, C.V., Innis, G.I. and Stewart, J.W.B. (1977) Simulation of phosphorus cycling in semi-arid grasslands. *Ecology* 58, 1–15.

Cole, C.V., Elliott, E.T., Hunt, H.W. and Coleman, D.C. (1978) Trophic interactions in soils as they affect energy and nutrient dynamics. V. Phosphorus transformations. *Microbial Ecology* 4, 381–387.

Crews, T.E., Kitayama, K., Fownes, J.H., Riley, R.H., Herber, D.A., Mueller-Dombois, D. and Vitousek, P.M. (1995) Changes in soil phosphorus fractions and ecosystem dynamics across a long chronosequence in Hawaii. *Ecology* 76, 1407–1424.

Daroub, S.H., Gerakis, A., Ritchie, J.T., Friesen, D.K. and Ryan, J. (2003) Development of a soil–plant phosphorus simulation model for calcareous and weathered tropical soils. *Agricultural Systems* 76, 1157–1181.

Gijsman, A.J., Oberson, A., Tiessen, H. and Friesen, D.K. (1996) Limited applicability of the CENTURY model to highly weathered tropical soils. *Journal of Agronomy* 88, 894–903.

Harrington, R.A., Fownes, J.H. and Vitousek, P.M. (2001) Production and resource use efficiencies in N- and P-limited tropical forests: a comparison of responses to long-term fertilization. *Ecosystems* 4, 646–657.

Hedin, L.O., Vitousek, P.M. and Matson, P.A. (2003) Nutrient losses over four million years of tropical forest development. *Ecology* 84, 2231–2255.

Hedley, M.J. and Stewart, J.W.B. (1982) Method to measure microbial phosphate in soils. *Soil Biology and Biochemistry* 14, 377–385.

Herbert, D.A. and Fownes, J.H. (1995) Phosphorus limitation of forest leaf area and net primary production on a highly weathered soil. *Biogeochemistry* 29, 223–235.

Hotchkiss, S.C., Vitousek, P.M., Chadwick, O.A. and Price, J.P. (2000) Climate cycles, geomorphological change, and the interpretation of soil and ecosystem development. *Ecosystems* 3, 522–533.

Huffman, S.A. (1996) Phosphorus solubility and availability in soils formed under different weathering regimes. PhD thesis, Colorado State University, Fort Collins, Colorado.

Jones, C.A., Cole, C.V., Sharpley, A.N. and Williams, J.R. (1984) A simplified soil and plant phosphorus model. 1. Documentation. *Soil Science Society of America Journal* 48, 800–805.

Kelly, R.H., Parton, W.J., Crocker, G.J., Grace, P.R., Klir, J., Korschens, M., Poulton, P.R. and Richter, D.D. (1997) Simulating trends in soil organic carbon in long-term experiments using the Century model. *Geoderma* 81, 75–90.

Knisel, W.G. (ed.) (1980) *CREAMS: a Field-Scale Model for Chemicals, Runoff, and Erosion from Agricultural Management Systems.* Conservation Research Report No. 26, Science and Education Administration, United States Department of Agriculture, Washington, DC, 643 pp.

Knisel, W.G., Leonard, R.A. and Davis, F.M. (1993) *GLEAMS (Groundwater Loading Effects of Agricultural Management Systems), Version 2.1, Part 1: Nutrient Component Documentation.* United States Department of Agriculture, Agricultural Research Service, Southeast Watershed Research Laboratory, Tifton, Georgia.

Kroes, J.G. and Rijtema, P.E. (1998) *ANIMO 3.5 Users Guide for the ANIMO Version 3.5 Nutrient Leaching Model.* Technical Document 46, DLO Winard Staring Centre, Wageningen, The Netherlands.

Lewis, D.R. and McGechan, M.B. (2002) A review of field-scale phosphorus dynamics models. *Biosystems Engineering* 82, 359–380.

McGill, W.B. and Cole, C.V. (1981) Comparative aspects of cycling of organic C, N, S and P through soil organic matter. *Geoderma* 26, 267–286.

McGuire, A.D., Melillo, J.M. and Schloss, A.L. (1997) Equilibrium responses of global net primary productivity and carbon storage to doubled atmospheric carbon dioxide: sensitivity to changes in vegetation nitrogen concentration. *Global Biogeochemical Cycles* 11, 173–189.

Metherell, A.K., Cole, C.V. and Parton, W.J. (1993) Dynamics and interactions of carbon, nitrogen, phosphorus and sulphur cycling in grazed pastures. In: *Proceedings of the XVII International Grassland Congress,* Palmerston North, New Zealand, pp. 1420–1421.

Olander, L.P. and Vitousek, P.M. (2000) Regulation of soil phosphatase and chitinase activity by N and P availability. *Biogeochemistry* 49, 175–190.

Olander, L.P. and Vitousek, P.M. (2004) Biological and geochemical sinks for phosphorus in soil from a wet tropical forest. *Ecosystems* 7, 404–419.

Parton, W.J., Stewart, J.W.B. and Cole, C.V. (1988) Dynamics of C, N, P, and S in grassland soils: a model. *Biogeochemistry* 5, 109–131.

Parton, W.J., Scurlock, J.M.O., Ojima, D.S., Gilmanov, T.G., Scholes, R.J., Schimel, D.S., Kirchner, T., Menaut, H.-C., Seastedt, T., Garcia Moya, E., Kamnalrut, A. and Kinyamario, J.L. (1993) Observations and modeling of biomass and soil organic matter dynamics for the grassland biome worldwide. *Global Biogeochemical Cycles* 7, 785–809.

Parton, W.J., Ojima, D.S., Cole, C.V. and Schimel, D.S. (1994) A general model for soil organic matter dynamics: sensitivity to litter chemistry, texture and management. In: *Quantitative Modeling of Soil Forming Processes.* Special Publication 39, Soil Science Society of America, Madison, Wisconsin, pp 147–167.

Paul, E.A. and van Veen, J. (1978) The use of tracers to determine the dynamic nature of organic matter. *Transactions of the 11th International Congress of Soil Science* 3, 61–102.

Paustian, K., Parton, W.J. and Persson, J. (1992) Influence of organic amendments and N fertilization on soil organic matter in long-term plots: model analysis. *Soil Science* 56, 476–488.

Raich, J.W., Parton, W.J., Russell, A.E., Sanford, R.L., Jr and Vitousek, P.M. (2000) Analysis of factors regulating ecosystem development on Mauna Loa using the Century model. *Biogeochemistry* 51, 161–191.

Running, S.W. (1994) Testing FOREST-BGC ecosystem process simulations across a climatic gradient in Oregon. *Ecological Applications* 4, 238–247.

Sanford, R.L., Jr, Parton, W.J., Ojima, D.S. and Lodge, D.J. (1991) Hurricane effect on soil organic matter dynamics and forest production in the Luquillo Experimental Forest, Puerto Rico: results of simulation modeling. *Biotropica* 23, 364–372.

Schimel, D.S., Emanuel, W., Rizzo, B., Smith, T., Woodward, F.I., Fisher, H., Kittel, T.G.F., McKeown, R., Painter, T., Rosenbloom, N., Ojima, D.S., Parton, W.J., Kicklighter, D.W., McGuire, A.D., Melillo, J.M., Pan, Y., Haxeltine, A., Prentice, C., Sitch, S., Hibbard, K., Nemani, R., Pierce, L., Running, S., Borchers, J., Chaney, J., Neilson, R. and Braswell, B.H. (1997) Continental-scale variability in ecosystem processes: models, data, and the role of disturbance. *Ecological Monographs* 67, 251–271.

Silver, W.L., Neff, J., Veldekamp, E., McGroddy, M., Keller, M. and Cosme, R. (2000) The effects of soil texture on belowground carbon and nutrient storage in a lowland Amazonian forest ecosystem. *Ecosystems* 3, 193–209.

Tate, K.R. (1984) The biological transformation of phosphorus in soil. *Plant and Soil* 76, 245–256.

Tiessen, H. and Moir, J.O. (1993) Characterization of available P by sequential extraction. In: Carter, M.R. (ed.) *Soil Sampling and Methods of Analysis.* Lewis, Chelsea, Michigan, pp. 75–86.

Tiessen, H. and Stewart, J.W.B. (1985) The biogeochemistry of soil phosphorus. In: Caldwell, D.E., Brierley, J.A. and Brierley, C.L. (eds) *Planetary Ecology.* van Nostrand and Reinhold, New York, pp. 463–472.

VEMAP members (1995) Vegetation/ecosystem modeling and analysis project: comparing biogeography and biogeochemistry models in a continental-scale study of terrestrial ecosystem response to climate change and CO_2 doubling. *Global Biogeochemical Cycles* 9, 407–437.

Vitousek, P.M. (2004) *Nutrient Cycling and Limitation: Hawaii as a Model System.* Princeton University Press, New Jersey, 232 pp.

Vitousek, P.M. and Farrington, H. (1997) Nutrient limitation and soil development: experimental test of a biogeochemical theory. *Biogeochemistry* 37, 63–75.

Vitousek, P.M. and Howarth, R.W. (1991) Nitrogen limitation on land and in the sea: how can it occur? *Biogeochemistry* 13, 87–115.

Vitousek, P.M. and Sanford, R.L. (1986) Nutrient cycling in moist tropical forest. *Annual Reviews of Ecology and Systematics* 17, 137–167.

Vitousek, P.M., Walker, L.R., Whiteaker, L.D. and Matson, P.A. (1993) Nutrient limitations to plant growth during primary succession in Hawaii Volcanoes National Park. *Biogeochemistry* 23, 197–215.

Walker, T.W. and Syers, J.K. (1976) The fate of phosphorus during pedogenesis. *Geoderma* 15, 1–19.

Warren, G.P. (1994a) Influence of soil properties on the response to phosphorus in some tropical soils. I. Initial response to fertilizer. *European Journal of Soil Science* 45, 337–344.

Warren, G.P. (1994b) Influence of soil properties on the response to phosphorus in some tropical

soils. II. Response to fertilizer P residues. *European Journal of Soil Science* 45, 345–351.
Wood, T. (1984) Phosphorus cycling in a northern hardwood forest: biological and chemical control. *Science* 223, 391–393.
Yanai, R.D. (1992) The phosphorus economy of a 70-year-old northern hardwood forest. *Biogeochemistry* 17, 1–22.
Yanai, R.D. (1998) The effect of whole-tree harvest on phosphorus cycling in a northern hardwood forest. *Forest Ecology and Management* 104, 281–295.

16 Modelling Organic Phosphorus Transformations in Aquatic Systems

Peter Reichert[1] and Bernhard Wehrli[2]

[1]*Department of Systems Analysis, Integrated Assessment and Modelling (SIAM), Swiss Federal Institute for Environmental Science and Technology (EAWAG), PO Box 611, 8600 Dübendorf, Switzerland;* [2]*Swiss Federal Institute of Technology (ETH), 8092 Zürich, Switzerland*

Introduction

In many aquatic systems, dissolved phosphate is a limiting nutrient for algal growth. Photosynthesis transforms dissolved phosphate into particulate organic phosphorus in algal biomass, from where it can be transferred to consumer biomass or released in dissolved inorganic or organic form through respiration or hydrolysis. Bacteria can use dissolved organic phosphorus as a substrate for growth. Again, phosphorus can then be released in dissolved form from bacterial or consumer biomass. This demonstrates that distinguishing different phosphorus fractions and investigating how they are affected by biogeochemical transformation processes is a central issue for improving our understanding of aquatic systems. Mechanistic models are useful tools to summarize and quantify our knowledge about the joint effect of these biogeochemical transformation processes.

This chapter reviews descriptions of organic phosphorus components and associated transformation processes as they are used in mechanistic models of aquatic systems. Different approaches have been developed for modelling organic phosphorus in lakes, rivers, sediments, marine systems and sewage treatment plants. A review of some of these approaches allows us to compare the concepts and process formulations developed for different systems. The comparison is facilitated by a unified process representation consisting of two elements: (i) chemical notation of organic substances is extended to a parameterized elemental composition; and (ii) processes are described by a process table formalism as used for models of activated sludge sewage treatment processes. This unified model formulation supports the exchange of knowledge between different fields and facilitates the development of integrative models which combine subsystems of different types (e.g. for modelling the effects of a sewage treatment plant upgrade on the receiving water body).

The chapter is structured as follows. The first section reviews descriptions of organic phosphorus components in biogeochemical models of aquatic systems. This is followed by a review of frequently used model formulations for important processes affecting organic phosphorus in lakes, rivers, sediments, marine systems and sewage treatment plants. The next section contains a brief review of how to embed organic phosphorus submodels into biogeochemical models of aquatic systems. Then we analyse which processes have been considered and which process formulations have been used for describing different aquatic systems. We discuss the selection of a suitable description with a view to building a more universal phosphorus model applicable to a variety of aquatic systems. Finally we conclude with a discussion of the historic development of models for different systems and draw conclusions about useful future directions of model development for different applications.

Formal Description of Organic Phosphorus Components

The term 'organic phosphorus' is defined by analytical procedures (see McKelvie, Chapter 1; Cade-Menun, Chapter 2; Cooper *et al.*, Chapter 3, this volume, for more details). Analytical protocols are available for the operational distinction between different chemical forms of phosphorus. Dissolved and particulate fractions are usually distinguished by filtration through a 0.45 μm membrane, a procedure that separates most bacteria, algae and mineral particles from the dissolved phase, but fails to separate colloidal particles. Free phosphate is estimated in the filtered fraction by reaction with molybdate, but some acid-labile organic and colloidal compounds can also be included. Total filterable phosphorus is measured after sample digestion, with filterable organic phosphorus calculated as the difference between total and reactive phosphorus. Filterable organic phosphorus can include several phosphorus pools with different bioavailability, such as nucleotides, simple phosphate esters and inorganic polyphosphate (Benitez-Nelson and Karl, 2002).

Differences in transport, bioavailability and reactivity mean that the distinctions between dissolved vs. particulate and organic vs. inorganic phosphorus are important for the interpretation of environmental data. In the absence of significant advective transport, dissolved forms have a much greater mobility due to their molecular diffusivity. On the other hand, particulate compounds are removed from the water column by sedimentation. Distinction of organic from inorganic compounds is relevant because the two fractions are affected by different processes. Dissolved phosphate is an essential nutrient for algal growth, but may adsorb to mineral surfaces and, to a lesser degree, to organic surfaces. Organic compounds are used as substrate by heterotrophic organisms. Organic compounds contain phosphorus in different chemical forms, but although measurement techniques support a distinction into different classes of compounds (e.g. into phosphate esters and phosphonates; Kolowith *et al.*, 2001), this has not been reflected in widely used biogeochemical models of organic phosphorus in aquatic systems. Such discrimination could be of relevance for modelling mineralization rates of organic compounds, because mineralization rates depend considerably on the compound. Nucleotides and phospholipids are mineralized quite rapidly, whereas phosphorus incorporated in humic acids is relatively inert (Brannon and Sommers, 1995). As a consequence of different mineralization rates of cytoplasm and cell walls, the carbon-to-phosphorus ratio increases substantially with depth in the solid phase in aquatic sediments. However, the bioavailability of organic compounds not only depends on the compound, but also on the organism using it as a substrate. In some models, classes of organic compounds are distinguished according to their degree of bioavailability. So far, these empirical classes have not been linked to analytically separable groups. The reasons for this are mainly the limited availability of data and the significant differences in bioavailability within some of the measurable compound classes.

Components of mechanistic biogeochemical models of aquatic systems are defined according to their function and not according to analytical procedures. To make model results comparable with measured data, it is important to link model components to analytically defined components. Due to the simplified representation of natural systems in mechanistic models and imperfect analytical procedures, this leads to systematic errors that must be minimized. Functional groups described by biogeochemical models include dissolved organic matter, particulate dead organic matter, algae and bacteria. The phosphorus content of dissolved organic matter is directly comparable to measurements of filterable organic phosphorus. For particulate phosphorus, the link between model components and measurements is more complicated. In addition to phosphorus that occurs as part of organic molecules, phosphorus can be associated with intracellular nutrient storage pools in algae (e.g. Droop, 1973, 2003), as polyphosphate accumulated as intracellular energy storage pools in bacteria (e.g. Henze *et al.*, 1999; Blackall *et al.*, 2002), and as phosphate adsorbed to mineral surfaces associated with cells (e.g. Hupfer *et al.*, 1995). If explicitly considered in the model, these components must be added to the phosphorus content of organic molecules to make model results comparable with particulate organic phosphorus measurements. The mathematical representations used to describe these contributions to organic phosphorus are described in the next two subsections.

Phosphorus as part of organic molecules

Organic matter consists of many compounds and it is impossible to account for this diversity in biogeochemical models of aquatic systems. However, the average composition of organic matter can be described by a similar notation as that used to describe chemical compounds. In the following subsections, the most frequently used descriptions of the composition of organic material are briefly reviewed.

Given elemental composition of organic matter

Based on measurements of elemental ratios of carbon, nitrogen and phosphorus in marine plankton of 106:16:1 (Redfield *et al.*, 1963) and on assumed contents of oxygen and hydrogen (based on typical compounds), the following chemical expression can be used to approximately describe the composition of organic matter

$$(CH_2O)_{106}(NH_3)_{16}(H_3PO_4)_1 = C_{106}H_{263}O_{110}N_{16}P \qquad (16.1)$$

(e.g. Stumm and Morgan, 1981). This expression specifies typical elemental mass fractions of carbon, oxygen, hydrogen, nitrogen and phosphorus in organic matter, but should not be interpreted as describing a 'molecule' of organic matter. Arguments for the long-term stability of these so-called Redfield ratios in marine systems are based on coupled feedback models of the biogeochemical cycles of nitrogen, phosphorus, carbon and oxygen (Redfield, 1958; van Cappellen and Ingall, 1994, 1996; Tyrrell, 1999; Benitez-Nelson, 2000; Lenton and Watson, 2000a). The most recent of these models was used to explain the long-term stability of oxygen concentration in the atmosphere (Lenton and Watson, 2000b).

Parameterized composition of organic matter

Although there are arguments for the long-term stability of the elemental composition of marine organic matter, measurements indicate significant short-term deviations of elemental mass fractions from the Redfield ratios given in equation 16.1. Notably, strong seasonal variations in carbon-to-phosphorus ratios occur in lakes in which algal growth is limited by the availability of phosphate during the summer (Gächter and Bloesch, 1985; Hupfer *et al.*, 1995; see Fig. 16.1).

Parameterization of the elemental ratios in organic matter can be used to consider site-specific information and as a basis for process formulation with variable stoichiometry. Different parameterizations have been proposed, some of which are briefly reviewed in the following paragraphs.

A straightforward generalization of the expression in equation 16.1 is to introduce the elemental ratios of carbon-to-phosphorus and nitrogen-to-phosphorus as model parameters x and y, while keeping the oxygen and hydrogen concentrations associated with carbon, nitrogen and phosphorus constant. This leads to the following

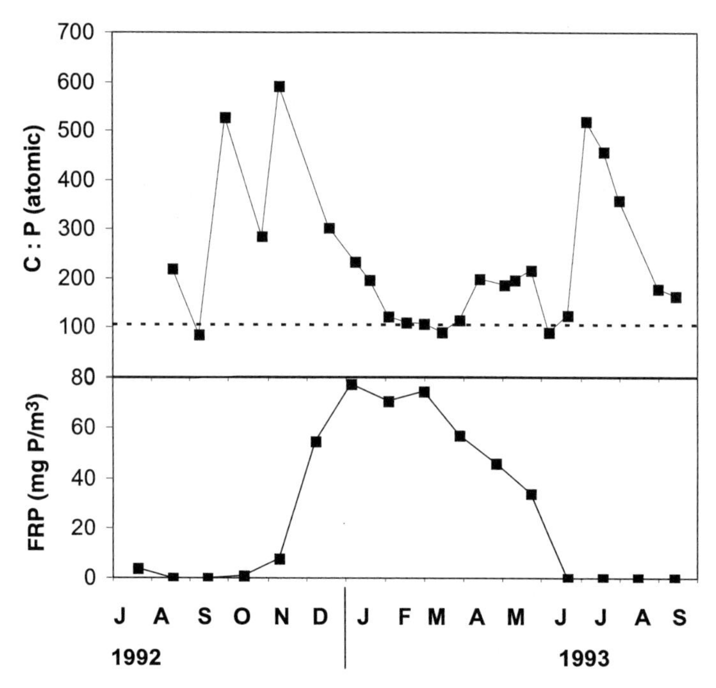

Fig. 16.1. Seasonal variation of carbon-to-phosphorus ratio in organic material collected in a sediment trap at 20 m depth in Lake Sempach, Switzerland (upper panel) and seasonal variation in filterable reactive phosphorus (FRP) concentration observed at 10 m depth in Lake Sempach (lower panel). The carbon-to-phosphorus ratio is significantly higher than the Redfield ratio of 106 : 1 (dashed line) during periods of low filterable reactive phosphorus. Reproduced from Hupfer *et al.* (1995) with permission from Birkhäuser Verlag.

expression for the composition of organic matter:

$$(CH_2O)_x(NH_3)_y(H_3PO_4) \qquad (16.2)$$

This approach has frequently been used in sediment modelling (Soetaert, *et al.*, 1996; Wijsman *et al.*, 2002). Thus, the Redfield approach (equation 16.1) is a special case of the generalized expression in equation 16.2 with $x=106$ and $y=16$.

In wastewater engineering, organic matter is usually quantified by the chemical oxygen demand (COD) required for mineralization. Consequently, in activated sludge sewage treatment models, the phosphorus and nitrogen contents of organic matter are expressed as mass of nitrogen or phosphorus per unit of chemical oxygen demand:

$$i_N = \frac{gN}{gCOD},\ i_P = \frac{gP}{gCOD} \qquad (16.3)$$

(Henze *et al.*, 1995, 1999).

Finally, it was proposed that the mass fractions of carbon, hydrogen, oxygen, nitrogen, phosphorus, and all other elements (*X*) could be introduced as model parameters Reichert *et al.*, 2001a,b):

$$\alpha_C, \alpha_H, \alpha_O, \alpha_N, \alpha_P, \alpha_X, \text{ with } \Sigma\, \alpha_i = 1 \qquad (16.4)$$

To link this approach to the previous one, we must formally describe the mineralization process (Reichert *et al.*, 2001b):

$$1\text{gOM} + \frac{1}{4}\left(\frac{\alpha_C}{3} + \alpha_H - \frac{\alpha_O}{8} - \frac{3\alpha_N}{14} + \frac{5\alpha_P}{31} + \alpha_X\left(\beta_+ - \beta_H + \frac{\beta_O}{8}\right)\right) \text{moles } O_2$$

$$+\frac{1}{2}\left(\frac{\alpha_C}{6} - \alpha_H + \frac{3\alpha_N}{14} + \frac{3\alpha_P}{31} + \alpha_X(\beta_H - \beta_+)\right) \text{moles } H_2O$$

$$\rightarrow \frac{\alpha_C}{12} \text{moles } HCO_3^-$$

$$+\left(\frac{\alpha_C}{12}-\frac{\alpha_N}{14}+\frac{2\alpha_P}{31}-\alpha_X\beta_+\right)\text{ moles H}^+$$
$$+\frac{\alpha_N}{14}\text{ moles NH}_4^+ + \frac{\alpha_P}{31}\text{ moles HPO}_4^{2-}$$
$$+\alpha_X\text{ gX XH}_{M\beta_H}\text{O}_{M\beta_O/16}{}^{M\beta_+} \qquad (16.5)$$

The coefficients of the compounds in this equation are uniquely determined by mass balances for carbon, hydrogen, oxygen, nitrogen and phosphorus, a mass balance for the elements summarized by *X*, and a charge balance. The term $XH_{M\beta_H}O_{M\beta_O/16}{}^{M\beta_+}$ describes the reference state of mineralization of the elements *X* parameterized by the molar mass, M, of *X*, and by β_H, β_O and β_+, which characterize hydrogen and oxygen content and charge (β_H and β_O specify the mass of hydrogen and oxygen, respectively, per unit mass of *X*, β_+ is the number of positive charge units per unit mass of *X*). From equation 16.5 we can derive conversion factors from chemical oxygen demand to dry mass or mass fractions. The conversion factor from chemical oxygen demand to organic carbon is given by:

$$\gamma_C=\frac{\text{gCOD}}{\text{gC}}=\frac{8}{\alpha_C}\left(\frac{\alpha_C}{3}+\alpha_H-\frac{\alpha_O}{8}-\frac{3\alpha_N}{14}+\frac{5\alpha_P}{31}+\alpha_X\left(\beta_+-\beta_H+\frac{\beta_O}{8}\right)\right) \qquad (16.6)$$

This leads to the following conversion formula between the phosphorus contents i_P and α_P:

$$i_P=\frac{\alpha_P}{8\left(\frac{\alpha_C}{3}+\alpha_H-\frac{\alpha_O}{8}-\frac{3\alpha_N}{14}+\frac{5\alpha_P}{31}+\alpha_X\left(\beta_+-\beta_H+\frac{\beta_O}{8}\right)\right)} \qquad (16.7)$$

The approach represented by equation (16.3) is difficult to apply if organic matter is not measured in chemical oxygen demand (COD) units. This is a problem, because some measurement techniques (e.g. counting of algae species and conversion to wet mass by an average species-specific volume) do not lead to a natural conversion to chemical oxygen demand units. The approach depicted in equation 16.2 uses a smaller number of parameters in comparison to equation 16.4, which in many cases is an advantage. However, the approach of 16.4 is very general, makes assumptions about hydrogen and oxygen content of organic matter more explicit, and is easier to extend to additional elements. The mass fractions of hydrogen and oxygen can still be chosen according to equation 16.1 if no specific information for these elements is available, but the additional parameters in approach (16.4) make it possible to analyse the sensitivity of model results with respect to the oxidation state of organic matter, which is essentially determined by its oxygen and hydrogen content.

Storage pools of phosphorus associated with particulate organic matter

So far, we have described phosphorus as part of organic molecules. Such molecules can accumulate additional phosphorus by several mechanisms, including storage as intracellular nutrients in algae, as polyphosphates in bacteria, or as phosphate adsorbed to mineral surfaces associated with cells. The significance and mathematical representation of these fractions are briefly discussed in the following subsections. The processes involved in building and consuming these phosphorus fractions will be reviewed in the next section.

Phosphorus in intracellular nutrient storage pools

Most models of algal growth combine the steps of nutrient uptake into the cell and growth on intracellular nutrients into a single step of uptake and growth. In the models that separate those two steps for phosphorus (e.g. Droop, 1973, 2003), an intracellular phosphorus pool must be included, in addition to the phosphorus that is part of the organic molecules. We denote the mass fraction of intracellular phosphorus contained in nutrient storage pools as α_{PS}.

Phosphorus in intracellular energy storage pools

Some bacteria are able to accumulate polyphosphate within their cells. This process is used in sewage treatment plants with enhanced biological phosphorus removal (Henze *et al.*, 1999; Blackall *et al.*, 2002). As polyphosphate is not part of organic molecules, it must be considered as an additional mass fraction of phosphorus denoted as α_{PP}.

Phosphorus adsorbed to mineral surfaces associated with cells

It has been observed that organic particles settling through the hypolimnion of a lake become enriched with phosphorus. The details of this enrichment process are not yet fully understood. It was suggested that the mechanism underlying this enrichment process could be adsorption of phosphate to surfaces of manganese oxides and iron hydroxides built by manganese-oxidizing and iron-oxidizing bacteria (Hupfer *et al.*, 1995). The mass fraction of phosphorus describing the result of this accumulation process is denoted as α_{PI}.

Phosphorus pools in dissolved organic matter

Particulate organic matter is not available for direct uptake by aquatic organisms. In a first step, called hydrolysis, it is transformed to dissolved substrates by enzymatic reactions that occur outside cells. In addition, cell lysis or 'sloppy feeding' (the breaking of cell walls by zooplankton without intake and digestion) can release dissolved organic substrate from living or dead particulate matter. For process modelling, the distinction between bioavailable and inert dissolved organic matter is essential. This requires two organic phosphorus pools to be distinguished, one that is accessible to biological processes and the other that is not accessible within the relevant time frame. In some cases it may be necessary to further divide the bioavailable organic matter fraction into rapidly and slowly degradable organic matter. At present the analytical distinction between such pools is not straightforward.

An additional level of complexity arises in some models involving processes which rely on fermentation products. To be able to describe such processes, Henze *et al.* (1999) divided degradable organic matter into fermentable material and fermentation products. Fermentation is the transformation of dissolved bioavailable organic matter in macromolecules into smaller substrates such as acetate and lactate. The range of dissolved organic matter available to fermenting bacteria is small compared with substrates for aerobic microorganisms. This may require the further division of degradable organic matter into fermentable and non-fermentable dissolved organic matter and fermentation products. The fermentation products are further used by anaerobic microorganisms, such as sulphate-reducers and methanogens, which can utilize an even narrower range of low-molecular-weight substrates. These fermentation products are typically phosphate-free. Detailed modelling of phosphorus turnover in such environments would therefore involve organic phosphorus in bioavailable, fermentable and inert organic fractions.

Description of Organic Phosphorus Transformation Processes

The processes affecting organic phosphorus concentrations that have been considered in widespread mechanistic models of aquatic systems are briefly described in the following subsections. To represent these processes, we use the process table formalism that has been popularized in wastewater engineering by the introduction of the activated sludge models (Henze *et al.*, 1986, 1995, 1999; Gujer *et al.*, 1999). This process table formalism is shown in Table 16.1.

The contribution of a process P_i to the conversion rate of substance j is given as $\nu_{i,j}r_i$ where r_i is the process rate and $\nu_{i,j}$ is the stoichiometric coefficient for the substance. One of the stoichiometric coefficients of each process can be selected to be plus or

Table 16.1. Process table formalism used to describe biogeochemical transformation processes. Each process, P_i, is characterized by its rate, r_i, and a series of stoichiometric coefficients, $\nu_{i,j}$.

Process	Components				Rate
	Dissolved		Particulate		
	S_1	S_2	X_1	X_2	
P_1	$\nu_{1,S1}$	$\nu_{1,S2}$	$\nu_{1,X1}$	$\nu_{1,X2}$	r_1
P_2	$\nu_{2,S1}$	$\nu_{2,S2}$	$\nu_{2,X1}$	$\nu_{2,X2}$	r_2
⋮	⋮	⋮	⋮	⋮	⋮

minus unity. This then defines the process rate as the (positive) contribution of the process to the transformation rate of this substance and makes the other stoichiometric coefficients of this process unique. The total transformation rate of substance j is given as:

$$r_j = \sum_i \nu_{i,j} r_i \quad (16.8)$$

This term is added to the transport terms in the differential equation for the temporal change of substance j.

The process table formalism gives an excellent overview of the system of transformation processes. Looking at column j of the stoichiometric matrix gives a quick overview of processes leading to production of substance j (positive stoichiometric coefficients), processes leading to consumption of substance j (negative stoichiometric coefficients) and processes that do not directly affect the concentration of substance j (stoichiometric coefficients equal to zero). In addition, looking at row i of the stoichiometric matrix gives a quick overview of substances affected by the process P_i. Finally, the last column of row i quantifies the dependence of the transformation rate of the process P_i on the concentration of modelled substances as well as on external influence factors.

In the following subsections, we discuss the qualitative stoichiometric matrix and an example of the rate formulation for the most relevant processes affecting organic phosphorus concentrations in aquatic systems. The stoichiometric matrix indicates the signs of the stoichiometric coefficients only; if the sign of a stoichiometric coefficient depends on the composition of the organic compound, this is indicated with a ± sign. The missing quantitative values of the stoichiometric coefficients can be calculated following the procedure described by Reichert *et al.* (2001a,b). References to more detailed model descriptions are also given. To describe variation in phosphorus concentrations with time, separate state variables for the concentration of phosphorus fractions are introduced. Following the practice in wastewater engineering, dissolved components are described by the symbol S and particulate components by the symbol X. Table 16.2 gives a short description of all model components used in the following process descriptions.

In addition to the notation for model components given in Table 16.2, we use the following symbols: T for temperature, K for half-saturation concentrations, k for specific rate constants, β for temperature dependence coefficients, and I for light intensity.

Primary production

Primary production is the key process that transforms dissolved phosphate into particulate organic phosphorus (as part of algae). In some models, this process is divided into two steps: nutrient uptake by cells to build intracellular nutrient pools and growth of cells on intracellular nutrients. Table 16.3 summarizes a typical model formulation of primary production considered as a one-stage process combining phosphate uptake with growth of algae. The rate formulation given in Table 16.3 combines approaches used by Omlin *et al.* (2001a) and Reichert *et al.* (2001a,b). However, it is representative of similar formulations in other simple models of primary production (e.g. Brown and Barnwell, 1987; Hamilton and Schladow, 1997). The terms in the rate formula represent a maximum specific growth rate at a given temperature, temperature-dependence of growth, light limitation, nitrogen limitation, switching between ammonium and nitrate as the nitrogen source, phosphorus limitation and proportionality to the concentration of algae.

Table 16.2. Summary of the state variables used for process descriptions.

Dissolved components (concentration of . . .)	
S_S	biodegradable dissolved organic substrate; in some models divided into S_F and S_A
$S_{P,S}$	phosphorus that is part of biodegradable dissolved organic substrate
S_F	fermentable, readily biodegradable dissolved organic substrate
$S_{P,F}$	phosphorus that is part of fermentable, readily biodegradable dissolved organic substrate
S_A	fermentation products
S_I	inert (within the relevant time) dissolved organic substrate
$S_{P,I}$	phosphorus that is part of the organic molecules of inert dissolved organic substrate
S_{NH_4}	ammonium
S_{NO_3}	nitrate
S_{HPO_4}	phosphate
S_{O_2}	dissolved oxygen
S_{HCO_3}	bicarbonate
S_H	protons
Particulate components (concentration of . . .)	
X_{ALG}	algae without their phosphorus content
$X_{P,ALG}$	phosphorus that is part of the organic molecules of algae
$X_{PS,ALG}$	phosphorus stored as a cell-internal nutrient storage pool in algae
X_{CON}	consumers without their phosphorus content
$X_{P,CON}$	phosphorus that is part of the organic molecules of consumers
X_{N1}	first-stage nitrifiers without their phosphorus content
$X_{P,N1}$	phosphorus that is part of the organic molecules of first-stage nitrifiers
X_{N2}	second-stage nitrifiers without their phosphorus content
$X_{P,N2}$	phosphorus that is part of the organic molecules of second-stage nitrifiers
X_H	heterotrophic bacteria without their phosphorus content
$X_{P,H}$	phosphorus that is part of the organic molecules of heterotrophic bacteria
X_{PAO}	phosphate-accumulating (heterotrophic) organisms without their phosphorus content
$X_{P,PAO}$	phosphorus that is part of the organic molecules of phosphate-accumulating organisms
$X_{PP,PAO}$	polyphosphate in phosphate-accumulating organisms
$X_{PHA,PAO}$	poly-hydroxy-alkanoates stored as a cell-internal energy storage pool in phosphate-accumulating organisms
X_S	dead biodegradable organic particles without their phosphorus content
$X_{P,S}$	phosphorus that is part of the organic molecules of dead biodegradable organic particles
$X_{PI,S}$	phosphate adsorbed to mineral surfaces associated with dead biodegradable organic particles
X_I	dead inert organic particles without their phosphorus content
$X_{P,I}$	phosphorus that is part of the organic molecules of dead inert organic particles

Table 16.4 shows a model for primary production that distinguishes uptake and growth. An important example of such a model is that of Droop (1973). Differential equations for this model are formulated in a more convenient way in Ducobu *et al.* (1998) and Ahn *et al.* (2002). The rate expressions given in Table 16.4 are from the Droop model. The dependence on environmental factors other than concentrations of phosphate and algae is absorbed by the variables u_m and μ_m. This includes switching between ammonium and nitrate as the nitrogen source, so for this reason we do not distinguish the rates for growth with ammonium and with nitrate. The processes as formulated in Table 16.4 assume that the base quota of Droop corresponds to the phosphorus concentration of the organic molecules ($Q_0 = \alpha_{P,ALG}$) and the total quota to the sum of the phosphorus concentration of the organic molecules and the storage product ($Q = \alpha_{P,ALG} + \alpha_{PS,ALG}$). The uptake rate is formulated by a Monod-type limitation term with respect to

Table 16.3. Qualitative process table of primary production modelled as a single-step process.

	Dissolved components						Particulate components		
Process	S_{NH_4}	S_{NO_3}	S_{HPO_4}	S_{O_2}	S_{HCO_3}	S_H	X_{ALG}	$X_{P,ALG}$	Rate
Growth of algae (ammonium)	−		−	+	−	−	+1	+	r_{gro,ALG,NH_4}
Growth of algae (nitrate)		−	−	+	−	−	+1	+	r_{gro,ALG,NO_3}

$$r_{gro,ALG,NH_4} = k_{gro,ALG,To}\exp(\beta(T-T_0))\frac{I}{K_I+I}\frac{S_{NH4}+S_{NO3}}{K_{N,ALG}+S_{NH_4}+S_{NO_3}}\frac{S_{NH4}}{K_{NH_4,ALG}+S_{NH_4}}\frac{S_{HPO4}}{K_{P,ALG}+S_{HPO_4}}X_{ALG}$$

$$r_{gro,ALG,NO_3} = k_{gro,ALG,To}\exp(\beta(T-T_0))\frac{I}{K_I+I}\frac{S_{NH4}+S_{NO3}}{K_{N,ALG}+S_{NH_4}+S_{NO_3}}\frac{K_{NH4,ALG}}{K_{NH_4,ALG}+S_{NH_4}}\frac{S_{HPO4}}{K_{P,ALG}+S_{HPO_4}}X_{ALG}$$

Table 16.4. Qualitative process table of primary production modelled as a two-stage process distinguishing phosphate uptake from growth on intracellular phosphate.

	Dissolved components						Particulate components			
Process	S_{NH_4}	S_{NO_3}	S_{HPO_4}	S_{O_2}	S_{HCO_3}	S_H	X_{ALG}	$X_{P,ALG}$	$X_{PS,ALG}$	Rate
Uptake of phosphate			−1						+1	r_{upt}
Growth of algae (ammonium) with internally stored phosphorus	−			+	−	−	+1	+	−	r_{gro}
Growth of algae (nitrate) with internally stored phosphorus		−		+	−	−	+1	+	−	r_{gro}

$$r_{upt} = u_m\frac{S_{HPO4}}{K_{HPO_4}+S_{HPO_4}}X_{ALG}$$

$$r_{gro} = \mu_m'\frac{X_{PS,ALG}}{X_{P,ALG}+X_{PS,ALG}}X_{ALG}$$

cell-external phosphate concentration and in proportion to the concentration of algae. The growth rate is again proportional to the concentration of algae and limited by cell internal phosphorus storage products.

Growth of chemolithotrophic bacteria

Growth of chemolithotrophic bacteria, such as nitrifiers, sulphide-oxidizers and methane-oxidizers, is an alternative means of primary production for the conversion of phosphate to particulate organic phosphorus. These bacteria may act as scavengers for phosphorus transported from the deep water in lakes upwards by eddy diffusion (Sinke *et al.*, 1993). The intensity of this secondary transformation of dissolved phosphorus into microbial biomass depends of the rate of dark photosynthesis. In deep anoxic basins, this can be substantial (Pimenov *et al.*, 2000).

Nitrification is the most important process involving chemolithotrophic bacteria in aerobic systems. It is an important process for the natural nitrogen cycle as well as for sewage treatment plants. Table 16.5 summarizes how nitrification can be formulated as a two-stage process converting ammonium to nitrite and nitrite to nitrate, parallel to growth of two classes of first- and second-stage nitrifiers. This process

Table 16.5. Qualitative process table of the two-stage nitrification process as an example for chemolithotrophic processes.

Process	Dissolved components							Particulate components				Rate
	S_{NH_4}	S_{NO_2}	S_{NO_3}	S_{HPO_4}	S_{O_2}	S_{HCO_3}	S_H	X_{N1}	$X_{P,N1}$	X_{N2}	$X_{P,N2}$	
First-stage nitrification	−	+		−	−	−	+	+1	+			$r_{gro,N1}$
Second-stage nitrification		−	+	−	−	−	−			+1	+	$r_{gro,N2}$

$$r_{gro,N1} = k_{gro,N1,To} \exp(\beta_{N1}(T - T_0)) \frac{S_{O2}}{K_{O_2,N1} + S_{O_2}} \frac{S_{NH4}}{K_{NH_4,N1} + S_{NH_4}} \frac{S_{HPO4}}{K_{HPO_4,N1} + S_{HPO_4}} X_{N1}$$

$$r_{gro,N2} = k_{gro,N2,To} \exp(\beta_{N2}(T - T_0)) \frac{S_{O2}}{K_{O_2,N2} + S_{O_2}} \frac{S_{NO2}}{K_{NO_2,N2} + S_{NO_2}} \frac{S_{HPO4}}{K_{HPO_4,N1} + S_{HPO_4}} X_{N2}$$

Table 16.6. Qualitative process table of the uptake/adsorption process.

Process	Dissolved component S_{HPO_4}	Particulate component $X_{PI,S}$	Rate
Uptake/adsorption	−	+1	$k_{upt}\left(\alpha_{PI,max} - \frac{X_{PI,S}}{X_S}\right) S_{HPO_4} X_S$

formulation is taken from Reichert *et al.* (2001a,b), who applied the same process formulation as used in the activated sludge sewage treatment plant models (Henze *et al.*, 1986, 1995, 1999; Gujer *et al.*, 1999). The terms in the rate expressions represent the maximum specific growth rate of nitrifiers, temperature dependence, oxygen limitation, limitation by ammonium (first stage) or nitrite (second stage), limitation by phosphate, and proportionality to the concentrations of nitrifiers in the corresponding stage. Growth of other chemolithotrophic bacteria can be formulated in a similar way.

Uptake/adsorption

It has been observed that organic particles settling through the hypolimnion of lakes become enriched with phosphorus (Gächter and Mares, 1985). The mechanism by which this occurs is still not completely clear, but seems to be related to adsorption of phosphate to mineral surfaces produced by iron- and manganese-oxidizing bacteria (Hupfer *et al.*, 1995). As this process can significantly affect phosphate concentrations in the water column and the phosphate flux to lake sediment, it was included and empirically parameterized in a biogeochemical lake model (Omlin *et al.*, 2001a).

Table 16.6 shows a simplified process table of uptake by adsorption. The parameterization of the rate is simplified from Omlin *et al.* (2001a). The rate is proportional to the concentration of phosphate in the water column, the concentration of biodegradable organic particles, and the difference between the mass fraction of adsorbed phosphate and a maximum mass fraction.

Growth of heterotrophic bacteria

Growth of heterotrophic organisms converts dissolved organic substrate partly into bacterial biomass and releases part of it in mineralized form. The process formulation for heterotrophic growth given in Table 16.7 (Reichert *et al.*, 2001a,b) is strongly influenced by how this process has been

Table 16.7. Qualitative process table of growth of heterotrophic bacteria.

	Dissolved components								Particulate components		
Process	S_S	S_{NH_4}	S_{NO_2}	S_{NO_3}	S_{HPO_4}	S_{O_2}	S_{HCO_3}	S_H	X_H	$X_{P,H}$	Rate
Aerobic growth (ammonium)	−	+/−			+/−	−	+	+	+1	+	r_{gro,H,aer,NH_4}
Aerobic growth (nitrate)	−			−	+/−	−	+	+	+1	+	r_{gro,H,aer,NO_3}
Anoxic growth (nitrate)	−		+	−	+/−		+	+	+1	+	$r_{gro,H,anox,NO_3}$
Anoxic growth (nitrite)	−		−		+/−		+	−	+1	+	$r_{gro,H,anox,NO_2}$

$$r_{gro,H,aer,NH_4} = k_{gro,H,aer,To} \exp(\beta_H(T-T_0)) \frac{S_S}{K_{S,H}+S_S} \frac{S_{O2}}{K_{O2,H}+S_{O2}} \left[\frac{S_{NH4}}{K_{N,H,aer}+S_{NH4}}\right] \left[\frac{S_{HPO4}}{K_{HPO4,H}+S_{HPO4}}\right] X_H$$

$$r_{gro,H,aer,NO_3} = \left[k_{gro,H,aer,To} \exp(\beta_H(T-T_0)) \frac{S_S}{K_{S,H}+S_S} \frac{S_{O2}}{K_{O2,H}+S_{O2}} \frac{K_{N,H,aer}}{K_{N,H,aer}+S_{NH4}} \frac{S_{NO3}}{K_{N,H,aer}+S_{NO3}} \times \left[\frac{S_{HPO4}}{K_{HPO4,H}+S_{HPO4}}\right] X_H \right]$$

$$r_{gro,H,amox,NO_3} = k_{gro,H,anox,To} \exp(\beta_H(T-T_0)) \frac{S_S}{K_{S,H}+S_S} \frac{K_{O2,H}}{K_{O2,H}+S_{O2}} \frac{S_{NO3}}{K_{NO3,H,anox}+S_{NO3}} \left[\frac{S_{HPO4}}{K_{HPO4,H}+S_{HPO4}}\right] X_H$$

$$r_{gro,H,anox,NO_2} = k_{gro,H,anox,To} \exp(\beta_H(T-T_0)) \frac{S_S}{K_{S,H}+S_S} \frac{K_{O2,H}}{K_{O2,H}+S_{O2}} \frac{S_{NO2}}{K_{NO2,H,anox}+S_{NO2}} \left[\frac{S_{HPO4}}{K_{HPO4,H}+S_{HPO4}}\right] X_H$$

described in the activated sludge models ASM 1, 2, 2d and 3 (Henze *et al.*, 1986, 1995, 1999; Gujer *et al.* 1999).

The stoichiometric coefficients with undetermined sign in the process table indicate the dependence of these signs on the nutrient content of the substrate, the composition of the bacteria, and the yield of the process. Bacteria growing on pure sugar need nutrients, whereas the mineralization of proteins leads to the release of excess nutrients. The corresponding limiting terms in the rate expression can be omitted for positive coefficients (release of ammonium or phosphate). These terms are indicated by square brackets.

Growth of phosphate-accumulating bacteria

Investigations stimulated by the discovery that bacteria actively store phosphorus in the form of intracellular polyphosphates during the activated sludge processes (Streichan *et al.*, 1990, Appeldoorn *et al.*, 1992), have led to a more detailed understanding of the underlying processes, the diversity and physiology of the microorganisms involved, and the optimization of activated sludge processes for biological phosphorus removal. Blackall *et al.* (2002) review the remarkable progress in this field in recent years. Activated sludge models have been adapted to model the details of this process (Henze *et al.*, 1995, 1999; Morgenroth and Wilderer, 1998; Manga *et al.*, 2001).

The phosphate accumulation process is described in current models as follows (Henze *et al.*, 1999). Some heterotrophic bacteria, so-called phosphate-accumulating organisms, grow under aerobic (dissolved oxygen available) as well as anoxic (no dissolved oxygen, but nitrate available) conditions on intracellular organic material (poly-hydroxy-alkanoates). These intracellular substrates are obtained from external fermentation products, with energy obtained from the hydrolysis of intracellular polyphosphate. Finally, polyphosphate is stored in the cell during consumption of extracellular phosphate by an aerobic or anoxic storage process under consumption of intracellular poly-hydroxy-alkanoates. Note that the growth of phosphate-accumulating organisms requires this polyphosphate pool, as growth only takes place on intracellular substrate, which can only be obtained through the release of phosphate from intracellular polyphosphate. Storage of intracellular substrate is based on fermentation products that are only formed under anaerobic conditions (neither dissolved oxygen nor nitrate available). Cell growth and storage of polyphosphate, however, only take place under aerobic or anoxic conditions. Therefore, phosphate-accumulating organisms grow only under cyclic anaerobic and aerobic or anoxic conditions. Table 16.8 illustrates this process as formulated in the Activated Sludge Model No. 2d (Henze *et al.*, 1999). The storage rate of poly-hydroxy-alkanoates is limited by extracellular fermentation products and by intracellular polyphosphate. The storage rate of polyphosphate is limited by extracellular phosphate, intracellular poly-hydroxy-alkanoates, by either dissolved oxygen (for aerobic storage) or nitrate (for anoxic storage), and by the approach of the mass fraction of intracellular polyphosphates to a maximum value. Finally, growth of phosphate-accumulating organisms is limited by ammonium, ambient phosphate, intracellular poly-hydroxy-alkanoates, and by either dissolved oxygen (for aerobic growth) or nitrate (for anoxic growth).

As a result of the research activities in the domain of biological wastewater treatment, several studies reassessed the role of bacteria in phosphorus removal in aquatic systems. The potential for phosphorus storage by general bacterial growth processes (not only by polyphosphate accumulation) has been identified in several analytical studies as an important mechanism for phosphorus retention in sediments and for phosphorus sedimentation in the water column (Gächter *et al.*, 1988; Sinke *et al.*, 1992, 1993; Hupfer *et al.*, 1995). Recently, it was suggested that polyphosphate accumulation could contribute to phosphorus accumulation in sediments (Davelaar, 1993; Khoshmanesh *et al.*, 2002). However, the role of microorganisms in the natural environment has still received little attention in modelling studies.

Growth of consumers

Consumption by macro-organisms is an important process that may limit particulate organic phosphorus concentrations in the lower trophic levels. From the point of view of elemental mass balances, growth of consumers is a similar process to growth of heterotrophic bacteria. However, due to the smaller yield and the presence of nitrogen and phosphorus, it can be assumed that the food contains sufficient nitrogen and phosphorus so that ammonium and phosphate are released and not consumed by this process. Table 16.9 gives an example of a process formulation for consumption (Reichert *et al.*, 2001a,b). Consumption is temperature-dependent and assumed to be proportional to the product of the concentrations of food and consumers. X_i represents any of the particulate organic components that can serve as food. Part of the consumed organic matter is transformed to consumer biomass, part is mineralized, and part is released as faecal pellets or as organic matter from sloppy feeding (the latter process could also result in the release of dissolved organic matter). Consumer growth processes, as described in Table 16.9, can be used to describe food webs with several trophic levels, which can be used to gain understanding of the complicated interactions between trophic levels (Carpenter *et al.*, 1985).

Respiration

Respiration leads to the oxidation of organic compounds and release of phosphate. Usually it is assumed that not all of the organic material is respired, but that a

Table 16.8. Qualitative process table for growth of phosphate-accumulating organisms.

	Dissolved components							Particulate components				
	S_A	S_{NH_4}	S_{NO_3}	S_{HPO_4}	S_{O_2}	S_{HCO_3}	S_H	X_{PAO}	$X_{P,PAO}$	$X_{PP,PAO}$	$X_{PHA,PAO}$	Rate
Storage of poly-hydroxy-alkanoates	−1			+		+	+			–	+	$r_{sto,PHA}$
Aerobic storage of polyphosphate				−1	–	+	+			+1	–	$r_{sto,aer,PP}$
Anoxic storage of polyphosphate			+	−1		+	+			+1	–	$r_{sto,anox,PP}$
Aerobic growth of phosphate-accumulating organisms		–		–	–	–	+	+1	+		–	$r_{gro,aer,PAO}$
Anoxic growth of phosphate-accumulating organisms		–	–	–		+	+	+1	+		–	$r_{gro,anox,PAO}$

$$r_{sto,PHA} = k_{sto,PHA} \frac{S_A}{K_A + S_A} \frac{S_{HCO_3}}{K_{HCO_3} + S_{HCO_3}} \frac{X_{PP,PAO}/X_{PAO}}{K_{PP} + X_{PP,PAO}/X_{PAO}} X_{PAO}$$

$$r_{sto,aer,PP} = k_{sto,aer,PP} \frac{S_{O_2}}{K_{O_2} + S_{O_2}} \frac{S_{HPO_4}}{K_{HPO_4} + S_{HPO_4}} \frac{S_{HCO_3}}{K_{HCO_3} + S_{HCO_3}} \frac{X_{PHA,PAO}/X_{PAO}}{K_{PHA} + X_{PHA,PAO}/X_{PAO}} \times \frac{K_{max} - X_{PP,PAO}/X_{PAO}}{K_{IPP} + K_{max} - X_{PP,PAO}/X_{PAO}} X_{PAO}$$

$$r_{sto,anox,PP} = k_{sto,aer,PP} \eta_{NO_3} \frac{K_{O_2}}{K_{O_2} + S_{O_2}} \frac{S_{NO_3}}{K_{NO_3} + S_{NO_3}} \frac{S_{HPO_4}}{K_{HPO_4} + S_{HPO_4}} \frac{S_{HCO_3}}{K_{HCO_3} + S_{HCO_3}} \frac{X_{PHA,PAO}/X_{PAO}}{K_{PHA} + X_{PHA,PAO}/X_{PAO}} \times \frac{K_{max} - X_{PP,PAO}/X_{PAO}}{K_{IPP} + K_{max} - X_{PP,PAO}/X_{PAO}} X_{PAO}$$

$$r_{gro,aer,PAO} = k_{gro,aer,PAO} \frac{S_{O_2}}{K_{O_2} + S_{O_2}} \frac{S_{NH_4}}{K_{NH_4} + S_{NH_4}} \frac{S_{HPO_4}}{K_{HPO_4} + S_{HPO_4}} \frac{S_{HCO_3}}{K_{HCO_3} + S_{HCO_3}} \frac{X_{PHA,PAO}/X_{PAO}}{K_{PHA} + X_{PHA,PAO}/X_{PAO}} X_{PAO}$$

$$r_{gro,anox,PAO} = k_{gro,aer,PAO} \eta_{NO_3} \frac{K_{O_2}}{K_{O_2} + S_{O_2}} \frac{S_{NO_3}}{K_{NO_3} + S_{NO_3}} \frac{S_{NH_4}}{K_{NH_4} + S_{NH_4}} \frac{S_{HPO_4}}{K_{HPO_4} + S_{HPO_4}} \frac{S_{HCO_3}}{K_{HCO_3} + S_{HCO_3}} \times \frac{X_{PHA,PAO}/X_{PAO}}{K_{PHA} + X_{PHA,PAO}/X_{PAO}}$$

small fraction is converted to inert (not degradable within the considered time scale) particulate organic material. Table 16.10 shows a typical formulation of this process limited by the concentration of dissolved oxygen. X_i represents any of the living organic fractions (algae, bacteria, consumers).

Death

In addition to respiration, a death process is introduced for living organic fractions (especially for algae and consumers). This process transforms living organic matter into biodegradable and inert organic particles.

Table 16.9. Qualitative process table of the growth process of consumers.

	Dissolved components					Particulate components						
Process	S_{NH_4}	S_{HPO_4}	S_{O_2}	S_{HCO_3}	S_H	X_i	$X_{P,i}$	X_S	$X_{P,S}$	X_{CON}	$X_{P,CON}$	Rate
Growth of consumers	+	+	−	+	+	−	−	+	+	+1	+	$k_{gro,CON,To} \times \exp(\beta_{CON}(T-T_0)) \times X_i X_{CON}$

Table 16.10. Qualitative process table of the respiration process.

	Dissolved components					Particulate components							
Process	S_{NH_4}	S_{HPO_4}	S_{O_2}	S_{HCO_3}	S_H	X_i	$X_{P,i}$	$X_{PS,i}$	$X_{PP,i}$	$X_{PI,i}$	X_I	$X_{P,I}$	Rate
Respiration	+	+	−	+	+	−1	−	−	−	−	+	+	$k_{resp,i,To}\exp(\beta_i(T-T_0)) \times \frac{S_{O_2}}{K_{O_2}+S_{O_2}} X_i$

The rate of this process is most frequently described as being proportional to the concentration of the organism. Table 16.11 shows a process table of this process.

Hydrolysis

The process of hydrolysis, as defined here, converts dead particulate organic matter into bioavailable dissolved substrate. It is usually described by a simple first-order transformation. The process involves extracellular enzyme reactions breaking down high-molecular-weight particulate material that cannot be directly metabolized by microorganisms. Table 16.12 shows a simple representation of this process. It is quantified by an empirical specific hydrolysis rate constant and a temperature-dependence coefficient.

Chemical equilibria

Chemical equilibria can be modelled as dynamic processes with large rate constants. It is sufficient to choose the rate constant large enough that the calculated concentrations are always close to their equilibrium values. It is then not necessary for these rate constants to describe chemical process kinetics appropriately. Table 16.13 shows some examples of acid–base equilibria involving phosphate and carbonate species and including the solubility of calcite as implemented in Reichert *et al.* (2001a,b).

Mineralization

In models that do not explicitly model bacteria, bacterially mediated mineralization of organic compounds is modelled without explicitly describing associated microbial growth. Depending on the redox conditions, mineralization with dissolved oxygen, nitrate, manganese oxide, iron hydroxide, and sulphate as electron acceptors, and mineralization by methanogenesis, must be considered in the models. This is of specific importance in sediment models (Rabouille and Gaillard, 1991a,b; Hunter *et al.*, 1998; Boudreau, 1999, Haeckel *et al.*, 2001; Wijsman *et al.*, 2002), because diffusion limitation often leads to the occurrence of different redox conditions within relatively small spatial regions. Some models distinguish the steps of hydrolysis of organic material from particulate to dissolved forms prior to mineralization; however, most models describe mineralization of particulate organic mater-

Table 16.11. Qualitative process table of the death process.

	Particulate components									
Process	X_i	$X_{P,i}$	$X_{PS,i}$	$X_{PP,i}$	$X_{PI,i}$	X_S	$X_{P,S}$	X_I	$X_{P,I}$	Rate
Death	−1	−	−	−	−	+	+	+	+	$k_{death,i,To}\exp(\beta_i(T-T_0))X_i$

Table 16.12. Qualitative process table of the hydrolysis process.

	Dissolved components		Particulate components		
Process	S_S	$S_{P,S}$	X_S	$X_{P,S}$	Rate
Hydrolysis	+1	+	−1	−	$k_{hyd,To}\exp(\beta_{hyd}(T-T_0))X_S$

ial directly and account for rate-limitation by hydrolysis through the choice of parameter values for the mineralization process. Table 16.14 gives an overview of the mineralization processes and of a possibility for formulating the rate expressions that account for the sequential preference for the different electron acceptors.

Oxidation

In models in which bacteria are not modelled explicitly, nitrification is modelled as oxidation of ammonium without explicit description of growth of the chemolithotrophic bacteria that perform the oxidation. This process is shown in Table 16.15. Other oxidation processes undertaken by chemolithotrophic bacteria are modelled similarly. When omitting the growth of chemolithotrophic bacteria, organic phosphorus is not affected by these processes.

Embedding Organic Phosphorus Models in Biogeochemical Models of Aquatic Systems

This section overviews the treatment of organic phosphorus transformation processes in biogeochemical models of lakes and rivers. Some milestones in the historical development are illustrated with representative models at each development stage. Typical parameter values for these models can be found in the cited literature.

Total phosphorus models

Early approaches to lake eutrophication modelling did not distinguish between inorganic, organic, particulate or dissolved phosphorus (Vollenweider, 1975, 1976). These models consist of a mass balance for total phosphorus in the lake by considering inflow, outflow and sedimentation. They are usually combined with a hydraulic model representing the lake as a well-mixed compartment and, despite their simplicity, are able to describe the major phosphorus mass fluxes in lakes.

Phosphate–particulate phosphorus models

Lake models of the next generation were still models for phosphorus only, but they distinguished dissolved phosphate and particulate phosphorus as part of algae. Table 16.16 shows the process table of such a model that considers algal growth and respiration/mineralization of algae (Imboden and Gächter, 1978). This type of model is usually combined with either a two-box (epilimnion–hypolimnion) lake model, or with a model for horizontally averaged

Table 16.13. Process table for selected chemical equilibria and precipitation/dissolution processes.

	Dissolved components								Particulate components	
Process	S_H	S_{OH}	S_{HPO_4}	$S_{H_2PO_4}$	S_{CO_2}	S_{HCO_3}	S_{CO_3}	S_{Ca}	X_{CaCO_3}	Rate
Dissociation of water	+1	+1								$k_{eq,w}(1 - S_H S_{OH}/K_{eq,w})$
Phosphate equilibrium	+		+1	−1						$k_{eq,P}(S_{H_2PO_4} - S_H S_{HPO_4}/K_{eq,p})$
Carbon dioxide equilibrium 1	+1				−1	+1				$k_{eq,1}(S_{CO_2} - S_H S_{HCO_3}/K_{eq,1})$
Carbon dioxide equilibrium 2	+1					−1	+1			$k_{eq,2}(S_{HCO_3} - S_H S_{CO_3}/K_{eq,2})$
Dissolution/precipitation of carbonate							+1	+1	−1	$k_{eq,s0}(1 - S_{Ca} S_{CO_3}/K_{eq,s0})$

Table 16.14. Qualitative process table of the mineralization processes.

Process	Dissolved components											Particulate components							Rate
	S_{NH_4}	S_{NO_3}	S_{HPO_4}	S_{O_2}	S_{HCO_3}	S_H	S_{Mn}	S_{Fe}	S_{SO_4}	S_{HS}	S_{CH_4}	X_i	X_{Pi}	$X_{PS,i}$	$X_{PP,i}$	$X_{PI,i}$	X_{MnO_2}	X_{FeOOH}	
Min. oxygen	+		+	−	+	+						−1	−	−	−	−			r_{min,O_2}
Min. nitrate	+	−	+		+	+						−1	−	−	−	−			r_{min,NO_3}
Min. manganese dioxide	+		+		+	−	+					−1	−	−	−	−	−		r_{min,MnO_2}
Min. FeOOH	+		+		+	−		+				−1	−	−	−	−		−	$r_{min,FeOOH}$
Min. sulphate	+		+		+	+			−	+		−1	−	−	−	−			r_{min,SO_4}
Methanogenesis	+		+		+	+					+	−1	−	−	−	−			r_{min,CH_4}

$$r_{min,O_2} = k_{min,O_2} \frac{S_{O_2}}{K_{min,O_2} + S_{O_2}} X_i$$

$$r_{min,NO_3} = k_{min,NO_3} \frac{K_{min,O_2}}{K_{min,O_2} + S_{O_2}} \frac{S_{NO_3}}{K_{min,NO_3} + S_{NO_3}} X_i$$

$$r_{min,MnO_2} = k_{min,MnO_2} \frac{K_{min,O_2}}{K_{min,O_2} + S_{O_2}} \frac{K_{min,NO_3}}{K_{min,NO_3} + S_{NO_3}} \frac{X_{MnO_2}}{K_{min,MnO_2} + X_{MnO_2}} X_i$$

$$r_{min,FeOOH} = k_{min,FeOOH} \frac{K_{min,O_2}}{K_{min,O_2} + S_{O_2}} \frac{K_{min,NO_3}}{K_{min,NO_3} + S_{NO_3}} \frac{K_{min,MnO_2}}{K_{min,MnO_2} + X_{MnO_2}} \frac{X_{FeOOH}}{K_{min,FeOOH} + X_{FeOOH}} X_i$$

$$r_{min,SO_4} = k_{min,SO_4} \frac{K_{min,O_2}}{K_{min,O_2} + S_{O_2}} \frac{K_{min,NO_3}}{K_{min,NO_3} + S_{NO_3}} \frac{K_{min,MnO_2}}{K_{min,MnO_2} + X_{MnO_2}} \frac{K_{min,FeOOH}}{K_{min,FeOOH} + X_{FeOOH}} \frac{S_{SO_4}}{K_{min,SO_4} + S_{SO_4}} X_i$$

$$r_{min,CH_4} = k_{min,CH_4} \frac{K_{min,O_2}}{K_{min,O_2} + S_{O_2}} \frac{K_{min,NO_3}}{K_{min,NO_3} + S_{NO_3}} \frac{K_{min,MnO_2}}{K_{min,MnO_2} + X_{MnO_2}} \frac{K_{min,FeOOH}}{K_{min,FeOOH} + X_{FeOOH}} \frac{K_{min,SO_4}}{K_{min,SO_4} + S_{SO_4}} X_i$$

Table 16.15. Process table for one-stage nitrification as an oxidation process of ammonium.

Process	Dissolved components				Rate
	S_{NH_4}	S_{NO_3}	S_{O_2}	S_H	
Nitrification	−1	+1	−2	+2	$k_{nitri}\frac{S_{NH4}}{K_{NH4,nitri}+S_{NH_4}}\frac{S_{O2}}{K_{O_2,nitri}+S_{O_2}}S_{NH4}$

Table 16.16. Process table of the lake eutrophication model by Imboden and Gächter (1978).

Process	Dissolved component S_{HPO_4}	Particulate component $X_{P,ALG}$	Rate
Growth of algae	−1	1	$\mu_{max}(I)\frac{S_{HPO4}}{K_{HPO_4}+S_{HPO_4}}X_{P,ALG}$
Respiration/mineralization of algae	1	−1	$k_{resp/min}X_{P,ALG}$

concentrations in a stratified lake, the depth of which is resolved continuously.

Early nutrient cycle models

Later, phosphorus models were extended to include dissolved oxygen, algae, and inorganic and organic nitrogen compounds. An example of such a model is the river water quality model QUAL2E (Brown and Barnwell, 1987; see Fig. 16.2 and Table 16.17). This was developed as an extension to the Streeter–Phelps model (Streeter and Phelps, 1925), which described the effect of organic pollutants on dissolved oxygen concentrations in a river. In the QUAL2E model (Fig. 16.2) dead organic matter is spread between three model components (CBOD, ORG-N and ORG-P) and living algae are modelled by an additional model component (so that ORG-P concentrations cannot be compared with measurements of organic phosphorus in the system). In this model, parameter values for organic phosphorus and nitrogen hydrolysis and degradation of organic matter can be independently selected. Similarly, sediment oxygen demand and nitrogen and phosphorus release from the sediment are independent of (past) organic matter sedimentation rates, organic nitrogen and organic phosphorus. This can lead to the choice of unreasonable parameter combinations by an inexperienced model user.

Closed mass balance nutrient cycle models

A next generation of lake water quality models removed inconsistencies present in the QUAL2E model by more explicitly connecting processes with model components and by closing material cycles (Riley and Stefan, 1988; Hamilton and Schladow, 1997; Schladow and Hamilton, 1997; Omlin *et al.*, 2001a). Figures 16.3–16.6 illustrate results of the model of Omlin *et al.* (2001a) for algae, phosphate, total particulate phosphorus sedimentation flux, and phosphate transformation rates. This model is representative of the above-mentioned class of models and contains a discussion of different model structures with respect to phosphorus fluxes. The algal profiles of Lake Zürich are reasonably well reproduced by the model (Fig. 16.3), with two distinguishable maxima in algal concentration in spring and in autumn. Phosphate profiles show the phosphate depletion in the epiliminion during the summer and the upward flux of phosphate released by mineralization processes from the sediment (Fig. 16.4). The

Table 16.17. Process table of the river water quality model QUAL2E.

	Dissolved components					Particulate components				
Process	S_{NH_4}	S_{NO_2}	S_{NO_3}	S_{HPO_4}	S_{O_2}	X_S	$X_{P,S}$	$X_{N,S}$	X_{ALG}	Rate
Growth of algae	$-F\alpha_1$		$-(1-F)\alpha_1$	$-\alpha_2$	$-\alpha_3$				$+1$	$\mu_{max}X_{ALG}f_L(I)f_{NP}(S_{NH_4},S_{NO_3},S_{HPO_4})$
Respiration of algae					$-\alpha_4$		α_2	α_1	-1	ρX_{ALG}
Sediment of algae									-1	$\frac{\sigma_1}{d} X_{ALG}$
Nitrification 1	-1	1			$-\alpha_5$					$\beta_1(1-e^{-K_{nit}X_{O_2}})S_{NH_4}$
Nitrification 2		-1	1		$-\alpha_6$					$\beta_2(1-e^{-K_{nit}X_{O_2}})S_{NO_2}$
Degradation of organic matter					-1	-1				K_1X_S
Sedimentation of organic matter						-1				$K_3X_{N,S}$
Hydrolysis of organic nitrogen	1							-1		$\beta_3X_{N,S}$
Sedimentation of organic nitrogen								-1		σ_4X_S
Hydrolysis of organic phosphorus				1			-1			$\beta_4X_{P,S}$
Sedimentation of organic phosphorus										$\sigma_5X_{P,S}$
Sediment oxygen demand					-1					$\frac{K_4}{d}$
Ammonium release from sediment	$+1$									$\frac{\sigma_3}{d}$
Phosphate release from sediment				$+1$						$\frac{\sigma_2}{d}$

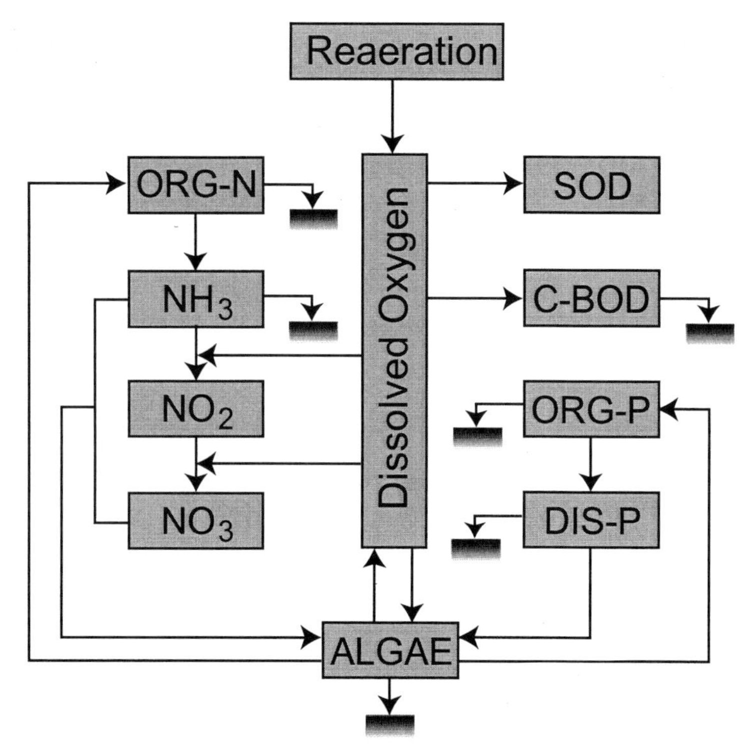

Fig. 16.2. Overview of model components, processes, and parameters of the river water quality model QUAL2E (Brown and Barnwell, 1987). (ORG-P = organic phosphorus; ORG-N = organic nitrogen; DIS-P = dissolved phosphorus; C-BOD = carbonaceous biochemical oxygen demand; SOD = sediment oxygen demand).

phosphorus sedimentation flux shows two yearly maxima corresponding to the maxima in algal concentrations (Fig. 16.5). Of special interest are the calculated phosphate transformation rates, because these cannot be measured directly (Fig. 16.6). In the top layers, phosphate in the water column is consumed by primary production. Below, phosphate release by mineralization in the water column is overcompensated by phosphate uptake on particles sinking through the hypolimnion. In the sediment, phosphate is released by mineralization.

Figure 16.7 shows an analysis of the effect of different model assumptions on phosphate profiles in the water column of the lake. Only the full model, with its increased phosphorus sedimentation rate due to phosphate uptake by sinking particles, is able to reproduce the observed profiles. The models omitting these processes predict a smaller flux of phosphate from the sediment (lower concentrations in the depth of the lake) and overestimate the concentrations immediately below the epilimnion. The difference between models with variable and constant stoichiometry of phosphorus uptake is observed in the concentration of nitrate (see Omlin *et al.*, 2001a). Figure 16.7 demonstrates the need for a variable stoichiometry of phosphorus uptake by growing algae and the need for a process involving phosphate uptake by particles sinking through the epilimnion.

With respect to rivers, more recent approaches to mechanistic biogeochemical modelling include bacterial populations in addition to algae and consumers (Reichert *et al.*, 2001a,b; Shanahan *et al.*, 2001; Vanrolleghem *et al.*, 2001). For small rivers, in which benthic transformation rates domi-

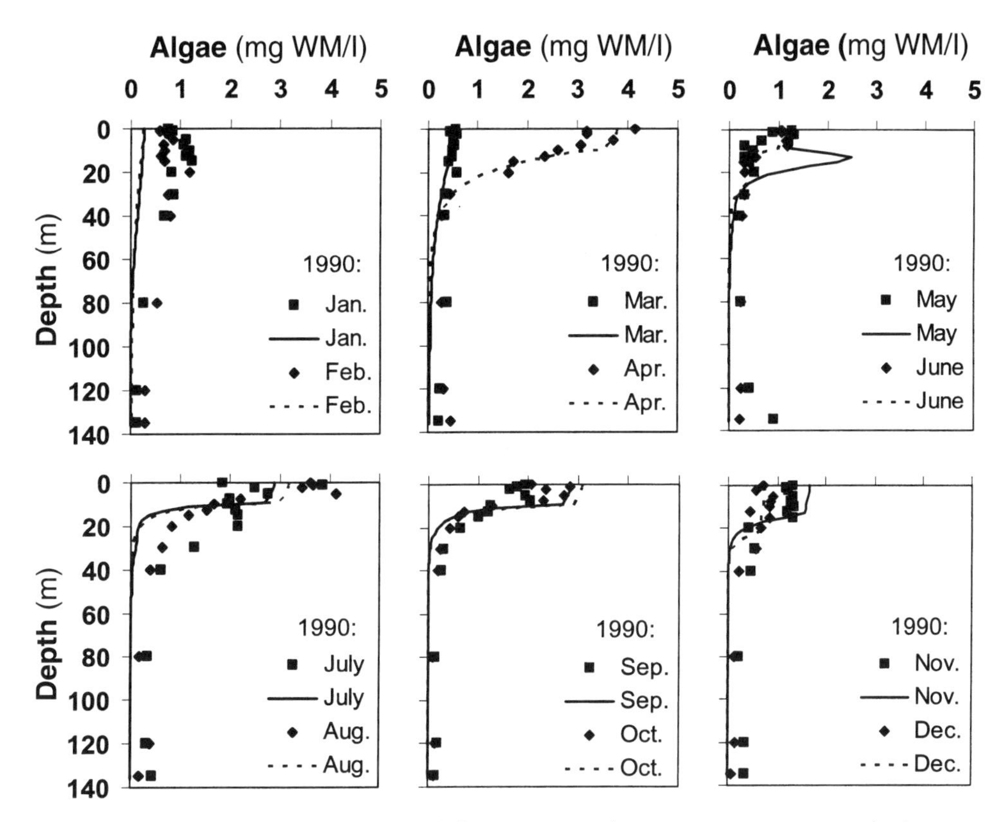

Fig. 16.3. Monthly algae profiles in Lake Zurich during 1990. Markers represent measurements by the water supply authority of Zurich (WVZ) and lines represent corresponding simulation results. Reproduced from Omlin *et al.* (2001a) with permission from Elsevier Science.

nate suspended transformation rates, models for the water column must be combined with models of the benthic population (McIntire, 1973; McIntire and Colby, 1978; Uehlinger *et al.*, 1996; Rutherford *et al.*, 1999).

Comparative Analysis of Organic Phosphorus Models

In this section, we compare how three aspects of models have been treated in different application areas and suggest which of the approaches is most useful for building more universal models in the future.

Description of biomass

As outlined in the second section, biomass has been described as organic matter with a constant elemental ratio (see equation 16.1), as a parameterized combination of CH_2O, ammonia and phosphate (equation 16.2), as chemical oxygen demand (COD), with nitrogen and phosphorus content per unit of chemical oxygen demand (equation 16.3), and by elemental mass fractions (equation 16.4). From the point of view of universality and generality, the last approach seems to be the most promising, as it is flexible and makes it possible to calculate the parameters of the other approaches uniquely. The disadvantage of a relatively large number of model parameters is compensated by the advantages of a clear declaration of (otherwise partially implicit) model assumptions and the potential for extended sensitivity analysis with respect to model assumptions.

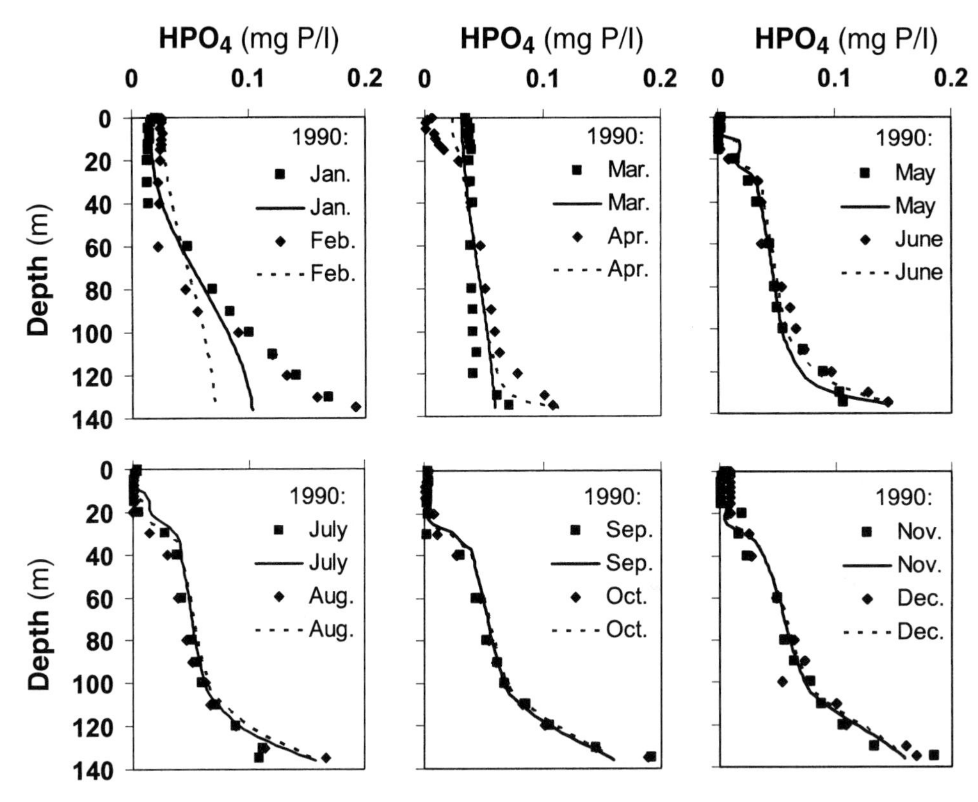

Fig. 16.4. Monthly phosphate profiles in Lake Zürich for the year 1990. Markers represent measurements by the water supply authority of Zurich (WVZ), lines represent corresponding simulation results. Reproduced from Omlin *et al.* (2001a), with permission from Elsevier Science.

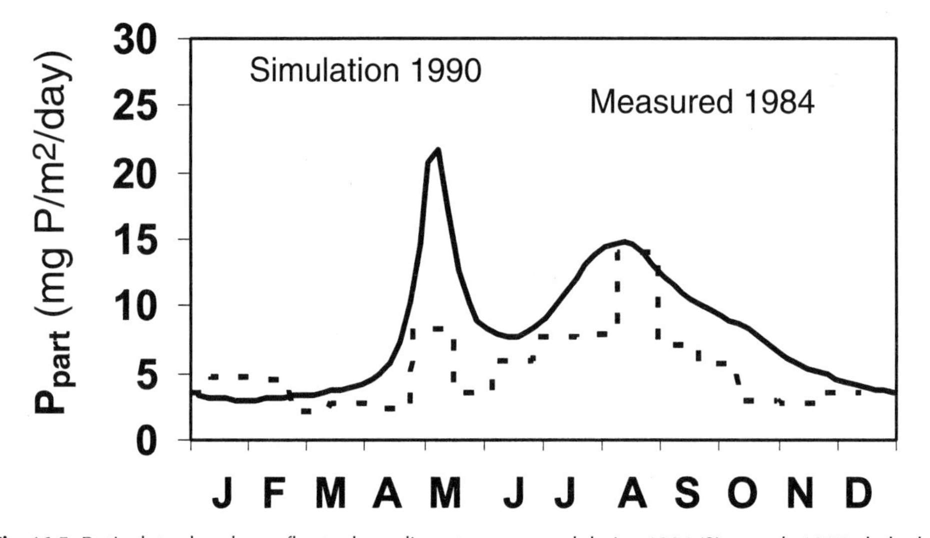

Fig. 16.5. Particulate phosphorus flux to the sediment as measured during 1984 (Sigg *et al.*, 1987; dashed line) and calculated for 1990 (solid line). Reproduced from Omlin *et al.* (2001a) with permission from Elsevier Science.

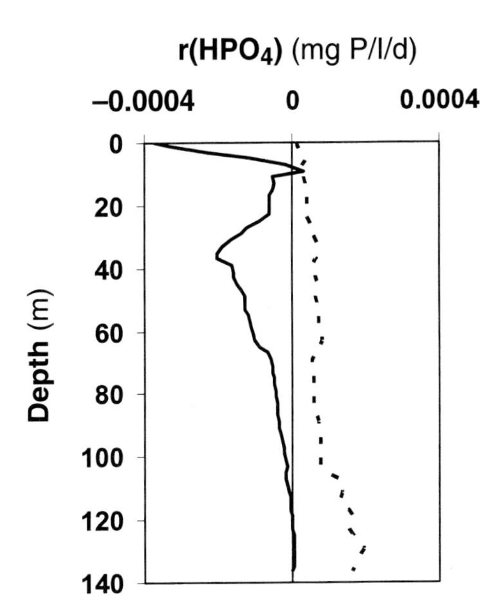

Fig. 16.6. Net phosphate transformation rates in September 1989 in the water column (solid line) and in the sediment (converted to water-column equivalent rates; dashed line). Reproduced from Omlin *et al.* (2001a) with permission from Elsevier Science.

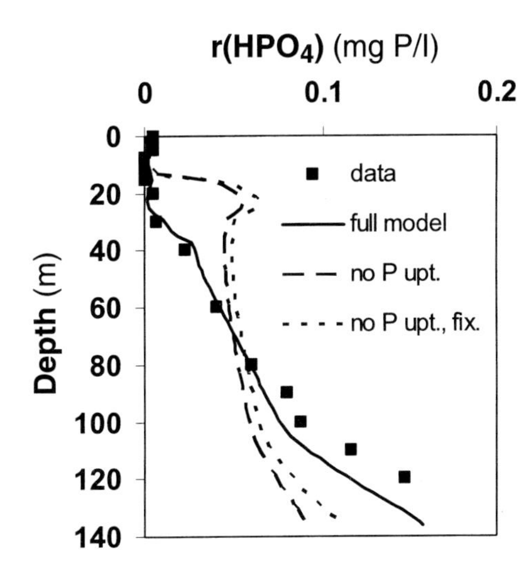

Fig. 16.7. Comparison of phosphate profiles in September 1989 of the full model (solid line) with the results of models with omission of the phosphate uptake process on sinking particles (long dashed line) and with omission of the phosphate uptake process and fixed Redfield stoichiometry of algal growth (short dashed line). Reproduced from Omlin *et al.* (2001a) with permission from Elsevier Science.

Model representation

The process table approach described above is used in wastewater engineering, whereas in ecological modelling it is more common to present a list of differential equations. The list of differential equations conceals a clear overview of processes by spreading information about a single process to terms in several differential equations. Therefore, we propose switching to the process table approach in ecological modelling.

Considered processes and process formulation

Table 16.18 gives an overview of the implementation of the processes described in the third section in typical biogeochemical models of rivers, lakes, oceans, sediments and wastewater treatment plants. Of the processes described in detail above (Tables 16.3–16.15), the two last processes of mineralization and oxidation are more aggregated than the ones described in the earlier subsections, because the rate constants of the latter depend on the bacterial populations responsible for the transformation. This means that models replacing the compound processes of mineralization and oxidation by growth (and respiration) of heterotrophic and chemolithotrophic bacteria have a higher degree of mechanistic resolution. Table 16.18 demonstrates that models for wastewater treatment plants have taken a leading role in the development of such a higher degree of mechanistic resolution.

Summary and Conclusions

There exist a large number of mechanistic models for organic phosphorus transformation processes in aquatic systems. However, different approaches have been chosen to describe different aquatic systems (rivers, lakes, ocean, sediment, sewage treatment plants), comparison of which is complicated

Table 16.18. Implementation of the processes described in the third section for typical biogeochemical models of rivers, lakes, oceans, sediments and wastewater treatment plants (WWTP). 'X' means that the process is usually implemented, 'x' means that it is implemented in some models, and an empty field means it is not usually implemented.

Table(s)	Process	River	Lake	Ocean	Sediment	WWTP
16.3, 16.4	Primary production	X	X	X		
16.5	Growth of chemolithotrophic bacteria	x				X
16.6	Uptake/adsorption		X	X		
16.7	Growth of heterotrophic bacteria	x				X
16.8	Growth of phosphate accumulating bacteria					X
16.9	Consumption	x	X	X		
16.10	Respiration	X	X	X		X
16.11	Death	X	X	X		
16.12	Hydrolysis	x				X
16.13	Chemical equilibria, precipitation and dissolution	x	x	x	x	x
16.14	Mineralization	X			X	
16.15	Oxidation	X	X	X	X	

by differences in notation. Some of the differences can be attributed to differences in dominant processes; for example, there is no photosynthesis in deep sediment or sewage treatment plants due to the absence of light, while anaerobic processes are not important in the water column of well-aerated rivers. However, some of the differences result from different process descriptions. Wastewater treatment plant models have played a pioneering role, not only in parameterizing the process rates of dissolved organic and inorganic components, but also in including population models of the microorganisms responsible for these transformations and modelling the complete cycles for the most important elements. This has been done historically to a much lesser degree in biogeochemical models of rivers and lakes, where the sediment is often described as a sink of organic particles and a source of nutrients without including a mechanistic relationship between these two aspects (an example is the QUAL2E river water quality model; Brown and Barnwell, 1987). This can lead to violations of mass conservation by inexperienced users, who do not consider these constraints when adjusting model parameters.

Although the approach used in sewage treatment plant models is conceptually more satisfactory, it cannot be transferred to natural systems without changes. First, the exclusive use of chemical oxygen demand for the quantification of organic material cannot be applied in ecological models, because some measurement techniques (such as counting organisms) are not easily converted to chemical oxygen demand, and because of a higher ecological significance of other measurement units. Secondly, it is much more difficult to measure populations of microorganisms in natural systems than in activated sludge sewage treatment plants. For this reason, the inclusion of these populations leads to over-parameterized models that are difficult to uniquely calibrate. Nevertheless, it is hoped that the comparison between models for different aquatic systems presented in this chapter will stimulate improvements of models for specific systems based on ideas already applied to other systems. In the following section, we summarize some ideas for future developments in three types of model applications: models for understanding, models for information exchange, and models for prediction.

Models for understanding

In models that are primarily used for quantitatively describing processes relevant to phosphorus cycling in the environment, it would be beneficial to include more mechanistic detail. The following extensions could be of interest:

- Distinguishing organic phosphorus compounds according to their bioavailability using measurable compound classes. This could lead to a fruitful interaction between newly available measurements and information gained from models in terms of the behaviour of organic phosphorus compound groups. Of special interest would be modelling the decrease in phosphorus concentrations of organic material during the degradation process due to the generally greater degradability of nutrient-rich compounds.
- Replacing empirical model parameters directly related to transformation rates of substances dissolved in the water phase by submodels of the bacterial population responsible for the transformations. In rivers and shallow lakes, this may lead to the need for coupling a microbial model with a benthic community model, because grazing can be one of the limiting factors determining the population density of algal and bacterial mats (McIntire, 1973; McIntire and Colby, 1978; Uehlinger *et al.*, 1996; Rutherford *et al.*, 1999).
- Due to the requirement of frequent changes between aerobic and anoxic conditions, polyphosphate accumulation in bacteria will probably be of minor importance for natural systems. However, sediments with periodically changing redox conditions, such as tidal flats, could be an exception (Davelaar, 1993; Khoshmanesh *et al.*, 2002). For such systems, it would be interesting to apply models for phosphate-accumulating bacteria developed for sewage treatment plants, in order to use knowledge from this field and at the same time gain information about the transferability of these models to other systems.
- The increase in mechanistic detail of process description is often accompanied by loss of parameter identifiability (Beck, 1999). This makes it extremely important to combine such model extensions with a careful analysis of parameter identifiability. Such an approach can clearly communicate not only the goodness of fit, but also the potential for gaining knowledge about parameter values, the conditionalities of parameters estimated from data, and the remaining degrees of freedom for parameter choices (Brun *et al.*, 2001; Omlin *et al.*, 2001b; Reichert and Vanrolleghem, 2001).

Models for information exchange

To better support the use of model formulations for exchange of information between different research fields, a common nomenclature would be advantageous. We propose combining the process table approach used in wastewater engineering (Henze *et al.*, 1986, 1999; Gujer *et al.*, 1999) with the elemental mass fraction approach recently proposed for river water quality modelling (Reichert *et al.*, 2001a,b). This formalism is understandable to biogeochemists and engineers and facilitates the conversion of mass units from this formalism to units used in different fields and to elemental analyses more readily than the chemical oxygen demand-based approach used in wastewater engineering. Such a common formalism would support closer collaboration between biogeochemists and engineers in the field of organic phosphorus transformation modelling.

Models for prediction

Models for prediction cannot be as detailed as models for understanding, because their calibration must usually be done with a much more limited data-set. On the other hand, the models should consider all processes relevant during the prediction time period. In general it can be assumed that oversimplified models lead to systematically wrong predictions and an underestimation of prediction accuracy (Reichert and Omlin, 1997). More complex models decrease systematic errors, but the uncertainty in their predictions is largely due to the large number of uncertain model parameters (Beck, 1999; Reckhow and Chapra, 1999). It is difficult to find an intermediate model complexity that minimizes prediction uncertainty, so the best approach is to develop several models of different complexity and compare their predictions in order to get a rough estimate of the effect of model structure on model predictions. It is a challenging research task to develop models of different complexity and to carefully compare their prediction uncertainty in well-investigated systems. This could lead to conclusions about reasonable model complexity as a function of the investigated system and the questions to be answered.

Acknowledgements

We thank Michael Hupfer, Mark Borsuk and an anonymous reviewer for helpful comments on the manuscript.

References

Ahn, C.-Y., Chung, A.-S. and Oh, H.-M. (2002) Diel rhythm of algal phosphate uptake rates in P-limited cyclostats and simulation of its effect on growth and competition. *Journal of Phycology* 38, 695–704.

Appeldoorn, K.J., Boom, K.J., Kortstee, G.J.J. and Zehnder, A.J.B. (1992) Biological phosphate removal by activated-sludge under defined conditions. *Water Research* 26, 453–460.

Beck, M.B. (1999) Coping with ever larger problems, models, and data bases. *Water Science and Technology* 39, 1–11.

Benitez-Nelson, C.R. (2000) The biogeochemical cycling of phosphorus in marine systems. *Earth-Science Reviews* 51, 109–135.

Benitez-Nelson, C.R. and Karl, D.M. (2002) Phosphorus cycling in the North Pacific subtropical gyre using cosmogenic ^{32}P and ^{33}P. *Limnology and Oceanography* 47, 762–770.

Blackall, L.L., Crocetti, G., Saunders, A.M. and Bond, P.L. (2002) A review and update of the microbiology of enhanced biological phosphorus removal in wastewater treatment plants. *Antonie Van Leeuwenhoek International Journal of General and Molecular Microbiology* 81, 681–691.

Boudreau, B.P. (1999) Metals and models: diagenetic modelling in freshwater lacustrine sediments. *Journal of Paleolimnology* 22, 277–251.

Brannon, C.A. and Sommers, L.E. (1985) Stability and mineralization of organic phosphorus incorporated into model humic polymers. *Soil Biology and Biochemistry* 17, 221–227.

Brown, L.C. and Barnwell, T.O. (1987) *The Enhanced Stream Water Quality Models QUAL2E and QUAL2E-UNCAS: Documentation and User Manual.* Environmental Research Laboratory Office of Research and Development, United States Environmental Protection Agency, Athens, Georgia.

Brun, R., Reichert, P. and Künsch, H.R. (2001) Practical identifiability analysis of large environmental simulation models. *Water Resources Research* 37, 1015–1030.

Carpenter, S.R., Kitchell, J.F. and Hodgson, J.R. (1985) Cascading trophic interactions and lake productivity. *BioScience* 35, 634–639.

Davelaar, D. (1993) Ecological significance of bacterial polyphosphate metabolism in sediments. *Hydrobiologia* 253, 179–192.

Droop, M.R. (1973) Some thoughts on nutrient limitation in algae. *Journal of Phycology* 9, 264–272.

Droop, M.R. (2003) In defence of the cell quota model of micro-algal growth. *Journal of Plankton Research* 25, 103–107.

Ducobu, H., Huisman, J., Jonker, R.R. and Mur, L.R. (1998) Competition between a prochlorophyte and a cyanobacterium under various phosphorus regimes: comparison with the Droop model. *Journal of Phycology* 34, 467–476.

Gächter, R. and Bloesch, J. (1985) Seasonal and vertical variation in the C:P ratio of suspended and settling seston of lakes. *Hydrobiologia* 128, 193–200.

Gächter, R. and Mares, A. (1985) Does settling seston release soluble reactive phosphorus in the hypolimnion of lakes? *Limnology and Oceanography* 30, 364–371.

Gächter, R., Meyer, J.S. and Mares, A. (1988) Contribution of bacteria to release and fixation of phosphorus in lake sediments. *Limnology and Oceanography* 33, 1542–1558.

Gujer, W., Henze, M., Mino, T. and van Loosdrecht, M. (1999) Activated sludge model No. 3. *Water Science and Technology* 39, 183–193.

Haeckel, M., Konig, I., Reich, V., Weber, M.E. and Suess, E. (2001) Pore water profiles and numerical modelling of biogeochemical processes in Peru Basin deep-sea sediments. *Deep-Sea Research Part II: Topical Studies in Oceanography* 48, 3713–3736.

Hamilton, D.P. and Schladow, S.G. (1997) Prediction of water quality in lakes and reservoirs. Part I. Model description. *Ecological Modelling* 96, 91–110.

Henze, M., Grady, C.P.L., Gujer, W., Marais, G.V.R. and Matsuo, T. (1986) *Activated Sludge Model No. 1.* Scientific and Technical Report No. 1, International Association for Water Pollution, Research and Control (IAWPRC), London.

Henze, M., Gujer, W., Takahashi, M., Matsuo, T., Wentzel, M.C. and Marais, G.v.R. (1995) *Activated Sludge Model No. 2.* Scientific and Technical Report No. 3, International Association for Water Pollution, Research and Control (IAWPRC), London.

Henze, M., Gujer, W., Mino, T., Matsuo, T., Wentzel, M.C., Marais, G.v.R. and van Loosdrecht, M.C.M. (1999) Activated sludge model No. 2d, ASM2d. *Water Science and Technology* 39, 165–182.

Hunter, K.S., Wang, Y. and van Cappellen, P. (1998) Kinetic modelling of microbially driven redox chemistry of subsurface environments: coupling transport, microbial metabolism and geochemistry. *Journal of Hydrology* 209, 53–80.

Hupfer, M., Gächter, R. and Giovanoli, R. (1995) Transformation of phosphorus species in settling seston and during early sediment diagenesis. *Aquatic Sciences* 75, 305–324.

Imboden, D.M. and Gächter, R. (1978) A dynamic lake model for trophic state prediction. *Ecological Modelling* 4, 77–98.

Khoshmanesh, A., Hart, B.T., Duncan, A. and Beckett, R. (2002) Luxury uptake of phosphorus by sediment bacteria. *Water Research* 36, 774–778.

Kolowith, L.C., Ingall, E.D. and Benner, R. (2001) Composition and cycling of marine organic phosphorus. *Limnology and Oceanography* 46, 309–320.

Lenton, T.M. and Watson, A.J. (2000a) Redfield revisited. 1. Regulation of nitrate, phosphate, and oxygen in the ocean. *Global Biogeochemical Cycles* 14, 225–248.

Lenton, T.M. and Watson, A.J. (2000b) Redfield revisited. 2. What regulates the oxygen content of the atmosphere? *Global Biogeochemical Cycles* 14, 249–268.

Manga, J., Ferrer, J., Garcia-Usach, F. and Seco, A. (2001) A modification to the Activated Sludge Model No. 2 based on the competition between phosphorus-accumulating organisms and glycogen-accumulating organisms. *Water Science and Technology* 43, 161–171.

McIntire, C.D. (1973) Algal dynamics in laboratory streams: a simulation model and its implications. *Ecological Monographs* 43, 399–420.

McIntire, C.D. and Colby, J.A. (1978) A hierarchical model of lotic ecosystems. *Ecological Monographs* 48, 167–190.

Morgenroth, E. and Wilderer, P.A. (1998) Modelling of enhanced biological phosphorus removal in a sequencing batch biofilm reactor. *Water Science and Technology* 37, 583–587.

Omlin, M., Brun, R. and Reichert, P. (2001a) Biogeochemical model of Lake Zürich: sensitivity, identifiability and uncertainty analysis. *Ecological Modelling* 141, 105–123.

Omlin, M., Reichert, P. and Forster, R. (2001b) Biogeochemical model of Lake Zürich: model equations and results. *Ecological Modelling* 141, 77–103.

Pimenov, N.V., Rusanov, I.I., Yusupov, S.K., Fridrich, J., Lein, A., Wehrli, B. and Ivanov, M.V. (2000) Microbial processes at the aerobic–anaerobic interface in the deep-water zone of the Black Sea. *Microbiology* 69, 436–448.

Rabouille, C. and Gaillard, J.-F. (1991a) Towards the edge: early diagenetic global explanation. A model depicting the early diagenesis of organic matter, O_2, NO_3, Mn, and PO_4. *Geochimica et Cosmochimica Acta* 55, 2511–2525.

Rabouille, C. and Gaillard, J.-F. (1991b) A coupled model representing the deep-sea organic carbon mineralization and oxygen consumption in surficial sediments. *Journal of Geophysical Research* 96, 2761–2776.

Reckhow, K.H. and Chapra, S.C. (1999) Modelling excessive nutrient loading in the environment. *Environmental Pollution* 100, 197–207.

Redfield, A.C. (1958) The biological control of chemical factors in the environment. *American Scientist* 46, 205–222.

Redfield, A.C., Ketchum, B.J. and Richards, F.A. (1963) The influence of organisms on the composition of seawater. In: Hill, M.R. (ed.) *The Sea,* Vol. 2. John Wiley & Sons, New York, pp. 26–77.

Reichert, P. and Omlin, M. (1997) On the usefulness of overparameterized ecological models. *Ecological Modelling* 95, 289–299.

Reichert, P. and Vanrolleghem, P. (2001) Identifiability and uncertainty analysis of the river water quality model No. 1 (RWQM1). *Water Science and Technology* 43, 329–338.

Reichert, P., Borchardt, D., Henze, M., Rauch, W., Shanahan, P., Somlyódy, L. and Vanrolleghem, P. (2001a) River Water Quality Model no. 1

(RWQM1). II. Biochemical process equations. *Water Science and Technology* 43, 11–30.

Reichert, P., Borchardt, D., Henze, M., Rauch, W., Shanahan, P., Somlyódy, L. and Vanrolleghem, P. (2001b) *River Water Quality Model No. 1*. Scientific and Technical Report No. 12, International Water Association (IWA), IWA Publishing, London.

Riley, M.J. and Stefan, H.G. (1988) MINLAKE: a dynamic lake water quality simulation model. *Ecological Modelling* 43, 155–182.

Rutherford, J.C., Scarsbrook, M.R. and Broekhuizen, N. (1999) Grazer control of stream algae: modelling temperature and flood effects. *Journal of Environmental Engineering* 126, 331–339.

Schladow, S.G. and Hamilton, D.P. (1997) Prediction of water quality in lakes and reservoirs. Part II. Model calibration, sensitivity analysis and application. *Ecological Modelling* 96, 111–123.

Shanahan, P., Borchardt, D., Henze, M., Rauch, W., Reichert, P., Somlyódy, L. and Vanrolleghem, P. (2001) River Water Quality Model No. 1 (RWQM1). I. Modelling approach. *Water Science and Technology* 43, 1–9.

Sigg, L., Sturm, M. and Kister, D. (1987) Vertical transport of heavy metals in settling particles in Lake Zurich. *Limnology and Oceanography* 32, 112–130.

Sinke, A.J.C., Cottaar, F.H.M., Buis, K. and Keizer, P. (1992) Methane oxidation by methanotrophs and its effects on the phosphate flux over the sediment–water interface in a eutrophic lake. *Microbial Ecology* 24, 259–269.

Sinke, A.J.C., Cottaar, F.H.M. and Keizer, P. (1993) A method to determine the contribution of bacteria to phosphate uptake by aerobic freshwater sediment. *Limnology and Oceanography* 38, 1081–1087.

Soetaert, K., Herman, P.M.J. and Middelburg, J.J. (1996) A model of early diagenetic processes from the shelf to abyssal depths. *Geochimica et Cosmochimica Acta* 60, 1029–1040.

Streeter, W.H. and Phelps, E.B. (1925) *A Study of the Pollution and Natural Purification of the Ohio River*. Public Health Bull. 146, United States Public Health Service, Washington, DC.

Streichan, M., Golecki, J.R. and Shon, G. (1990) Polyphosphate-accumulating bacteria from sewage plants with different processes for biological phosphorus removal. *FEMS Microbial Ecology* 73, 113–124.

Stumm, W. and Morgan, J.J. (1981) *Aquatic Chemistry*. John Wiley & Sons, New York.

Tyrrell, T. (1999) The relative influence of nitrogen to phosphorus on oceanic primary production. *Nature* 400, 525–531.

Uehlinger, U., Bührer, H. and Reichert, P. (1996) Periphyton dynamics in a flood prone pre-alpine river: evaluation of significant processes by modelling. *Freshwater Biology* 36, 249–263.

van Cappellen, P. and Ingall, E.D. (1994) Benthic phosphorus regeneration, net primary production, and oceanic anoxia: a model of the coupled marine biogeochemical cycles of carbon and phosphorus. *Palaeooceanography* 9, 677–692.

van Cappellen, P. and Ingall, E.D. (1996) Redox stabilizations of the atmosphere and oceans by phosphorus-limited marine productivity. *Science* 271, 493–496.

Vanrolleghem, P., Borchardt, D., Henze, M., Rauch, W., Reichert, P., Shanahan, P. and Somlyódy, L. (2001) River Water Quality Model No. 1 (RWQM1). III. Biochemical submodel selection. *Water Science and Technology* 43, 31–40.

Vollenweider, R.A. (1975) Input–output models with special reference to the phosphorus loading concept in limnology. *Cweizerische Zeitschrift fur Hydrologie* 37, 53–83.

Vollenweider, R.A. (1976) Advances in defining critical loading levels for phosphorus in lake eutrophication. *Memorie dell'Istituto Italiano di Idrobiologia* 33, 53–83.

Wijsman, J.W.M., Herman, P.M., Middelburg, J.J. and Soetaert, K. (2002) A model for early diagenetic processes in sediments of the continental shelf of the Black Sea. *Estuarine, Coastal and Shelf Science* 54, 403–421.

17 Synthesis and Recommendations for Future Research[1]

Organic phosphorus is intimately involved in processes that influence some of the most important issues facing the modern world. Indeed, successful solutions to challenging global issues as diverse as food production, environmental pollution and climate change require knowledge of the factors that determine the movement and transformations of organic phosphorus compounds in the environment. Therefore, it is not unrealistic to suggest that our global future may depend on a detailed understanding of the biogeochemistry of organic phosphorus.

Among the greatest challenges facing humans in the 21st century is to increase agronomic productivity to feed the world population (Rosegrant and Cline, 2003). Recent predictions of global population growth indicate the strong probability that the world population will increase from the current figure of just over 6 billion to around 9 billion by the end of this century (Lutz *et al.*, 2001). Most of this growth will occur in 'developing' world regions that contain some of the most fragile agroecosystems, including the arid and semi-arid tropics. Managing agriculture in these regions to achieve the required levels of productivity is difficult, especially considering the marked decline in crop yield in many tropical agroecosystems (Stocking, 2003). It is estimated that approximately 80% of the future increase in crop production in developing countries must come from intensification (Alexandratos, 1995), yet standard practices used in much of the 'developed' world, including the widespread use of mineral fertilizers, remain beyond the financial capacity of farmers in developing regions. Phosphorus is generally the nutrient that most limits agricultural productivity in tropical agroecosystems, notably those on Oxisols and Ultisols (well-weathered, iron-rich soils with a strong capacity to fix phosphorus) in South America, Africa and Southeast Asia (Fairhurst *et al.*, 1999). Optimizing organic phosphorus use by crops in tropical systems may therefore hold the key to providing food for the growing population, particularly in regions where a decline in soil fertility is the main constraint on food security.

Systems that provide sufficient phosphorus to sustain crop growth may risk becoming a source of phosphorus pollution to waterbodies, which can contribute to eutrophication and the growth of toxic algae (Sharpley *et al.*, 1996). Eutrophication

[1] This chapter is in part the result of a discussion session involving all delegates at the Organic Phosphorus 2003 meeting, held on 13–18 July 2003 in Ascona, Switzerland.

resulting from diffuse agricultural pollution is often considered to be an exclusively developed-world phenomenon, but many examples exist in developing countries, and these are likely to become increasingly common following any future agricultural intensification. For example in Lake Victoria in East Africa, the second largest freshwater lake in the world, nutrient pollution in runoff from the surrounding catchment during recent decades has contributed to severe eutrophication and the associated disappearance of indigenous fish communities (Verschuren *et al.*, 2002). Therefore, a critical issue concerns the potential environmental impact of the required increases in agricultural productivity. Reconciling increased productivity with a healthy environment demands robust 'sustainable' systems, defined here as production systems that meet food requirements in an economically and environmentally sustainable manner (Runge-Metzner, 1995). Such systems must optimize natural processes to improve fertility, while simultaneously minimizing 'leakage' in the form of pollutant transfer to waterbodies and gaseous emissions to the atmosphere.

Threats to natural ecosystems from human impacts, including loss of habitat and biodiversity, are major current environmental concerns. From tropical rainforests to Arctic tundra, threats to ecosystems from climate change, pollution, and other human activities are well documented (Matson *et al.*, 2002; Jenkins, 2003; Root *et al.*, 2003). Protecting natural environments requires a comprehensive understanding of biogeochemical cycles, of which organic phosphorus is a key component. Phosphorus availability often limits biological productivity in both terrestrial and aquatic ecosystems, so organic phosphorus turnover may regulate key processes involved in the soil carbon cycle. Such processes include those regulating phytoplankton productivity in the oceans (Tyrell, 1999) and the decomposition of organic matter in tundra soils (Hobbie *et al.*, 2002). Regulation of ecosystem processes by phosphorus availability is also likely to be exacerbated in the future following continued deposition of reactive nitrogen from the atmosphere. This promotes or enhances biological phosphorus limitation in many ecosystems, and can contribute to marked shifts in species composition (Aber *et al.*, 1998; Bobbink *et al.*, 1998).

How can we address these issues, and what are the knowledge gaps? Our understanding of abiotic reactions and mechanisms regulating inorganic phosphorus in environmental processes is relatively strong, and backed by many decades of research. However, despite some recent advances, our understanding of the role and fate of organic phosphorus in the environment remains rudimentary. Four principal areas of research are necessary to address this shortfall:

1. The development of simple, robust techniques for identifying and quantifying organic phosphorus in environmental samples.
2. An improved knowledge of the mechanisms and rates of organic phosphorus transformations, so that we can better predict the consequences of management interventions on organic phosphorus dynamics.
3. Like other areas of natural resource management and ecology, the development of scaling rules to allow knowledge developed to address issues at one temporal or spatial scale to be applied at smaller or larger scales.
4. Detailed investigation of mechanisms by which organisms access organic phosphorus compounds in the environment, which is critical to understanding almost all the key issues involving organic phosphorus dynamics.

The driving force for many advances in environmental science is often the development of novel analytical techniques, and organic phosphorus research is no exception. Indeed, investigation of organic phosphorus in the environment has been hampered by the lack of suitable techniques for detecting the diverse range of chemical compounds present in soil, sediment and water samples. Organic phosphorus compounds are chemically diverse, and most environmental samples contain compounds ranging from simple phosphate esters, with molecular weights of a few hundred daltons, to complex molecules with molecular weights of several hundred thousand daltons. Extracting organic phosphorus from soil or sediment without altering its chemical structure can also be difficult, and there are few reliable methods for assessing such changes. Furthermore, quantitative detection of organic phosphorus species is a highly specialized process that generally requires expensive and sophisticated equipment. Access to many of the powerful analytical instruments necessary for such studies is frequently limited for environmental researchers – a situation that must change if the study of organic phosphorus in the environment is to become a truly 'molecular' science. At the same time, the development of robust, inexpensive, yet accurate, techniques for speciating the often low concentrations of organic phosphorus found in environmental samples would facilitate a wider array of studies and hasten a more mechanistic understanding of organic phosphorus in the environment.

Understanding ecosystem processes at a mechanistic level is a fundamental aim of environmental scientists, but there is a clear lack of such information for organic phosphorus. Interest in mechanisms unites scientists working at various scales in diverse disciplines, providing a basis for interdisciplinary collaboration. Mechanisms of key interest include those involved in the use of organic phosphorus by organisms, but our understanding of such mechanisms remains limited, despite their widespread occurrence in the environment. For example, processes involving the synthesis of phosphatase enzymes by plants and microbes have been widely studied in both terrestrial and aquatic environments, yet we remain relatively ignorant of some of the most basic processes involved. The nutritional role of inositol phosphates remains largely unknown despite their ubiquity in the environment, and although it is now known that some high-latitude plants can directly take up organic nitrogen compounds (Näsholm *et al.*, 1998), no such information exists for organic phosphorus. Furthermore, information is urgently required on rates and mechanisms of organic phosphorus transformations, notably those involved in the synthesis and hydrolysis of organic phosphorus. It remains unclear how these processes influence the biogeochemical cycling of other elements, yet such information is central to understanding key issues like ecosystem response to climate change. At the largest scale, we remain ignorant of the quantitative importance of organic phosphorus in the global phosphorus cycle and its implications for the biogeochemical cycles of other elements, notably in the oceans. Advancing our knowledge of these processes requires interdisciplinary effort, but such information almost certainly holds the key to understanding the role of organic phosphorus in major ecosystem processes.

Finally, in contemporary science it is often necessary to attach fundamental importance to research proposals to secure funding, yet research into organic phosphorus remains conspicuously underfunded, despite its clear importance. How can we ensure that organic phosphorus remains at the forefront of research in the environmental sciences? There is a clear need to articulate the importance of organic phosphorus research (for example in terms of global food production, pollution of water resources, and carbon sequestration) to senior managers and government policy advisers. It is our hope that this book goes some way towards informing that discussion.

References

Aber, J.D., Nadelhoffer, K.J., Steudler, P. and Melillo, J.M. (1989) Nitrogen saturation in northern forest ecosystems. *Bioscience* 39, 378–386.

Alexandratos, N. (ed.) (1995) *World Agriculture Towards 2010: an FAO Study.* John Wiley & Sons, Chichester, UK, 488 pp.

Bobbink, R., Hornung, M. and Roelofs, J.G.M. (1998) The effects of air-borne nitrogen pollutants on species diversity in natural and semi-natural European vegetation. *Journal of Ecology* 86, 717–738.

Fairhurst, T., Lefroy, R.D.B., Mutert, E. and Batjes, N. (1999) The importance, distribution and causes of phosphorus deficiency as a constraint to crop production in the tropics. *Agroforestry Forum* 9, 2–8.

Hobbie, S.E., Nadelhoffer, K.J. and Högberg, P. (2002) A synthesis: the role of nutrients as constraints on carbon balances in boreal and arctic regions. *Plant and Soil* 242, 163–170.

Jenkins, M. (2003) Prospects for biodiversity. *Science* 302, 1175–1177.

Lutz, W., Sanderson, W. and Scherbov, S. (2001) The end of world population growth. *Nature* 412, 543–545.

Matson, P.A., Lohse, K. and Hall, S. (2002) The globalization of nitrogen deposition: consequences for terrestrial ecosystems. *Ambio* 31, 113–119.

Näsholm, T., Ekblad, A., Nordin, A., Giesler, R., Högberg, M. and Högberg, P. (1998) Boreal forest plants take up organic nitrogen. *Nature* 392, 914–916.

Root, T.L., Price, J.T., Hall, K.R., Schneider, S.H., Rosenzweig, C. and Pounds, J.A. (2003) Fingerprints of global warming on wild animals and plants. *Nature* 421, 57–60.

Rosegrant, M.W. and Cline, S.A. (2003) Global food security: challenges and policies. *Science* 302, 1917–1919.

Runge-Metzner, A. (1995) Closing the cycle: obstacles to efficient P management for improved global food security. In: Tiessen, H. (ed.) *Phosphorus in the Global Environment: Transfers, Cycles and Management.* John Wiley & Sons, New York, pp. 27–42.

Sharpley, A.N., Hedley, M.J., Sibbesen, E., Hillbricht-Ilkowska, A., House, W.A. and Ryszkowski, L. (1996) Phosphorus transfers from terrestrial to aquatic systems. In: Tiessen, H. (ed.) *Phosphorus in the Global Environment: Transfers, Cycles and Management.* John Wiley & Sons, New York, pp. 171–199.

Stocking, M.A. (2003) Tropical soils and food security: the next 50 years. *Science* 302, 1356–1359.

Tyrrell, T. (1999) The relative influence of nitrogen and phosphorus on oceanic primary productivity. *Nature* 400, 572–531.

Verschuren, D., Johnson, T.C., Kling, H.J., Edgington, D.N., Leavitt, P.R., Brown, E.B., Talbot, M.R. and Hecky, R.E. (2002) History and timing of human impact on Lake Victoria, East Africa. *Proceedings of the Royal Society London, Series B* 269, 289–294.

Appendix: Organic Phosphorus Compounds in the Environment

This appendix provides a list of some of the most common organic phosphorus compounds found in the environment. It also includes complex inorganic phosphates (condensed phosphates) and synthetic organic phosphates commonly used in assays of phosphatase activity. It is not an exhaustive list, but rather provides a handy reference to most of the compounds referred to in the book. Numerous compounds are not included, and details of these can be found in several comprehensive texts, notably Corbridge (2000). For solution phosphorus-31 nuclear magnetic resonance chemical shift values of these compounds see Turner *et al.* (2003).

Compound	Chemical formula	Structure	Functional class	Comments	References
Adenosine bisphosphate	$C_{10}H_{15}N_5O_{10}P_2$		Phosphate monoester	Occurs in either 2′,5′ or 3′,5′ form.	Anderson (1970)
Adenosine 3′,5′-cyclic monophosphate	$C_{10}H_{12}N_5O_6P$		Phosphate diester	Guanosine 3′,5′-cyclic monophosphate also occurs in the environment.	Francko and Webzel (1982)
Adenosine 5′ diphosphate	$C_{10}H_{15}N_5O_{10}P_2$	(β) (α)	Organic polyphosphate	Intermediate in energy transfer from AMP to ATP.	

Adenosine 5′-monophosphate	$C_{10}H_{14}N_5O_7P$	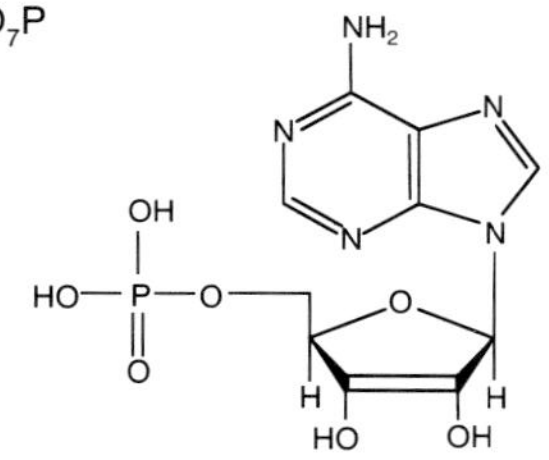	Phosphate monoester	Occurs mainly in 5′ form, although the 2′ and 3′ forms also occur. The ribonucleotides guanosine, cytodine, and uridine 5′-monophosphate also occur, and may also be phosphorylated in the 2′ and 3′ positions. In addition, 2-deoxyribonucleodes occur (e.g 2-deoxyadenosine 5′-monophosphate), which may be phosphorylated in the 3′ or 5′ positions.	Anderson (1958) Corbridge (2000)
Adenosine 5′-triphosphate	$C_{10}H_{16}N_5O_{13}P_3$	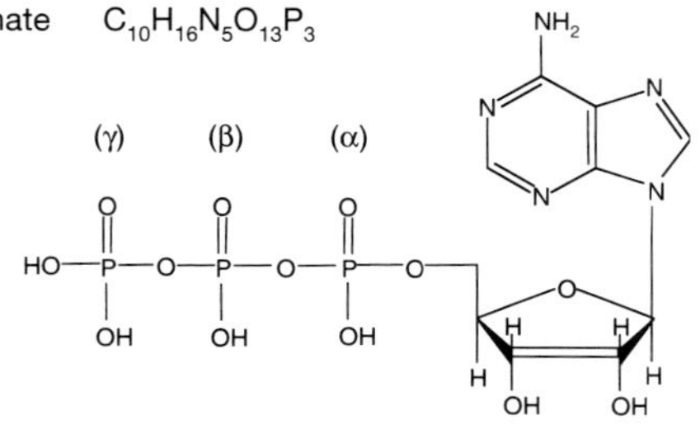	Organic phosphoanhydride (organic condensed phosphate)	Involved in energy transfer in biochemical reactions. Uridine, cytidine, guanosine and thymine triphosphates are also common in biological systems. Nicotinamide adenine dinucleotide phosphate is also an important energy carrier.	Corbridge (2000)
2-Aminoethyl phosphonic acid	$C_2H_8NO_3P$	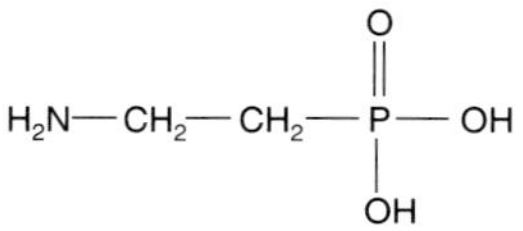 $H_2N—CH_2—CH_2—P(=O)(OH)—OH$	Phosphonate	Most common naturally occurring phosphonate, found in a variety of organisms.	Newman and Tate (1980) Hilderbrand (1983) Clark *et al.* (1999)
2-(5′-chloro-2′-phosphoryloxy-phenyl)-6 chloro-4-(3H)-quinazolinone	$C_{14}H_7Cl_2N_2O_5P$	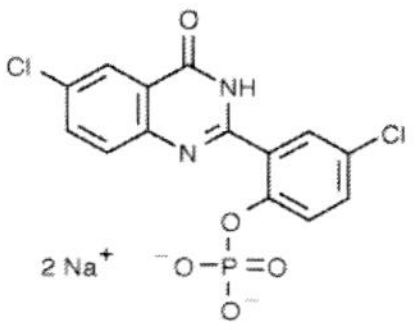	Phosphate monoester	Enzyme-labelled fluorescence substrate (ELF 97 phosphate, also known as ELFP) used to detect phosphatase activity by staining the site of cleavage.	Huang *et al.* (1992) Štrojsová *et al.* (2003)

Compound	Chemical formula	Structure	Functional class	Comments	References
Choline phosphate (phosphorylcholine)	$C_5H_{14}NO_4P$	$^-O-P(OH)(=O)-O-CH_2-CH_2-N^+(CH_3)_3$	Phosphate monoester	Originates from the enzymatic cleavage of phosphatidyl choline by phospholipase C.	
Deoxyribonucleic acid (DNA)	$C_2H_8NO_4P$		Phosphate diester	Polynucleotide containing cellular genetic information.	
Ethanolamine phosphate (o-phosphorylethanolamine)		$HO-P(OH)(=O)-O-CH_2-CH_2-NH_2$	Phosphate monoester	Originates from the enzymatic cleavage of phosphatidyl ethanolamine by phospholipase C.	
α-Glucose 1-phosphate	$C_6H_{13}O_9P$		Phosphate monoester	Sugar phosphate found widely in plants, where it is the precursor of starch.	
D-Glucose 6-phosphate	$C_6H_{13}O_9P$		Phosphate monoester	Common sugar phosphate. Other sugar phosphates, such as fructose 6-phosphate also occur.	
β-Glycerophosphate	$C_3H_9O_6P$	$(HO-CH_2)_2CH-O-P(OH)(=O)-OH$	Phosphate monoester	Originates from alkaline hydrolysis of phosphatidyl choline following loss of the fatty acyl chains and the choline moiety. Thus common in alkaline soil extracts.	Turner *et al*. (2003)

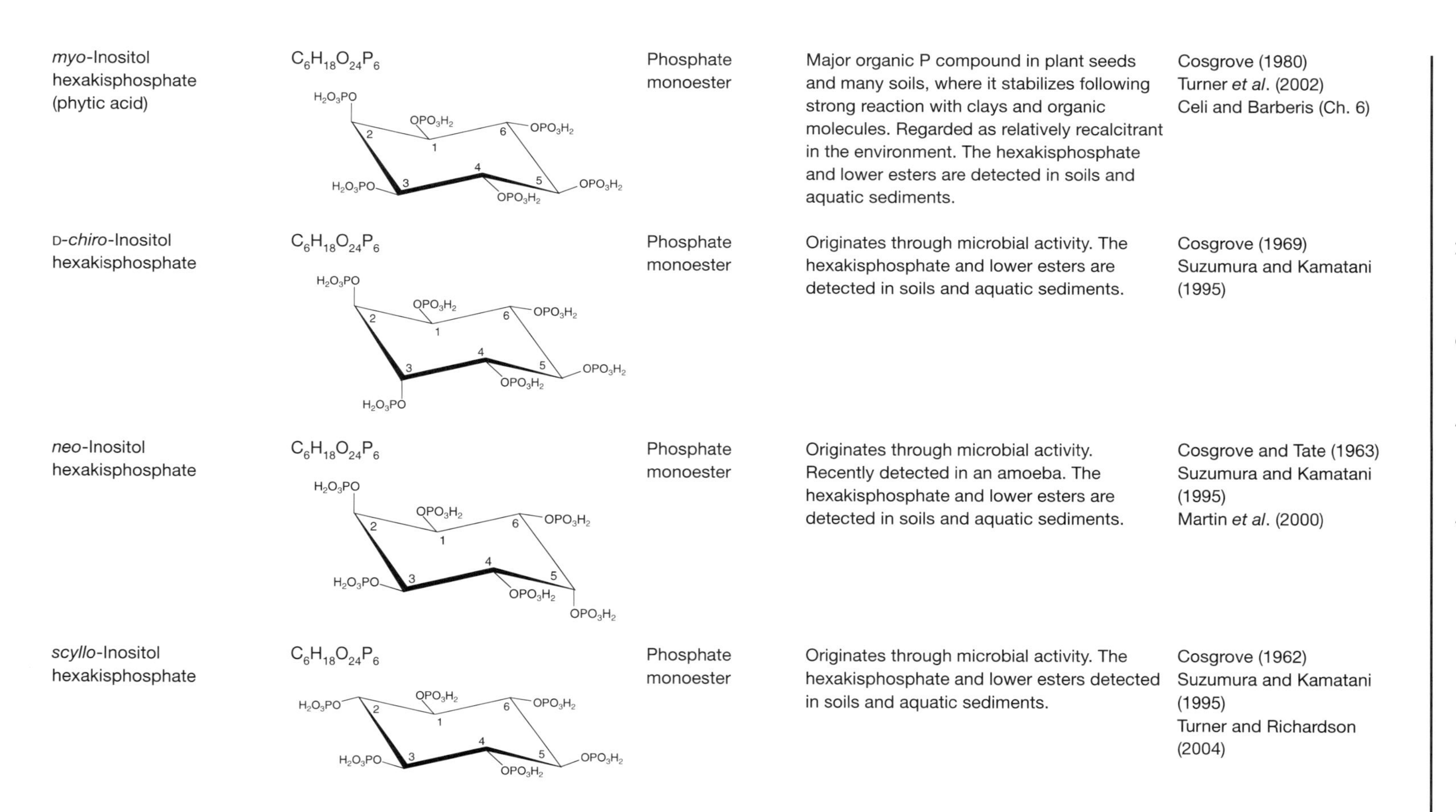

myo-Inositol hexakisphosphate (phytic acid)	$C_6H_{18}O_{24}P_6$	Phosphate monoester	Major organic P compound in plant seeds and many soils, where it stabilizes following strong reaction with clays and organic molecules. Regarded as relatively recalcitrant in the environment. The hexakisphosphate and lower esters are detected in soils and aquatic sediments.	Cosgrove (1980) Turner *et al*. (2002) Celi and Barberis (Ch. 6)
D-*chiro*-Inositol hexakisphosphate	$C_6H_{18}O_{24}P_6$	Phosphate monoester	Originates through microbial activity. The hexakisphosphate and lower esters are detected in soils and aquatic sediments.	Cosgrove (1969) Suzumura and Kamatani (1995)
neo-Inositol hexakisphosphate	$C_6H_{18}O_{24}P_6$	Phosphate monoester	Originates through microbial activity. Recently detected in an amoeba. The hexakisphosphate and lower esters are detected in soils and aquatic sediments.	Cosgrove and Tate (1963) Suzumura and Kamatani (1995) Martin *et al*. (2000)
scyllo-Inositol hexakisphosphate	$C_6H_{18}O_{24}P_6$	Phosphate monoester	Originates through microbial activity. The hexakisphosphate and lower esters detected in soils and aquatic sediments.	Cosgrove (1962) Suzumura and Kamatani (1995) Turner and Richardson (2004)

Compound	Chemical formula	Structure	Functional class	Comments	References
Methylumbelliferyl phosphate	$C_{10}H_9O_6P$	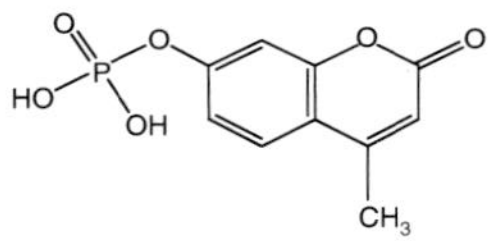	Phosphate monoester	Synthetic fluorimetric monoester used as a sensitive substrate in assays of phosphomonoesterase activity.	Kang and Freeman (1999)
bis-Methylumbelliferyl phosphate	$C_{20}H_{15}O_8P$	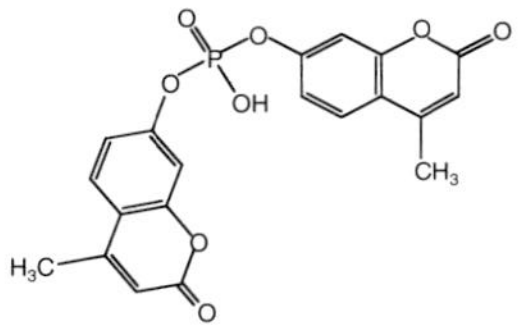	Phosphate diester	Synthetic fluorimetric diester used as a sensitive substrate in assays of phosphodiesterase activity.	
para-nitrophenyl phosphate	$C_6H_6NO_6P$	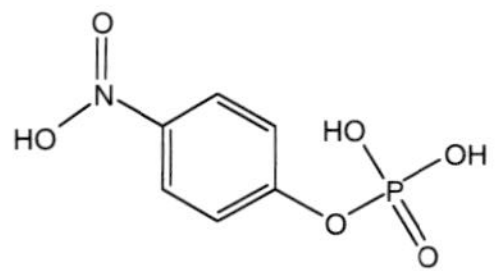	Phosphate monoester	Synthetic colorimetric monoester widely used to determine phosphomonoesterase activity in soils and plant material.	Tabatabai (1994) Turner *et al.* (2001)
bis-*para*-Nitrophenyl phosphate	$C_{12}H_9N_2O_8P$	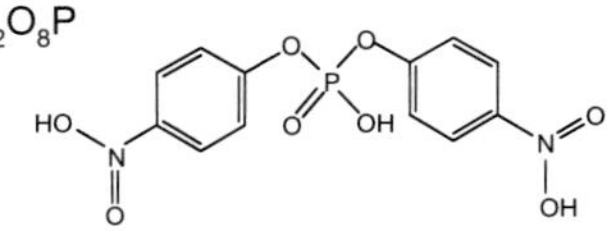	Phosphate diester	Synthetic colorimetric monoester widely used to determine phosphodiesterase activity in soils and plant material.	Tabatabai (1994) Turner *et al.* (2001)
L-α-Phosphatidyl choline (lecithin)	$C_{10}H_{19}NO_8P(2R)$[a] 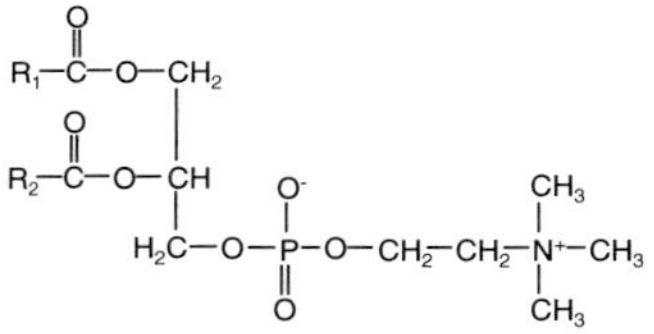		Phosphate diester	Phospholipid commonly found in plants, but less so in microorganisms.	Kowalenko and McKercher (1971) Hanahan (1997)

L-α-Phosphatidyl ethanolamine (cephalin)	$C_7H_{12}NO_8P(2R)$[a]	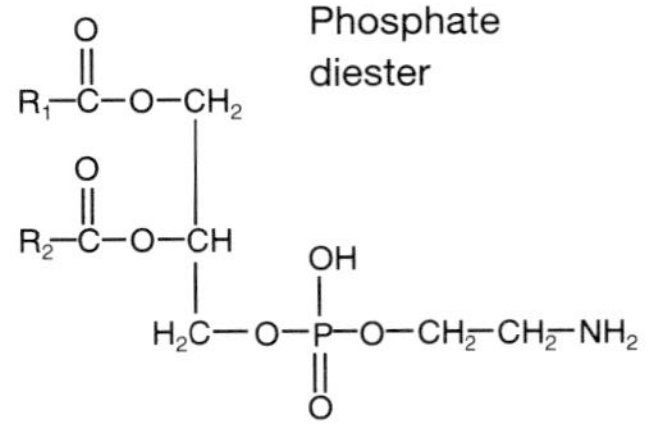	Phosphate diester	Phospholipid commonly found in microorganisms, but largely absent in plants.	Kowalenko and McKercher (1971) Hanahan (1997)
L-α-Phosphatidyl-L-serine	$C_8H_{12}NO_{10}P(2R)$[a]	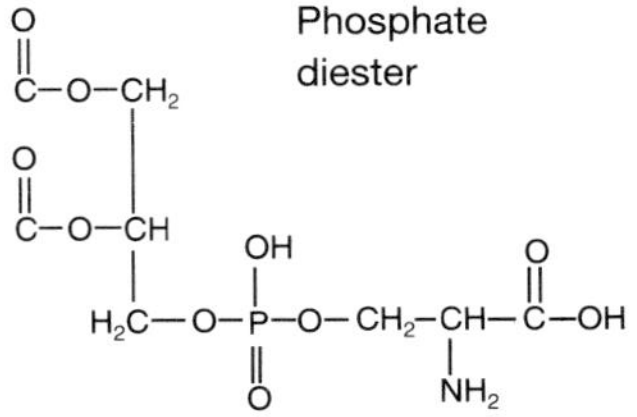	Phosphate diester	Phospholipid commonly occurring in mammalian tissue.	Hanahan (1997)
Phosphoenolpyruvate	$C_3H_5O_6P$	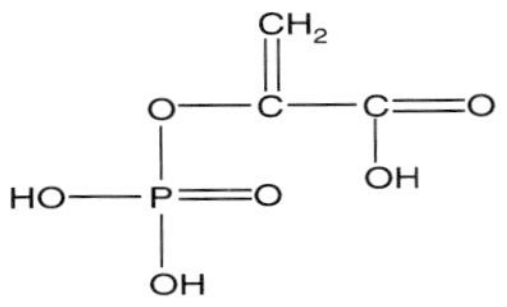	Phosphate monoester	Commonly found in plants. Contains a high energy bond involved in biochemical reactions.	
Phosphocreatine	$C_4H_{10}N_3O_5P$	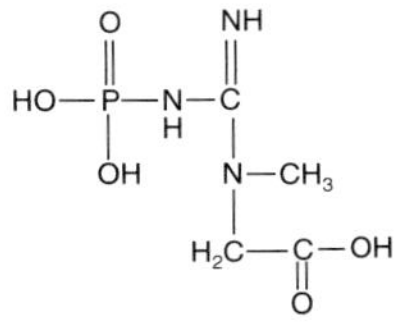		Common in vertebrate muscle. Contains a high energy direct N–P bond, as does phosphoarginine.	

Compound	Chemical formula	Structure	Functional class	Comments	References
N-(Phosphonomethyl) glycine (Glyphosate)	$C_3H_8NO_5P$	$OH-C(=O)-CH_2-NH-H_2C-P(=O)(OH)-OH$	Phosphonate	Common herbicide marketed under the name 'Round Up'. Numerous other similar compounds occur.	
Polyphosphate (linear)	$H_{2n}O_{3n+1}P_n$	$HO-P(=O)(OH)-O\left[-P(=O)(OH)-O\right]_x-P(=O)(OH)-OH$	Phosphoanhydride (condensed phosphate)	Product of microbial activity found in soils and sediments. Also occurs in cyclic form (metaphosphate) in some bacteria.	Ghonsikar and Miller (1973) Kulaev (1979) Hupfer *et al*. (1995)
Pyrophosphate	$H_4O_7P_2$	$HO-P(=O)(OH)-O-P(=O)(OH)-OH$	Phosphoanhydride (condensed phosphate)	Polyphosphate with two phosphate moieties. Common in soils and sediments.	Kulaev (1979) Sundareshwar *et al*. (2001)
Ribonucleic acid (RNA)			Phosphate diester	Polynucleotide involved in protein synthesis.	
Serine phosphate	$C_3H_8NO_6P$	$HO-P(=O)(OH)-O-CH_2-CH(NH_2)-C(=O)-OH$	Phosphate monoester	Derived from cleavage of phosphatidyl serine by phospholipase C.	
Teichoic acids	Various	See Corbridge (2000) for some typical structures.	Phosphate diester	Common component of cell walls of plants and bacteria, notably Gram-positive bacteria. Typically contain phosphate, glycerol and sugar units.	Corbridge (2000) Naumova *et al*. (2001)

[a] R represents hydrophobic fatty acyl chains which may not be identical.

References

Anderson, G. (1958) Identification of derivatives of deoxyribonucleic acid in humic acid. *Soil Science* 86, 169–174.

Anderson, G. (1970) The isolation of nucleoside diphosphates from alkaline extracts of soil. *Soil Science* 21, 96–104.

Clark, L.L., Ingall, E.D. and Benner, R. (1999) Marine organic phosphorus cycling: novel insights from nuclear magnetic resonance. *American Journal of Science* 299, 724–737.

Corbridge, D.E.C. (2000) *Phosphorus 2000: Chemistry, Biochemistry and Technology*. Elsevier Science, Amsterdam, 1267 pp.

Cosgrove, D.J. (1962) Forms of inositol hexaphosphate in soils. *Nature* 194, 1265–1266.

Cosgrove, D.J. (1969) The chemical nature of soil organic phosphorus. II. Characterization of the supposed DL-*chiro*-inositol hexaphosphate component of soil phytate as D-*chiro*-inositol hexaphosphate. *Soil Biology and Biochemistry* 1, 325–327.

Cosgrove, D.J. (1980) *Inositol Phosphates: Their Chemistry, Biochemistry and Physiology*. Elsevier Science, Amsterdam, 197 pp.

Cosgrove, D.J. and Tate, M.E. (1963) Occurrence of *neo*-inositol hexaphosphate in soils. *Nature* 200, 568–569.

Francko, D.A. and Wetzel, R.G. (1982) The isolation of cyclic adenosine 3′:5′ monophosphate (cAMP) from lakes of different trophic status: correlation with planktonic variables. *Limnology and Oceanography* 27, 27–38.

Ghonsikar, C.P. and Miller, R.H. (1973) Soil inorganic polyphosphates of microbial origin. *Plant and Soil* 38, 651–655.

Hanahan, D.J. (1997) *A Guide to Phospholipid Chemistry*. Oxford University Press, Oxford, 214 pp.

Hilderbrand, R.L. (1983) *The Role of Phosphonates in Living Systems*. CRC Press, Boca Raton, Florida.

Huang, Z., Terpetschnig, E., You, W. and Haugland, R.P. (1992) 2-(2′-phosphoryloxyphenyl)-4(3H)-quinazolinone derivates as fluorogenic precipitating substrates of phosphatases. *Analytical Biochemistry* 207, 32–39.

Hupfer, M., Gächter, R. and Rüegger, H. (1995) Polyphosphate in lake sediments: ^{31}P NMR spectroscopy as a tool for its identification. *Limnology and Oceanography* 40, 610–617.

Kang, H. and Freeman, C. (1999) Phosphatase and arylsulphatase activities in wetland soils: annual variation and controlling factors. *Soil Biology and Biochemistry* 31, 449–454.

Kowalenko, C.G. and McKercher, R.B. (1971) Phospholipid components extracted from Saskatchewan soils. *Canadian Journal of Soil Science* 51, 19–22.

Kulaev, I.S. (1979) *The Biochemistry of Inorganic Polyphosphates*. John Wiley & Sons, New York, 255 pp.

Martin, J.-B., Laussmann, T., Bakker-Grunwald, T., Vogel, G. and Klein, G. (2000) *neo*-Inositol polyphosphates in the amoeba *Entamoeba histolytica*. *Journal of Biological Chemistry* 275, 10134–10140.

Naumova, I.B., Shashkov, A.S., Tul'skaya, E.M., Streshinskaya, G.M., Kozlova, Y.I., Potekhina, N.V., Evtushenko, L.I. and Stackebrandt, E. (2001) Cell wall teichoic acids: structural diversity, species specificity in the genus *Nocardiopsis*, and chemotaxonomic perspective. *FEMS Microbiology Reviews* 25, 269–283.

Newman, R.H. and Tate, K.R. (1980) Soil phosphorus characterisation by ^{31}P nuclear magnetic resonance. *Communications in Soil Science and Plant Analysis* 11, 835–842.

Štrojsová, A., Vrba, J., Nedoma, J., Komárková, J. and Znachor, P. (2003) Seasonal study of extracellular phosphatase expression in the phytoplankton of a eutrophic reservoir. *European Journal of Phycology* 38, 295–306.

Sundareshwar, P.V., Morris, J.T., Pellechia, P.J., Cohen, H.J., Porter, D.E. and Jones, B.C. (2001) Occurrence and ecological implications of pyrophosphate in estuaries. *Limnology and Oceanography* 46, 1570–1577.

Suzumura, M. and Kamatani, A. (1995) Origin and distribution of inositol hexaphosphate in estuarine and coastal sediments. *Limnology and Oceanography* 40, 1254–1261.

Tabatabai, M.A. (1994) Soil enzymes. In: Weaver, R.W., Angle, S., Bottomley, P., Bezdicek, D., Smith, S., Tabatabai, A. and Wollum, A. (eds), *Methods of Soil Analysis, Part 2. Microbiological and Biological Properties*. Soil Science Society of America, Madison, Wisconsin, pp. 775–833.

Turner, B.L. and Richardson, A.E. (2004) Identification of *scyllo*-inositol phosphates in soil by solution phosphorus-31 nuclear magnetic resonance spectroscopy. *Soil Science Society of America Journal* 68, 802–808.

Turner, B.L., Baxter, R., Ellwood, N.T.W. and Whitton, B.A. (2001) Characterization of the phosphatase activity of mosses in relation to their environment. *Plant, Cell and Environment* 24, 1165–1176.

Turner, B.L., Papházy, M., Haygarth, P.M. and McKelvie, I.D. (2002) Inositol phosphates in the environment. *Philosophical Transactions of the Royal Society London, Series B* 357, 449–469.

Turner, B.L., Mahieu, N. and Condron, L.M. (2003) Phosphorus-31 nuclear magnetic resonance spectral assignments of phosphorus compounds in soil NaOH-EDTA extracts. *Soil Science Society of America Journal* 67, 497–510.

Index

Page numbers in **bold** refer to illustrations and tables